Algebra I

An Incremental Development

Algebra I

An Incremental Development

JOHN H. SAXON, JR.

Rose State College

GRASSDALE PUBLISHERS, INC.

Algebra I: An Incremental Development
Teachers Edition

Grassdale Publishers, Inc.
1002 Lincoln Green
Norman, Oklahoma 73069

To The Student

Algebra is not difficult. Algebra is just different, and time is required in order for different things to become familiar. In this book we provide the necessary time by reviewing all concepts in every Problem Set. Also, the parts of a particular concept are introduced in small units so that they may be practiced for a period of time before the next part of the same concept is introduced. Understanding the first part makes it easier to understand the second part. If you find that a particular problem is troublesome, get help at once because the problem won't go away. It will appear again and again in future Problem Sets.

The Problem Sets contain all the review that is necessary. **Your task is to work all the problems in every Problem Set.** The answers to the Odd-Numbered Problems are in the Appendix. It will be necessary to check the answers to the even problems with a classmate. Don't be discouraged when you continue to make mistakes. Everyone makes these mistakes, and makes them often, and for a long period of time. A large part of learning algebra is devising defense mechanisms to protect you from yourself. If you work at it, you can find ways to prevent these mistakes. Your teacher is an expert because your teacher has made the same mistakes many times and has finally found ways to prevent them. You must do the same. Each person must devise his or her own defense mechanisms.

The repetition is necessary to permit all students to master all of the concepts, and then the application must be practiced for a long time to insure retention. This practice has an element of drudgery to it, but it has been demonstrated that people who are not willing to practice fundamentals often find success elusive. Ask your favorite athletic coach for his opinion on the necessity of practicing fundamental skills.

To The Teacher

The effectiveness of this book was demonstrated during the 1980–1981 school year in twenty Oklahoma public schools. Over 1,360 ninth grade Algebra I students participated. In each school one teacher taught one section from a prototype of this book and one or more sections from the Algebra I book normally used. During the spring, sixteen 10–15 minute tests were given. Each test was on one fundamental skill of beginning algebra and the tests were constructed from problems submitted by the teachers. The topics tested were: signed numbers, evaluation of expressions, solutions of equations in one unknown, adding like terms, number word problems, natural number exponents, factoring, percent word problems, value word problems, addition of rational expressions, simplification of radicals, linear equations, simultaneous equations, and uniform motion word problems. Overall, the students who used this book more than doubled the scores of the students who had the same teacher but used a standard book. For tests on scientific notation and negative exponents, eight classes of Algebra II students were used as controls. The 9th graders who had used this book more than tripled the scores of the Algebra II students on both of these tests. The test program was monitored by the Oklahoma Federation of Teachers, and they have certified the results.

Students should work every problem in every problem set. This book provides only the problems that are necessary and the problems are not in pairs so that either the odd problems or the even problems may be assigned. Experience with this book will demonstrate that no extra problems of the new kind are necessary or desirable. In this book the learning process is spread out and comprehension will come in time. The emphasis is on review and not on an all out attack on the new concept. This development spreads out the learning process, increases the depth of understanding and improves long term retention. Experience with books that use the traditional development of topics has proved that an intensive initial attack on a new concept lends little to long term retention. Remember that this book never drops a topic once it has been introduced and that the student will continue to wrestle with every concept in every homework problem set.

Acknowledgments

I thank the following Oklahoma teachers for their help in conducting the test program: Max Coon in Carnegie; Jim Shields in Cushing; Mickey Yarberry in Del City; Gwen Gaskins in Holdenville; Margaret Spann in Lindsay; Linda Smiley in Madill; Emet Callaway in Marlow; Paula Busking and Elizabeth Drew in Moore; Jack Harding and Annette Gravitt in Oklahoma City; Joyce Fisher in Okemah; Corinne Pitts in Okmulgee; Dennis Lebeda, Anne Boesch, Helen Hull, and Lloyd Gelmers in Ponca City; Hobert Higgs in Purcell; George Beaty in Seminole; David Wade in Stillwater; Virgina Weir in Tecumseh: Calvin Tooley in Wetumka; and Robert Beck in Yukon.

I thank Johnny, Selby, Bruce, and Sarah for their support, suggestions, and contributions. I thank Oscar Rose Junior College for the freedom to experiment to develop new teaching techniques.

I thank Julie Filer for help in reading proof. I thank Frank Wang for reading proof, checking answers, and for his constructive comments. I thank William F. Buckley, Jr. and Bob Worth for their encouragement and help. Lastly, I thank Detia Roe, graduate physicist and expert typist for her typing, patience, and support.

Norman, Oklahoma   John Saxon
August 1981

Contents

Basic

Course

REVIEW LESSON A   *Addition and subtraction of fractions*

A.A
addition
and subtraction
of fractions

In order to add or subtract fractions that have the same denominators, the numerators are added or subtracted and the result is recorded over a single denominator, as shown here.

$$\frac{5}{11} + \frac{2}{11} = \frac{7}{11} \qquad \frac{5}{11} - \frac{2}{11} = \frac{3}{11}$$

If the denominators are not the same, it is necessary to rewrite the fractions so that they have the same denominators.

	PROBLEM	REWRITTEN WITH EQUAL DENOMINATORS	ANSWER
(a)	$\frac{1}{3} + \frac{2}{5}$	$\frac{5}{15} + \frac{6}{15}$	$\mathbf{\frac{11}{15}}$
(b)	$\frac{2}{3} - \frac{1}{8}$	$\frac{16}{24} - \frac{3}{24}$	$\mathbf{\frac{13}{24}}$

A mixed number is the sum of a whole number and a fraction. Thus the notation

$$13\frac{3}{5}$$

does not mean 13 multiplied by $\frac{3}{5}$ but instead 13 plus $\frac{3}{5}$.

$$13 + \frac{3}{5}$$

When we add and subtract mixed numbers, we handle the fractions and the whole numbers separately. In some subtraction problems it is necessary to borrow, as shown in (e).

	PROBLEM	REWRITTEN WITH EQUAL DENOMINATORS	ANSWER
(c)	$13\frac{3}{5} + 2\frac{1}{8}$	$13\frac{24}{40} + 2\frac{5}{40}$	$\mathbf{15\frac{29}{40}}$
(d)	$13\frac{3}{5} - 2\frac{1}{8}$	$13\frac{24}{40} - 2\frac{5}{40}$	$\mathbf{11\frac{19}{40}}$
		BORROWING	
(e)	$13\frac{3}{5} - 2\frac{7}{8}$	$13\frac{24}{40} - 2\frac{35}{40} = 12\frac{64}{40} - 2\frac{35}{40}$	$\mathbf{10\frac{29}{40}}$

problem set A   Add or subtract as indicated. Write answers as proper fractions reduced to lowest terms or as mixed numbers.

1. $\dfrac{1}{5} + \dfrac{2}{5}$ **2.** $\dfrac{3}{8} - \dfrac{2}{8}$ **3.** $\dfrac{4}{3} - \dfrac{1}{3} + \dfrac{8}{3}$

4. $\dfrac{2}{7} + \dfrac{13}{7} - \dfrac{5}{7}$ **5.** $\dfrac{18}{11} - \dfrac{4}{11} + \dfrac{1}{11}$

Different denominators:

6. $\dfrac{1}{3} + \dfrac{1}{5}$ **7.** $\dfrac{3}{8} - \dfrac{1}{5}$ **8.** $\dfrac{2}{3} - \dfrac{1}{8}$

9. $\dfrac{1}{13} + \dfrac{1}{5}$ **10.** $\dfrac{17}{15} - \dfrac{2}{3}$ **11.** $\dfrac{5}{9} + \dfrac{2}{5}$

12. $\dfrac{7}{8} - \dfrac{4}{5}$ **13.** $\dfrac{12}{13} - \dfrac{3}{4}$ **14.** $\dfrac{17}{20} - \dfrac{4}{5}$

15. $\dfrac{14}{17} - \dfrac{6}{34}$ **16.** $\dfrac{5}{13} + \dfrac{1}{26}$ **17.** $\dfrac{4}{7} - \dfrac{2}{5}$

18. $\dfrac{4}{7} + \dfrac{1}{8} + \dfrac{1}{2}$ **19.** $\dfrac{3}{5} + \dfrac{1}{8} + \dfrac{1}{8}$ **20.** $\dfrac{5}{11} - \dfrac{1}{6} + \dfrac{2}{3}$

21. $\dfrac{3}{8} + \dfrac{1}{2} - \dfrac{2}{5}$ **22.** $\dfrac{3}{5} + \dfrac{15}{3} + \dfrac{17}{10}$

Addition and subtraction of mixed numbers:

23. $2\dfrac{1}{2} + 3\dfrac{1}{5}$ **24.** $7\dfrac{3}{8} + 4\dfrac{7}{3}$ **25.** $1\dfrac{1}{8} + 7\dfrac{2}{5}$

26. $6\dfrac{1}{3} + 1\dfrac{2}{5}$ **27.** $8\dfrac{13}{3} - 2\dfrac{2}{5}$ **28.** $411\dfrac{1}{3} - 24\dfrac{2}{15}$

Subtraction with borrowing:

29. $15\dfrac{1}{3} - 7\dfrac{4}{5}$ **30.** $42\dfrac{3}{8} - 21\dfrac{3}{4}$ **31.** $22\dfrac{2}{5} - 13\dfrac{7}{15}$

32. $421\dfrac{1}{11} - 17\dfrac{4}{3}$ **33.** $78\dfrac{2}{5} - 14\dfrac{7}{10}$ **34.** $43\dfrac{1}{13} - 6\dfrac{5}{8}$

35. $21\dfrac{1}{5} - 15\dfrac{7}{13}$ **36.** $21\dfrac{2}{19} - 7\dfrac{7}{10}$ **37.** $43\dfrac{3}{17} - 21\dfrac{9}{10}$

LESSON 1   *Real numbers and the number line · Multiplication and division of fractions*

1.A
numerals and numbers   A number is an idea. A numeral is a single symbol or a collection of symbols that we use to express the idea of a particular number.

The three drawings above all have the quality of *threeness*. The three children and the three pencils both bring to mind the idea of *three*. The drawing at the right also brings to mind the idea of *three*, although all the things in the drawing are not of the same kind.

If we wish to use a symbol to designate the idea of three, we could write any of the following:

$$\text{III}, \quad 3, \quad \frac{30}{10}, \quad \frac{27}{9}, \quad \frac{33}{11}, \quad 2 + 1, \quad 6 \div 2, \quad 11 - 8$$

Each of these is a symbolic representation of the idea of 3. Throughout the book when we use the word *number*, we are describing the idea; and we will use numerals to designate the numbers. But we will remember that none of the marks we make on paper are numbers because

A number is an idea!

Since the symbols

$$3 \quad \text{and} \quad \frac{30}{10}$$

are both numerals that represent the same number, we say that they have the same value. **Thus, the value of a numeral is the number represented by the numeral, and we see that the words value and number have the same meaning.**

1.B
natural or counting numbers

The system of numeration that we use to designate numbers is called the decimal system. It was invented by the Hindus of India, passed to their Arab neighbors, and finally transmitted to Europe circa 1200 A.D. The decimal system uses 10 symbols that we call **digits.** These digits are

$$0, \quad 1, \quad 2, \quad 3, \quad 4, \quad 5, \quad 6, \quad 7, \quad 8, \quad 9$$

We use these digits by themselves or in combination with one another to form the numerals that we use to designate decimal numbers. **We call the numbers that we use to count objects or things the natural numbers or the counting numbers.** When we begin counting, we always begin with the number 1 and follow it with the number 2, etc.

$$1, \quad 2, \quad 3, \quad 4, \quad 5, \quad 6, \quad 7, \quad 8, \quad 9, \quad 10, \quad 11, \quad 12, \quad 13, \ldots$$

It would not be natural to try to count by using numbers such as $\frac{1}{2}$ or 0 or $\frac{3}{4}$, so these numbers are not called natural or counting numbers. We designate the natural or counting numbers with the listing above. The three dots after the number 13 indicate that this listing continues without end.

1.C
real numbers

The numbers of arithmetic are zero and the positive real numbers. **We say that a positive real number is any number that can be used to describe a physical distance greater than zero.**

Thus, all of the numbers shown here

$$\frac{3}{4} \qquad .000163 \qquad 363 \qquad 3\frac{3}{8} \qquad 46 \qquad \frac{11}{7} \qquad 400.1623232323$$

are positive real numbers, for all of them can be used to describe physical distances when used with descriptive units such as feet, yards, etc.

$$\frac{3}{4} \text{ mile} \qquad .000163 \text{ yard} \qquad 363 \text{ feet} \qquad 3\frac{3}{8} \text{ meters}$$

$$46 \text{ inches} \qquad \frac{11}{7} \text{ kilometers} \qquad 400.16232323 \text{ centimeters}$$

The number zero is not a positive number, but it can be used to describe a physical distance of no magnitude, and we say that zero is also a real number. In addition to the positive numbers and zero, in algebra we use numbers that we call **negative numbers** and these numbers are also called real numbers. The ancients did not understand or use negative numbers. A man could not own negative 10 sheep. If he owned any sheep at all, the number of sheep had to be designated by a number greater than zero. The ancients could subtract 4 from 6 and get 2, but they felt that it was impossible to subtract 6 from 4 because that would result in a number that was less than zero itself. To their way of thinking, this was clearly impossible.

While some might tend to agree with the ancients, to the modern mathematician, physicist, or chemist, the idea or concept of negative numbers does exist, and it is a useful concept. **We say that every positive real number has a negative counterpart, and we call these numbers the negative real numbers.** We must always use a minus sign when we designate a negative number as we see here by writing negative seven.

$$-7$$

We may use a plus sign to designate a positive number as we see by writing positive seven,

$$+7$$

or we may leave off the plus sign as we did in arithmetic and just write the numerical part with no sign.

$$7$$

We must remember that when we write a numeral with no sign, we designate a positive number. When we are talking about negative numbers as well as positive numbers, we say that we are talking about **signed numbers.** As we shall see later, the use of signed numbers will enable us to lump the operations of addition and subtraction into a single operation which we will call algebraic addition.

1.D
number lines

In the 1950s the so-called new math appeared, and among other things it introduced the **number line** at the elementary algebra level. The number line can be used as a graphic aid when discussing signed numbers, and it is especially useful when discussing the addition of signed numbers.

To construct a number line, we first draw a line and divide it into equal units of length. The units may be any length as long as they are all the same length.

Many books show small arrows on the ends of number lines to emphasize that the lines continue without end in both directions. The arrowheads are not necessary and may be omitted.

Now we choose a point on the line as our base point. We call this base point the **origin,** and we associate the number zero with this point.

0

Then we associate the positive real numbers with the points to the right of the origin and the negative real numbers with the points to the left of the origin.

-7   -6   -5   -4   -3   -2   -1   0   1   2   3   4   5   6   7

On the number line above we have indicated the location of zero, the counting numbers, and the negative counterpart of each counting number. As required, we can indicate the position of any real number by locating it in relation to the numbers shown. For example, on the number line below we indicate the position of $+\frac{3}{4}$, $-1\frac{1}{2}$, and $+2.6$ by placing a dot at the approximate location of these numbers.

When we place a dot on the number line to indicate the location of a number, we say that we have graphed the number and that the dot is the graph of the number. Conversely, the number is said to be the coordinate of the point that we have graphed. We use the number line to tell if one number is greater than another number by saying that a number is greater than another number if its graph lies to the right of the graph of the other number. Thus $\frac{3}{4}$ is greater than $-1\frac{1}{2}$ because the graph of $\frac{3}{4}$ lies to the right of the graph of $-1\frac{1}{2}$. This topic will be discussed in considerable detail in later lessons.

On the line above we have graphed $+2$ and -2. The number $+2$ (usually the $+$ sign is omitted) lies 2 units to the right of the origin and the number -2 lies 2 units to the left of the origin. Since the graphs of these numbers are equidistant from the origin but in opposite directions, it is sometimes helpful to think of each of these numbers as being the **opposite** of the other number. In this example, **we say that -2 is the opposite of 2 and that 2 is the opposite of -2.**

1.E
review of multiplication and division of fractions

Fractions are multiplied by multiplying the numerators to get the new numerator, and by multiplying the denominators to get the new denominator.

PROBLEM

(a) $\dfrac{4}{3} \times \dfrac{7}{5}$

SOLUTION

$\dfrac{4 \times 7}{3 \times 5} = \dfrac{28}{15} = 1\dfrac{\mathbf{13}}{\mathbf{15}}$

We divide fractions by inverting the divisor and then multiplying.

PROBLEM

(b) $\dfrac{4}{3} \div \dfrac{15}{8}$

INVERTING

$\dfrac{4}{3} \times \dfrac{8}{15}$

SOLUTION

$\dfrac{4 \times 8}{3 \times 15} = \dfrac{\mathbf{32}}{\mathbf{45}}$

If cancellation is possible, it is easier if we cancel before we multiply.

PROBLEM	CANCELLATION	SOLUTION
(c) $\dfrac{7}{3} \times \dfrac{30}{9}$	$\dfrac{7}{\cancel{3}} \times \dfrac{\overset{10}{\cancel{30}}}{9}$	$\dfrac{70}{9} = 7\dfrac{7}{9}$
(d) $\dfrac{3}{5} \times \dfrac{5}{6} \times \dfrac{21}{23}$	$\dfrac{\overset{1}{\cancel{3}}}{\cancel{5}} \times \dfrac{\overset{1}{\cancel{5}}}{\underset{2}{\cancel{6}}} \times \dfrac{21}{23}$	$\dfrac{21}{46}$

We change mixed numbers to improper fractions first and then multiply or divide as indicated.

PROBLEM	IMPROPER FRACTION	SOLUTION
(e) $2\dfrac{1}{2} \times 5\dfrac{1}{3}$	$\dfrac{5}{\cancel{2}} \times \dfrac{\overset{8}{\cancel{16}}}{3}$	$\dfrac{40}{3} = 13\dfrac{1}{3}$
(f) $12\dfrac{1}{3} \div 2\dfrac{1}{6}$	$\dfrac{37}{\cancel{3}} \times \dfrac{\overset{2}{\cancel{6}}}{13}$	$\dfrac{74}{13} = 5\dfrac{9}{13}$
(g) $\dfrac{3\dfrac{1}{3}}{2\dfrac{1}{5}}$	$\dfrac{\dfrac{10}{3}}{\dfrac{11}{5}}$	$\dfrac{10}{3} \cdot \dfrac{5}{11} = \dfrac{50}{33} = 1\dfrac{17}{33}$

In every problem set in this book the asterisks identify problems that were worked out as example problems in the lesson.

problem set 1

1. What is the difference between a number and a numeral?

2. What do we call our system of numeration?

3. Who invented this system?

4. List the digits that we use in this system.

5. What numbers are called the counting numbers?

6. What numbers are called natural numbers?

7. The numbers of arithmetic are zero and the positive real numbers. How does the book define positive real numbers?

8. Is zero a real number?

9. What is the graph of a number?

10. What do we call the point on the number line with which we associate the number zero?

11. What is the coordinate of a point on the number line?

12. Does every real number have a sign?

13. If a numeral (other than zero) is not preceded by a sign, does it designate a positive number or a negative number?

14. Which real numbers cannot be graphed on the number line?

15. Are $+5$ and $-.006$ both real numbers?

16. What number is the opposite of -2?

17. Draw a number line and graph the numbers:

   (a) $+5$ (b) 6 (c) $\dfrac{5}{2}$

   (d) $-3\dfrac{1}{2}$ (e) $-\dfrac{5}{2}$

18. When we use the words *signed numbers*, what do we mean?

19. How do we tell if one number is greater than another number?

20. Is -10 greater than -9? Why?

Add, subtract, multiply, or divide as indicated:

21. $15\dfrac{1}{3} - 7\dfrac{4}{5}$ 22. $18\dfrac{1}{5} - 3\dfrac{7}{8}$ *23. $\dfrac{4}{3} \times \dfrac{7}{5}$

*24. $\dfrac{4}{3} \div \dfrac{15}{8}$ *25. $\dfrac{7}{3} \times \dfrac{30}{9}$ 26. $\dfrac{3}{5} \times \dfrac{10}{7} \times \dfrac{21}{8}$

*27. $2\dfrac{1}{2} \times 5\dfrac{1}{3}$ *28. $12\dfrac{1}{3} \div 2\dfrac{1}{6}$ 29. $17\dfrac{3}{4} - 6\dfrac{7}{8}$

30. $21\dfrac{3}{4} - 5\dfrac{4}{5}$ 31. $\dfrac{7}{4} \times \dfrac{28}{35}$ 32. $3\dfrac{1}{5} \div 4\dfrac{2}{3}$

33. $3\dfrac{1}{4} \div 2\dfrac{1}{3}$ *34. $\dfrac{3\frac{1}{3}}{2\frac{1}{5}}$ 35. $2\dfrac{3}{4} - 1\dfrac{9}{16}$

36. $\dfrac{7\frac{1}{2}}{2\frac{1}{8}}$ 37. $7\dfrac{1}{3} - 2\dfrac{3}{8}$ 38. $415\dfrac{9}{13} - 17\dfrac{38}{39}$

39. $\dfrac{5}{3} \times \dfrac{1}{3} \times \dfrac{9}{15}$ 40. $7\dfrac{1}{3} \div 2\dfrac{1}{5}$

LESSON 2   *Sets and operations · Decimal numbers*

2.A
sets   We use the word **set** to designate a well-defined collection of numbers, objects, or things. We say that the individual objects or things that make up the set are the elements of the set or the members of the set. It is customary to designate a set by enclosing the members of the set within braces.

$$A = \{1, 2, 3, 4, 5\}$$

We have designated set A as the set whose members are the counting numbers 1 through 5 inclusive. We could also designate this set by placing a verbal phrase within the braces as

$$A = \{\text{the counting numbers 1 through 5 inclusive}\}$$

or we could designate the members of this set by writing a sentence.

"Set A is the set whose members are the counting numbers 1 through 5 inclusive."

Since we can designate which numbers are natural or counting numbers so that there is no doubt as to whether a number is or is not a natural or counting number, we say that these numbers constitute a set. Both *natural* and *counting* are names for the same set, and we normally use one name or the other. Thus we say

$$\text{Natural numbers} = \{1, 2, 3, 4, 5, \ldots\}$$

The three dots indicate that the listing of numbers continues without end.
   If we include the number zero with the set of natural numbers, we say that we have designated the set of whole numbers.

$$\text{Whole numbers} = \{0, 1, 2, 3, 4, 5, \ldots\}$$

We designate that the listing of the members of the set of whole numbers continues without end by using the three dots after the last digit recorded. **Now if we include in our list the negative of every member of the set of counting numbers, we have designated the set that we call the integers.**

$$\text{Integers} = \{\ldots, -3, -2, -1, 0, 1, 2, 3, \ldots\}$$

The three dots on each end indicate that the listing continues without end in both directions.

2.B
symbols of equality and inequality

We use the equals ($=$) sign to designate that two quantities are equal. Thus we can write

$$5 + 2 = 7$$

because the number represented by the notation $5 + 2$ is the same number as that represented by the numeral 7. In the same way we use the symbol $\neq$ to designate that two quantities are not equal. Thus we can write that

$$5 + 2 \neq 11$$

because 7 is not equal to 11.

2.C
basic operations

The four basic operations of arithmetic are also the basic operations of algebra. The operations are addition, subtraction, multiplication, and division. We will review these operations here and will restrict our discussion to the numbers of arithmetic which are the positive real numbers and zero.

addition

When we wish to add two numbers to get a result, we use the plus sign ($+$) to indicate the operation of addition. We call each of the numbers an **addend,** and we call the result a **sum.**

$$2 + 3 = 5$$

In this example, we use the plus sign to indicate addition; we say that the numbers 2 and 3 are addends, and we say that 5 is the sum.

We note that the sum of zero and any particular real number is the particular real number itself.

$$4 + 0 = 4 \quad \text{and} \quad 15 + 0 = 15$$

subtraction   When we wish to subtract one number from another number, we use the minus sign $(-)$ to indicate the operation of subtraction. We call the first number the **minuend**; the second number, the **subtrahend**; and the result, the **difference**.

$$9 - 5 = 4$$

In this example, 9 is the minuend, 5 is the subtrahend, and 4 is the difference.

multiplication   If two numbers are to be multiplied to achieve a result, each of the numbers is called a **factor** and the result is called a **product**. There are several ways to indicate the operation of multiplication.

$$4 \cdot 3 = 12, \qquad 4(3) = 12, \qquad (4) \cdot (3) = 12, \qquad (4)(3) = 12, \qquad 4 \times 3 = 12$$

In each of the five examples shown here, the notation indicates that 4 is to be multiplied by 3 and the result is 12. In algebra, we will avoid the last notation because the cross can be confused with the letter x, a symbol we will use for other purposes. In each of the above, we say that 4 and 3 are factors, and we say that 12 is the product.

We note that the product of a particular real number and the number 1 is the particular real number itself.

$$4 \cdot 1 = 4 \quad \text{and} \quad 15 \cdot 1 = 15$$

The number zero also has a unique multiplicative property. **The product of any real number and the number zero is the number zero.**

$$4 \cdot 0 = 0 \quad \text{and} \quad 15 \cdot 0 = 0$$

division   If one number is to be divided by another number to achieve a result, the first number is called the **dividend**, the second number is called the **divisor**, and the result is called the **quotient**.

$$\frac{10}{5} = 2 \qquad 10 \div 5 = 2$$

Both of the notations shown here indicate that 10 is to be divided by 5 and that the result is 2. We call 10 the dividend, call 5 the divisor, and say that the quotient is 2. When the indicated division is expressed in the form of a fraction such as $\frac{10}{5}$, we say that 10 is the **numerator** of the fraction and that 5 is the **denominator** of the fraction.

2.D
review of operations with decimal numbers

We must align the decimal points vertically when we add and subtract decimal numbers as we show here.

$$\begin{array}{r} 1.005 \\ +\,300.012 \\ \hline \mathbf{301.017} \end{array}$$

example 2.1   Add 4.0016 and .02163.

solution   We remember to place the numbers so that the decimal points are aligned.

$$\begin{array}{r} 4.0016 \\ +\ \ .02163 \\ \hline \mathbf{4.02323} \end{array}$$

example 2.2 Subtract .02163 from 4.0016.

solution Again we align the decimal points.

$$\begin{array}{r} 4.0016 \\ -.02163 \\ \hline \textbf{3.97997} \end{array}$$

example 2.3 Multiply 4.06 × .016.

solution We do not align the decimal points when we multiply.

$$\begin{array}{r} 4.06 \\ .016 \\ \hline 2436 \\ 406 \\ \hline \textbf{.06496} \end{array}$$

example 2.4 Divide 6.039 by .03.

solution As the first step, we adjust the decimal points as necessary. Then we divide.

$$.03\overline{)6.039} \qquad 3\overline{)\begin{array}{c} \textbf{201.3} \\ 603.9 \end{array}}$$

$$\begin{array}{r} 6 \\ \hline 3 \\ 3 \\ \hline 9 \\ 9 \\ \hline \end{array}$$

problem set 2

1. Use braces and digits to designate the set of integers.

2. Use braces and digits to designate the set of whole numbers.

3. Designate the set of natural numbers.

4. Designate the set of counting numbers.

5. What meaning does the symbol ≠ have?

6. Elements of a set are also called what?

7. When two numbers are added to get an answer, what do we call the numbers and what do we call the answer?

8. Define a positive real number.

9. Is zero a real number?

10. How can we tell if one number is greater than another number?

11. Is −4 greater than −1? Why?

12. Does every real number have a sign that is plus or minus?

13. If a numeral is not preceded by a sign, does the numeral represent a positive number or a negative number?

14. Who invented the number system we use?

15. Are $-\dfrac{3}{2}$ and $+2.6$ both real numbers? How can you tell?

16. If we multiply two numbers to get an answer, what do we call the numbers and what do we call the answer?

17. What do we call the answer to a division problem?

18. Tell about the two special properties of the number zero.

19. What do we call the top of a fraction and what do we call the bottom of the fraction?

Add, subtract, multiply, or divide as indicated:

20. $4.0016 + .02163 - .016$

*21. $4.06 \times .016$

22. $\dfrac{5\frac{3}{8}}{7\frac{1}{4}}$

*23. $6.039 \div .03$

24. $\dfrac{3\frac{1}{5}}{2\frac{1}{2}}$

25. $17\frac{1}{3} - 1\frac{7}{8}$

26. $5\frac{2}{3} \times 3\frac{1}{4}$

27. $5\frac{2}{3} - 3\frac{1}{4}$

28. $\dfrac{61.9737}{5.01}$

29. $5\frac{2}{3} \div 3\frac{1}{4}$

LESSON 3   *Absolute value · Addition on the number line*

3.A
absolute value

The number zero is neither positive nor negative and can be designated with the single symbol 0. Every other real number is either positive or negative and thus requires a two-part numeral. One of the parts is the plus or minus sign, and the other part is the numerical part. If we look at the two numerals

$$+7 \qquad \text{and} \qquad -7$$

we note that the numerical part of each one is the same and that the numerals differ only in their signs. We can think of the numerical part as designating the quality of "bigness" of the number, and we use the words **absolute value** to describe this quality. **However, when we try to write the absolute value of one of these numbers by just writing the numerical part**

$$7$$

we find that we have written a positive number because we have agreed that a numeral written with no sign designates a positive number. Because of this agreement, we are forced to define the absolute value of any nonzero real number to be a positive number. We define the absolute value of zero to be zero. If we enclose a number[†] within vertical

---

[†] We really should use the word *numeral* here, but from now on we will often use the words *number* and *numeral* interchangeably because excessive attention to the difference between these words is counterproductive.

lines, we are designating the absolute value of the number. We will demonstrate this notation by designating the absolute value of zero, the absolute value of positive 7, and the absolute value of negative 7.

$$|0| = 0 \qquad \text{read ``the absolute value of zero equals zero.''}$$
$$|+7| = +7 \qquad \text{read ``the absolute value of 7 equals 7.''}$$
$$|-7| = +7 \qquad \text{read ``the absolute value of } -7 \text{ equals 7.}$$

Since the plus sign in front of a positive number is customarily omitted, the above can be written without recording the plus signs:

$$|7| = 7 \qquad \text{read ``the absolute value of 7 equals 7.''}$$
$$|-7| = 7 \qquad \text{read ``the absolute value of } -7 \text{ equals 7.''}$$

Thus we have two rules for stating the absolute value of a real number.

1. **The absolute value of zero is zero.**
2. **The absolute value of any nonzero real number is a positive number.**

Here we designate the absolute value of zero and several other real numbers.

(a) $|0| = \mathbf{0}$ (b) $|-7.12| = \mathbf{7.12}$ (c) $|7.12| = \mathbf{7.12}$
(d) $|-5| = \mathbf{5}$ (e) $|5| = \mathbf{5}$ (f) $|\frac{3}{4}| = \frac{3}{4}$

example 3.1 Simplify (a) $|-5|$, (b) $|11 - 2|$, (c) $-|20 - 2|$.

solution (a) The absolute value of -5 is 5.

$$|-5| = \mathbf{5}$$

(b) First we simplify within the vertical lines:

$$|11 - 2| = |9|$$

and the absolute value of 9 is 9.

$$|9| = \mathbf{9}$$

(c) Again we simplify within the vertical lines:

$$-|20 - 2| = -|18|$$

The absolute value of 18 is 18, but we want the negative of this, so our answer is -18.

$$-|18| = \mathbf{-18}$$

3.B
addition of signed numbers

In arithmetic the minus sign always means to subtract, but in algebra we also use the minus sign to designate that a number is a negative number. This can be confusing at first as we see if we look at the expression

$$3 - 2$$

and ask if the minus sign means to subtract or if it means that -2 is a negative number. It turns out that we will find the same answer with either thought process, but in algebra we normally prefer the second process in which we think of the negative sign as designating that -2 is a negative number. If we do this, we say that we are using algebraic addition.

        To help explain the rules for algebraic addition, we will use diagrams drawn on a number line.

We will represent signed numbers with arrows and say that the arrows indicate directed numbers. We represent positive numbers with arrows that point to the right and negative numbers with arrows that point to the left. The length of each arrow corresponds to the absolute value of the number represented. For instance, $+3$ and -2 can be represented with the following arrows.

To use these arrows to add $+3$ and -2, we begin at the origin and draw the $+3$ arrow pointing to the right.

Then from the head of this arrow we draw the -2 arrow, which points to the left. The head of the -2 arrow is over $+1$ on the number line.

This is a graphical solution to the problem

$$(+3) + (-2) = \textbf{+1}$$

Please note that we obtain the same answer when we add signed numbers algebraically as we do when we use only the positive numbers of arithmetic and subtract!

$$3 - 2 = \textbf{1}$$

It may seem that we are trying to turn an easy problem into a difficult problem, but such is not the case. **In algebra the operations of addition and subtraction are lumped together in the one operation of algebraic addition, and this enables a straightforward solution to problems that would be very confusing if the concepts of signed numbers and algebraic addition were not used.**

example 3.2   Use directed numbers and the number line to add $+3$ and -5.

solution

We begin at the origin and draw an arrow 3 units long that points to the right to represent the number $+3$. From the head of this arrow we draw an arrow 5 units long that points to the left to represent the number -5. The head of the second arrow is just above the number -2 on the number line. Thus we see that

$$(+3) + (-5) = -2$$

example 3.3   Use directed numbers and the number line to add -5 and $+3$.

solution

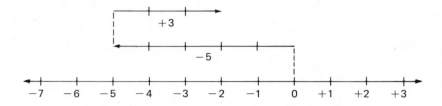

We will use the same arrows, but this time we will draw the -5 arrow first and then draw the $+3$ arrow. We note that again we get an answer of -2. This demonstrates that we may exchange the order of the numbers in an addition problem without changing the answer to the problem. Since the numbers to be added are called addends, we can make a formal statement of this peculiarity† of real numbers as follows:

> EXCHANGE OF ADDENDS IN ADDITION
>
> The order of addends in a real number addition problem does not affect the sum.

When the signed numbers to be added have the same signs, the arrows will point in the same direction as we see in the next two examples.

example 3.4   Use directed numbers and the number line to add $+2$ and $+1$.

solution

We see from the graph that the solution is $+3$.

$$(+2) + (+1) = +3$$

example 3.5   Use directed numbers and the number line to add -2 and -1.

solution

---

† The Latin word for exchange is *commutare*, and in later algebra courses this definition will be restricted to two numbers and called the commutative property. A full discussion of the properties of the set of real numbers is in the enrichment lessons in the appendix.

We see from the graph that the solution is -3.

$$(-2) + (-1) = -3$$

The numbers to be added may also be exchanged when three or more numbers are being added. To demonstrate this we will add four signed numbers, and then exchange the order of the numbers and work the problem again. The sum will be the same.

example 3.6   Use directed numbers and the number line to add these numbers:

$$(-4) + (+2) + (-1) + (+5)$$

solution   We will use arrows and add the numbers in the order they are written.

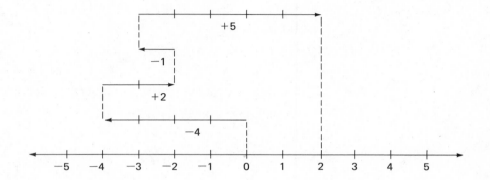

We began at the origin and moved 4 units to the left for -4, then 2 units to the right for $+2$, then 1 unit to the left for -1, and finally 5 units to the right for $+5$ and find that we end up directly above the number $+2$ on the number line. Thus

$$(-4) + (+2) + (-1) + (+5) = 2$$

the answer will be the same regardless of the order in which we draw the arrows. To show this we will work the problem again with the order of the numbers changed.

$$(-1) + (+2) + (-4) + (+5)$$

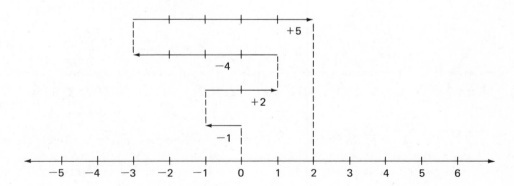

Again we find that the sum of the numbers is **2**.

problem   Simplify:
set 3   *1. $|-7|$ *2. $|+7|$ *3. $|-5|$

*4. $-|20 - 2|$ 5. $-|15 - 5|$

Draw a number line for each of the following problems and use directed numbers (arrows) to add the signed numbers.

*6. $(+3) + (-5)$ ***7.** $(-1) + (-2)$

8. $(+4) + (+3)$ ***9.** $(-4) + (+2) + (-1) + (+5)$

10. $(+3) + (-5) + (+7) + (-1)$ **11.** $(+7) + (-5) + (-3) + (-4)$

12. Designate the set of natural numbers.   **13.** Designate the set of integers.

14. What do we call the answer to a division problem?

15. What is the coordinate of a point on the number line?

16. What is the graph of a number?

17. What do we call the answer to a multiplication problem?

18. What is a factor?

19. When we divide 10 by 5 and get an answer of 2, we call 2 the quotient. What do we call 10 and 5?

20. What real number cannot be graphed on the number line?

Add, subtract, multiply or divide as indicated:

21. $1472\frac{1}{2} - 1432\frac{15}{16}$ **22.** $\frac{1}{2} + \frac{7}{4} + \frac{9}{8} - \frac{1}{16}$ **23.** $\frac{14}{32} \times \frac{8}{21}$

24. $5\frac{1}{3} + 7\frac{3}{8} - 1\frac{1}{4}$ **25.** $8.48636 \div 2.12$

26. $42.003 + 4.1 + 2.0606 - 1.3$ **27.** $.0402 \times 3.1604$

28. $2\frac{1}{4} \div 3\frac{1}{8}$ **29.** $7\frac{2}{5} \times 3\frac{5}{7}$ **30.** $7\frac{3}{8} + 7\frac{3}{5} - 3\frac{3}{10}$

31. $12.16608 \div 3.04$ **32.** $.00143 + .012 + 443.6 + .0007$

33. $3.628 \times .0404$ **34.** $4\frac{1}{4} \div 3\frac{2}{5}$ **35.** $\dfrac{2\frac{1}{8}}{3\frac{4}{3}}$

LESSON 4   *Rules for addition · Definition of subtraction*

4.A
rules for addition

In the last lesson we learned to add signed numbers by using a number line and arrows to represent the numbers. This procedure allows us to have a graphical picture of what we are doing. Unfortunately this method is slow and time-consuming. We do not have time to go through all algebra drawing number lines and arrows so we must develop rules that will allow us to do algebraic addition quickly. We need two rules—one to use when the numbers to be added have the same signs and one to use when the numbers have different signs. In the next problem, we will draw two diagrams that will help us state the first rule.

example 4.1 Use directed numbers and the number line to add $+1$ and $+3$ algebraically, and use directed numbers and the number line to add -1 and -3 algebraically.

solution $(+1) + (+3)$

$(-1) + (-3)$

We see from these figures that

$$(+1) + (+3) = \mathbf{+4} \quad \text{and that} \quad (-1) + (-3) = \mathbf{-4}$$

Now we will generalize. **To add algebraically two signed numbers that have the same sign, we add the absolute values of the numbers and give the result the same sign as the sign of the numbers.**

Now we will use two problems in which numbers with different signs are added algebraically to help us state the second rule.

example 4.2 Use directed numbers and the number line to add -2 and $+5$ algebraically, and use directed numbers and the number line to add $+2$ and -5 algebraically.

solution $(-2) + (+5)$ $(+2) + (-5)$

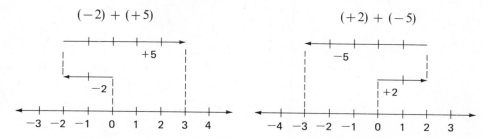

From the figure we see that the absolute value of each answer is 3 but that one of the answers is $+3$ and that one of the answers is -3.

$$(-2) + (+5) = \mathbf{+3} \qquad (+2) + (-5) = \mathbf{-3}$$

In the first case the number $+5$ had the larger absolute value and thus the sign of the result was positive. In the second case the number -5 had the larger absolute value and thus the sign of the result was negative. In both cases the absolute value of the answer was the difference in the absolute values of the numbers.

Now we will generalize. **To add algebraically two signed numbers that have opposite signs, we take the difference in the absolute values of the numbers and give to this result the sign of the original number whose absolute value is the greatest.**

When two numbers have the same absolute value but different signs, their sum is zero. For instance, the sum of (-5) and $(+5)$ is zero.

$$(-5) + (+5) = \mathbf{0}$$

We say that -5 is the opposite of $+5$ and that $+5$ is the opposite of -5. **Every real number except zero has an opposite, and the sum of any number and its opposite is zero. Another name for the opposite of a number is the additive inverse of the number, so we can also say that the sum of any number and its additive inverse is zero.**

Test your understanding of the rules by covering the answers to the following problems and seeing if your answers are the same.

(a) $(+7) + (-3) = \mathbf{+4}$ (e) $(+2) + (+6) = \mathbf{+8}$
(b) $(-7) + (-3) = \mathbf{-10}$ (f) $(-2) + (-8) = \mathbf{-10}$
(c) $(-7) + (+3) = \mathbf{-4}$ (g) $(-2) + (+8) = \mathbf{+6}$
(d) $(-4) + (-1) = \mathbf{-5}$ (h) $(+2) + (-8) = \mathbf{-6}$

4.B
adding more than two numbers

We have noted that signed numbers may be added in any order and the answer will not change. Many people add from left to right, and others begin by first adding numbers that have the same sign.

example 4.3   Add $(-5) + (4) + (-3) + (+2)$.

solution   This time we will add from left to right.

$$(-5) + (4) + (-3) + (+2) \qquad \text{original problem}$$
$$(-1) + (-3) + (+2) \qquad \text{added } (-5) \text{ and } (4)$$
$$(-4) + (+2) \qquad \text{added } (-1) \text{ and } (-3)$$
$$-2 \qquad \text{added } (-4) \text{ and } (+2)$$

example 4.4   Add $(-3) + (+2) + (-2) + (+4)$.

solution   We see that we have two negative numbers and two positive numbers. As the first step, we will add (-3) to (-2) and $(+2)$ to $(+4)$ and then add these sums.

$$(-3) + (+2) + (-2) + (+4) \qquad \text{original problem}$$
$$(-5) + (+6) \qquad \text{added } (-3) \text{ to } (-2) \text{ and } (+2) \text{ to } (+4)$$
$$+1 \qquad \text{added } (-5) \text{ to } (+6)$$

4.C
inserting parentheses mentally

Most signed number problems are written without having parentheses enclosing the signed numbers. We must insert the parentheses mentally before we can add. **We will let the sign preceding the number designate whether the number is a positive number or a negative number, and we will mentally insert a plus sign in front of each number to indicate algebraic addition.** If we use this process,

$$4 - 3 + 2 \qquad \text{can be read as} \qquad (+4) + (-3) + (+2)$$

and

$$-6 - 3 - 2 + 5 \qquad \text{can be read as} \qquad (-6) + (-3) + (-2) + (+5)$$

Thus, to simplify an expression such as

$$-4 + 2 - 3 - 3 - 2 + 6$$

we mentally enclose each of the numbers in parentheses, insert the extra plus signs, and then add.

$$(-4) + (+2) + (-3) + (-3) + (-2) + (+6) = -4$$

Care must be used to avoid associating the signs with the wrong numbers. If the mental parentheses are not used, some would incorrectly read the original expression from right to left as "6 plus 2 minus 3," etc. Guard against this.

example 4.5   Simplify $-4 - 3 + 2 - 4 - 3 - 2$.

solution   We mentally enclose each number in parentheses and use plus signs so that we can read the problem as

$$(-4) + (-3) + (+2) + (-4) + (-3) + (-2)$$

Now we add the numbers and get a sum of -14.

$$(-4) + (-3) + (+2) + (-4) + (-3) + (-2) = \mathbf{-14}$$

example 4.6   Simplify $-2 + 11 - 4 + 3 - 2$.

solution   We mentally enclose the numbers in parentheses and add algebraically to get a sum of $+6$.

$$(-2) + (+11) + (-4) + (+3) + (-2) = \mathbf{+6}$$

example 4.7   Simplify $(-4) + |-2| + 3 - 7 - 2$.

solution   We mentally insert parentheses so that the problem reads as follows.

$$(-4) + (|-2|) + (3) + (-7) + (-2)$$

Now we simplify and get an answer of -8.

$$(-4) + (2) + (3) + (-7) + (-2) = \mathbf{-8}$$

4.D
algebraic subtraction

As we have seen, if we use algebraic addition, we can handle minus signs without using the word *subtraction*. We let the signs tell whether the numbers are positive or negative, and we mentally insert parentheses and extra plus signs as necessary. Thus the subtraction problem on the left

$$7 - 4 = 3 \qquad 7 + (-4) = 3$$

can be turned into the algebraic addition problem on the right. A definition of algebraic subtraction does exist, however, and some people prefer to use it rather than using mental parentheses. The result is exactly the same, but the definition uses the word *subtraction*. **To subtract algebraically, we change the sign of the subtrahend and add.**

$$7 - 4 = 3 \qquad 7 + (-4) = 3$$

The formal definition of the operation of algebraic subtraction is as follows.

ALGEBRAIC SUBTRACTION

If a and b are real numbers, then

$$a - b = a + (-b)$$

where $-b$ is the opposite of b.

Thus there are two thought processes that may be used to simplify expressions that contain minus signs such as

$$7 - 4$$

Since we prefer to consider that the minus sign designates a negative number, we will discuss only algebraic addition in this book and will avoid the use of the word subtraction whenever we can.

problem set 4

1. State the rule for adding two numbers whose signs are alike.

2. State the rule for adding two numbers whose signs are different.

3. What is (a) a factor? (b) a quotient? (c) a product?

Use the rules for addition to simplify:

*4. $(-2) + (+5)$ ***5.** $(-1) + (-3)$

6. $(-3) + (-7)$ **7.** $(+3) + (-18)$

8. $(-11) + (-2)$ **9.** $(+3) + (-14)$

10. $(-3) + (-14)$ **11.** $(-14) + (-21)$

12. $(-32) + (+4)$ **13.** $(-7) + (-24)$

*14. $(-5) + (4) + (-3) + (+2)$ ***15.** $(-3) + (+2) + (-2) + (+4)$

16. $(-2) + (-5) + (3) + (-5)$ **17.** $(+2) + (-5) + (-3) + (-7)$

18. $(-5) + (-3) + (11) + (-2)$ **19.** $(-14) + (-3) + (-7) + (-14)$

Insert parentheses mentally and simplify:

*20. $-4 - 3 + 2 - 4 - 3 - 2$ ***21.** $-2 + 11 - 4 + 3 - 2$

22. $-11 - 3 + 14 - 2 - 5 + 7$ **23.** $-5 - 11 + 20 - 14 + 5$

24. $-2 - 8 + 3 - 2 + 5 - 7$ **25.** $7 - 3 - 2 - 11 + 4 - 5 + 3$

26. $-7 - 4 - 13 + 4 - 2 + 7$ **27.** $-8 + 13 - 4 + 13 - 2 - 5 - 7$

Both the preceding notations and absolute value are included in the following problems:

28. $-7 + (-8) + 3$ **29.** $+|-2| + (-5)$

30. $|-2 - 3| + (-5)$ **31.** $-7 + (-3) + 4 - 3 + (-2)$

*32. $-4 + |-2| + (-3) - 5$ **33.** $(-8) + (5) + (-10) - 4 - |-2|$

34. $(-5) + (-2) - 3 - 2 + 14$ **35.** $-2 + |-2|$

36. $|-2 - 3| - 2$ **37.** $-4 - 2 + (-8) + |-5|$

38. $+|-2 - 3| - 4 + (-8)$ **39.** $|-2| + (-2) - 2$

40. $+3 + |+7| - 3 + (-4)$

Simplify:

41. $17\frac{3}{8} - 9\frac{4}{5}$ **42.** $\frac{48}{7} \times \frac{21}{6} \times \frac{42}{33} \times \frac{5}{3}$

43. $(.0105)(.1417)$ **44.** $.0891 \div 4.05$

LESSON 5 *Opposites and multiple signs*

5.A

the opposite
of a number

We can use the thought "opposite of a number" to help us understand and simplify expressions such as $-(-2)$, $-(-(-2))$, $-(-(-(2)))$, etc. We begin by graphing the the number 2 and the number -2.

From the figure we see that the number -2 lies on the **opposite** side of the origin from 2 and is exactly the same distance from the origin. Thus we can think of $+2$ as being the **opposite of** -2 and -2 as being the opposite of $+2$. Often it is helpful to read a negative sign as "the opposite of." If we use this wording, it is easy to locate $-(-2)$, for we read this as the opposite of the opposite of 2. Well, the opposite of 2 is -2 so the opposite of that must be 2 itself.

Thus 2 and $-(-2)$ are different numerals or symbols for the same number. If this is true, then where does $-(-(-2))$ lie? We can read this as the opposite of the opposite of the opposite of 2.

If we begin at 2, we can find the opposite of 2 at -2. Then the opposite of the opposite of 2 is back on the right side, so the opposite of the opposite of the opposite of 2 is on the left and is another way to write -2. Of course, we could go on forever with this process—but we won't.

5.B
simplifying more difficult notations

Complicated expressions such as

$$-(-4) + (-2) + [-(-6)]$$

can be simplified by using algebraic addition and the concept of opposites. We begin by noting that algebraic addition of three numbers is indicated. We emphasize this by enclosing the numbers that are to be added and writing plus signs between the enclosures.

$$\boxed{-(-4)} + \boxed{+(-2)} + \boxed{+[-(-6)]}$$

The number in the first enclosure is $+4$, in the second is -2, and in the third is $+6$. So we can write

$$(4) + (-2) + (6) = \mathbf{8}$$

example 5.1 Simplify $-(+4) - (-5) + 5 - (-3) + (-6)$.

solution This problem indicates addition of five numbers.

$$\boxed{-(+4)} + \boxed{-(-5)} + \boxed{(+5)} + \boxed{-(-3)} + \boxed{+(-6)}$$

We simplify within each enclosure and add algebraically.

$$(-4) + (+5) + (+5) + (+3) + (-6) = \mathbf{3}$$

example 5.2 Simplify $-(-3) - [-(-2)] + [-(-3)]$.

solution We see three numbers are to be added. We begin by enclosing each number and inserting the necessary plus signs.

$$\boxed{-(-3)} + \boxed{-[-(-2)]} + \boxed{+[-(-3)]}$$

Now we simplify within each enclosure and then add

$$3 + (-2) + (3) = \mathbf{4}$$

example 5.3   Simplify $-(-4) + (-2) - [-(-6)]$.

solution   This time we will picture the enclosures mentally but won't write them down. If we do this, we can simplify the given expression as

$$(4) + (-2) + (-6) = -4$$

example 5.4   Simplify $-(+4) - (-5) + 5 - (-3) + (-6)$.

solution   This time we won't even use parentheses but will write the simplification directly as

$$-4 + 5 + 5 + 3 - 6 = 3$$

It will take a lot of practice to become adept in doing simplifications such as this one. Don't get discouraged if you find these problems to be troublesome.

problem set 5

1. The opposite of 4 is -4 and the opposite of -4 is 4. What is the opposite of $-45,654$?

2. What is the sum of a number and its opposite?

Use the concept of opposites to simplify:

3. $-(+4)$ 4. $-(-4)$

5. $-[-(-4)]$ 6. $-\{-[-(-4)]\}$

Use the concept of opposites and algebraic addition to simplify the following. Use additional plus signs and brackets as required.

*7. $-(-4) + (-2) + [-(-6)]$

*8. $-(+4) - (-5) + 5 - (-3) + (-6)$

*9. $-(-4) + (-2) - [-(-6)]$

10. $+7 - (-3) + (-2)$ 11. $-3 + (-2) - (-3)$ 12. $4 - (-3) - 7 + (-2)$

13. $3 - (+4) - (-2)$ 14. $-6 - (-8) - (-6)$ 15. $-2 - |-2|$

16. $6 + |-2|$ 17. $-3 - (-3) + |-3|$

18. $-2 - (-3) - \{-[-(-4)]\}$

19. $-2 + 5 - (-3) + |-3|$ 20. $-|-10| - (-10)$

21. $-2 - (-(-6)) + |-5|$ 22. $-7 + (-5) - (+5) - |2|$

23. $|-2 - 5 - 7| - |-4|$ 24. $-8 - 3 - 4 - (-10) + |12|$

25. $|7 - 3| - (-2) + 7 - 4 + |-11|$ 26. $-4 - (-3) - 7 + (-3)$

27. $-(-5) + (-2) - 3 - |-14|$ 28. $-(-2) - (+2) - 3 - (-3)$

29. $-|-3 - 2| - (-3) - 2 - 5$ 30. $-(-3) - [-(-4)] - 2 + 7$

Simplify:

31. $31\frac{3}{8} - 4\frac{7}{15}$ 32. $3\frac{2}{5} \div 3\frac{1}{4}$ 33. $\dfrac{7\frac{1}{3}}{3\frac{9}{7}}$

34. $.416 + 5.007$ 35. $.00402 \div .01$ 36. $.3004 \times 21.02$

37. $\dfrac{.0612}{1.02}$

LESSON 6   *Rules for multiplication of signed numbers*

6.A
multiplication of signed numbers

The sum of three 2s is 6. Also the sum of two 3s is 6.

$$2 + 2 + 2 = 6 \qquad 3 + 3 = 6$$

We can get the same results from multiplication by writing

$$3 \cdot 2 = 6 \qquad \text{or} \qquad 2 \cdot 3 = 6$$

because multiplication is just a shorthand notation for repeated addition. Thus, if we wish to use the number line to explain the multiplication of $3 \cdot 2$, we can do it two ways. We can show the sum of two 3s or the sum of three 2s.

Now let's find the product of 2 and -3 on the number line. We can obtain the same answer by adding two -3s.

Thus we see that

$$-3 - 3 = -6 \qquad \text{so} \qquad 2 \text{ times } -3 = -6$$

But now if we attempt to use the number line to show -3 times 2 by trying to draw -3 arrows that are $+2$ units long, we find that the task is impossible because we don't know how to draw -3 arrows, for any number of arrows we draw will be a number equal to or greater than 1.

The number line was a useful graphic aid in understanding the concept of signed numbers and the algebraic addition of signed numbers but is of less help when discussing the multiplication of signed numbers (and also the division of signed numbers), so we will not try to use it further for this purpose.

We could now just give the rules for multiplication and be done with it, but let's try to get some feeling for why the rules are as they are.

We need three rules for multiplication:

1. A rule for use when both factors are positive.
2. A rule for use when one factor is negative and the other factor is positive.
3. A rule for use when both factors are negative.

6.B
multiplication with both factors positive

We use the first figure in this lesson to justify the following rule.

The product of two positive real numbers is a positive real number whose absolute value is the product of the absolute values of the two numbers.

EXAMPLES

(a) $(+4)(+5) = 20$ (b) $4(3) = 12$ (c) $2 \cdot 9 = 18$

6.C
multiplication with one negative factor and one positive factor

Now we use the same figure again and note that 3 times 2 equals 6.

$$3(2) = 6$$

So it seems reasonable that the product of the opposite of 3 times 2 would be the opposite of 6, or -6. It is!

$$(-3)(2) = -6 \quad \text{and also} \quad (3)(-2) = -6$$

The product of two signed real numbers that have opposite signs is a negative real number whose absolute value is the product of the absolute values of the numbers.

EXAMPLES

(a) $-3(5) = -15$ (b) $(4)(-2) = -8$ (c) $(-2)(6) = -12$

6.D
multiplication with both factors negative

We know that 3 times 2 equals 6,

$$3 \cdot 2 = 6$$

and in the last paragraph we said that 3 times the opposite of 2 equals the opposite of 6.

$$3(-2) = -6$$

Now it seems reasonable to guess that the product of the opposite of 3 times the opposite of 2 would be the opposite of the opposite of 6, which is 6 itself. It is!

$$(-3)(-2) = +6$$

The product of two negative real numbers is a positive real number whose absolute value is the product of the absolute values of the two numbers.

EXAMPLES

(a) $-3(-2) = 6$ (b) $(-5)(-4) = 20$ (c) $-5(-3) = 15$

problem set 6

1.   What is the coordinate of a point on the number line?

2.   What is the additive inverse of -2?

3.   How do we tell if one number is greater than another number?

4.   When we multiply numbers, what do we call the answer?

5.   What do we call the answer to a division problem?

6.   Indicate the set of integers.

Simplify:

7. $(-4)(-3)$ 8. $-4(-3)$

9. $4(-3)$ *10. $(-3)(2)$

11. $(-3)(-5)$ 12. $5(-2)$

13. $-5(2)$ 14. $-[-(-4)]$

15. $-|-2| - (-2)$ 16. $4 + (-2) + (-4)$

17. $-(-4) + (-2) - (-3)$ 18. $|-4| + 5 - 6 - |-2 - 4|$

J. C. Warren

19. $-3 + 7 - 8 - 5 - (4)$

20. $-|-2| + |2| - (-2)$

21. $-7 + 3 - 2 - 5 + (-6)$

22. $-3 + (-3) + (-6) - 2$

23. $-5 + 3 - 2 - 5 - (-2)$

24. $5 - 3 - (-2) - (-(-3))$

25. $-3 - (-3) + (-2) - (3)$

26. $|-4 - 3| - 2 + 7 - (-3)$

27. $7 - 4 - 5 + 12 - 2 - |-2|$

28. $|-3 - 2| - (-3) - 4 - 6$

29. $5 - |-2 + 5| - (-3) + 2$

30. $-8 + 5 - 3 - (-2) + (-3)$

31. $-3 + 7 - (-2) + (-3) - 2$

32. $4 - 3 - (-2) - |12 - 3 + 4|$

33. $41.263 + .002$

34. $21\frac{2}{5} - 7\frac{7}{8}$

35. $\dfrac{4\frac{2}{5}}{6\frac{1}{8}}$

36. $\dfrac{5\frac{1}{2}}{6\frac{2}{3}}$

37. $15\frac{1}{2} - 6\frac{2}{3}$

38. $\dfrac{15}{7} \times \dfrac{21}{5} \times \dfrac{2}{49}$

39. $-3\frac{1}{5} + 2\frac{1}{8}$

40. $\dfrac{3\frac{1}{5}}{2\frac{1}{8}}$

41. $4.05 \div .0005$

LESSON 7 *Inverse operations · Rules for multiplication and division*

7.A
inverse operations

If one operation will *undo* another operation, the two operations are called inverse operations. If we take a particular number and then add and subtract the same number, the result is the particular number itself. For example, if we begin with the number 7 and add 3 and then subtract 3, the result is 7.[†]

$$7 + 3 - 3 = \mathbf{7}$$

Thus addition and subtraction are inverse operations.

Multiplication and division are also inverse operations. If we multiply 7 by 2 and then divide by 2, the result is 7.

$$\frac{7 \cdot 2}{2} = \mathbf{7}$$

There has been no change since dividing by 2 *undoes* the effect of multiplying by 2.

Since multiplication and division are inverse operations, our rules for the multiplication and division of signed numbers must be stated in such a way that these operations are inverse operations.

[†] Now we use algebraic addition instead of subtraction.

7.B
division of one positive number by another positive number

In Lesson 6 we said that the product of two positive real numbers is a positive real number. For example, 4 times 3 equals 12.

$$4 \cdot 3 = \mathbf{12}$$

Since division undoes multiplication, we can divide the product by one of the factors to give the other factor:

$$\frac{12}{3} = \mathbf{4} \qquad \text{and} \qquad \frac{12}{4} = \mathbf{3}$$

If one positive real number is divided by another positive real number, the quotient is a positive real number whose absolute value is the quotient of the absolute values of the original numbers

7.C
division of negative numbers

In Lesson 6 we said that the product of two real numbers with opposite signs is a negative real number. Therefore, we must define the division of two numbers of opposite signs so that the division will be an inverse operation. Hence, 2 times the opposite of 3 is the opposite of 6.

$$(2)(-3) = \mathbf{-6}$$

Division must be defined in such a way that it will *undo* the multiplication.

$$\frac{-6}{-3} = \mathbf{2} \qquad \text{and} \qquad \frac{-6}{2} = \mathbf{-3}$$

Thus we see that we will have to define division so that the quotient of two negative real numbers is a positive real number and that the quotient of a negative real number divided by a positive real number is a negative real number.

If one negative real number is divided by another negative real number, the quotient is a positive real number whose absolute value is the quotient of the absolute values of the original numbers.

If a negative real number is divided by a positive real number, the quotient is a negative real number whose absolute value is the quotient of the absolute values of the original numbers.

We need to cover one more possibility. In Lesson 2 we found that the product of the opposite of 2 times the opposite of 3 equals 6.

$$(-2)(-3) = \mathbf{6}$$

Thus, since division is the inverse operation of multiplication and must *undo* the multiplication, the following must be true.

$$\frac{6}{-2} = \mathbf{-3} \qquad \text{and also} \qquad \frac{6}{-3} = \mathbf{-2}$$

This gives us the last case. The quotient of a positive real number divided by a negative real number is a negative real number.

If a positive real number is divided by a negative real number, the quotient is a negative real number whose absolute value is the quotient of the absolute values of the original numbers.

The following examples demonstrate all the rules for the division of signed numbers.

(a) $\dfrac{-12}{4} = \mathbf{-3}$ (b) $\dfrac{-10}{-2} = \mathbf{5}$ (c) $\dfrac{20}{-2} = \mathbf{-10}$ (d) $\dfrac{20}{2} = \mathbf{+10}$

7.D
rules for the division and multiplication of signed real numbers

We can consolidate all we have learned about the multiplication and division of signed numbers into two rules.

like signs

The product or the quotient of two signed numbers that have the same sign is a positive number whose absolute value is the absolute value of the product or the quotient of the absolute value of the original numbers.

unlike signs

The product or the quotient of two signed numbers that have opposite signs is a negative number whose absolute value is the absolute value of the product or the quotient of the absolute value of the original numbers.

We can state the rules above in a less rigorous but more easily remembered way if we say

---

IN BOTH MULTIPLICATION AND DIVISION

1. *Like* signs $\xrightarrow{\text{yield}}$ a positive number

2. *Unlike* signs $\xrightarrow{\text{yield}}$ a negative number

---

problem set 7

1. What operation is the inverse operation of multiplication?

2. What operation is the inverse operation of division?

3. What are inverse operations?

Simplify:

*4. $\dfrac{-12}{4}$

5. $4(-12)$

*6. $\dfrac{-10}{-2}$

7. $-10(-2)$

*8. $\dfrac{20}{2}$

9. $-4(-3)$

10. $\dfrac{100}{-5}$

11. $(-3)(8)$

12. $-|-5 + 3 - 2| + 2$

13. $-4 + 5 - 7 + (-3) - 2 + 7$

14. $-[-3(-2)] + (-5)$

15. $-3 - (+6) + (-6) - 2$

16. $-|-6| - [-(-2)] + 5$

17. $-5 - (3) - 2 + (-3)$

18. $-7 - 4 - (-3) + |-3|$

19. $-2 + (-3) - (-4) + 2$

20. $-|-3 - 3| - 2$

21. $-3 + (-3) - (-5) - |7|$

22. $-4 - (6) - (-3) - 2$

23. $-|-2 - 5 + 7| - (-3)$

24. $7 - |-3 - 5 + 1| - (-2)$

25. $-6 - 4 - (3) - (-3) + 3$

26. $5 - |-2 - 3| + |+7 - 3|$

27. $-6 + (-3) - [-(-2)] + 7$

28. $|-3| + (-2) - 4 + 7$

29. $-6 - (-6) - 4 + (-2) - 1$

30. $-5 + (-2) - (-7) - 4$

31. $8 - 7 + (-6) - 3 + (-5) - |-2|$

32. $3\dfrac{2}{5} - 7\dfrac{6}{15}$

33. $\dfrac{3\frac{2}{5}}{7\frac{6}{15}}$

34. $\dfrac{.0832}{4.16}$

35. $\dfrac{4\frac{2}{3}}{3\frac{1}{9}}$

LESSON 8 · *Division by zero · Exchange of factors in multiplication*

8.A
division by zero

The operation of division is the inverse operation of the operation of multiplication, for division is defined as the process that will *undo* multiplication. Thus if $3 \times 2 = 6$, it is necessary that

$$\frac{6}{3} = 2 \qquad \text{and} \qquad \frac{6}{2} = 3$$

We will now use the same thought process to try to decide what the result will be if we divide a nonzero number by zero. We will use the example of 6 divided by 0.

$$\frac{6}{0} = ?$$

Since we say that division *undoes* multiplication, the multiplication that is to be undone by the above division must be

$$6 = ? \cdot 0$$

But the product of zero and any real number is zero—it is not 6. There is no number that we can substitute for ? so that the product of ? and 0 equals 6. Therefore, we say that since the multiplication does not exist that is to be undone by the division, the expression

$$\frac{6}{0}$$

has no meaning, or is undefined. A similar reasoning process is used to show that we can't divide zero by zero, and we say that zero divided by zero is indeterminate rather than saying it is undefined. Thus indicated divisions such as

$$\frac{0}{0} \qquad \frac{142}{0} \qquad \frac{6}{0} \qquad \frac{-5}{0}$$

have no meaningful simplifications and must be avoided.

8.B
exchange of factors in multiplication

In Lesson 3, we noted that we can change the order in which signed numbers are added without changing the answer. Now we note that the order of multiplying signed numbers does not affect the answer. We will repeat the statement for addition here and make the companion statement about the exchange of factors in multiplication.

> EXCHANGE OF ADDENDS IN ADDITION
>
> The order of addends in a real number addition problem does not affect the sum.

> EXCHANGE OF FACTORS IN MULTIPLICATION
>
> The order of factors in a real number multiplication problem does not affect the product.

The order in which signed numbers are multiplied does not affect the value of the product!

example 8.1   Find the product of $(-4)(3)(-6)(-2)$.

solution

$$(-4)(3)(-6)(-2) \quad \text{given}$$
$$-12(-6)(-2) \quad \text{multiplied } -4 \text{ by } 3$$
$$72(-2) \quad \text{multiplied } -12 \text{ by } -6$$
$$-144 \quad \text{multiplied } 72 \text{ by } -2$$

In the first step we multiplied -4 by 3 and got -12. In the second step, we multiplied -12 by -6 and got $+72$, which we multiplied by -2 to get the final result of -144.

example 8.2   Find the product of $(-6)(-2)(3)(-4)$.

solution

$$(-6)(-2)(3)(-4) \quad \text{given}$$
$$12(3)(-4) \quad \text{multiplied } -6 \text{ by } -2$$
$$36(-4) \quad \text{multiplied } 12 \text{ by } 3$$
$$-144 \quad \text{multiplied } 36 \text{ and } -4$$

example 8.3   Find the product of $(-6)(-4)(3)(-2)$.

solution

$$(-6)(-4)(3)(-2) \quad \text{given}$$
$$24(3)(-2) \quad \text{multiplied } -6 \text{ by } -4$$
$$72(-2) \quad \text{multiplied } 24 \text{ by } 3$$
$$-144 \quad \text{multiplied } 72 \text{ by } -2$$

In each of the three examples above the same factors were multiplied, but the order of multiplication was different. The product was -144, however, regardless of the order of the factors.

problem set 8

1. Is the product of two negative numbers always a positive number?

2. Is the sum of a positive number and a negative number always a negative number?

3. Why is division by zero said to be undefined or indeterminate?

Simplify:

4. $(-4)(3)(-2)$

5. $4(-3)(-4)$

6. $(-2)(3)(4)$

7. $\dfrac{-15}{-3}$

8. $\dfrac{-4}{2}$

9. $\dfrac{4}{-2}$

10. $(-3)(2)(-1)(3)$

11. $(-2)(-3)(-2)(2)$

12. $-4(-2)(-3)$

13. $4(-2)(-3)(2)$

14. $-3 - 6 + 5 - 2 + 4 - 3$

15. $-3 + (-2) + 3 - (-4)$

16. $|-2| + |-4 - 5| + 2$

17. $-2 - (-3) + (-4) - |-3|$

18. $-|-2 + 3 - 5| - |-3 - 6|$

19. $-2 - (-3) - |-4 - 3| + 2$

20. $-\{-[-(-2)]\} - |-4 - 2|$

21. $-5 + (-3) - (-2) + 2$

22. $-|-3 - 2| + (-5)$

23. $7 + 5 - 3 - 2 + (-5)$

24. $-[-(-4)] - (-3) + 2$

25. $|-6| + |-3| - 5 + (-3)$

26. $-4 - 3 - (+3) + (-3)$

27. $|-2 - 5 + 7| - (-3) + 2$

28. $3 - (-4) + (-3) - (-4)$

29. $3 - |-2 - 3| + (-6) - (-3)$

30. $(4) + (-3) + (-2) - (-5)$

31. $4 - |-2| + |5| - 3 - (-2)$

32. $\dfrac{3\frac{2}{5}}{7\frac{1}{3}}$

33. $3\frac{2}{5} - 7\frac{1}{3}$

34. $7\frac{1}{3} \div 3\frac{2}{5}$

35. $4.00165 - 1.00072$

36. $.008484 \div .0028$

37. $-2\frac{3}{5} + 1\frac{2}{3}$

38. $-7\frac{4}{11} + 2\frac{7}{8}$

LESSON 9 *Reciprocal and multiplicative inverse · Order of operations*

9.A
reciprocal or multiplicative inverse

If one fraction is the inverted form of another fraction, each of the fractions is said to be the **reciprocal** of the other fraction.

$$\frac{2}{3} \quad \text{is the reciprocal of} \quad \frac{3}{2}$$

$$\frac{3}{2} \qquad \text{is the reciprocal of} \qquad \frac{2}{3}$$

$$-\frac{4}{11} \qquad \text{is the reciprocal of} \qquad -\frac{11}{4}$$

$$-\frac{11}{4} \qquad \text{is the reciprocal of} \qquad -\frac{4}{11}$$

Since numbers such as 4 can also be written in a form such as $\frac{4}{1}$, these numbers also have reciprocals.

$$\frac{1}{4} \qquad \text{is the reciprocal of} \qquad 4$$

$$4 \qquad \text{is the reciprocal of} \qquad \frac{1}{4}$$

$$-5 \qquad \text{is the reciprocal of} \qquad -\frac{1}{5}$$

$$-\frac{1}{5} \qquad \text{is the reciprocal of} \qquad -5$$

The number zero does not have a reciprocal because if we try to write the reciprocal of zero we get

$$\frac{1}{0} \qquad \text{(meaningless)}$$

which we say is a meaningless notation because division by zero is undefined. **Zero is the only real number that does not have a reciprocal.** The **reciprocal** of a number is often called the **multiplicative inverse** of the number.

DEFINITION OF RECIPROCAL OR MULTIPLICATIVE INVERSE

For any nonzero real number *a*, the reciprocal or multiplicative inverse of the number is $\frac{1}{a}$.

If a number is multiplied by its multiplicative inverse or its reciprocal, the product is the number 1. Thus

$$4 \cdot \frac{1}{4} = \mathbf{1}, \qquad -5 \cdot \frac{1}{-5} = \mathbf{1}, \qquad \text{and} \qquad -\frac{1}{13} \cdot (-13) = \mathbf{1}$$

This simple fact is of great importance and will be very useful in the solutions of equations, a topic that will be discussed later in the book.

9.B
order of operations

If we wish to compute the value of

$$4 + 3 \cdot 2$$

we have a problem. It appears that there are two possible solutions.

(a) $4 + 3 \cdot 2$

Here we will first multiply 3 by 2 to get 6

$$4 + 6$$

and then add to get 10

$$4 + 6 = \mathbf{10}$$

(b) $4 + 3 \cdot 2$

Here we will first add 4 and 3 to get 7

$$7 \cdot 2$$

and then multiply to get 14

$$7 \cdot 2 = \mathbf{14}$$

We worked the problem two ways and got two different answers. **Neither way was basically more correct than the other, but since there are two possible ways to work the problem, mathematicians have found it necessary to agree on one way so that everyone will get the same answer. They have agreed to do the multiplications first and then to do the additions.** Thus, to simplify an expression such as

$$4 \cdot 3 + 5 - 6 + 4 - 3 \cdot 5 + 6 - 4 \cdot 2$$

we will use a two-step process. First we will perform all the multiplications and get

$$12 + 5 - 6 + 4 - 15 + 6 - 8$$

Now we will do the algebraic additions.

$$17 - 6 + 4 - 15 + 6 - 8 \qquad \text{added 12 and 5}$$
$$11 + 4 - 15 + 6 - 8 \qquad \text{added 17 and } -6$$
$$15 - 15 + 6 - 8 \qquad \text{added 11 and 4}$$
$$0 + 6 - 8 \qquad \text{added 15 and } -15$$
$$\mathbf{-2} \qquad \text{added 6 and } -8$$

example 9.1   Simplify $4 \cdot 3 + 2 + (-3)5$.

solution   We perform the multiplications first and get

$$12 + 2 - 15$$

which we now add algebraically.

$$12 + 2 - 15 = \mathbf{-1}$$

example 9.2   Simplify $-5(2) - 3 + 6(3)$.

solution   We perform the two multiplications first

$$-10 - 3 + 18$$

and now add to get

$$\mathbf{5}$$

example 9.3   Simplify $4 \cdot 3 - 2 \cdot 5 + 6 - 5 \cdot 2$.

solution
$$12 - 10 + 6 - 10 \qquad \text{performed multiplications}$$
$$\mathbf{-2} \qquad \text{added algebraically}$$

9.C
identifying multiplication and addition

When confronted with an expression such as

$$4 - 3(5) - 7(-6) - (4)(-5)$$

the beginner often has difficulty telling whether quantities are to be added or multiplied.

There is an easy way to identify indicated multiplication. **If there is no + or − sign between symbols, multiplication is indicated.** Let's simplify the expression above from left to right by first performing the indicated multiplications. The first place where there is no sign between the symbols is between the 3 and the parentheses enclosing the 5. The second place is between the 7 and the parentheses enclosing the −6, and the third is between the parentheses enclosing both the 4 and the −5.

$$4 - 3(5) - 7(-6) - (4)(-5)$$

The places where multiplication is indicated are designated by arrows. If we perform the indicated multiplications, we have

$$4 - 15 - (-42) - (-20)$$

Now we can simplify this expression and add.

$$4 - 15 + 42 + 20 = \mathbf{51}$$

problem set 9

1. Which real number does not have a reciprocal and why?

2. (a) What is the product of any nonzero real number and its reciprocal? (b) Is this product sometimes a negative number?

3. (a) What is a quotient? (b) What is a product?

4. What is the additive inverse of -8?

Simplify. Remember that multiplication is done before addition.

*5. $4 \cdot 3 + 2 + (-3)5$ 　　　　　　　　***6.** $-5(2) - 3 + 6(3)$

*7. $4 \cdot 3 - 2 \cdot 5 + 6 - 5 \cdot 2$ 　　　　***8.** $4 - 3(5) - 7(-6) - 4(-5)$

9. $6 - 8 + 2(3)$ 　　　　　　　　　　**10.** $3 - 2 \cdot 4 + 3 \cdot 2$

11. $-5 - 7 - 3 \cdot 2$ 　　　　　　　　**12.** $4 - 5(-5) + 3$

13. $-2(-2) - 2 - 2$ 　　　　　　　　**14.** $-3(-2)(-3) - 2$

15. $-3 - 6 + 2 \cdot 5$ 　　　　　　　　**16.** $-6(-2) - 3(-2)$

17. $-4(-3) + (-2)(-5)$ 　　　　　　**18.** $-2 - 3(+6)$

19. $\dfrac{-18}{-9}$ 　　　　　　　　　　　**20.** $(-2)(-2) - 2$

21. $-3 + (-2) - (+2)(2)$ 　　　　　**22.** $-2 - 2(-2) + (-2)(-2)$

23. $(-2)(-2)(-2) - |-8|$ 　　　　　　**24.** $-(-5) + (-2) + (-5)|-3|$

25. $(-7)(2) - 2(-3) + 6$ 　　　　　　**26.** $(-5) - (-5) + 2(-2) + 4$

27. $-3 - (-2) + (-3) - 2(-2)$ 　　　**28.** $3(-2) - |(-3)(+3)| + 9 - 7(-2)$

29. $(-5)(-5) - 5(2) + 3$ 　　　　　　**30.** $(-3)(-3) - 3 - 2|(-3)(2) + 5|$

31. $3\dfrac{2}{7} - 7\dfrac{6}{15}$ 　　　　　　　**32.** $\dfrac{3\dfrac{1}{5}}{7\dfrac{6}{15}}$

33. $\dfrac{4.16}{.52}$ 　　　　　　　　　　**34.** $2\dfrac{3}{8} \cdot 5\dfrac{3}{5}$

LESSON 10   *Symbols of inclusion · Order of operations*

10.A
symbols of inclusion

In Lesson 9 we found that the simplification of

$$4 + 3 \cdot 2$$

is 10 because we have agreed to do the multiplication first and then do the addition. So,

$$4 + 3 \cdot 2 = 4 + 6 = \mathbf{10}$$

Parentheses, brackets, braces, and bars are called **symbols of inclusion,** and can be used to help us emphasize the meaning of our notation. Using these symbols, the notation above could be written in any of the following ways:

(a) $4 + (3 \cdot 2)$ (b) $4 + [3 \cdot 2]$ (c) $4 + \{3 \cdot 2\}$ (d) $4 + \overline{3 \cdot 2}$

Each of the notations emphasizes that 3 is to be multiplied by 2 and that 4 is to be added to this product. A further benefit of the use of symbols of inclusion is that a nonstandard order of operations can be indicated. For example, we can use parentheses to indicate that 4 is to be added to 3 and the result multiplied by 2 by writing

$$2(3 + 4) \qquad \text{or} \qquad (3 + 4)2$$

While bars and braces can be used as indicated above, we normally reserve the use of braces to indicate a set, and bars are most often used as fraction lines as shown here.

$$\frac{4 + (3 \cdot 2)}{5(2 - 3)} = \frac{4 + 6}{-5} = \frac{10}{-5} = \mathbf{-2}$$

The parentheses in the numerator are used to emphasize that 4 is to be added to the product of 3 and 2, and the parentheses in the denominator are used to designate that 5 is to be multiplied by the algebraic sum of 2 and -3.

10.B
order of operations

To simplify numerical expressions that contain symbols of inclusion, we begin by simplifying within the symbols of inclusion. Then we simplify the resulting expression, remembering that multiplication is performed before addition.

example 10.1   Simplify $4(3 + 2) - 5(6 - 3)$.

solution   First we will simplify within the parentheses.

$$4(5) - 5(3) \qquad \text{simplified within parentheses}$$
$$20 - 15 \qquad \text{multiplied}$$
$$\mathbf{5} \qquad \text{added algebraically}$$

example 10.2   Simplify $-3(2 - 3 + 5) - 6(4 + 2) - 3$.

solution   First we simplify within the parentheses, then multiply and finish by adding.

$$-3(4) - 6(6) - 3 \qquad \text{simplified within parentheses}$$
$$-12 - 36 - 3 \qquad \text{multiplied}$$
$$\mathbf{-51} \qquad \text{added algebraically}$$

example 10.3   Simplify $-2(-3 - 3)(-2 - 4) - (-3 - 2) + 3(4 - 2)$.

solution   We begin by simplifying within the parentheses.

$$-2(-6)(-6) - (-5) + 3(2) \qquad \text{simplified within parentheses}$$

$$-72 + 5 + 6 \qquad \text{multiplied}$$

$$\mathbf{-61} \qquad \text{added algebraically}$$

When the expression is in the form of a fraction, we begin by simplifying both the numerator and the denominator. Then we have our choice of dividing or leaving the result in the form of a fraction.

example 10.4   Simplify $\dfrac{5(-5 + 3) + 7(-5 + 9) + 2}{(4 - 2) + 3 + 5}$.

solution   First we will simplify the numerator and the denominator.

(a) $\dfrac{5(-2) + 7(4) + 2}{2 + 3 + 5}$ simplified within parentheses

(b) $\dfrac{-10 + 28 + 2}{2 + 3 + 5}$ multiplied

(c) $\dfrac{20}{10}$ added algebraically

(d) **2**   divided

example 10.5   Simplify $\dfrac{-3(4 - 2) - (-5)}{4 - (3)(-3)}$.

solution   First we simplify above and below.

(a) $\dfrac{-3(2) + 5}{4 - (3)(-3)}$ simplified within parentheses

(b) $\dfrac{-6 + 5}{4 + 9}$ multiplied

(c) $\dfrac{-1}{13}$ added algebraically

(d) $-\dfrac{1}{13}$ this doesn't divide evenly, and we will leave it in fractional form

problem set 10

1. Designate the set of whole numbers.

2. Designate the set of integers.

3. (a) What is a factor? (b) What is a quotient? (c) What is a sum?

Simplify:

*4. $4(3 + 2) - 5(6 - 3)$ **5.** $-3(-6 - 2) + 3(-2 + 5)$

6. $(-2 - 2)(-3 - 4)$ **7.** $(-4 + 7) + (-3 - 2)$

8. $(-3 - 2) - (-6 + 2)$ **9.** $(-3 - 2)(-2)(-2 - 2)$

*10. $\dfrac{5(-5 + 3) + 7(-5 + 9) + 2}{(4 - 2) + 3 + 5}$ ***11.** $\dfrac{-3(4 - 2) - (-5)}{4 - 3(-3)}$

12. $5(9 + 2) - (-4)(5 + 1)$ **13.** $8(12 + 4) + (-4)(8 + 4)$

14. $\dfrac{-150}{-25}$

15. $\dfrac{75}{-3}$

16. $-2(-5 - 7) - 3(-8 + 2)$

17. $(-2 - 7 + 4) - (-3 - 2)$

18. $(2 - 3)(-8 + 2) + |-3 + 5|$

19. $-4 - 6 - (-3) - (-3 - 8)$

20. $\dfrac{1}{4}(8 - 4) - 5(8 - 2) - 2$

21. $(6 - 2)(-3 - 5) - (-5)$

22. $-|-2 - 5 + 3|(5 - 2)$

23. $2(2 - 4) - 8 - 6(7 + 3) - |-2|$

24. $4(8 + 4) + 7(10 - 8)$

25. $-8 - 4 - (-2) - (+2)(-3)$

26. $6(10 + 3) + 2(-3 - 2)(-2 - 2)$

27. $5(12 + 2) - 6(-3 + 8) - (2 + 3)$

28. $-6 - (+3) + (-3) - 5(4 - 3)$

29. $4 - 6 - 2(-3) - 5(6) + 7$

30. $7(14 - 7) - 6(-12 - 4)$

31. $2 - 4 - 5(-2) + 5(-2) - 4$

32. $-8(+2) + 3(-2)(4 - 3)$

33. $\dfrac{.01608}{-.004}$

34. $\dfrac{3\frac{2}{3}}{-5\frac{1}{6}}$

35. $-4\frac{1}{5} + 2\frac{1}{3}$

LESSON 11 *Multiple symbols of inclusion*

11.A
multiple symbols of inclusion

Often we encounter expressions such as

$$-3[(-2 - 4) - 3] - 2$$

where symbols of inclusion are within other symbols of inclusion. We simplify these expressions by beginning with the innermost symbol of inclusion and working our way out. Here we will simplify within the parentheses and then within the brackets.

$-3[(-6) - 3] - 2$	simplified within parentheses
$-3[-9] - 2$	simplified within brackets
$27 - 2$	multiplied
25	added

example 11.1   Simplify $4\{2[(-3 - 2)(-7 + 4) - 5]\} - 2$.

solution   We will begin on the inside with the parentheses and work our way out.

$4\{2[(-5)(-3) - 5]\} - 2$	simplified within parentheses
$4\{2[10]\} - 2$	simplified within brackets
$4\{20\} - 2$	simplified within braces
$80 - 2$	multiplied
78	added

example 11.2   Simplify $\dfrac{-3\{[(-2-3)][-2]\}}{-3(4-2)}$.

solution   First we will simplify the numerator and the denominator. Then we will divide as the last step.

$$\dfrac{-3\{[(-5)][-2]\}}{-3(2)} \quad \text{simplified within parentheses}$$

$$\dfrac{-3\{10\}}{-3(2)} \quad \text{simplified within braces}$$

$$\dfrac{-30}{-6} \quad \text{multiplied}$$

$$5 \quad \text{divided}$$

problem set 11

1. Designate the set of whole numbers.

2. How do we define real numbers?

3. (a) What is a factor? (b) A product? (c) A quotient?

Simplify:

4. $-2(-6-3)$

5. $(-7)-(-(-2))5$

6. $-2+3|-4|$

7. $(-3+2)(-6)$

8. $(-3-2)-(5+2)$

9. $(-3+5)(2-3)$

10. $(5-3)(-3)+(-2)7$

11. $-2-3-(-2)+(-3)$

12. $-2-(-2)-|-2|(2)$

13. $-|-4-2|(-2)+3$

14. $-3(-3)(-2-5+|-11|)$

15. $-2+3-2(-2)3$

16. $-|-11|+(-3)|-3+5|$

*17. $-3[(-2-4)-3]-2$

*18. $4\{2[(-3-2)(-7+4)-5]\}-2$

*19. $\dfrac{-3\{[(-2-3)][-2]\}}{-3(4-2)}$

20. $\dfrac{-3(4-2)-(-5)}{4-(3)(-3)}$

21. $\dfrac{-(-2-6)}{(-2)(-1-3)}$

22. $\dfrac{-(-3+5)+7}{4-(-3)}$

23. $-3-(-2)+(3-5)(-2)-5$

24. $\dfrac{3(-4-2)}{2(-3)(-4)}$

25. $-3-(-(-2))+(-3)(5)$

26. $-6-2-(-3)+2-6$

27. $\dfrac{-6(2)(-2)}{-(-5-3)}$

28. $\dfrac{-4(-2-2)}{-3-(-2)}$

29. $\dfrac{-3(-2+5)}{-5(-6+4)}$

30. $\dfrac{3.1563}{3.006}$

31. $\dfrac{3\frac{2}{5}}{-3\frac{1}{3}}$

32. $-3\frac{1}{5}+2\frac{3}{7}$

LESSON *12* *More on order of operations · Products of signed numbers*

12.A

more on order of operations

In the discussion of the order of operations in Lesson 9 we said that mathematicians have agreed that when they write

$$4 \cdot 3 + 2$$

the **multiplication should be done first** and then the addition.

$$12 + 2 \qquad \text{multiplied}$$

$$\mathbf{14} \qquad \text{added}$$

We did not discuss division because if symbols of inclusion are properly used, the order in which division is to be performed is apparent.

If we write

$$\frac{4 \cdot 3 + 2}{-7 + 5}$$

we find the value of the numerator and the value of the denominator and then divide.

$$\frac{14}{-2} = -7$$

If the following problem is encountered, however,

$$4 + \frac{14}{2} - 3 \cdot 6$$

the notation clearly indicates that only 14 is to be divided by 2, and if we do this first, we get

$$4 + 7 - 3 \cdot 6$$

Now we do the multiplication and conclude with algebraic addition.

$$4 + 7 - 18$$

$$11 - 18$$

$$-7$$

At this point in an algebra book, however, it is customary to give a rule for finding the number represented by

$$6 + 3 \cdot 6 \div 2 - 6 \cdot 2$$

The rule is to perform the operations from left to right in the following order.

1. Multiplication and division
2. Algebraic addition

First we will go through the problem from **left to right** performing the multiplications and divisions **in the order in which they are encountered.**

$6 + 3 \cdot 6 \div 2 - 6 \cdot 2$	original problem
$6 + 18 \div 2 - 6 \cdot 2$	multiplied 3 times 6
$6 + 9 - 6 \cdot 2$	divided 18 by 2
$6 + 9 - 12$	multiplied 6 times 2

Now we go through the problem again from left to right performing the algebraic additions as they are encountered.

$$6 + 9 - 12 \qquad \text{from above}$$

$$15 - 12 \qquad \text{added 6 and 9}$$

$$3 \qquad \text{added 15 and } -12$$

If symbols of inclusion had been properly used, instead of stating the problem as

$$6 + 3 \cdot 6 \div 2 - 6 \cdot 2$$

the problem would have been written as follows

$$6 + \frac{(3 \cdot 6)}{2} - (6 \cdot 2)$$

and here the method of solution is clearly indicated. We simplify within the parentheses as the first step.

$$6 + \frac{18}{2} - 12 \qquad \text{simplified within parentheses}$$

$$6 + 9 - 12 \qquad \text{divided 18 by 2}$$

$$15 - 12 \qquad \text{added 6 and 9}$$

$$3 \qquad \text{added 15 and } -12$$

We will use symbols of inclusion to include the use of a bar as a fraction line when stating problems. Thus problems such as the one just discussed will not be encountered again in this book.

12.B
products of signed numbers

Let's review the concept of the opposite of a number by watching the pattern that develops here.

	READ AS	WHICH IS
2	2	2
-2	the opposite of 2	-2
$-(-2)$	the opposite of the opposite of 2	2
$-[-(-2)]$	the opposite of the opposite of the opposite of 2	-2
$-\{-[-(-2)]\}$	the opposite of the opposite of the opposite of the opposite of 2	2

The expressions in the left-hand column are all equivalent expressions for 2 or for -2. If we look at the right-hand column, we see that every time an additional $(-)$ is included in the left-hand expression, the right-hand expression changes sign.

A similar alternation in sign occurs whenever a particular number is multiplied by a negative number. For instance,

$$-2 = -2$$

$$(-2)(-2) = +4$$

$$(-2)(-2)(-2) = -8$$

$$(-2)(-2)(-2)(-2) = +16$$

$$(-2)(-2)(-2)(-2)(-2) = -32$$

The numbers on the right have different absolute values, but they *alternate in sign*. We note that

The product of **two** negative factors is **positive**.

The product of **three** negative factors is **negative**.

The product of **four** negative factors is **positive**.

The product of **five** negative factors is **negative**.

Without proof we will generalize these observations.

1. **The product of an even number of negative real numbers is a positive real number.**
2. **The product of an odd number of negative real numbers is a negative real number.**

We can use these observations to determine the sign of the product of several signed numbers. Let's consider

$$(4)(-3)(-4)(-2)(11) = \;?$$

Here we have $+4$ and $+11$ as two of the five factors. Since multiplication by a positive number does not affect the sign of the product, we will not consider these numbers. The other three factors are negative. We can look at the rules developed above and see that the sign of the product of three negative numbers is negative. Thus our answer can be expressed as

$$-(4 \cdot 3 \cdot 4 \cdot 2 \cdot 11) = \mathbf{-1056}$$

example 12.1   Determine the signs of the following products and give the reasons. Do *not* do the multiplications.

solution

		SIGN	REASON
(a)	$(-4)(-3)(2)(+5)(+6)$	positive	Even number of negative factors
(b)	$(3)(+2)(6)$	positive	No negative factors
(c)	$(-3)(-2)(6)(4)(-2)$	negative	Odd number of negative factors
(d)	$(-3)(-2)(-5)(-7)(-2)$	negative	Odd number of negative factors
(e)	$(-3)(-4)(-2) + 2(-3)$	?	Rule does not apply as this is an indicated *sum*. We will do the problem in three steps.
(f)	$(-3)(-4)(-2)$	negative	Odd number of negative factors
(g)	$2(-3)$	negative	Odd number of negative factors
(h)	$-(3)(4)(2) - (2)(3)$	negative	Algebraic sum of two negative numbers is a negative number.

problem set 12

1. Is the product of 33 negative numbers and 2 positive numbers a positive number or a negative number?

2. Is the product of 33 negative numbers and 3 positive numbers a positive number or a negative number?

3. What is the reciprocal of -5?

4. What is another name for the reciprocal of a number?

Simplify:

5. $6 + \dfrac{(3 \cdot 6)}{2} - (6 \cdot 2)$ ***6.** $-[-(-2)]$

*7. $-\{-[-(-2)]\}$

9. $4 - \dfrac{(+12)}{(-3)} + 2$

11. $-3(-2 - 5)4$

13. $\dfrac{-6 - 3}{-4 + (4 - 3)}$

15. $\dfrac{-2(-6) - 2}{-3 + (-7 + 2)}$

17. $-3 - (2) + (-2) - (-3)(-2)$

19. $\dfrac{(-8 - 2)(-2)}{-2 - 6(2)}$

21. $-2(-6 - 1 - 2) - (-2 + 7)$

23. $(-5 - 6) - 2(3 - 6) + |-4|$

25. $-2[3 - 2 - (-3)] - [(3 - 2)(-2)]$

27. $-7 - (2) + (-2) - 3|-4|$

29. $-3(-2|-11|) + 5(-3 + 2)$

31. $-3\dfrac{1}{5} + 2\dfrac{1}{6}$

33. $-3\dfrac{1}{5} \div 2\dfrac{1}{6}$

*8. $(-2)(-2)(-2)(-2)$

10. $\dfrac{-6}{-1} + (-3)(-2) + 3|-4 - 2|$

12. $-5(-3 - 2) + (-2) - (-3 - 4)$

14. $\dfrac{-5(-2) - 4}{(-2 - 1)(-1)}$

16. $\dfrac{-2 - 4(-3 - 2)}{3 + (-2)(+7)}$

18. $\dfrac{(-3)|-1 - 4|}{-3 - |-2|}$

20. $-2 + (-2) - (-4)5$

22. $-3 - (-2 - 6)(-2 + 4)$

24. $-2 - 3 - (2)(-2) + (-1)(-3)$

26. $(-2)(-3)(-4 + 2) - (3 + 1)$

28. $5 - 6 - 4 + 3 - (-2)(-3 - 2)$

30. $3 - 6 - 2 - (-3)(-4) + 2$

32. $-.1386 \div .063$

LESSON 13   *Evaluation of algebraic expressions*

13.A
algebraic expressions

In Lesson 1 we said that a **number** is an **idea** and that when we wish to write down something to represent this **idea,** we use a **numeral.** If we wish to bring to mind the number 7, we could write any of the following:

$$7, \qquad \frac{14}{2}, \qquad 4 + 3, \qquad \frac{-21}{-3}, \qquad 2 + 2 + 2 + 1$$

We call each of these notations a **numerical expression** or just a **numeral.** Every numerical expression represents only one number and we call this number the **value** of the expression. Each of the numerical expressions shown above has a **value** of 7.

In algebra we often use letters to represent numbers. When letters as well as numbers are used in an expression, we don't call the expression a numerical expression but we call it by the more general name of **algebraic expression or mathematical expression.** These words are used to describe expressions that contain only numbers or only letters or contain both numbers and letters.

If we write the algebraic expression

$$4 + x$$

the expression has a **value** that depends on the value that we assign to x. If we give x a value of 5, then the expression has a value of 9 because

$$4 + 5 = \mathbf{9}$$

If we give x a value of 11, then the expression has a value of 15 because

$$4 + 11 = \mathbf{15}$$

Because the value assigned to x can be changed or varied, we call letters such as x **variables.**

We also call the letters **unknowns** since they represent unknown or unspecified numbers. The numeral 4 in this example does not change value and has a constant value of 4. For this reason the symbol that we use to denote a number is called a **constant.**

When we use variables in algebraic expressions, the notations that we use to indicate the operations of division and algebraic addition are the same as the notations that we use to indicate the division and algebraic addition of real numbers. The notation for the multiplication of variables is sometimes slightly different. We can denote that we wish to multiply 4 by the variable x by writing any of the following:

(a) $4x$ (b) $4(x)$ (c) $(4)(x)$ (d) $4 \cdot x$ (e) $(4) \cdot (x)$

The last four notations are the same as the notations that we use for real number multiplication, but the notation shown in (a) is different from the real number notation.

$4x$ indicates that 4 is to be multiplied by the value of x

whereas

45   does not indicate that 4 is to be
multiplied by 5 but instead is a
numeral that represents the number 45.

Thus the expression xym indicates that the values of x, y, and m are to be multiplied. If we give x a value of 1, y a value of 2, and m a value of 3, the value of the expression xym can be found.

$$1 \cdot 2 \cdot 3 = \mathbf{6}$$

If we write the algebraic expression

$$4x + mx$$

we indicate that 4 is to be multiplied by the value of x and that the value of m is to be multiplied by the value of x and that the two products are to be added. If we give x a value of 3 and m a value of 5, then we can find the value of the expression.

$$4 \cdot 3 + 5 \cdot 3$$
$$12 + 15 \quad = \mathbf{27}$$

Thus the value of the expression when x equals 3 and m equals 5 is 27.

If we give x the value of 2 and m the value of 6, then the expression will have a different value.

$$4x + mx = 4 \cdot 2 + 6 \cdot 2 = 8 + 12 = \mathbf{20}$$

In this case the value of the expression is 20. It is of **utmost importance** to note that in the first case when we gave x a value of 3, the value of x everywhere in the expression was 3. When we gave x a value of 2, the value of x everywhere in the expression was 2.

While the values assigned to variables may change or be changed, under any set of conditions the value assigned to a particular variable in an expression is the same value throughout the expression. Also, when we begin solving equations and working problems, we must remember that the value assigned to any particular variable under any set of conditions must be the same value regardless of where the particular variable appears in the equation or the problem.

example 13.1 Find the value of xmp if $x = 4$, $m = 5$, and $p = 2$

solution We replace x with 4, m with 5, and p with 2.

$$xmp = 4 \cdot 5 \cdot 2 = 20 \cdot 2 = \mathbf{40}$$

example 13.2 Evaluate (find the value of) $4yz - 5$ if $y = 2$ and $z = 10$.

solution We replace y with 2 and z with 10 and then simplify.

$$4yz - 5 = 4(2)(10) - 5 = 80 - 5 = \mathbf{75}$$

example 13.3 Evaluate $y - z$ if $y = -2$ and $z = -6$.

solution We replace y with -2 and $-z$ with $+6$ since $-z$ represents the opposite of z.

$$y - z = -2 + 6 = \mathbf{+4}$$

example 13.4 Evaluate $-a - b - ab$ if $a = -3$ and $b = -4$.

solution The value of a is -3 so the opposite of a is 3. The value of b is -4 so the opposite of b is $+4$. Finally, $ab = 12$ and the opposite of this is -12. Thus, we get -5 for an answer.

$$3 + 4 - 12 = \mathbf{-5}$$

example 13.5 Evaluate $-x - (-a + b)$ if $x = 2$, $a = -4$, and $b = -6$.

solution Some people find that it is helpful to replace each variable with parentheses. Then the proper number is written inside the parentheses.

$$-(\ ) - [-(\ ) + (\ )] \qquad \text{replaced variables with parentheses}$$
$$-(2) - [-(-4) + (-6)] \qquad \text{numbers inserted}$$

The first entry can be read as the opposite of 2, or -2. Inside the brackets we have $-(-4)$, read the opposite of the opposite of 4, which is 4 itself. The last entry inside the brackets is $+(-6)$, read plus the opposite of 6, which is the same as -6. Thus we have

$$-2 - (4 - 6) = -2 - (-2) = -2 + 2 = \mathbf{0}$$

example 13.6 Evaluate $x - y(-a + x)$ if $x = -2$, $y = +3$, and $a = -4$.

solution We will replace each variable with parentheses.

$$(\ ) - (\ )[-(\ ) + (\ )] \qquad \text{replaced variables with parentheses}$$
$$(-2) - (+3)[-(-4) + (-2)] \qquad \text{numbers inserted}$$
$$= -2 - (3)(4 - 2) = -2 - (3)(2) \qquad \text{simplified}$$
$$= -2 - 6 = \mathbf{-8} \qquad \text{simplified}$$

example 13.7   Evaluate $-(m + x)(-a + mx)$ if $m = 2, x = -3, a = -4$.

solution   We will replace each variable with parentheses.

$$-[(\ ) + (\ )][-(\ ) + (\ )(\ )]$$ replaced variables with parentheses
$$-[(2) + (-3)][-(-4) + (2)(-3)]$$ numbers inserted
$$= -(2 - 3)[4 + (-6)] = -(-1)(4 - 6)$$ simplified
$$= -(-1)(-2) = -(2) = \mathbf{-2}$$ simplified

example 13.8   Evaluate $-xa(x - a) + a$ if $a = -2$ and $x = 4$.

solution   We will replace each variable with parentheses.

$$-(\ )(\ )[(\ ) - (\ )] + (\ )$$ replaced variables with parentheses
$$-(4)(-2)[(4) - (-2)] + (-2)$$ numbers inserted
$$= -(-8)(4 + 2) - 2 = 8(6) - 2$$ simplified
$$= 48 - 2 = \mathbf{46}$$ simplified

problem set 13

1.   What is the difference between a numerical expression and an algebraic expression?

2.   What do we mean by the value of an expression?

3.   What is (a) a variable? (b) an unknown?

Evaluate:

*4. xmp if $x = 4, p = 2, m = 5$ *5. $4yz - 5$ if $y = 2$ and $z = 10$

*6. $y - z$ if $y = -2$ and $z = -6$ 7. $xy - y$ if $x = -2$ and $y = 3$

8. $xm - 2m$ if $x = -2$ and $m = -3$

9. $ma - m - a$ if $m = -2$ and $a = -4$

10. $2abc - 3ab$ if $a = 2, b = -3$, and $c = 4$

11. $xy - 3y$ if $x = 2$ and $y = 4$

*12. $-x - (-a + b)$ if $x = 2, a = -4$, and $b = -6$

*13. $-a - b - ab$ if $a = -3$ and $b = -4$

*14. $x - y(-a + x)$ if $x = -2, y = 3$, and $a = -4$

*15. $-(m + x)(-a + mx)$ if $m = 2, x = -3$, and $a = -4$

*16. $-xa(x - a) + a$ if $a = -2$ and $x = 4$

17. $|-b - a| + a$ if $a = -4$ and $b = 2$

18. $-a + (-a + b)$ if $a = -3$ and $b = -5$

19. $-xy - (-x + y)$ if $x = -3$ and $y = -4$

20. $-c - (p - c)$ if $p = -5$ and $c = 2$

21. $-xy - x(x - y)$ if $x = -4$ and $y = -1$

Simplify:

22. $-2[-3(-2 - 5)(3)]$ 23. $-3 - (-2) - 3(-2 + 5) + 2|-3|$

24. $-|-3|(2 - 5) - [-(-3)]$ 25. $-4 - 2(3 - 2) - (-2 - 5)$

26. $-3(-2 - 3)(5 - 7) - 2$

27. $5 - 3(-2 + 6) - (5 - 7) - 2$

28. $-5(-3 + 7)(-2)(-3 + 2)$

29. $-3(5 - 3) - (-2)(-6 - 1)$

30. $\dfrac{(-5 - 2) + (-3 - 2)}{-3 - (-2)}$

31. $\dfrac{-2[-(-3)]}{(-2)(-4 + 3)}$

32. $-3\dfrac{1}{4} + 2\dfrac{3}{11}$

33. $-5\dfrac{1}{3} \div 6\dfrac{2}{3}$

LESSON 14 *More complicated evaluations*

14.A
parentheses within brackets

The procedures discussed in Lesson 13 are also used to evaluate more complicated expressions. The use of parentheses, brackets, and braces is often helpful in preventing mistakes. We will use all of these symbols of inclusion in the following examples.

example 14.1 Evaluate $-a[-a(p - a)]$ if $p = -2$ and $a = -4$.

solution We use parentheses, brackets, and braces as required.

$$-(\ \ )\{-(\ \ )[(\ \ ) - (\ \ )]\}$$

Now we will insert the numbers inside the parentheses.

$$-(-4)\{-(-4)[(-2) - (-4)]\}$$

Lastly, we simplify:

$$4\{4[2]\} = 4\{8\} = \mathbf{32}$$

example 14.2 Evaluate $ax[-a(a - x)]$ if $a = -2$ and $x = -6$.

solution This time we will not use parentheses. We will replace a with -2, $-a$ with 2, x with -6, and $-x$ with 6.

$$12[2(-2 + 6)]$$

Now we simplify, remembering to begin with the innermost symbol of inclusion.

$$12[2(-2 + 6)] = 12[2(4)] = 12[8] = \mathbf{96}$$

example 14.3 Evaluate $-b[-b(b - c) - (c - b)]$ if $b = -4$ and $c = -6$.

solution We replace b with -4, $-b$ with 4, c with -6, and $-c$ with 6.

$$4[4(-4 + 6) - (-6 + 4)]$$

Now we simplify, remembering to begin within the innermost symbols of inclusion and to multiply before adding.

$$4[4(2) - (-2)]$$
$$4[8 + 2]$$
$$4[10]$$

**problem
set 14**

1. What is (a) a factor? (b) a quotient? (c) a sum?

Evaluate:

2. $x - xy$ if $x = -2$ and $y = -3$

3. $x(x - y)$ if $x = -2$ and $y = -3$

4. $(x - y)(y - x)$ if $x = 2$ and $y = -3$

5. $(x - y) - (x - y)$ if $x = -2$ and $y = 3$

6. $(-x) + (-y)$ if $x = -2$ and $y = 3$

7. $-xa(x - a)$ if $a = -2$ and $x = 4$

8. $(-x + a) - (x - a)$ if $x = -4$ and $a = 5$

9. $-x(a - xa)$ if $x = 3$ and $a = -5$

10. $-mp(p - m)$ if $m = -5$ and $p = 2$

11. $(p - x)(a - px)$ if $a = -3, p = 2,$ and $x = -4$

12. $(p - px) + (a + p)$ if $a = -3, p = 2,$ and $x = -4$

13. $(p - px) + (a + p)$ if $a = -5, p = -3,$ and $x = 4$

*14. $-a[-a(p - a)]$ if $p = -2$ and $a = -4$

*15. $ax[-a(a - x)]$ if $a = -2$ and $x = -6$

*16. $-b[-b(b - c) - (c - b)]$ if $b = -4$ and $c = -6$

17. $-a[(-x - a) - (x - y)]$ if $a = -3, x = 4,$ and $y = -5$

Simplify:

18. $-3[-2 - 5(3 - 7)]$

19. $-3 - (-2) - \{-[-(5)]\}$

20. $-2 + (-3) - |-5 + 2|3$

21. $-8 - 6(-2 - 1) + (-5)$

22. $-3[(2 - 5) - (3 - 1)]$

23. $\dfrac{-3(-6 - 2) + 5}{-3(-2 + 1)}$

24. $\dfrac{3(-2 - 1)}{-7(2 - 4)}$

25. $\dfrac{3(-2) - 5}{-3(-2)}$

26. $-2(-4) - \{-[-(-6)]\}$

27. $-3 - 2[(5 - 3)2 - (2 - 3)]$

28. $-2(-5 - 2) + (-3)(-6) - 2$

29. $5(-2 - 3) - 3(-2 + 5)$

30. $-3 - 2 - 5(-2 - 1) + (-3)$

31. $4[2(3 - 2) - (6 - 4)]$

32. $-2 - |-2 - 5| + (-3)(-6 - 2)$

LESSON 15   *Terms and the distributive property*

**15.A
factors
and
coefficients**

If the form in which variables and constants are written in an expression indicates that the variables and constants are to be multiplied, we say that the expression is an **indicated product.** If we write

$$4xy$$

we indicate that 4 is to be multiplied by the product of x and y. Each of the symbols is said to be a factor of the expression. Any one factor of an expression or any product of factors of an expression can also be called the **coefficient** of the rest of the expression. Thus in the expression $4xy$ we can say that

(a) 4 is the coefficient of xy $4(xy)$

(b) x is the coefficient of $4y$ $x(4y)$

(c) y is the coefficient of $4x$ $y(4x)$

(d) xy is the coefficient of 4 $xy(4)$

(e) $4y$ is the coefficient of x $4y(x)$

(f) $4x$ is the coefficient of y $4x(y)$

As mentioned earlier, the value of a product is not affected by the order in which the multiplication is performed, so we may arrange the factors in any order without affecting the value of the expression. You note that we change the order at will in (a) through (f) above.

If the coefficient is a number as in (a) above, we call it a **numerical coefficient,** and if the coefficient consists entirely of variables or letters as in (b), (c), and (d) above, we call it a **literal coefficient.** We need to speak of numerical coefficients so often that we usually drop the adjective *numerical* and use the single word *coefficient*. Thus in the following expressions

$$4xy, \qquad -15pq, \qquad 81xmz$$

4 is the coefficient of xy, -15 is the coefficient of pq, and 81 is the coefficient of xmz.

15.B
terms

A **term** is an algebraic expression that

1. Consists of a single variable or constant.
2. Is the indicated product or quotient of variables and/or constants.
3. Is the indicated product or quotient of expressions that contain variables and/or constants.

$$4, \qquad x, \qquad 4x, \qquad \frac{4xy(a+b)}{p}, \qquad \frac{3x+2y}{m}$$

All the expressions above can be called **terms.** The first two are terms because they consist of a single symbol. The third is a term because it is an indicated product of symbols. The fourth and fifth are terms because they are considered to be indicated quotients even though the numerator of the fourth term is an indicated product and the numerator of the fifth term is an indicated sum. **A term is thought of as a single entity that represents or has the value of one particular number.** The word *term* is very useful in allowing us to identify or talk about the parts of a larger expression. For instance, the expression

$$x + 4xym - \frac{6p}{y+2} - 8$$

is an expression that has four terms. We can speak of a particular term of this expression, say the third term, without having to write out the term in question. The terms of an expression are numbered from left to right beginning with the number 1. Thus, for the expression above:

The first term is $+x$ The third term is $-\dfrac{6p}{y+2}$.

The second term is $+4xym$.   The fourth term is -8.

If we consider that the sign preceding a term indicates addition or subtraction, then the sign is not a part of the term. In this book we prefer to use the thought of algebraic addition, and thus most of the time we will consider the sign preceding a term to be a part of the term. But we must be careful.

Let's look at the third term in the expression we are considering.

$$-\frac{6p}{y + 2}$$

If p and y are given values such that $\frac{6p}{y + 2}$ is a negative number, then $-\frac{6p}{y + 2}$ will be positive. For example, if p is equal to -4 and y is equal to 1, then the expression has a value of $+8$.

$$-\frac{6p}{y + 2} = -\frac{6(-4)}{1 + 2} = -(-8) = +8$$

15.C
the distributive property

We have noted that the order of adding signed numbers does not change the answer. Also, the order of multiplying signed numbers does not change the answer. We call these properties or peculiarities of the real numbers the **exchange property for addition** and the **exchange property for multiplication.**

Now we will discuss another property of real numbers that is of considerable importance, the **distributive property of real numbers.** If we write

$$4(5 - 3)$$

we indicate that we are to multiply 4 by the algebraic sum of the numbers 5 and -3. A property (or peculiarity) of real numbers permits the value of this product to be found two different ways.

$4(5 - 3)$	$4(5 - 3)$
$4(2)$	$4 \cdot 5 + 4(-3)$
8	$20 - 12$
	8

On the left we first added 5 and -3 to get 2, and then multiplied by 4 to get 8. On the right we first multiplied 4 by both 5 and -3, and then added the products 20 and -12 to get 8. Both methods of simplifying the expression led to the same result. **We call this property or peculiarity of real numbers the distributive property because we get the same result if we distribute the multiplication over the algebraic addition.**[†]

DISTRIBUTIVE PROPERTY

For any real numbers $a, b, c,$

$$a(b + c) = ab + ac$$

[†] Note that while multiplication can be distributed over addition, the reverse is not true, for addition cannot be distributed over multiplication. For example,

$$2 + (3 \cdot 5) \neq (2 + 3) \cdot (2 + 5)$$

because $17 \neq 35$.

It is possible to extend the distributive property so that the extension is applicable to the indicated product of a number or a variable and the algebraic sum of any number of real numbers or variables.

EXTENSION OF THE DISTRIBUTIVE PROPERTY

For any real numbers $a, b, c, d, \ldots$

$$a(b + c + d + \cdots) = ab + ac + ad + \cdots$$

---

example 15.1   Use the distributive property to find the value of $4(6 - 2 + 5 - 7)$.

solution   We begin by multiplying 4 by each of the terms within the parentheses and then we add the resultant products.

$$4(6 - 2 + 5 - 7) = 4(6) + 4(-2) + 4(5) + 4(-7)$$
$$= 24 - 8 + 20 - 28 = \mathbf{8}$$

example 15.2   Use the distributive property to expand $mn(x + y + 2p)$.

solution   We will multiply mn by each of the terms within the parentheses.

$$mn(x + y + 2p) = \mathbf{mnx + mny + 2mnp}$$

example 15.3   Use the distributive property to expand $(x - 3y + xz)mp$.

solution   The order is different, but we use the same procedure. Thus, mp is multiplied by each term within the parentheses.

$$(x - 3y + xz)mp = \mathbf{mpx - 3ymp + mpxz}$$

example 15.4   Use the distributive property to expand $-3(2x - 4)$.

solution   **This can be read two ways! The first way is to read it as the opposite of** $3(2x - 4)$. Since $3(2x - 4)$ can be expanded as

$$3(2x - 4) = 6x - 12$$

we can write the opposite of this as

$$\mathbf{-6x + 12}$$

The second way is to multiply -3 **by both** $2x$ **and** -4 **and write the result as an algebraic sum. If we do this, we get the same answer.**

$$-3(2x - 4) = \mathbf{-6x + 12}$$

problem set 15

1.   What is a coefficient?

2.   What is a literal coefficient?

*3.   Use the letters a, b, and c to state the distributive property.

Evaluate by using the distributive property:

*4. $4(5 - 3)$ *5. $4(6 - 2 + 5 - 7)$ 6. $3(-2 + 5)$

Expand by using the distributive property:

*7. $mn(x + y + 2p)$ 8. $(mn - 3)4x$ 9. $-3(2x - 4)$

10. $(a + bc)2x$ **11.** $3a(x + 2y)$

Evaluate:

12. $-a(a - b)$ if $a = -2$ and $b = -7$

13. $(-a + b) + (-a)$ if $a = -2$ and $b = 5$

14. $(a - x)(-x)$ if $a = 2$ and $x = -5$

15. $(x - y) - (y - x)$ if $x = -2$ and $y = -4$

16. $x - 2a(-a)$ if $x = 4$ and $a = -3$

17. $-x(a - xa)$ if $x = -4$ and $a = -3$

18. $-y[-ay - (xy)]$ if $a = -2$, $x = 2$, and $y = -3$

Simplify:

19. $-|-2| + (-3) - 3 - (-4 - 2)$ **20.** $4[(2 - 4) - (6 - 3)]$

21. $5[(3 - 2)(-5 - 3)]$ **22.** $-3 - 5(-2) - 4 + (-6)(3)$

23. $-2[-5 - 3(-2)][(-4) + 2]$ **24.** $-5(-2)(-2 - 3) - (-|-2|)$

25. $-[-(-3)] - 2(-2) + (-3)$ **26.** $3(-2)(-3 - 2) - (-4 - 2)$

27. $\dfrac{3 - (-2)(4)}{5 - (-3)}$ **28.** $-|-3| - 2(-3) + (-3) - 5 - 2$

29. $\dfrac{3 + 7(-3)}{-6 - 2(-4)}$ **30.** $(-2 - 5 + 3)(-2) - [-6 + 3(-2)]$

31. $\dfrac{-.06561}{4.05}$ **32.** $\dfrac{-3\frac{1}{3}}{2\frac{1}{5}}$

33. $-3\frac{1}{5} + 1\frac{3}{8}$

LESSON *16* *Like terms · Addition of like terms*

16.A
like terms

Like terms are terms that have the same variables in the same form or in equivalent forms so that the terms (excluding numerical coefficients) **represent the same number regardless of the nonzero values assigned to the variables.** Let's look at the indicated sum of terms

$$4xmp - 2pmx + 6mxp$$

Now whether terms are **like terms** or not does not depend on the signs of the terms or on the values of the numerical coefficients. So we won't consider the + and − signs or the numbers 4, 2, and 6. We just need to know if

$$xmp, \qquad pmx, \qquad \text{and} \qquad mxp$$

are in the same form or equivalent forms and if each expression represents the same number regardless of the nonzero values that are assigned to the variables.

1. They are in equivalent forms, for they have the same variables in the form of an indicated product, and the order of multiplication of the factors does not affect the value of the product.
2. They do represent the same number regardless of the values assigned to the variables. We will not attempt to prove this but will demonstrate this with one set of values for the variables. If we let $x = 4$, $p = 6$, and $m = 2$, we have

$$
\begin{array}{ccc}
xpm & pmx & mxp \\
4 \cdot 6 \cdot 2 & 6 \cdot 2 \cdot 4 & 2 \cdot 4 \cdot 6 \\
\mathbf{48} & \mathbf{48} & \mathbf{48}
\end{array}
$$

Thus $4xmp$, $2pmx$, and $6mpx$ are like terms because the variables represent the same number regardless of the real number replacements used for the variables.

16.B
addition of like terms

The extension of the distributive property of Lesson 15 can be rewritten as

$$ba + ca + da + \cdots = (b + c + d + \cdots)a$$

We note that a is a common factor of each of the terms on the left and is written outside the parentheses on the right. If we look at the indicated sum of terms

$$4xpm - 2pmx + 6mxp$$

we see that the factor xpm is a factor of all three terms and can be treated in the same manner as the a factor on the left side of the statement of the distributive property.

Thus we can write the sum of three terms as a product of $(4 - 2 + 6)$ and pmx as shown here.

$$4xpm - 2pmx + 6mxp = (4 - 2 + 6)pmx = \mathbf{8pmx}$$

The factors of the three variables in the expression $8pmx$ can be written in any order without changing the *value* of the expression. Thus any of the following would be equally acceptable.

$$8pmx, \qquad 8pxm, \qquad 8xmp, \qquad 8xpm, \qquad 8mpx, \qquad 8mxp$$

The above is a rather detailed approach to justify the following statement:
To add like terms we algebraically add the numerical coefficients. Thus to add

$$4xpm - 2pmx + 6mxp$$

we simply add the numerical coefficients 4, -2, and $+6$ to get $8pxm$.

$$4xpm - 2pmx + 6mxp = \mathbf{8pxm}$$

If the expression contains signed numbers, these are added separately, as shown in the following examples.

example 16.1   Simplify by adding like terms: $3x + 5 - xy + 2yx - 5x$.

solution   The first term and the fifth term are like terms, and the third and fourth are like terms. If we add these terms, we get

$$\mathbf{-2x + xy + 5}$$

example 16.2   Simplify by adding like terms: $3xy + 2xyz - 10yx - 5yzx$.

solution   The first term and the third term are like terms and may be added. Also the second and fourth terms are like terms and may be added. If we add these terms, we get

$$\mathbf{-7yx - 3xyz}$$

example 16.3   Simplify by adding like terms: $4 + 7mxy + 5 + 3yxm - 15$.

   solution   We add like terms and get

$$-6 + 10mxy$$

example 16.4   Simplify by adding like terms:

$$3x - x - y + 5 - 2y - 3x - 10 - 8y$$

   solution   We add the x terms, the y terms, and the numbers and get

$$-x - 11y - 5$$

example 16.5   Simplify by adding like terms:

$$-3 + xmy - y - 5 + 8ymx - 3y - 14$$

   solution   We add the y terms, the ymx terms, and the numbers in any order and get

$$-22 - 4y + 9myx$$

Of course, the letters myx could be in any order.

**problem
set 16**

1.   What is a term?

2.   What kind of terms may be added?

3.   What is a (a) factor? (b) product? (c) quotient?

Simplify by adding like terms:

*4. $3x + 5 - xy + 2yx - 5x$ *5. $3xy + 2xyz - 10yx - 5yzx$

*6. $4 + 7mxy + 5 + 3yxm - 15$

*7. $-3 + xmy - y - 5 + 8ymx - 3y - 14$

8. $3xyz + 2zxy - 7zyx + 2xy$ 9. $4x + 3 - 2xy - 5x - 7 + 4yx$

Expand by using the distributive property:

10. $(4 + 2y)x$ 11. $3x(y - 2m)$ 12. $2p(xy - 3k)$

Evaluate:

13. $-a(x - a)$ if $a = -3$ and $x = 6$

14. $-x - (-a)(a - x)$ if $x = -2$ and $a = 4$

15. $(m - p)p$ if $m = 3$ and $p = -2$

16. $-p(-x) - px$ if $p = -3$ and $x = 4$

17. $(x - y) - (y - x)$ if $x = -3$ and $y = -2$

18. $-x(-y) - xy$ if $x = 3$ and $y = -2$

19. $-px(x - p)$ if $x = -4$ and $p = 5$

20. $(-a)(b)(-a + b)$ if $a = 6$ and $b = -3$

Simplify:

21. $-3 - 2(-4 + 7) - 5 - |-2 - 5|$ 22. $-\{3(-2)(-4 + 2) - [3 - (-2)]\}$

23. $-4 - (-2) - [-(-2)] - |-3|$ 24. $-6 - 2(-3)(-1) - 5(3 - 2 - 2)$

25. $3 - (-6 + 8)2 - 4(-3) + (-3)$

26. $-5 - 2 - 6(-3 + 7)2 - 2(-3)$

27. $-2[(-3 + 5)(-2) - (3 - 2)]$

28. $\dfrac{-2(-3 + 7)}{(-2)(-3)}$

29. $\dfrac{-7(-2 + 3)}{-2(-3)}$

30. $\dfrac{-3\frac{1}{5}}{5\frac{7}{10}}$

31. $-5\frac{1}{5} + 7\frac{2}{3}$

32. $\dfrac{.09338}{-.046}$

LESSON 17 *Exponential notation · Integers as bases*

17.A
exponents

Often we find it necessary to indicate that a number is to be multiplied by itself a given number of times. For instance, if we wish to indicate that 5 is to be used as a factor seven times, we could write $5 \cdot 5 \cdot 5 \cdot 5 \cdot 5 \cdot 5 \cdot 5$. This is a cumbersome expression and mathematicians have developed a sort of mathematical shorthand called **exponential notation** that allows the expression to be written more concisely. The exponential notation for 5 used as a factor seven times is 5^7, read **"five to the seventh."** The general form of the expression is x^n and indicates that x is to be used as a factor n times and is read **"x to the n."**

DEFINITION OF EXPONENTIAL NOTATION[†]

$$\underbrace{x \cdot x \cdot x \cdot \ldots \cdot x}_{n \text{ factors}} = x^n$$

The symbol that is to be multiplied by itself is called the **base** and the symbol that indicates how many times it is to be used as a factor is called the **exponent** or the **power**. The base together with the exponent is called an **exponential**.

For example,

$x^4 = x \cdot x \cdot x \cdot x$ The base is x and the exponent is 4.

$(-4)^3 = (-4)(-4)(-4)$ The base is (-4) and the exponent is 3.

$\left(\dfrac{1}{3}\right)^4 = \left(\dfrac{1}{3}\right)\left(\dfrac{1}{3}\right)\left(\dfrac{1}{3}\right)\left(\dfrac{1}{3}\right)$ The base is $\dfrac{1}{3}$ and the exponent is 4.

17.B
integers
as bases

When a positive number is raised to a positive power, the result is always a positive number.

example 17.1   Simplify: (a) 3^2 (b) 3^3 (c) 3^4 (d) -3^4

[†] In this definition n is restricted to the set of positive integers.

solution   Each of these notations tells us that $+3$ is the base. The exponents 2, 3, and 4 tell us to use 3 as a factor twice, three times, and four times.

$$\text{(a)} \quad 3^2 = (3)(3) = \mathbf{9} \qquad \text{(b)} \quad 3^3 = (3)(3)(3) = \mathbf{27}$$
$$\text{(c)} \quad 3^4 = (3)(3)(3)(3) = \mathbf{81}$$

(d)   We must be careful here because -3^4 means **the opposite of** 3^4 and not $(-3)^4$.

$$-3^4 = -(3)(3)(3)(3) = \mathbf{-81}$$

When a negative number is raised to an even power, the result is always positive; and when raised to an odd power, the result is always negative as demonstrated in the next example.

example 17.2   Simplify: (a) $(-3)^2$ (b) $(-3)^3$ (c) $(-3)^4$ (d) $-(-3)^4$

solution   The first three are straightforward.

$$\text{(a)} \quad (-3)^2 = (-3)(-3) = \mathbf{9} \qquad \text{(b)} \quad (-3)^3 = (-3)(-3)(-3) = \mathbf{-27}$$
$$\text{(c)} \quad (-3)^4 = (-3)(-3)(-3)(-3) = \mathbf{81}$$

(d)   Be careful here. We want the opposite of $(-3)^4$.

$$-(-3)^4 = -(-3)(-3)(-3)(-3) = \mathbf{-81}$$

example 17.3   Simplify $-3^3 - (-3)^2 + (-2)^2$.

solution   Be careful with the first one.

$$-(3)(3)(3) - (-3)(-3) + (-2)(-2) = -27 - 9 + 4 = \mathbf{-32}$$

example 17.4   Simplify $-2^2 - 4(-3)^3 - 2(-2)^2 - 2$.

solution -2^2 is -4, $(-3)^3$ is -27, and $(-2)^2$ is 4, so we get

$$-4 - 4(-27) - 2(4) - 2 = -4 + 108 - 8 - 2 = \mathbf{94}$$

problem set 17

1.   Use the numbers 2, 3, and 4 to demonstrate the distributive property.

2.   Which integer is not a real number?

Simplify:

*3. $(-3)^2$ *4. -3^4

*5. $-3^3 - (-3)^2 + (-2)^2$ *6. $-2^2 - 4(-3)^3 - 2(-2)^2 - 2$

7. $-3^3 - (-3)^2 - (-3)^2 + (-2)^2$

Simplify by adding like terms:

8. $xym - 3ymx + 4xmy - 3my + 2ym$ 9. $-3pxk + pkx - 3kpx - kp - 3kx$

10. $m + 4 + 3m - 6 - 2m + mc - 4mc$ 11. $a - 3 - 7a + 2a - 6ax + 4xa - 5$

12. $-p - 5 - 3p - 6 - 2p + 7 - ax + 3xa$

Expand by using the distributive property to multiply:

13. $x(4 - ap)$ 14. $(5p - 2c)4xy$ 15. $4k(2c - a + 3m)$

Evaluate:

16. $|x - a| - a(-x)$ if $a = -3$ and $x = 4$

17. $(-x - a) - a(x - a)$ if $a = -3$ and $x = -4$

18. $-a(b - a)$ if $a = -4$ and $b = -3$

19. $-(a - x)(x - a)$ if $a = -5$ and $x = 3$

20. $(-p) - a(p - a)$ if $a = -4$ and $p = 5$

21. $-a[(x - a) + (2x + a)]$ if $a = -4$ and $x = 3$

22. $-(x + xy)$ if $x = -3$ and $y = 2$

23. $m[(x + 2xm) - (3x - mx)]$ if $x = -4$ and $m = 2$

Simplify:

24. $-3(4 - 3) - 3 - |-3|$

25. $-2^2 + (-3 - 5) - (-2)$

26. $-4(-3 + 7) - (-2) - 3$

27. $-2(-5 - 2)(-2)(-2 - 3)$

28. $-5 + (-5) - (3) + (2)$

29. $\dfrac{-4(2 - 4)}{(-2)(-4)}$

30. $-2(-4) - 3^2 + 2 - 5$

31. $(-2)^3 - 2(2) - 3(-5)$

32. $-7 + (6) - (-3) + (-2)^2$

LESSON 18   *Evaluation of exponentials*

18.A
evaluation of exponential expressions

Evaluation of expressions with exponents is straightforward when the replacements of the variables are all positive numbers. To evaluate

$$yx^2m^3$$

with $y = 3$, $x = 4$, and $m = 2$, we proceed as follows:

$$(3)(4)^2(2)^3 = (3)(16)(8) = \mathbf{384}$$

We must be careful, however, when the expressions contain minus signs or when some replacement values of the variables are negative numbers.

example 18.1   If $a = -2$, what is the value of each of the following:

$$\text{(a)} \quad a^2 \qquad \text{(b)} \quad -a^2 \qquad \text{(c)} \quad -a^3 \qquad \text{(d)} \quad (-a)^3$$

solution   (a) a^2 means a times a, or $(-2)(-2) = \mathbf{+4}$
(b) $-a^2$ asks for the opposite of a^2, or $-(-2)(-2) = \mathbf{-4}$
(c) $-a^3$ means the opposite of a times a times a, or

$$-(-2)(-2)(-2) = \mathbf{+8}$$

(d) $(-a)^3$ means $(-a)(-a)(-a) = (2)(2)(2) = \mathbf{+8}$

example 18.2   Evaluate x^2z^3y if $x = 2$, $z = 3$, and $y = -2$.

solution   We replace x with 2, z with 3, and y with -2.
$$(2)^2(3)^3(-2) = (4)(27)(-2) = \mathbf{-216}$$

example 18.3   Evaluate $pm^2 - z^3$ if $p = 1$, $m = -4$, and $z = -2$.

solution   We replace p with 1, m with -4, and z with -2.
$$(1)(-4)^2 - (-2)^3 = (1)(16) - (-8) = 16 + 8 = \mathbf{24}$$

problem set 18

1.   Designate (a) the set of whole numbers; (b) the set of integers.

2.   What is another name for the multiplicative inverse?

3.   Is the product of 142 negative numbers and 3 positive numbers a positive number or a negative number?

Simplify:

4. $3^2 + (-3)^2$

5. $-2^3 + (-2)^3$

6. $-2^2 + (-4)^2$

7. $-(-3)^2 - (-2)^3$

8. $-3^2(-2)^2 - 2$

9. $-3 - 2^3 - (-3)^3$

10. $-2 - 2^2 - 2^3 - (-2)^3$

Evaluate:

11. $x^2 + y^3$ if $x = 2$ and $y = -3$

*12. $pm^2 - z^3$ if $p = 1$, $m = -4$, and $z = -2$

13. $-x^2 - y^3$ if $x = -3$ and $y = -2$

14. $a^2 - b^2a$ if $a = -2$ and $b = 3$

Simplify:

15. $-3^2 - 2(-3 - 4)$

16. $-2^2 - 4(-3)$

17. $-2(-2)(-2 - 3)$

18. $-|-2| - 3 + (-3 - 2)$

19. $5(3 - 4)(-2) + (-5 - 2)$

20. $2[-3(-2 - 4)(3 - 2)]$

Simplify by adding like terms:

21. $5 - x + xy - 3yx - 2 + 2x$

22. $-3bpx - 3bp - 3 + 5pb$

23. $7 - 3k - 2k + 2kx - xk + 8$

24. $-8 - py + 2yp + 4 - y$

Expand by using the distributive property:

25. $(4 - 2p)4x$

26. $-3(-x - 4)$

27. $-2x(a - 3p)$

Evaluate:

28. $-x(a - 3x) + x$ if $a = 3$ and $x = 4$

29. $-p(-a + 2p) + p$ if $p = -3$ and $a = 2$

30. $k(ak - 4a) + k$ if $k = -3$ and $a = 2$

Simplify

31. $\dfrac{2\frac{3}{5}}{-4\frac{7}{8}}$

32. $-1\frac{1}{8} + 2\frac{7}{16}$

33. $-.000012 \div .003$

LESSON 19   *Product theorem for exponents · Like terms with exponents*

19.A
product theorem for exponents

The product theorem for exponents can be deduced from the definition of exponential notation. We know that

$$\text{(a)} \quad 3^5 \qquad \text{means} \qquad 3 \cdot 3 \cdot 3 \cdot 3 \cdot 3$$

and thus

$$\text{(b)} \quad 3^2 \cdot 3^3 \qquad \text{means} \qquad (3 \cdot 3) \cdot (3 \cdot 3 \cdot 3), \text{ or } 3^5$$

and also

$$\text{(c)} \quad 3 \cdot 3^4 \qquad \text{means} \qquad (3) \cdot (3 \cdot 3 \cdot 3 \cdot 3), \text{ or } 3^5$$

We see that when we multiply exponentials whose bases are the same, the exponent of the product is obtained by adding the exponents of the factors. Thus,

(a) $x^5 \cdot x^7 \cdot x^2 = \boldsymbol{x^{14}}$
(c) $p^5 \cdot p^{12} = \boldsymbol{p^{17}}$

(b) $5^2 \cdot 5^3 \cdot 5^2 = \boldsymbol{5^7}$
(d) $4^2 \cdot 4^3 \cdot 4^{25} = \boldsymbol{4^{30}}$

We call this rule the product theorem for exponents and give the formal definition as follows.

PRODUCT THEOREM FOR EXPONENTS

If *m* and *n* are real numbers and $x \neq 0$,

$$x^m \cdot x^n = x^{m+n}$$

We can use this theorem to help simplify expressions that contain exponents.

example 19.1   Simplify $x^2 y^2 x^5 y^3$.

solution   Since rearranging the order of the factors of a product does not change the value of the product, we may write $x^2 y^2 x^5 y^3$ as

$$x^2 x^5 y^2 y^3 = \boldsymbol{x^7 y^5}$$

example 19.2   Simplify $x^2 y^3 m^5 x^3 y^2$.

solution   First we rearrange the factors and then we simplify.

$$x^2 x^3 y^2 y^3 m^5 = \boldsymbol{x^5 y^5 m^5}$$

Now we define the notation x^1.

DEFINITION

$$x^1 = x$$

This says that x means the same thing as x^1 and that x^1 means the same thing as x. **If any variable or constant is written without an exponent, it is understood to have an exponent of 1. Thus,**

$$5 \text{ equals } 5^1, \qquad x^1 \text{ equals } x, \qquad \text{and} \qquad 7^1 \text{ equals } 7$$

example 19.3 Simplify xyy^2x^3.

solution First, we rearrange the letters. Then we simplify, remembering that x means x^1 and y means y^1.

$$xx^3yy^2 = x^4y^3$$

example 19.4 Simplify $m^3pmxm^2x^3p^5$.

solution We rearrange and simplify to get the result.

$$m^3mm^2xx^3pp^5 = m^6x^4p^6$$

19.B
adding like terms that contain exponents

Above we noted that we multiply exponentials with like bases by adding the exponents. Unfortunately, the task of adding like terms that contain exponents appears similiar, but the rule is different. When we add like terms that contain exponents, we do not add the exponents. Thus

$$3x^2 + 2x^2 = 5x^2$$

and does not equal $5x^4$. Addition and multiplication are often confused, so we discuss them in the same lesson so that we can point out the difference.

When we add, we can only add like terms. We recall that letters stand for unspecified numbers and that the order of multiplication of real numbers can be changed. This means that

$$x^2yp^5 \qquad \text{and} \qquad p^5x^2y$$

are like terms, and that

$$xy^2p^5 \qquad \text{and} \qquad y^2xp^5$$

are like terms because the literal factors of the terms are the same.

example 19.5 Simplify by adding like terms:
$$x^2yp^5 + 2xy^2p^5 + 3p^5x^2y - 7y^2xp^5$$

solution The first and third terms are like terms, and the second and fourth terms are also like terms. We add the numerical coefficients of these terms and get

$$4x^2yp^5 - 5xy^2p^5$$

example 19.6 Simplify by adding like terms:
$$2m^3xy^2p + 3pxy^2m^3 - 10xy^2m^3p + yx^2m^3p$$

solution The first three terms are like terms and may be added.

$$-5xy^2m^3p + x^2ym^3p$$

example 19.7 Simplify by adding like terms
$$2x^2y + 3yx^2 + x^2y^2 - x^2y - 4x^2y^2$$

solution The first, second, and fourth terms are like terms and so are the third and fifth terms. Thus we add the first, second, and fourth terms and the third and fifth terms to obtain

$$4x^2y - 3x^2y^2$$

problem   **1.**   What is a variable?
set 19
                2.   When can terms be called like terms?

Simplify:

*3. $x^2y^3m^5x^3y^2$ ***4.** xyy^2x^3 ***5.** $m^3pmxm^2x^3p^5$

6. $xm^2xm^3x^3m$ **7.** $ky^2k^3k^2y^5$ **8.** $a^2ba^2b^3ab^4$

Simplify by adding like terms:

*9. $x^2yp^5 + 2xy^2p^5 + 3p^5x^2y - 7y^2xp^5$ ***10.** $2x^2y + 3yx^2 + x^2y^2 - x^2y - 4x^2y^2$

11. $3ab^2 - 2ab + 5b^2a - ba$ **12.** $x^2 - 3yx + 2yx^2 - 2xy + yx$

13. $xym^2p - 3m^2yxp + 7pm^2xy - 3y^2mxp$

Expand by using the distributive property:

14. $x(3p - 2y)$ **15.** $(3 - 2b)a^2$ **16.** $5(2 - 4p)$

Evaluate:

17. $x^2 - y^2$ if $x = -3, y = -2$ **18.** $x(y^2 - x^2)$ if $x = -3, y = 2$

19. $a^3 - y^3$ if $a = -3, y = 2$ **20.** $a(b^3 - a)$ if $a = -2, b = -4$

21. $(a - x)(x - a)$ if $a = -3, x = 4$ **22.** $x^2(a - x)$ if $a = -5, x = 3$

23. $m(x - m) + x$ if $m = -3, x = -4$ **24.** $|-p + a| - a^2$ if $p = 4, a = -2$

Simplify:

25. $-4 - (-2)(-2 + 5) - 3$ **26.** $\dfrac{-3 - [-(-3)]}{-(-2)}$

27. $-6 - [-3 + 5(-2)] - 2$ **28.** $-3^2 - 2^2 - (-3)^3$

29. $-5 - (-5)^2 - 3 + (-2)$ **30.** $-2[(-3 - 5) + (-7 + 2)]$

31. $\dfrac{1\frac{3}{4}}{-2\frac{1}{3}}$ **32.** $5\frac{2}{3} - 7\frac{9}{10}$

33. $(.004)(.012)$

LESSON 20 *Statements and sentences · Conditional equations*

20.A
**numerical
and algebraic
expressions
(again)**

Each of the following expressions

(a) 4   (b) $6 + 3$ (c) $4(2 + 3)$ (d) $\dfrac{7(8 + 4)}{5 + 2}$

is called a **numerical expression** because it consists of **a meaningful arrangement of numerals and symbols that designate specific operations.** Every numerical expression represents a particular number, and we say that this number is the **value** of the expression. The **values** of expressions (a), (b), (c), and (d) are 4, 9, 20, and 12, respectively.

We use the words **algebraic expression** to describe **numerical expressions** and also to describe expressions that contain variables.

(e) 6   (f) $x + 4$ (g) $x^2 - 6$ (h) $x(x + 4)$

Each of the expressions shown here is an algebraic expression. The value of expression (e) is 6, but the values of expressions (f), (g), and (h) depend on the value that we assign to the variable x. If we assign to x the value of 3, then the values of expressions (f), (g), and (h) are 7, 3, and 21, respectively.

20.B
statements and sentences

If we wish to make a statement that certain quantities are equal or are not equal, we can do so by writing a grammatical **sentence** in English.

(a)   The number of peaches equals the number of apples.
(b)   The number of peaches does not equal the number of apples.
(c)   The number of peaches is greater than the number of apples.
(d)   The number of peaches is less than the number of apples.

If we use the variables

$$N_p \quad \text{and} \quad N_a$$

to represent the number of peaches and the number of apples, and if we use the symbols

$=$ to mean   equal

$\neq$ to mean   not equal

$>$ to mean   greater than (read left to right)

$<$ to mean   less than (read left to right)

we can make the same statements, by writing algebraic sentences.

(a) $N_p = N_a$

(b) $N_p \neq N_a$

(c) $N_p > N_a$

(d) $N_p < N_a$

We see that all four are called statements or sentences but only the first one, (a) uses the equals sign. This algebraic statement is called an *equation*. The other three statements, (b), (c), and (d), do not use the equals sign and are called *inequalities*. We will discuss equations here and hold the topic of inequalities for a later lesson.

20.C
equations

An equation is an algebraic statement consisting of two algebraic expressions connected by an equals sign. Thus all the following are statements and all are also equations.

(a) $4 = 3 + 1$ (b) $4 + x = 2 + 2 + x$ (c) $4 = 6$
(d) $4 + x = 6 + x$ (e) $x + 4 = 8$

Equations are not always true equations as we see here. Two of these equations are true equations, two are false equations, and the truth or falsity of one of them depends on the number we use as a replacement for x in the equation.

(a)   This is a true equation because 4 does equal the sum of 3 and 1.

(b)   This is a true equation regardless of the number we use as a replacement for x. We will demonstrate this by replacing x with -3 and then by replacing x with $+7$.

WITH (-3)	WITH $(+7)$
(b) $4 + (-3) = 2 + 2 + (-3)$	(b) $4 + (+7) = 2 + 2 + (+7)$
$4 - 3 = 4 - 3$	$4 + 7 = 4 + 7$
$1 = 1$ True	$11 = 11$ True

(c)   This is a false equation because 4 is not equal to 6.

(d)   This is a false equation regardless of the number we use as a replacement for x. We will demonstrate this by replacing x with -3 and then by replacing x with $+7$.

WITH (-3)	WITH $(+7)$
(d) $4 + (-3) = 6 + (-3)$	(d) $4 + (+7) = 6 + (+7)$
$4 - 3 = 6 - 3$	$4 + 7 = 6 + 7$
$1 = 3$ False	$11 = 13$ False

(e)   We call this equation a **conditional equation** because its truth or falsity is conditioned by the number used as a replacement for x. If we use -2 as the replacement for x, we get a false equation; but if we use 4 as the replacement for x, we get a true equation.

WITH (-2)	WITH (4)
(e) $(-2) + 4 = 8$	(e) $(4) + 4 = 8$
$-2 + 4 = 8$	$4 + 4 = 8$
$2 = 8$ False	$8 = 8$ True

Replacement values of the variable that turn the equation into a true equation are called solutions of the equation or roots of the equation and are said to satisfy the equation. Thus in the equation

$$x + 4 = 8$$

we say that the number 4 is a **solution** or **root** of the equation, and we also say that the the number 4 **satisfies** the equation.

example 20.1   Does -2 or $+5$ satisfy the equation $x^2 = -5x - 6$?

solution   First we try -2.

$$(-2)^2 = -5(-2) - 6 \qquad \text{replaced } x \text{ with } -2$$
$$4 = 10 - 6 \qquad \text{simplified}$$
$$4 = 4 \qquad \text{True}$$

Now we try $+5$.

$$(5)^2 = -5(5) - 6 \qquad \text{replaced } x \text{ with } 5$$
$$25 = -25 - 6 \qquad \text{simplified}$$
$$25 = -31 \qquad \text{False}$$

Thus -2 is a solution and $+5$ is not a solution.

example 20.2   Is -2 or $+5$ a root of the equation $x^2 - 3x = 10$?

solution   First we try -2.

$$(-2)^2 - 3(-2) = 10$$
$$4 + 6 = 10 \qquad \text{True}$$

Now we try $+5$.

$$(5)^2 - 3(5) = 10$$
$$25 - 15 = 10 \qquad \text{True}$$

Thus, both -2 and $+5$ are solutions or roots to the given equation, and we say that both -2 and $+5$ satisfy this equation.

problem   **1.** Give an example of (a) a true equation, (b) a false equation, (c) a conditional
set 20   equation.

Does -2 or $+5$ satisfy any of the following equations?

*2. $x^2 = -5x - 6$ ***3.** $x^2 - 3x = 10$ **4.** $-4x^2 = -x - 95$

Simplify:

5. $x^2xxy^2xy^3$ **6.** $p^2m^5ypp^3my^2$ **7.** $8k^5nn^2kn^3k^5k$

8. $a^2aba^3b^2a^5$ **9.** $m^2pap^2ma^2aa^3$ **10.** $4p^2x^2kpx^3k^2k$

Simplify by adding like terms:

11. $3x + 2 - x^2 + 2x^2 - 4$ **12.** $xy - 3xy^2 + 5y^2x - 4xy$

13. $-3x^2ym + 7x - 5ymx^2 + 16x$ **14.** $5mp^2y - 6myp^2 + 3ymp^2 - 2p^2my$

15. $x + 2x^2 - 3 + 5x - 6x^2 - 10$ **16.** $m^2y - 6ym^2 + 2y - 3m^2y + 4y$

17. $5 - 3x + 7 - 4 + 4x^2 - 2x - x^2$

Expand by using the distributive property:

18. $a(3x - 2)$ **19.** $4xy(5 - 2a)$ **20.** $(3a - 4)6x$

Simplify:

21. $(-3)^2 - 2^3$ **22.** $-3^2 - (-2)^2$

23. $(-3)^3 + (-2)^3 - |-2|$ **24.** $-3[(-3 + 5)(-2 - 6)] - 3$

25. $-2[(5 - 3) - (5 - 8)]$

Evaluate:

26. $a^3 - b^3$ if $a = -2$ and $b = 3$

27. $a - b(a^2 - b)$ if $a = -2$ and $b = 3$

28. $cy[(cx - y)]$ if $x = -3, y = 3$, and $c = -2$

29. $-b^2a(a - b)$ if $a = -3, b = 2$ **30.** $b(b^2) - a^2$ if $a = 3, b = -2$

Simplify:

31. $\dfrac{3\frac{1}{4}}{-2\frac{5}{7}}$ **32.** $-3\frac{1}{4} + 2\frac{5}{7}$ **33.** $\dfrac{.003636}{.0303}$

LESSON 21 *Equivalent equations · Additive property of equality*

21.A
equivalent
equations

Two equations are said to be equivalent equations if *every* solution of either one of the equations is also a solution of the other equation.

$$\text{(a)} \quad x + 6 = 9 \qquad \text{(b)} \quad x + 10 = 13$$

The two equations shown are equivalent equations, for the number 3 will satisfy both equations and 3 is the only number that will satisfy either equation.

21.B
additive
property of
equality

If we begin with the true statement that

$$2 = 2$$

and add $+4$ to both sides of the equality, we get the true statement that 6 equals 6.

$$2 + 4 = 2 + 4 \qquad 4 + 2 = 4 + 2$$
$$6 = 6 \qquad\qquad 6 = 6$$

On the left we placed the 4s after the 2s and on the right we placed the 4s before the 2s. We note that both procedures yield the same result. We emphasize this fact in the formal definition given in the box by writing the definition twice, the second time with the order of the addends reversed.

ADDITIVE PROPERTY OF EQUALITY

If a, b, and c are any real numbers and if $a = b$, then

$$a + c = b + c \qquad \text{and also} \qquad c + a = c + b$$

The additive property of equality can be used to find the solution of conditional equations such as

$$x + 4 = 6$$

This equation is a conditional equation and is neither true nor false because no value has been assigned to x so the additive property of equality does not apply. Thus we must hedge a little. We assume that some real number exists that when substituted for x will make the equation a true equation. We further assume that x in the equation represents this number. Now we have assumed that

$$x + 4 \text{ equals } 6$$

is a true statement, and thus we can use the additive property of equality. We will do this to eliminate the $+4$ that is now on the left side with x. We will add -4 to both sides of the equation.

$$
\begin{array}{rl}
x + 4 = & 6 \\
-4 & -4 \qquad \text{add } -4 \text{ to both sides} \\
\hline
x + 0 = & 2 \\
x \quad = & 2
\end{array}
$$

Now since we made an assumption to permit the use of the additive property of equality we must check our solution in the original equation.

$$x + 4 = 6 \qquad \text{original equation}$$

$$(2) + 4 = 6 \qquad \text{substitute 2 for } x$$

$$6 = 6 \qquad \text{True}$$

Since using the number 2 for x in the equation makes the equation a true equation, we say that the number 2 is a root or solution of the equation and that the number 2 satisfies the equation. It can be shown that the use of the additive property of equality will not change the numbers that satisfy the equation, so we say that the use of the additive property of equality results in an equation that is an **equivalent equation** to the original equation.

> If the same quantity is added to both sides of an equation, the resulting equation will be an **equivalent equation** to the original equation and thus every solution of one of these equations will also be a solution of the other equation.

The rule says that we may add the same quantity to both sides of an equation, but it does not specify a particular format to be used and one format is usually just as acceptable as another. Below are shown three possible formats for the problem worked above.

(a) $\begin{array}{rl} x + 4 = & 6 \\ -4 = & -4 \\ \hline x \quad\;\; = & 2 \end{array}$ (b) $\begin{array}{rl} x + 4 = & 6 \\ -4 & -4 \\ \hline x \quad\;\; = & 2 \end{array}$ (c) $\begin{array}{l} x + 4 + (-4) = 6 + (-4) \\ \quad x + 0 = 2 \\ \qquad\quad x = 2 \end{array}$

In (a) we added -4 to both sides of the equation and placed an equals sign between the -4s to *emphasize* that they are equal. In (b) we added -4 to both sides but omitted the equals sign since it really isn't necessary. In (c) we added the -4s on the same line with the rest of the numbers and variables. This form is adequate for very simple problems such as this one but is less desirable for more complicated problems. By the end of the book one should be able to perform this calculation mentally, without writing anything down, so that none of these formats will be necessary.

example 21.1   Solve $x - 3 = 12$.

solution   To solve the equation we want to isolate x on one side of the equals sign. We can do this if we eliminate the -3. Thus we will add $+3$ to both sides of the equation.

$$\begin{array}{rl} x - 3 = & 12 \\ +3 & +3 \\ \hline x \quad\;\; = & 15 \end{array}$$

This same procedure can be used when the equation contains fractions or mixed numbers as we see in the next two examples.

example 21.2   Solve $x + \dfrac{1}{4} = -\dfrac{3}{8}$.

solution   To isolate x we must eliminate the $\frac{1}{4}$. Thus we add $-\frac{1}{4}$ to both sides of the equation.

$$x + \frac{1}{4} = -\frac{3}{8}$$

$$\underline{-\frac{1}{4} \qquad -\frac{1}{4}}$$

$$x \qquad = -\frac{3}{8} - \frac{1}{4} \longrightarrow x = -\frac{5}{8}$$

example 21.3   Solve $k + 2\frac{1}{3} = \frac{2}{9}$.

solution   This time we add $-2\frac{1}{3}$ to both sides and then simplify the result.

$$k + 2\frac{1}{3} = \frac{2}{9}$$

$$\underline{-2\frac{1}{3} \qquad -2\frac{1}{3}}$$

$$k \qquad = \frac{2}{9} - 2\frac{1}{3} \longrightarrow k = \frac{2}{9} - \frac{21}{9} \longrightarrow k = -\frac{19}{9}$$

problem set 21

1.   What does "solve an equation" mean?

2.   What are equivalent equations?

Solve:

*3. $x + 4 = 6$ 4. $x - 15 = 30$ 5. $y - 13 = 23$

6. $p + 5 = -17$ 7. $k + 7 = 93$ 8. $m - 2 = 17$

9. $4 + k = -7$ *10. $x + \frac{1}{4} = -\frac{3}{8}$ *11. $k + 2\frac{1}{3} = \frac{2}{9}$

Simplify:

12. $x^2ym^5x^2y^4$ 13. x^3y^2myxm 14. xxx^2yyy^3xy

Simplify by adding like terms:

15. $x^2y + 3yx - 2y^2x - 4yx^2$ 16. $3x - 3 - x^2 - 2x^3 + 7 - 2x + 6x^2$

17. $p^2xy - 3yp^2x + 2xp^2y - 5$ 18. $-4 + 7x - 3x - 5 + 2x - 4x^2$

Expand by using the distributive property:

19. $4x(a + 2b)$ 20. $(2x + 4)3$ 21. $4px(my - 3ab^2)$

Evaluate:

22. $a^3 - b^3$ if $a = -2$ and $b = 3$

23. $(a - b)(b - x)$ if $x = 2$, $a = 3$, and $b = -3$

24. $x^3 - a(a - b)$ if $x = -2$, $a = 3$, and $b = -3$

25. $x(x - y) - y$ if $x = -2$ and $y = 3$

Simplify:

26. $(-2)^3 - 2^3$ 27. $-3^2 - (-3)^2$

28. $-4(-3 - 2) - 5(-2) - 2|-4|$ 29. $\dfrac{5(-2^2 - 4)}{-4 - 6(-2)}$

30. $\dfrac{-3[5(-2 - 1) - (6 - 2)]}{2(-3 - 2)}$

LESSON 22 *Multiplicative property of equality*

22.A
multiplicative
property of
equality

To demonstrate the **multiplicative property of equality** we will begin with the true equation

$$2 = 2 \qquad \text{true}$$

and multiply both sides of the equation by 3.

$$3 \cdot 2 = 3 \cdot 2$$

The result is the true equation that 6 equals 6.

$$6 = 6 \qquad \text{still true}$$

The formal statement of the multiplicative property of equality in the box below is made twice to emphasize the fact that the order of the factors does not affect the product.

MULTIPLICATIVE PROPERTY OF EQUALITY

If *a*, *b*, and *c* are any real numbers and if

$$a = b,$$

then

$$ca = cb \qquad \text{and also} \qquad ac = bc$$

It is possible to use the multiplicative property of equality to prove that multiplying or dividing every term on both sides of an equation by the same nonzero number does not change the solution(s) to the equation. This means that the new equation is an equivalent equation to the original equation.

If **every term** on both sides of an equation is multiplied or divided by the same nonzero quantity, the resulting equation will be an equivalent equation to the original equation, and thus every solution of one of these equations will be a solution of the other equation.

We can use this rule with either of two thought processes to solve equations such as

$$4x = 12$$

The first way is to remember that division by 4 will undo multiplication by 4 because division and multiplication are inverse operations. Thus, we solve by dividing both sides of the equation by 4.

$$4x = 12 \quad \longrightarrow \quad \frac{4x}{4} = \frac{12}{4} \quad \longrightarrow \quad x = 3$$

The second way is to remember that the product of 4 and its multiplicative inverse (reciprocal) is 1. To solve by using this thought, we will multiply both sides of the equation by $\frac{1}{4}$, which is the reciprocal of 4:

$$4x = 12 \quad \longrightarrow \quad \frac{1}{4} \cdot 4x = \frac{1}{4} \cdot 12 \quad \longrightarrow \quad x = 3$$

Both of the preceding thought processes are correct and either one can be used at any time.

example 22.1   Solve $5x = 20$.

solution   We can solve by (a) multiplying both sides by $\frac{1}{5}$, or by (b) dividing both sides by 5.

$$\text{(a)} \quad \frac{1}{5} \cdot 5x = \frac{1}{5} \cdot 20 \qquad\qquad \text{(b)} \quad \frac{5x}{5} = \frac{20}{5}$$

$$x = 4 \qquad\qquad\qquad\qquad x = 4$$

example 22.2   Solve $\frac{2}{5}x = 7$.

solution   We can solve by (a) multiplying both sides by $\frac{5}{2}$, or by (b) dividing both sides by $\frac{2}{5}$.

$$\text{(a)} \quad \frac{5}{2} \cdot \frac{2}{5}x = \frac{5}{2} \cdot 7 \qquad\qquad \text{(b)} \quad \frac{\frac{2}{5}x}{\frac{2}{5}} = \frac{7}{\frac{2}{5}}$$

$$x = \frac{35}{2} \qquad\qquad\qquad\qquad x = 7 \cdot \frac{5}{2} \quad \longrightarrow \quad x = \frac{35}{2}$$

example 22.3   Solve $2\frac{1}{4}x = 3$.

solution   We can undo multiplication by $2\frac{1}{4}$ by dividing by $2\frac{1}{4}$,

$$2\frac{1}{4}x = 3 \quad \longrightarrow \quad \frac{2\frac{1}{4}x}{2\frac{1}{4}} = \frac{3}{2\frac{1}{4}} \quad \longrightarrow \quad x = \frac{3}{\frac{9}{4}} \quad \longrightarrow \quad x = \frac{12}{9} = \frac{4}{3}$$

or by rewriting $2\frac{1}{4}$ as the improper fraction $\frac{9}{4}$ and then multiplying both sides by $\frac{4}{9}$, the the reciprocal of $\frac{9}{4}$.

$$\frac{9}{4}x = 3 \quad \longrightarrow \quad \frac{4}{9} \cdot \frac{9}{4}x = \frac{4}{9} \cdot 3 \quad \longrightarrow \quad x = \frac{4}{3}$$

example 22.4   Solve $\frac{x}{3} = 9$.

solution   We can undo division by 3 by multiplying by 3. Thus, we multiply both sides of the equation by 3 and cancel.

$$3 \cdot \frac{x}{3} = 9 \cdot 3 \qquad \text{and thus} \qquad x = 27$$

We can also use the concept of inverse operations when x is divided by a fraction or a mixed number.

example 22.5   (a) Solve $\dfrac{x}{2\frac{1}{2}} = 7$.   (b) Solve $\dfrac{p}{\frac{3}{2}} = 4\frac{1}{3}$.

solution   (a) We can undo division by $2\frac{1}{2}$ by multiplying by $2\frac{1}{2}$.

$$\frac{x}{2\frac{1}{2}} = 7 \longrightarrow \frac{\cancel{2\frac{1}{2}}x}{\cancel{2\frac{1}{2}}} = 7 \cdot 2\frac{1}{2} \longrightarrow x = 7 \cdot \frac{5}{2} \longrightarrow x = \frac{35}{2}$$

(b) We can undo division by $\frac{3}{2}$ by multiplying by $\frac{3}{2}$.

$$\frac{p}{\frac{3}{2}} = 4\frac{1}{3} \longrightarrow \frac{\cancel{\frac{3}{2}}p}{\cancel{\frac{3}{2}}} = 4\frac{1}{3} \cdot \frac{3}{2} \longrightarrow p = \frac{13}{3} \cdot \frac{3}{2} = \frac{13}{2}$$

problem set 22

1. What is another name for multiplicative inverse?

2. What is the product of any number and its reciprocal?

Solve:

*3. $5x = 20$ *4. $\dfrac{x}{3} = 9$ 5. $7x = 49$

6. $\dfrac{x}{7} = 5$ 7. $2x = 20$ 8. $\dfrac{x}{3} = 5$

*9. $\dfrac{2}{5}x = 7$ *10. $2\frac{1}{4}x = 3$ *11. $\dfrac{x}{2\frac{1}{2}} = 7$

12. $x + 5 = 7$ 13. $x + \dfrac{1}{2} = 2\frac{1}{5}$ 14. $y - 3 = 2$

15. $y - \dfrac{1}{2} = -2\frac{1}{2}$ 16. $\dfrac{x}{3} = 19$ 17. $\dfrac{x}{\frac{1}{2}} = 4$

18. $3x = 4\frac{1}{2}$ 19. $x - \dfrac{1}{4} = \dfrac{7}{8}$ 20. $k - \dfrac{2}{3} = 3\frac{1}{3}$

21. $3\frac{2}{5}m = \dfrac{1}{2}$

Simplify:

22. $m^2xyp^2x^3y^5$ 23. $3p^2xxypp^3xy^2$

Simplify by adding like terms:

24. $a + 3 - 2a - 5a^2 + 5 - a$ 25. $-3x^2ym + 5myx^2 - 2my^2x$

26.   Expand $(3a - 5p)xy$ by using the distributive property.

Evaluate:

27. $a^3 - (a - b) - |-a^3|$ if $a = -2$ and $b = 3$

28. $(-x)^3 - y$ if $x = -2$ and $y = 4$

Simplify:

29. $(-2)^3 - (-2)^2 - 5$ **30.** $-2\{-5(-3 + 4) - [3 - (-4)]\}$

LESSON 23 *Solution of equations*

In Lessons 21 and 22, we were introduced to the two rules for solving equations. These are two of the most important rules in algebra. In order to emphasize this fact, they are repeated here in bold print.

1. **The same quantity can be added to both sides of an equation without changing the answers[†] to the equation.**
2. **Every term on both sides of an equation can be multiplied or divided by the same nonzero quantity without changing the answers to the equation.**

In many equations, it is necessary to use both of these rules to find the solution. We will always use the addition rule first and then the multiplication/division rule. We do this because we are undoing a normal order of operations problem. To demonstrate we will begin with the number 4, multiply by 3, and then add -2 for a result of 10.

$$3(4) - 2 = 10$$

Now, to undo what we have done and get back to 4, we must undo the addition of -2 first and then undo the multiplication. **This is the reason that in solving equations, we reverse the normal order of operations and undo addition first and then undo multiplication or division.** We demonstrate this procedure by replacing 4 with x in the above expression and getting the equation $3x - 2 = 10$. Now we solve to find that x equals 4.

$$
\begin{array}{ll}
\begin{aligned}
3x - 2 &= 10 \\
\underline{+2 \quad\; +2} & \\
3x \quad\;\; &= 12
\end{aligned}
& \text{replaced 4 with } x \\
& \text{add } +2 \text{ to both sides}
\end{array}
$$

$$\frac{\cancel{3}x}{\cancel{3}} = \frac{12}{3} \longrightarrow x = 4 \qquad \text{divided both sides by 3}$$

example 23.1   Solve $4x + 5 = 17$.

solution   We must use the addition rule to eliminate the $+5$ and then use the multiplication/division rule to eliminate the 4. We always use the addition rule first.

$$
\begin{array}{ll}
\begin{aligned}
4x + 5 &= 17 \\
\underline{-5 \quad\; -5} & \\
4x \quad\;\; &= 12
\end{aligned}
& \text{original equation} \\
& \text{add } -5 \text{ to both sides}
\end{array}
$$

$$\frac{4x}{4} = \frac{12}{4} \longrightarrow x = 3 \qquad \text{divided both sides by 4}$$

[†] We remember that the answers to an equation are formally called the roots or solutions of the equation.

example 23.2 Solve $-5m + 6 = 8$.

solution To isolate m, we must first eliminate the 6 and then eliminate the -5.

$$
\begin{array}{r}
-5m + 6 = 8 \quad \text{original equation} \\
\underline{\; -6 \quad -6} \quad \text{add } -6 \text{ to both sides} \\
\text{(a)} \quad -5m = 2
\end{array}
$$

Now we complete the solution by dividing both sides by -5.

$$
\frac{-5m}{-5} = \frac{2}{-5} \longrightarrow \boldsymbol{m = -\frac{2}{5}}
$$

Dividing by a negative number sometimes leads to errors. In this problem division by a negative number can be avoided by mentally multiplying both sides of equation (a) by -1. This changes the signs on both sides.

$$
-5m = 2 \xrightarrow{\text{mentally multiplying both sides by } -1} 5m = -2
$$

Now we can finish by dividing both sides by $+5$.

$$
\frac{5m}{5} = \frac{-2}{5} \longrightarrow \boldsymbol{m = -\frac{2}{5}} \quad \text{divided by 5}
$$

example 23.3 Solve $-7k - 4 = -21$.

solution We begin by adding $+4$ to both sides.

$$
\begin{array}{r}
-7k - 4 = -21 \quad \text{original equation} \\
\underline{\; +4 \quad +4} \quad \text{add } +4 \text{ to both sides} \\
-7k = -17
\end{array}
$$

Now we change the signs by mentally multiplying both sides of the equation by -1 and get

$$
7k = 17
$$

We finish by dividing both sides by $+7$.

$$
\frac{7k}{7} = \frac{17}{7} \longrightarrow \boldsymbol{k = \frac{17}{7}} \quad \text{divided by 7}
$$

example 23.4 Solve $-11p + 5 = 17$.

solution We first eliminate the $+5$ and then the -11.

$$
\begin{array}{r}
-11p + 5 = 17 \quad \quad \text{original equation} \\
\underline{\; -5 \quad -5} \quad \quad \text{add } -5 \text{ to both sides} \\
-11p = 12
\end{array}
$$

$$
11p = -12 \quad \quad \text{multiplied by } -1
$$

$$
\frac{11p}{11} = \frac{-12}{11} \longrightarrow \boldsymbol{p = -\frac{12}{11}} \quad \text{divided by 11}
$$

example 23.5 Solve $\frac{1}{5}m - \frac{1}{2} = \frac{3}{4}$.

solution We first eliminate the $-\frac{1}{2}$.

$$\frac{1}{5}m - \frac{1}{2} = \frac{3}{4} \qquad \text{original equation}$$

$$\frac{+\frac{1}{2} \quad +\frac{1}{2}}{\frac{1}{5}m \qquad = \frac{5}{4}} \qquad \text{add } \frac{1}{2} \text{ to both sides}$$

Now we finish by multiplying both sides by $\frac{5}{1}$

$$\frac{5}{1} \cdot \frac{1}{5}m = \frac{5}{4} \cdot \frac{5}{1} \qquad \text{multiplied by } \frac{5}{1}.$$

$$\boldsymbol{m = \frac{25}{4}}$$

example 23.6 Solve $.4x - .2 = -.16$.

solution We first add $+.2$ and then we divide by $.4$.

$$.4x - .2 = -.16 \qquad \text{original equation}$$
$$\frac{+.2 \quad +.2}{.4x \qquad = .04} \qquad \text{add .2 to both sides}$$

Now we divide both sides by $.4$.

$$\frac{\cancel{.4}x}{\cancel{.4}} = \frac{.04}{.4} \longrightarrow \quad x = .1 \qquad \text{divided by .4}$$

problem set 23

1. $3x = 4\frac{1}{2}$

2. $x + \frac{1}{2} = \frac{2}{3}$

*3. $3x - 2 = 10$

4. $5k - 4 = -30$

5. $-2p + 3 = -29$

6. $\frac{2}{3}y = 5\frac{1}{2}$

7. $-2x - 2 = 5$

8. $\dfrac{x}{3\frac{1}{2}} = 5$

*9. $\frac{1}{5}m - \frac{1}{2} = \frac{3}{4}$

10. $x + 3\frac{1}{3} = 5$

11. $y - \frac{1}{4} = \frac{5}{7}$

12. $-3y + \frac{1}{2} = \frac{5}{7}$

*13. $.4x - .2 = -.16$

14. $.02x - 1.3 = .003$

Simplify:

15. $x^2 k x k^5 x^2 y k x^2$

16. $aaa^3 bxa^2 b^3 abx^4$

Simplify by adding like terms:

17. $3a^2 xy + 5xa^2 y - 2ya^2 x$

18. $6c - 6 - 2c - 5 - 3c + 7$

19. $a^2 xx + a^2 x^2 - 3x^2 aa$

Expand by using the distributive property:

20. $4x(2y - 3 + 2a)$

21. $(x - 4)3$

Evaluate:

22. $a(a^3 - b) - b$ \quad if $a = -2$ and $b = 3$

23. $x(y - a) + a(y - x)$ \quad if $x = -3, y = 2$, and $a = -1$

24. $-b(b - a)$ if $a = -2$ and $b = 1$

25. $(-d)(d - c)$ if $c = -3$ and $d = -5$

Simplify:

26. $-4[(-5 + 2) - 3(-2 - 1)]$

27. $-3^2 + (-3)^3 - 4$

28. $\dfrac{2^2 - 3^3 - 4^2}{(-2)^3}$

29. $(-2)^2 - 3^3 - (-4)^2 - |-2 - 3|$

LESSON 24 *More complicated equations*

24.A
more complicated equations

Often we will encounter equations that have variables on both sides of the equation. When this occurs, we will begin the solution by using the addition rule to eliminate the variable on one side or the other. It makes no difference which side we choose, as will be demonstrated by working the same problem both ways.

example 24.1 Solve $3x - 4 = 5x + 7$.

solution We begin by eliminating the variable on the right side by adding $-5x$ to both sides.

$$
\begin{array}{ll}
3x - 4 = \quad 5x + 7 & \text{original equation} \\
\underline{-5x \qquad\quad -5x} & \text{add } -5x \text{ to both sides} \\
-2x - 4 = \qquad\quad 7 &
\end{array}
$$

We finish by adding $+4$ and then dividing by -2.

$$
\begin{array}{l}
-2x - 4 = \quad 7 \\
\underline{+4 \qquad +4} \qquad \text{add } +4 \text{ to both sides} \\
-2x \qquad = \; 11 \;\longrightarrow\; \dfrac{-2x}{-2} = \dfrac{11}{-2} \;\longrightarrow\; x = -\dfrac{11}{2} \quad \text{divided by } -2
\end{array}
$$

example 24.2 Solve $3x - 4 = 5x + 7$.

solution This time we begin by eliminating the variable on the left side by adding $-3x$ to both sides.

$$
\begin{array}{ll}
3x - 4 = \quad 5x + 7 & \text{original equation} \\
\underline{-3x \qquad\quad -3x} & \text{add } -3x \text{ to both sides} \\
-4 = \quad 2x + 7 &
\end{array}
$$

Now we finish by adding -7 and dividing by 2. We will get $-\frac{11}{2}$, the same answer that we got in the last example.

$$
\begin{array}{l}
-4 = 2x + 7 \\
\underline{-7 \qquad\quad -7} \qquad \text{add } -7 \text{ to both sides} \\
-11 = 2x \;\longrightarrow\; \dfrac{-11}{2} = \dfrac{2x}{2} \;\longrightarrow\; -\dfrac{11}{2} = x \quad \text{divided by } 2
\end{array}
$$

In many problems we must begin by simplifying on both sides by adding like terms.

example 24.3　Solve $3x + 2 - x + 4 = -5 - x - 4$.

solution　We begin by adding like terms on both sides of the equation to get

$$2x + 6 = -9 - x$$

This time we decide to eliminate the x term on the right side so we add $+x$ to both sides.

$$
\begin{array}{rl}
2x + 6 = -9 - x & \\
\underline{+x \qquad\qquad +x} & \quad \text{add } +x \text{ to both sides} \\
3x + 6 = -9 &
\end{array}
$$

Now we can finish by adding -6 to both sides and then dividing both sides by 3.

$$
\begin{array}{rl}
3x + 6 = \quad -9 & \\
\underline{\quad -6 \quad\quad -6} & \quad \text{add } -6 \text{ to both sides} \\
3x \quad\quad = -15 \quad\longrightarrow\quad \dfrac{3x}{3} = \dfrac{-15}{3} \quad\longrightarrow\quad x = -5 & \quad \text{divided by 3}
\end{array}
$$

example 24.4　Solve $k + 3 - 4k + 7 = 2k - 5$.

solution　We will begin by adding like terms on both sides to get

$$-3k + 10 = 2k - 5$$

Next, we eliminate the $-3k$ on the left by adding $+3k$ to both sides and then finish the solution.

$$
\begin{array}{rl}
-3k + 10 = \quad 2k - 5 & \quad \text{added like terms} \\
\underline{+3k \qquad\qquad +3k} & \quad \text{add } 3k \text{ to both sides} \\
10 = \quad 5k - 5 & \\
\underline{+5 \qquad\qquad +5} & \quad \text{add } +5 \text{ to both sides} \\
15 = \quad 5k & \\
\dfrac{15}{5} = \dfrac{5k}{5} & \quad \text{divide both sides by 5} \\
3 = \quad k &
\end{array}
$$

example 24.5　Solve $-7n + 3 + 2n = 4n - 5 + n$.

solution　We begin by simplifying and then eliminating the $5n$ term on the right side.

$$
\begin{array}{rl}
-5n + 3 = \quad 5n - 5 & \quad \text{simplified both sides} \\
\underline{-5n \qquad\qquad -5n} & \quad \text{add } -5n \text{ to both sides} \\
-10n + 3 = -5 & \\
\underline{\qquad\quad -3 = -3} & \quad \text{add } -3 \text{ to both sides} \\
-10n \qquad = -8 & \\
10n = 8 & \quad \text{multiplied both sides by } (-1)
\end{array}
$$

$$\dfrac{10n}{10} = \dfrac{8}{10} \quad\longrightarrow\quad n = \dfrac{4}{5} \qquad \begin{array}{l}\text{divide each side by 10 and} \\ \text{simplify answer}\end{array}$$

We note that we used four steps in the solution above. These are the last four steps out of a five-step process. The first step will be discussed in Lesson 30. If x is the variable, the steps are

1. (Lesson 30)
2. **Simplify by adding like terms.**

3. **Eliminate x on one side.**
4. **Eliminate the constant term.**
5. **Eliminate the coefficient of x.**

Check the order of these steps in the next problem.

example 24.6   Solve $2x - 5 + 7x = 5 + 3x + 10$.

solution   After we simplify, we will eliminate the $3x$ term on the right side.

$$
\begin{array}{rcl}
9x - 5 = & 3x + 15 & \text{simplified} \\
-3x & -3x & \text{add } -3x \text{ to both sides} \\
\hline
6x - 5 = & 15 & \\
+5 & +5 & \text{add } +5 \text{ to both sides} \\
\hline
6x \quad = & 20 & \\
\end{array}
$$

$$\frac{6x}{6} = \frac{20}{6} \qquad \text{divide both sides by 6 and simplify answer}$$

$$x = \frac{10}{3}$$

Many beginning algebra students would prefer to write the preceding answer as the mixed number $3\frac{1}{3}$. Both forms of the answer are equally correct, but we prefer the improper fraction $\frac{10}{3}$ to the mixed number $3\frac{1}{3}$ because the improper fraction is easier to use. Suppose the instructions in this problem had been to, say, solve and then multiply the answer by $\frac{11}{4}$. If we had written the answer as the mixed number $3\frac{1}{3}$, we would have to change it back to the improper fraction $\frac{10}{3}$ before the multiplication could be performed. Instructors at this level and in higher courses usually prefer the improper fraction to the mixed number.

**problem
set 24**

1. What are equivalent equations?

2. Are $x + 2 = 6$ and $3x - 14 + x - 5 = -3$ equivalent equations? Why?

Solve:

3. $2\frac{1}{2}x = \frac{3}{7}$ 4. $-x = 3$

5. $3x - 4 = 7$ 6. $.02p - 2.4 = .006$

*7. $3x + 2 - x + 4 = -5 - x - 4$ *8. $k + 3 - 4k + 7 = 2k - 5$

*9. $-7n + 3 + 2n = 4n - 5 + n$ 10. $3p + 7 - (-3) = p + (-2)$

11. $2x + x + 3 = x + 2 - 5$ 12. $5x - 3 - 2 = 3x - 2 + x$

13. $m + 4m - 2 - 2m = 2m + 2 - 3$ 14. $-m - 6m + 4 = -2m - 5$

15. $y + 2y - 4 - y = 3y - 2 + y$

Simplify:

16. $m^2y^5myy^3m^3$ 17. $k^5mmm^2k^2m^2k^3aa^2$

Simplify by adding like terms:

18. $xym^2z - 3m^2zxy + 2xm^2yz - 5xym^2z$

19. $a^2bc + 2bc - bca^2 + 5ca^2b - 3cb$ 20. $a - ax + 2xa - 3a - 3$

Expand by using the distributive property:

21. $4(7 - 3x)$

22. $(m + 2p)3axy$

Evaluate:

23. $a(-a^2 + b)$ if $a = -2$ and $b = 4$

24. $x(x - y)$ if $x = -3$ and $y = 5$

25. $p(a - 2p^2)$ if $a = -2$ and $p = -4$

26. $(y - m^2) - m$ if $m = 2$ and $y = -3$

Simplify:

27. $-3^2 - (-3)^3 + (-2)$

28. $-4(-3 + 2) - 3 - (-4) - |-3 + 2|$

29. $\dfrac{-3(-2 - 3 - 4)}{-(-3 + 2)}$

30. $\dfrac{-2(-3) + (-6)}{-4(-2)}$

LESSON 25 *More on the distributive property · Simplifying decimal equations*

25.A
the distributive property

Remember that we can simplify expressions such as

$$4(2 + 7)$$

by adding first or by using the distributive property and multiplying first.

ADDING FIRST	MULTIPLYING FIRST
$4(2 + 7)$	$4(2 + 7)$
$4(9)$	$8 + 28$
36	**36**

Thus far, we have restricted our use of this property to expanding simple expressions such as $4p(x + 3y)$.

$$4p(x + 3y) = 4px + 12py$$

In the following examples, we will use the distributive property to expand expressions that are more complicated. We remember that in each case the expression on the outside is multiplied by every term inside the parentheses.

example 25.1   Expand $xy(y^2 - x^2z)$.

solution   The xy is multiplied by y^2 and also by $-x^2z$.

$$xy(y^2 - x^2z) = (xy)(y^2) + (xy)(-x^2z) = xy^3 - x^3yz$$

example 25.2   Expand $4xy^3(x^4y - 5x)$.

solution $4xy^3$ is to be multiplied by both x^4y and $-5x$.

$$4xy^3(x^4y - 5x) = (4xy^3)(x^4y) + (4xy^3)(-5x) = 4x^5y^4 - 20x^2y^3$$

example 25.3   Expand $(ay - 4y^5)2x^2y$.

solution   It is not necessary to write down two steps. We can do the multiplications in our head if we are careful.

$$(ay - 4y^5)2x^2y = \mathbf{2ax^2y^2 - 8x^2y^6}$$

example 25.4   Expand $8m^2x(5m^3x - 3x^5 + 2x)$.

solution   This time $8m^2x$ must be multiplied by all three terms inside the parentheses.

$$(8m^2x)(5m^3x) + (8m^2x)(-3x^5) + (8m^2x)(2x) = \mathbf{40m^5x^2 - 24m^2x^6 + 16m^2x^2}$$

25.B

simplifying decimal equations   Finding the solutions of equations such as

$$(a)\quad .4 + .02m = 4.6\quad \text{and}\quad .002k + .02 = 4.02$$

can be facilitated if we begin by multiplying every term on both sides of the equation by the power of 10 that will make every decimal coefficient an integer. The value of the smallest decimal number in the problem often determines whether we multiply by 10 or 100 or 1000 or 10,000, etc.

example 25.1   Solve $.4 + .02m = 4.6$

solution   The smallest decimal number in the problem is .02. We can convert .02 to 2 if we multiply by 100. Thus, we will multiply every term on both sides of the equation by 100 and then solve.

$$
\begin{array}{ll}
.4 + .02m = 4.6 & \text{original equation} \\
40 + 2m = 460 & \text{multiplied every term by 100} \\
2m = 420 & \text{added } -40 \text{ to both sides} \\
\mathbf{m = 210} & \text{divided both sides by 2}
\end{array}
$$

example 25.2   Solve $.002k + .02 = 4.02$

solution   This time the smallest number is .002, so we will use 1000 as our multiplier.

$$
\begin{array}{ll}
.002k + .02 = 4.02 & \text{original equation} \\
2k + 20 = 4020 & \text{multiplied every term by 1000} \\
2k = 4000 & \text{added } -20 \text{ to both sides} \\
\mathbf{k = 2000} & \text{divided both sides by 2}
\end{array}
$$

problem set 25

1.   Use letters a, b, and c to state the distributive property.

2.   What is the reciprocal of $-\dfrac{1}{4}$?

Solve:

3. $3x + 2 = 7$

*4. $.4 + .02m = 4.6$

*5. $.002k + .02 = 4.02$

6. $\dfrac{1}{2}x + \dfrac{3}{4} = -\dfrac{3}{7}$

7. $3\frac{1}{3}x - \frac{1}{2} = 5$ 8. $-\frac{1}{7} + \frac{4}{3}x = \frac{1}{5}$

9. $x + 3 - 5 - 2x = x - 3 - 7x$ 10. $4x - 5 + 2x = 3x - 4 + x$

11. $-2y - 4 - y = -y + 2 + 3y$ 12. $-5 - x - 2 + 7x = 3 - 4x$

13. $-a - 2a + 4 - 4a = 7 - 3a$ 14. $p - 2p - 5 + 7p = 3 - 2p$

Simplify:

15. $p^2xyy^2x^2yx^2x$ 16. $3p^2x^4yp^5xxyy^2$

Simplify by adding like terms:

17. $-3x^2y + 5 - yx^2 - 13 + xy$ 18. $-4x + x^2 - 3x - 5 + 7x^2$

19. $xyp^2 - 4p^2xy + 5xp^2y - 7yxp^2$

Expand by using the distributive property:

*20. $xy(y^2 - x^2z)$ 21. $4x^2(ax - 2)$

*22. $4xy^3(x^4y - 5x)$ *23. $(ay - 4y^5)2x^2y$

*24. $8m^2x(5m^3x - 3x^5 + 2x)$

Evaluate:

25. $-(-a - x) - x^2$ if $x = -3$ and $a = 4$

26. $-y^2(y - 2x)$ if $y = -2$ and $x = 3$

27. $-(-p)^2 + (p - x)$ if $p = -2$ and $x = 5$

28. $(a - b) + (-a)^2$ if $a = -3$ and $b = 6$

Simplify:

29. $-3^2 - 3(3^2 - 4) - 2 - |-7 + 2|$ 30. $\dfrac{-6 - (-2 - 3)}{4 - (-3)}$

LESSON 26 *Fractional parts of numbers*

26.A
fractional
parts of
numbers

When we multiply a number by a fraction, we say that the result is a fractional part of the number. If we multiply $\frac{7}{8}$ by 48 we get 42. We say this mathematically by writing

$$\frac{7}{8} \times 48 = 42$$

and if we use words we say that

(seven-eighths)   (of 48)   (is 42)

We can generalize this problem into an equation that has three parts.

$$(F) \times (\text{of}) = (\text{is})$$

The letter F stands for fraction, and the words *of* and *is* associate with parts of the statement as we note in the following problems. We will use the variable *WN* to represent what number and *WF* to represent what fraction. We will avoid the use of the meaningless variable *x*.

example 26.1 $\frac{3}{4}$ of what number is 69?

solution In this problem, the fraction is $\frac{3}{4}$, the word *of* associates with what number (*WN*), and the word *is* associates with 69. We make these replacements and get.

$$(F) \times (\text{of}) = (\text{is}) \quad \longrightarrow \quad \left(\frac{3}{4}\right) \times (WN) = 69$$

We can undo multiplication by $\frac{3}{4}$ by multiplying by $\frac{4}{3}$. Thus we solve by multiplying both sides of the equation by $\frac{4}{3}$.

$$\frac{4}{3} \cdot \frac{3}{4} WN = 69 \cdot \frac{4}{3} \quad \longrightarrow \quad \boldsymbol{WN = 92}$$

example 26.2 What fraction of 40 is 24?

solution This time the fraction is unknown, *of* associates with 40, and *is* associates with 24. We make these replacements. Then to solve, we divide both sides of the equation by 40.

$$(F) \times (\text{of}) = (\text{is}) \quad \longrightarrow \quad (WF)(40) = (24) \quad \longrightarrow \quad \frac{WF \cdot 40}{40} = \frac{24}{40} \quad \longrightarrow \quad \boldsymbol{WF = \frac{3}{5}}$$

example 26.3 $2\frac{1}{2}$ of 240 is what number?

solution This time the fraction is written as the mixed number $2\frac{1}{2}$. We see that *of* associates with 240 and *is* with what number (*WN*). We make these substitutions and solve by multiplying $2\frac{1}{2}$ and 240.

$$(F) \times (\text{of}) = (\text{is}) \quad \longrightarrow \quad (2\tfrac{1}{2})(240) = WN \quad \longrightarrow \quad \boldsymbol{600 = WN}$$

problem set 26 Solve:

*1. $\frac{3}{4}$ of what number is 69?

*2. What fraction of 40 is 24?

*3. $2\frac{1}{2}$ of 240 is what number?

4. $3\frac{1}{5}$ of what number is 32?

5. What fraction of 324 is 270?

6. $\frac{7}{3}$ of 42 is what number?

Solve:

7. $\frac{1}{2}x + \frac{1}{2} = 2\frac{1}{5}$

8. $-\frac{7}{8} + \frac{1}{2}x = \frac{3}{4}$

9. $.005p + 1.4 = .005$

10. $x - 2 - 2x = 3 - x + 4x$

11. $3y - y + 2y - 5 = 7 - 2y + 5$

12. $p - 2p + 4 - 7 = p + 2$

13. $k - 4 - 2k = 7 + 3k - k + 5$

14. $x - 5 - (-2) + 2x = 7$

15. $3x - 4x + 7 = 5 + x - 6$

Simplify:

16. $y^5x^2y^3yxy^2$ **17.** $m^2myy^2m^3ym$

Simplify by adding like terms:

18. $xym^2 + 3xy^2m - 4m^2xy + 5mxy^2$ **19.** $pc - 4cp + c - p + 7pc - 7c$

20. $a^2xy + 4xa^2y - yxa^2 + 3yx^2a$

Expand by using the distributive property:

21. $x^2y^3(3xy - 5y)$ **22.** $3x^4a(x^3 - 2x^4a^3)$ **23.** $(xyp - 3xp)p^2xy$

Evaluate:

24. $x^3y(x - y)$ if $x = -3, y = 1$ **25.** $p^2 - a^2(p - a)$ if $p = 3, a = 5$

26. $ka(-a) - k + a$ if $a = -3, k = 4$

27. $p(a) - xp(-a)$ if $p = -2, a = 3, x = 4$

Simplify:

28. $-3^2 + (-2)^3$ **29.** $-4(-7 + 5)(-2) - |-2 - 5|$

30. $\dfrac{-5(-3 + 7)}{-5 + (-2)}$

LESSON 27   *Negative exponents*

27.A

negative exponents   It is convenient to have an alternate way to write the reciprocal of an exponential. Here we show the alternate way to write 1 over 5^2 and 1 over 5^{-2}.

$$\frac{1}{5^2} = 5^{-2} \qquad \frac{1}{5^{-2}} = 5^2$$

In the formal definition we will use x and n to represent the base and the exponent.

DEFINITION OF x^{-n}

If n is any real number and x is any real number that is not zero,

$$\frac{1}{x^n} = x^{-n}$$

(a) $\dfrac{1}{3^4} = 3^{-4}$ (b) $7^{-2} = \dfrac{1}{7^2}$ (c) $\dfrac{1}{5^{-8}} = 5^8$ (d) $6^{-3} = \dfrac{1}{6^3}$

In (a) we moved 3^4 from the denominator to the numerator and **changed the sign of the exponent** from plus to minus. In (b) we moved 7^{-2} from the numerator to the denominator **and changed the sign of the exponent** from minus to plus. In (c) we moved 5^{-8} from the denominator to the numerator **and changed the sign of the exponent** from minus to plus. In (d) we moved 6^{-3} to the denominator and **changed the sign of the exponent.** The formal definition of x^{-n} is stated in the box above. We will now state the definition informally.

> A number or a variable that is written as an exponential can be written in reciprocal form if the sign of the exponent is changed.

If the exponent was positive, it is negative in the reciprocal form. If the exponent was negative, it is positive in the reciprocal form.

example 27.1   Simplify 3^{-2}.

solution   The negative exponent is meaningless as an operation indicator. Thus the first step in the solution is to write 3^{-2} in reciprocal form and change the negative exponent to a positive exponent.

$$3^{-2} = \frac{1}{3^2}$$

Now we can complete the simplification because a positive exponent is an operation indicator because 3^2 means $3 \cdot 3$.

$$3^{-2} = \frac{1}{3^2} = \frac{1}{3 \cdot 3} = \frac{1}{9}$$

example 27.2   Simplify $\dfrac{1}{3^{-3}}$.

solution   Again, as the first step we write the expression in reciprocal form so that the negative exponent can be changed to a positive exponent.

$$\frac{1}{3^{-3}} = 3^3$$

Now 3^3 is meaningful as $3 \cdot 3 \cdot 3$ and 3^3 equals 27, so

$$\frac{1}{3^{-3}} = 3^3 = 3 \cdot 3 \cdot 3 = 27$$

example 27.3   Simplify $(-3)^{-2}$.

solution   We first change the negative exponent to a positive exponent by writing the exponential in reciprocal form

$$(-3)^{-2} = \frac{1}{(-3)^2} = \frac{1}{9}$$

We have defined negative exponents so that their use will not conflict with the use of the product theorem, which is repeated here.

> PRODUCT THEOREM FOR EXPONENTS
>
> If m and n are real numbers and $x \neq 0$,
>
> $$x^m \cdot x^n = x^{m+n}$$

When the bases are the same, we multiply exponentials by adding the exponents. This is true even if some of the exponents are negative numbers.

(a) $x^{-5}x^2 = x^{-3}$ (b) $y^7y^5y^{-2} = y^{10}$ (c) $p^{10}p^{-15} = p^{-5}$

example 27.4   Simplify $x^4m^2x^{-2}m^{-5}$.

solution   We first change the order of multiplication

$$x^4x^{-2}m^2m^{-5}$$

and then add the exponents of exponentials whose bases are the same.

$$x^2m^{-3}$$

example 27.5   Simplify $x^{-2}y^{-6}y^5x^4zxz^5$

solution   We change the orders of the factors and add the exponents of exponentials that have the same bases and get

$$x^{-2}x^4xy^{-6}y^5zz^5 = x^3y^{-1}z^6$$

problem set 27

1.   Designate (a) the set of integers, (b) the set of whole numbers.

2. $2\frac{1}{4}$ of what number is 750?   3.   What fraction of 72 is 63?

4. $3\frac{1}{8}$ of 72 is what number?

Solve:

5. $-5 + 2\frac{1}{2}x = 17$ 6. $3\frac{1}{2}x + 2 = 9$

7. $\frac{1}{4}x + \frac{1}{2} = \frac{7}{8}$ 8. $3x + 5 - x = x + 5$

9. $3y - 5 = 7 - 2y + 8$ 10. $7p - 14 = 4p - 5 + p$

11. $.0025k + .06 = 4.003$ 12. $3m - 2 - m = -2 + m - 5$

13. $x - 3x - 5 - 2x = 7x + 3 - 5 + 2x$

Simplify:

*14. 3^{-2} *15. $\dfrac{1}{3^{-3}}$ *16. $(-3)^{-2}$ 17. $\dfrac{1}{(-2)^{-2}}$

Simplify:

*18. $x^4m^2x^{-2}m^{-5}$ *19. $x^{-2}y^{-6}y^5x^4zxz^5$

20. $x^{-5}y^5axy^5y^{-8}a^{-4}$ 21. $m^2p^{-4}m^{-2}p^6m^4m^{-5}$

Simplify by adding like terms:

22. $4x^2yp - px^2y + 3ypx^2 - 4$ 23. $5m^2x^2y - 2x^2m^2y + 8m^2y^2x$

Expand:

24. $(6x^2yp - 4p + 2)x^2y^3p$ 25. $4mz(m^3cz^2 - 5mz^5)$

Evaluate:

26. $-xa^2(a + x)$ if $x = -2, a = 3$ 27. $m^2p - p(m - p)$ if $m = -3, p = 5$

28. $4x(a + x)(-x)$ if $x = -3, a = 2$ 29. $m(k^2 - m^3)$ if $k = -2, m = -1$

Simplify:

30. $\dfrac{4(2 - 5)}{-2(4 - 2) - (-2)}$

LESSON 28 *Zero exponents · Decimal parts of a number*

28.A
zero
exponents

We know that a nonzero number divided by itself equals 1. For instance,

$$1 = \frac{4^2}{4^2}$$

We can simplify this expression by moving the 4^2 on the bottom to the top and changing the exponent from 2 to -2. Then we multiply 4^2 by 4^{-2} and get 4^0.

$$1 = \frac{4^2}{4^2} = 4^2 \cdot 4^{-2} = 4^0$$

Now, 4^0 must equal 1 because 4^2 divided by 4^2 equals 1. In the same way, we see that any nonzero quantity raised to the zero power must have a value of 1.

$$(x + y + z^2)^0 = 1 \qquad (pm)^0 = 1 \qquad (px^{-4})^0 = 1$$

Each of the above has a value of 1 if the expression in parentheses does not have a value of zero.

DEFINITION

If x is any real number that is not zero,

$$x^0 = 1$$

---

example 28.1   Simplify the following expressions:

$$\text{(a)} \quad x^2y^5y^{-2}x^{-2} \qquad \text{(b)} \quad m^5b^2mb^{-2}$$

solution   (a) $x^2y^5y^{-2}x^{-2} = y^3x^0 = \boldsymbol{y^3}$ (because $x^0 = 1$ if $x \neq 0$)
(b) $m^5b^2mb^{-2} = m^6b^0 = \boldsymbol{m^6}$ (because $b^0 = 1$ if $b \neq 0$)

Since we must not use the expression 0^0, it is necessary in problems such as (a) and (b) that we assume that the variable with the zero exponent is not zero. Further, in the problem sets, we will assume a nonzero value for any variable that has zero for its exponent.

example 28.2   Expand $x^5y^0z(p^{-3}z^0 - 4x^{-5}z^{-1})$.

solution   We choose to begin by simplifying x^5y^0z and $p^{-3}z^0$, remembering that $y^0 = 1$ and that $z^0 = 1$. Now we have

$$x^5z(p^{-3} - 4x^{-5}z^{-1})$$

We finish by doing the two multiplications and get

$$x^5zp^{-3} - 4x^0z^0 \qquad \text{which equals} \qquad \boldsymbol{x^5zp^{-3} - 4}$$

example 28.3   Expand $x^{-2}y^{-2}(x^2y^2 + 4x^4y^2)$.

solution   We do the two multiplications and get

$$x^{-2}y^{-2}x^2y^2 + x^{-2}y^{-2}4x^4y^2$$

Now we simplify by remembering that both x^0 and y^0 equal 1.

$$x^0 y^0 + 4x^2 y^0 = (1)(1) + 4x^2(1) = \mathbf{1 + 4x^2}$$

28.B
decimal parts of a number

Many people call decimal numbers decimal fractions because terminating decimal numbers can be written as fractions. For example,

(a) $28.6132 = \dfrac{286{,}132}{10{,}000}$ (b) $.000463 = \dfrac{463}{1{,}000{,}000}$

We have been working problems concerning fractional parts of a number by using the relationship

$$(F) \times (\text{of}) = (\text{is})$$

We can solve statements about decimal parts of a number by using the slightly different relationship

$$(D) \times (\text{of}) = (\text{is})$$

where D stands for the decimal (decimal fraction) part of the number and *of* and *is* have the same meanings as before.

example 28.4   .32 of what number is 24.32?

solution   We will use

$$(D) \times (\text{of}) = (\text{is})$$

We replace D with .32, *of* with WN, and *is* with 24.32 and then solve.

$$.32WN = 24.32 \quad \longrightarrow \quad \frac{.32WN}{.32} = \frac{24.32}{.32} \quad \longrightarrow \quad \mathbf{WN = 76}$$

example 28.5   What decimal part of 42 is 26.04?

solution   In $(D) \times (\text{of}) = (\text{is})$, we replace D with WD, *of* with 42, and *is* with 26.04. Then we solve.

$$WD(42) = 26.04 \quad \longrightarrow \quad \frac{WD(42)}{42} = \frac{26.04}{42} \quad \longrightarrow \quad \mathbf{WD = .62}$$

example 28.6   .42 of 86 is what number?

solution   This time .42 replaces D and 86 replaces *of*. Then we multiply to find WN.

$$(.42)(86) = WN \quad \longrightarrow \quad \mathbf{36.12 = WN}$$

problem set 28

*1.   .32 of what number is 24.32?   *2.   What decimal part of 42 is 26.04?

*3.   .42 of 86 is what number?

Solve:

4. $3x - 7 = 42$ 5. $2\frac{1}{2}x - 5 = 17$

6. $\dfrac{3}{4} + \dfrac{1}{2}x + 2 = 0$ 7. $.4k + .4k - .02 = 4.02$

8. $-2x - 5 - x - 8 - 5x - 3 = 0$ **9.** $5m - m - 2m + 5 = -3m - 2$

10. $-3x - 2 = 3x - 5 + 8$ **11.** $-p - 2p - 4 + 7p = 5 + 2p - 6$

12. $-2k + k - 3k = 7 + 2k - 2$

Simplify:

13. $(-3)^{-2}$ **14.** $\dfrac{1}{4^{-2}}$ **15.** $\dfrac{1}{(-4)^{-3}}$

16. $\dfrac{1}{2^{-3}}$ **17.** $(-5)^{-3}$

Expand:

*18. $x^5 y^0 z (p^{-3} z^0 - 4x^{-5} z^{-1})$ ***19.** $x^{-2} y^{-2} (x^2 y^2 + 4x^4 y^2)$

20. $(x^2 - 4x^5 y^{-5}) 3p^0 x^{-2}$ **21.** $2x^{-2} y^0 (x^2 y^0 - 4x^{-6} y^4)$

22. $x^{-4} y^0 (x^4 - 3y^2 x^5 p^0)$ **23.** $m^0 x (x^{-1} y^0 - y^2 m^0)$

Simplify by adding like terms:

24. $x^2 y m^3 - 3x^2 y + 6y m^3 x^2 + y x^2$ **25.** $abc^2 - 2ab^2 c + 6c^2 ab - 4b^2 ac$

Evaluate:

26. $-a^3 (a^0 - b)$ if $a = -2, b = 4$ **27.** $-c(c - b)$ if $c = -2, b = 4$

28. $x(x^0 - y)(y - 2x)$ if $x = -3, y = 5$

Simplify:

29. $-3^2 + (-3)^3 - 3^0 - |-3 - 3|$ **30.** $\dfrac{-4[2 - (-2)]}{-7(5 - 2)}$

LESSON 29   *Algebraic phrases*

29.A
algebraic phrases

In algebra we learn to answer verbal questions by turning these questions into algebraic equations. Then we solve the equations to get the desired answers. The equations that we write contain algebraic phrases that have the same meanings as the verbal phrases used in the questions. There are several keys to writing these phrases. The word *sum* means that things are added, and the word *product* means that things are multiplied. Seven more than, or increased by 7, means to add 7; while 7 less than, or decreased by 7, means to subtract 7. If we use N to represent an unknown number, then we will use $-N$ to represent the opposite of the unknown number. In the same way, twice a number would be represented by $2N$ and 5 times the opposite of a number would be represented by $5(-N)$. If we write the sum of twice a number and negative 10 as $2N - 10$, we could write 7 times this sum by writing $7(2N - 10)$. Cover the answers in the right hand column below and see if you can write the algebraic phrase that is indicated.

The sum of a number and 7	$N + 7$
Seven less than a number	$N - 7$
The opposite of a number decreased by 5	$-N - 5$

The sum of the opposite of a number and -5	$-N - 5$
The product of twice a number and 8	$8(2N)$
The sum of twice a number and -5	$2N - 5$
Five times the sum of twice a number and -5	$5(2N - 5)$
Six times the sum of twice the opposite of a number and -8	$6[2(-N) - 8]$
The product of 7 and a number increased by 10	$7(N + 10)$
The sum of 3 times a number and -4, multiplied by 5	$5(3N - 4)$
The sum of -10 and 6 times the opposite of a number	$6(-N) - 10$

problem set 29

1. 1.6 of what number is 3200?

2. What decimal part of 80 is 8400?

3. $4\dfrac{5}{6}$ of 4596 is what number?

Write the algebraic phrases that correspond to these word phrases.

*4. Five times the sum of twice a number and negative 5.

*5. The product of 7 and a number increased by 10.

6. The sum of 5 times a number and -8.

7. Three times the sum of the opposite of a number and negative 7.

Solve:

8. $\dfrac{1}{2}x + 2 = -\dfrac{3}{4}$

9. $.3k + .85k - 2 = 2.6$

10. $\dfrac{1}{7}k - \dfrac{4}{7} = -7$

11. $2k - 5 + k - 3 = 2 + 2k + 5$

12. $-3 - 6p + p - 2 = 7 - p$

13. $3m - m = 5 - 4 + 2m + 5 - 5m$

Simplify:

14. $(-6)^{-2}$

15. 3^{-3}

16. $\dfrac{1}{(-4)^{-2}}$

17. $\dfrac{1}{5^{-3}}$

Expand:

18. $p^0 x^{-1}(x - 2x^0)$

19. $4x^2 p^0(3xp^5 - 2x^{-2})$

20. $(y^{-2}x^{-1} + 3p^2 y^{-2}k^0)xp^0 y^2$

21. $4m^2 x^{-5}(2m^{-2}x^{-5} - 3m^{-2}k^0)$

22. $2p^{-4}x^2 y^{-2}(p^4 x^2 y^2 k^0 - 3p^2 x)$

23. $4x^2 yp^{-3}(x^{-2}y^4 p^6 - 5x^4 yp^{-3})$

Simplify by adding like terms:

24. $xmp^{-2} - 4p^{-2}xm + 6p^{-2}mx - 5mx$

25. $k^2 p^{-4}y - 5k^2 yp^{-4} + 2yk^2 p^{-4} - 5k^2 yp^{-4}$

Evaluate:

26. $a^3 - (b^0 - a)$ if $a = -3, b = 4$

27. $p - a(p - ap)$ if $a = -2, p = -3$

28. $-k^0 - (-km)$ if $m = 3, k = -5$

Simplify:

29. $-2(-3) - (-4)^0(-3)|-5 - 2|$

30. $\dfrac{-7(-4 - 6)}{-(-4) - [-(-6)]}$

LESSON 30 *Equations with parentheses*

30.A
equations
with
parentheses

When equations contain parentheses, we begin by eliminating the parentheses. If the parentheses are preceded by a number, we use the distributive property. We multiply the number by every term inside the parentheses and discard the parentheses.

example 30.1   Solve $2(3 - b) = b - 5$.

solution   As the first step we will use the distributive property on the left side to expand $2(3 - b)$. Then we will complete the solution.

$$2(3 - b) = b - 5 \qquad \text{original equation}$$
$$6 - 2b = b - 5 \qquad \text{multiplied}$$
$$\underline{\quad 2b \quad 2b \quad} \qquad \text{add } 2b \text{ to both sides}$$
$$6 \quad = 3b - 5$$
$$\underline{5 \qquad\qquad 5} \qquad \text{add 5 to both sides}$$
$$11 \quad = 3b \longrightarrow \frac{11}{3} = b \qquad \text{divided by 3}$$

example 30.2   Solve $3(1 + 2x) + 7 = -4(x + 2)$.

solution   This equation has parentheses on both sides. Thus we begin by using the distributive property on the left side and again on the right side to eliminate both sets of parentheses.

$$3(1 + 2x) + 7 = -4(x + 2) \qquad \text{original equation}$$
$$3 + 6x + 7 = -4x - 8 \qquad \text{used distributive property}$$
$$10 + 6x = -4x - 8 \qquad \text{added like terms}$$
$$\underline{+ 4x \qquad +4x} \qquad \text{add } 4x \text{ to both sides}$$
$$10 + 10x = -8$$
$$\underline{-10 \qquad\qquad -10} \qquad \text{add } -10 \text{ to both sides}$$
$$10x = -18$$
$$x = -\frac{18}{10} \longrightarrow x = -\frac{9}{5} \qquad \text{divided by 10 and simplified}$$

In this problem, we used all of the five steps that we will use to solve equations. Sometimes one of the steps is not necessary as in Example 30.1 above where addition of like terms was not required. If the variable is x, the five steps are:

1. **Eliminate parentheses.**
2. **Add like terms on both sides.**
3. **Eliminate x on one side or the other.**
4. **Eliminate the constant term.**
5. **Eliminate the coefficient of x.**

example 30.3   Solve $15(4 - 5x) = 16(4 - 6x) + 10$.

solution   As the first step we will use the distributive property as required on both sides of the equation.

$$60 - 75x = 64 - 96x + 10 \qquad \text{used distributive property}$$

$$\begin{array}{rl} 60 - 75x = & 74 - 96x \qquad \text{added like terms} \\ +\,96x & +\,96x \qquad\quad \text{add } 96x \text{ to both sides} \\ \hline 60 + 21x = & 74 \\ -60 & -60 \qquad\quad\ \text{add } -60 \text{ to both sides} \\ \hline 21x = & 14 \end{array}$$

$$x = \frac{14}{21} \longrightarrow x = \frac{2}{3} \qquad \begin{array}{l}\text{divided both sides by 21}\\ \text{and simplified}\end{array}$$

In the last three examples we began by using the distributive property. To solve the next two problems, we need to have two rules for eliminating parentheses preceded by a plus sign or a minus sign. The rules are:

1. **When parentheses are preceded by a plus sign, both the parentheses and the sign may be discarded as demonstrated here.**

$$+(-4 + 3x) = -4 + 3x$$

2. **When parentheses are preceded by a minus sign, both the minus sign and the parentheses may be discarded if the signs of all terms within the parentheses are changed. This rule is used because the minus sign indicates the negative of, or the opposite of, the quantity within the parentheses.**

$$-(x - 3y + 6 - k) = -x + 3y - 6 + k$$

example 30.4   Solve $12 - (2x + 5) = -2 + (x - 3)$.

solution   As the first step we drop the parentheses, remembering that if the parentheses are preceded by a minus sign we must change all signs inside the parentheses.

$$12 - 2x - 5 = -2 + x - 3$$

Now we simplify on both sides of the equation

$$7 - 2x = x - 5$$

and solve for x:

$$\begin{array}{rl} 7 - 2x = & x - 5 \\ +5 + 2x & +2x + 5 \qquad \text{add } +5 + 2x \text{ to both sides} \\ \hline 12 \quad = & 3x \longrightarrow \frac{12}{3} = \frac{3x}{3} \longrightarrow 4 = x \qquad \text{divided by 3} \end{array}$$

example 30.5   Solve $-(4y - 17) + (-y) = (2y - 1) - (-y)$.

solution   Again we remember that when we discard parentheses preceded by a minus sign, the signs of all terms within the parentheses are changed.

$$-4y + 17 - y = 2y - 1 + y$$

First we add like terms and then we solve.

$$\begin{array}{rl} -5y + 17 = & 3y - 1 \\ +5y + \ 1 & +5y + 1 \qquad \text{add } +5y + 1 \text{ to both sides} \\ \hline 18 = & 8y \longrightarrow \frac{18}{8} = \frac{8y}{8} \longrightarrow \frac{9}{4} = y \qquad \text{divided by 8} \end{array}$$

problem
set 30

Write the algebraic phrases that correspond to these word phrases.

1. Seven times the sum of a number and negative 5.

2. Seven less than twice the opposite of a number.

3. The sum of 7 times a number and -51.

4. A number is multiplied by 4 and this product decreased by 15.

5. 3.25 of what number is 585?

6. What fraction of $6\frac{7}{8}$ is $\frac{1}{4}$?

7. $2\frac{5}{8}$ of 21 is what number?

Solve:

8. $.1p - .2p + 2 = -4.6$

9. $\frac{1}{4} + 4\frac{1}{2}k = \frac{1}{8}$

*10. $2(3 - b) = b - 5$

11. $3(1 + 2x) + 7 = -4(x - 2)$

*12. $15(4 - 5x) = 16(4 - 6x) + 10$

13. $(3 - 5)m - 2m = 4(-2 + 5)$

14. $12 - 2x + 5 = -2 + (x - 3)$

15. $-4(4y - 17) + (-y) = (2y - 1) - (-y)$

Simplify:

16. $\dfrac{1}{(-4)^{-3}}$

17. 4^{-3}

Expand:

18. $y^0 x^{-4}(x^4 - 5y^4 x^4)$

19. $(y^{-5} + 3x^5 y^2)x^0 y^5$

20. $-2x^2(3x^4 - 6x^{-2}y^0 p)$

21. $5x^0 y^2(y^4 x^6 - 5x^0 y^{-4})$

22. $3m^2 n^2(p^0 m^4 n - m^{-2}n^{-2})$

23. $(x^0 p^5 - 3x^0 p^{-5})2p^0 x^5$

Simplify by adding like terms:

24. $3xym^2 z^3 + 2x^2 xy^2 y^{-1}m^2 z^2 - xym^5 m^{-3}z^2 x^2$

25. $3xy - 2x^2 yx^{-1} + 5x^3 x^{-2}y^3 y^{-2} + 5xxxx^{-2}y$

Evaluate:

26. $-a^2 - 3a(a - b)$ if $a = -2, b = -1$

27. $-c(ac - a^0)$ if $a = -3, c = 4$

28. $-n(n^0 - m) - |m^2|$ if $n = -4, m = 6$

Simplify:

29. $-2^0 - 2^2 - (-2)^3$

30. $-3 + (-3)(-3)^2 + (-3)^3$

LESSON 31 *Word problems*

31.A

word problems

To solve word problems, we look for statements in the problems that describe equal quantities. Then we use algebraic phrases and equals signs to write equations that make the same statements of equality. We will begin by solving problems that contain only one statement of equality. These problems require that we write only one equation. Later, we will encounter types of problems that contain more than one statement of equality. These problems will require more than one equation for their solution.

We will avoid the use of the letters x and y in writing these equations. We will try to use variables whose meaning is easy to remember. The problems in this lesson discuss some unknown number. We will use the letter N to represent the unknown number.

example 31.1 The sum of twice a number and 13 is 75. Find the number.

solution **The word *is* means equals to. Thus, the sum of twice a number and 13 equals 75.**

$$2N + 13 = 75 \qquad \text{equation}$$

We can solve this equation by adding -13 to both sides and then dividing by 2.

$$
\begin{array}{rl}
2N + 13 = & 75 \qquad \text{equation} \\
-13 \quad & -13 \qquad \text{add } -13 \text{ to both sides} \\
\hline
2N \quad = & 62 \\
N = & \mathbf{31} \qquad \text{divided both sides by 2}
\end{array}
$$

Solutions to word problems should always be checked to see if they really do solve the problem.

Check: $2(31) + 13 = 75 \longrightarrow 62 + 13 = 75 \longrightarrow 75 = 75 \qquad$ Check

example 31.2 Find a number such that 13 less than twice the number is 137.

solution We will use N to represent the unknown number. Then twice the unknown number is $2N$ and 13 less than that is $2N - 13$.

$$
\begin{array}{rl}
2N - 13 = & 137 \qquad \text{equation} \\
+13 \quad & +13 \qquad \text{add } +13 \text{ to both sides} \\
\hline
2N \quad = & 150 \\
N = & \mathbf{75} \qquad \text{divided both sides by 2}
\end{array}
$$

Check: $2(75) - 13 = 137 \longrightarrow 150 - 13 = 137 \longrightarrow 137 = 137 \qquad$ Check

example 31.3 If 5 times a number is decreased by 14, the result is twice the opposite of the number.

solution If we use N for the number, then $2(-N)$ will represent twice the opposite of the number.

$$
\begin{array}{ll}
5N - 14 = 2(-N) & \text{equation} \\
5N - 14 = -2N & \text{multiplied} \\
2N + 14 \qquad 2N + 14 & \text{add } 2N + 14 \text{ to both sides} \\
\hline
7N \quad = 14 \longrightarrow N = \mathbf{2} & \text{divided both sides by 7}
\end{array}
$$

Check: $5(2) - 14 = 2(-2) \longrightarrow 10 - 14 = -4 \longrightarrow -4 = -4 \qquad$ Check

example 31.4 Find a number which decreased by 18 equals 5 times its opposite.

solution Again we use N for the number and $-N$ for its opposite.

$$
\begin{array}{ll}
N - 18 = 5(-N) & \text{equation} \\
N - 18 = -5N & \text{multiplied} \\
5N + 18 \qquad 5N + 18 & \text{add } 5N + 18 \text{ to both sides} \\
\hline
6N \quad = \qquad 18 & \\
N = \mathbf{3} & \text{divided both sides by 6}
\end{array}
$$

Check: $3 - 18 = 5(-3) \longrightarrow 3 - 18 = -15 \longrightarrow -15 = -15 \qquad$ Check

example 31.5   We get the same result if we multiply a number by 3 *or* if we multiply the number by 5 and then add 2. Find the number.

solution   The statement of the problem leads to the following equation.

$$3N = 5N + 2 \qquad \text{equation}$$
$$\underline{-5N \quad -5N} \qquad\quad \text{add } -5N \text{ to both sides}$$
$$-2N = 2 \qquad\qquad$$
$$N = -1 \qquad\qquad \text{divided both sides by } -2$$

Check: $3(-1) = 5(-1) + 2 \longrightarrow -3 = -5 + 2 \longrightarrow -3 = -3$ Check

problem set 31

*1.   The sum of twice a number and 13 is 75. Find the number.

*2.   Find a number such that 13 less than twice the number is 137.

*3.   If 5 times a number is decreased by 14, the result is twice the opposite of the number. Find the number.

*4.   Find a number which decreased by 18 equals 5 times its opposite.

*5.   We get the same result if we multiply a number by 3 *or* if we multiply the number by 5 and then add 2. Find the number.

6. $2\frac{1}{9}$ of what number $= 3\frac{4}{5}$?   7.   What decimal part of .05 is 1.25?

Solve:

8. $\dfrac{p}{7} - 2 = \dfrac{15}{3}$ 9. $2\frac{1}{4}k + \dfrac{1}{4} = \dfrac{1}{8}$

10. $1.3p + .3p - 2 = 1.2$ 11. $3(p - 2) = p + 7$

12. $+2(3x - 5) = 7x + 2$ 13. $-(x - 3) - 2(x - 4) = 7$

14. $-5(p - 4) - 3(-2 - p) = p - 2$ 15. $2(3p - 2) - (p + 4) = 3p$

Simplify:

16. 6^{-2} 17. $\dfrac{1}{(-3)^{-3}}$ 18. $(-2)^{-2} - 2^2$ 19. $\dfrac{1}{(-4)^{-3}}$

Expand:

20. $-3x^0p(-5xp^0 - 2p^{-1})$ 21. $2xp^{-4}(x^{-4}p - 3x^2p^{-2})$

22. $2p^4x^0y(p^5m^4x - 5x^{-2}y^{-4})$ 23. $(x^4 - 2p^2)3x^0p^{-4}$

Simplify by adding like terms:

24. $-3x^2x^0xy^2 + 2x^3y^{-3}y^5 + 5x^{-3}x^{-6}yy^2y^{-5}$

25. $2xym^2 + 3x^2ym - 4y^2my^{-1}m^0x^4x^{-2}$

Evaluate:

26. $a - a(b^0 - a)$ if $a = -2, b = 5$

27. $x^2 - xy^2(x - y)$ if $x = -3, y = 4$

28. $m - m^2 - (m - n)$ if $m = -3, n = -5$

Simplify:

29. $-3^3 - 3^2 - (-3)^4 - |-3^2 - 3|$ 30. $\dfrac{-5(-5 - 4)}{-2^0(-8 - 1)}$

LESSON 32 *Products of prime factors · Statements about unequal quantities*

32.A
prime numbers

The number 6 can be composed by multiplying the two counting numbers 3 and 2.

$$3 \cdot 2 = 6$$

Because 6 can be composed in this way, we say that 6 is a composite number. The number 35 is also a composite number because it can be composed as the product of the counting numbers 5 and 7.

$$5 \cdot 7 = 35$$

The number 1 must be one of the factors if we wish to compose 17 by multiplying.

$$17 \cdot 1 = 17$$

The number 1 must also be a factor if we wish to compose either 3 or 11 or 23.

$$1 \cdot 3 = 3 \qquad 1 \cdot 11 = 11 \qquad 1 \cdot 23 = 23$$

Since these numbers can be composed only if 1 is one of the factors, we do not call these numbers composite numbers. We call them prime numbers.

A prime number is a counting number greater than 1 whose only counting number factors are 1 and the number itself.

The number 12 can be written as a product of integral factors in four different ways.

(a) $12 \cdot 1$ (b) $4 \cdot 3$ (c) $2 \cdot 6$ (d) $2 \cdot 2 \cdot 3$

In (a), (b), and (c), one of the factors is not a prime number, but in (d) all three of the factors are prime numbers. **A prime factor is a factor that is a prime number.** To find the prime factors of a counting number we divide by prime numbers as we see in the following examples.

example 32.1   Express 80 as a product of prime factors.

solution   We will divide by prime numbers.

$$\frac{80}{2} = 40, \qquad \frac{40}{2} = 20, \qquad \frac{20}{2} = 10, \qquad \frac{10}{2} = 5$$

Using the five factors we have found, we can express 80 as a product of prime factors as $2 \cdot 2 \cdot 2 \cdot 2 \cdot 5$.

example 32.2   Express 147 as a product of prime factors.

solution   147 is not divisible by 2 or by 5, so let's try 3.

$$\frac{147}{3} = 49 \qquad \text{and} \qquad \frac{49}{7} = 7$$

So 147 expressed as a product of prime factors is $3 \cdot 7 \cdot 7$.

32.B
statements about unequal quantities

Often a word problem makes a statement about quantities that differ by a specified amount. Thus, the statement tells us that the quantities are not equal, and our task is to write an equation about quantities that are equal. To perform this task we must add as required so that both sides of the equation represent equal quantities.

example 32.3   Twice a number is 42 less than -102. Find the number.

solution   **We must be careful because the problem tells us about things that are not equal. We begin by writing an equation that we know is incorrect.**

$$2N = -102 \qquad \text{incorrect}$$

The problem said that $2N$ was 42 less than -102, so we must add 42 to $2N$ or we must add -42 to -102.

ADDING 42 TO $2N$ or   ADDING -42 TO -102

$$
\begin{array}{ll}
2N + 42 = -102 \quad \text{correct} \\
\underline{ -42 \qquad -42} \\
2N = -144 \\
 N = \mathbf{-72}
\end{array}
$$

$$
\begin{array}{l}
2N = -102 - 42 \quad \text{correct} \\
2N = -144 \\
N = \mathbf{-72}
\end{array}
$$

Check: $2(-72) + 42 = -102 \longrightarrow -144 + 42 = -102 \longrightarrow -102 = -102$ Check

example 32.4   Five times a number is 72 greater than the opposite of the number. Find the number.

solution   **This statement is tricky because it describes quantities that are unequal. As the first step in writing the desired equation, we will write an equation that we know is incorrect.**

$$5N = -N \qquad \text{incorrect}$$

This equation is incorrect because $5N$ is really 72 greater. We can make the equation correct by adding -72 to $5N$ or by adding $+72$ to $-N$.

ADDING -72 TO $5N$ or   ADDING $+72$ TO $-N$

$$
\begin{array}{l}
5N - 72 = -N \\
\underline{+N + 72 \quad +N + 72} \\
6N = 72 \\
 N = \mathbf{12}
\end{array}
$$

$$
\begin{array}{l}
5N = -N + 72 \\
\underline{+N \quad +N} \\
6N = 72 \\
N = \mathbf{12}
\end{array}
$$

Check: $5(12) - 72 = -12 \longrightarrow 60 - 72 = -12 \longrightarrow -12 = -12$ Check

example 32.5   If the sum of twice a number and -14 is multiplied by 2, the result is 12 greater than the opposite of the number. Find the number.

solution   **Again we begin by writing an equation that we know is incorrect.**

$$2(2N - 14) = -N \qquad \text{incorrect}$$

We know that the left side is greater by 12. We can write a correct equation by adding -12 to the left side or by adding $+12$ to the right side.

ADDING -12 TO THE LEFT SIDE   or   ADDING $+12$ TO THE RIGHT SIDE

$$
\begin{array}{l}
2(2N - 14) - 12 = -N \\
4N - 28 - 12 = -N \\
5N = 40 \\
N = \mathbf{8}
\end{array}
$$

$$
\begin{array}{l}
2(2N - 14) = -N + 12 \\
4N - 28 = -N + 12 \\
5N = 40 \\
N = \mathbf{8}
\end{array}
$$

Check: $2(2 \cdot 8 - 14) - 12 = -8 \longrightarrow 2(2) - 12 = -8 \longrightarrow -8 = -8$ Check

example 32.6   Five times a number is 21 less than twice the opposite of the number. What is the number?

solution   We must be careful because 5 times the number is 21 less. Thus we will add 21 so that it will be equal.

$$5N + 21 = 2(-N)$$
$$5N + 21 = -2N$$

$$\begin{array}{rcl} \underline{2N - 21} & & \underline{2N - 21} \\ 7N & = & -21 \end{array}$$

$$N = -3$$

Check: $5(-3) + 21 = 2(3) \longrightarrow -15 + 21 = 6 \longrightarrow 6 = 6$ Check

problem set 32

*1. Five times a number is 72 greater than the opposite of the number. Find the number.

*2. If the sum of twice a number and -14 is multiplied by 2, the result is 12 greater than the opposite of the number. What is the number?

*3. Five times a number is 21 less than twice the opposite of the number. What is the number?

4. $\frac{3}{7}$ of what number is $2\frac{1}{5}$?   5. What fraction of 40 is 90?

6. 1.025 of 50 is what number?

Solve:

7. $\frac{5}{8}x - 3 = \frac{1}{2}$ 8. $\frac{1}{7}y + 10 = 14\frac{1}{4}$

9. $.3 + .06p + .02 - .02p = 4$ 10. $3p - 4 - 6 = -2(p - 5)$

11. $k + 4 - 5(k + 2) = 3k - 2$ 12. $x - 4(x - 3) + 7 = 6 - (x - 4)$

13. $p - 3(p + 4) = 2(p + 1)$

Write the following numbers as products of prime factors:

*14. 80   *15. 147   16. 250   17. 450

Simplify:

18. $(-3)^{-3}$ 19. 2^{-3}

Expand:

20. $2x^{-2}(x^{-2}y^0 + x^2y^5p^0)$ 21. $x^{-3}p^0(x^6p^5 - 3x^3p^0)$

22. $4x^2y^0(x^0y^2 - 3x^2y^{-2})$ 23. $(4p^{-2} - 3x^{-3}p^5)p^2x^0$

Simplify by adding like terms:

24. $-3x^{-2}y^2x^5 + 6x^3y^{-2}y^4 - 3x^3y^2 + 5x^2y^3$

25. $-xyz^5z^{-4} + 5xy^{-4}y^5z - 3zxy^7y^{-6}$

Evaluate:

26. $m - (-m)(m^0 - a)$ if $m = -2, a = 3$

27. $k^2 - k^3(km^0)$ if $k = -3, m = 2$ 28. $a^3x - x^3$ if $a = -3, x = -2$

Simplify:

29. $-3^3 - 2^2 - 4^3 - |-2^2 - 2|$ 30. $\dfrac{-(-3 + 7) - 4^0}{(-2)(-3 + 5)}$

LESSON 33 *Greatest common factor*

The number 210 has four prime number factors as shown here:

$$2 \cdot 3 \cdot 5 \cdot 7 = 210$$

We call factors that are numbers **numerical factors.** Some expressions have factors that are letters, and some expressions have both numbers and letters as factors as does $210xy^2z^3$.

$$210xy^2z^3 = 2 \cdot 3 \cdot 5 \cdot 7 \cdot x \cdot y \cdot y \cdot z \cdot z \cdot z$$

We call the letter factors **literal factors,** and we use the words **algebraic factor** as a general term to describe factors that are either numbers or letters or both numbers and letters.

DEFINITION

The greatest common factor (GCF) of two or more terms is the product of all prime algebraic factors common to every term, each to the highest power that it occurs in all of the terms.

The expression $6x^2y^2m^2 + 3xy^3m^2 + 3x^3y^2$ can be written as

$$2 \cdot 3 \cdot x \cdot x \cdot y \cdot y \cdot m \cdot m + 3 \cdot x \cdot y \cdot y \cdot y \cdot m \cdot m + 3 \cdot x \cdot x \cdot x \cdot y \cdot y$$

Now only the first term has 2 as a factor, so 2 is not a factor of the greatest common factor. Each term has 3 as a factor at least once, so 3 is a factor of the greatest common factor of all the terms.

$$3$$

Each term has x as a factor at least once in every term so x is a factor of the GCF.

$$3x$$

In the same way, y is used as a factor at least twice in every term so the greatest common factor of the three given terms is

$$3xy^2$$

The variable m is not included because it is not a factor of the third term of the original expression.

example 33.1   Find the GCF of $8z^4m^2p - 12z^3m^4p^2$.

solution   The greatest common factor of the terms is $4z^3m^2p.$

example 33.2   Find the greatest common factor of $4x^2y^3z - 8y^2xz^3$.

solution   The GCF is $4xy^2z.$

example 33.3   Find the GCF of $16x^2yp^3 - 4x^3y^2p + 2x^2y^2p^{15}$.

solution   The GCF is $\mathbf{2x^2yp}$.

problem set 33

1. Precia had a secret number. She found that the sum of 3 times her number and 60 equaled -50. What was her number?

2. Twice the sum of 3 times a number and 60 is 155 greater than the opposite of the number. Find the number.

3. .125 of what number is 5.25?

4. What fraction of 4 is $\frac{1}{4}$?

5. $\frac{3}{5}$ of $6\frac{2}{3}$ is what number?

Solve:

6. $3\frac{1}{2} + 2\frac{1}{4}x = \frac{5}{4}$

7. $3(x - 2) + (2x + 5) = x + 7$

8. $-4.2 + .02x - .4 = .03x$

9. $-p - 4 - (2p - 5) = 4 + 2(p + 3)$

10. $8 - k + 2(4 - 2k) = k + 2k$

Find the greatest common factor of:

11. $6x^2y^2m^2 + 3xy^3m^2 + 3x^3y^2$

*12. $8z^4m^2p - 12z^3m^4p^2$

*13. $16x^2yp^3 - 4x^3y^2p + 2x^2y^2p^{15}$

14. $4ab^2c^4 - 2a^2b^3c^2 + 6a^3b^4c$

15. $5x^2y^5m^2 - 10xy^2m^2 + 15x^2y^4m^2$

Write as products of prime factors:

16. 630

17. 600

Simplify:

18. 2^{-4}

19. $\frac{1}{4^{-3}}$

Expand:

20. $3x^2y^0(x^{-2} - 3y^2x^4)$

21. $2p^{-5}(p^2x^5 - 3x^0p^5)$

22. $4x^{-3}y^2(x^{-3}y^{-2} - 2x^4y^4)$

23. $(y^{-5} - 2y^7x^5)x^0y^5$

Simplify by adding like terms:

24. $3xyz^2 - 4z^2xy + 7yx^2z - 5zx^2y$

25. $3x^2xyy^3y^{-1} + 2x^2xyyy - 4x^{-2}yx^5y^2 + 7x^2$

Evaluate:

26. $-x^0 - a(x - 2a)$ if $x = -5, a = 3$

27. $p^3 - a^2 + ap$ if $p = -3, a = 2$

28. $a^2 - a^3 - a^4$ if $a = -2$

Simplify:

29. $-3^3 - 3^2 - (-3)^2 - |-2^2|$

30. $\dfrac{-4(3^0 - 6)(-2)}{-4 - (-3)(-2) - 3}$

LESSON 34 *Factoring the greatest common factor*

34.A

factoring When we use the distributive property, we change an expression from a product to a sum. The expression $2a(x + c)$ tells us to multiply $2a$ by $x + c$. If we do this multiplication, we get the algebraic sum $2ax + 2ac$:

$$2a(x + c) = 2ax + 2ac$$

If we reverse the process and write $2ax + 2ac$ as the product of the two factors $2a$ and $(x + c)$, we say that we are factoring.

Factoring is the process of writing an indicated sum as a product of factors.

example 34.1 Factor $2ax + 2ac$.

solution We will factor in three steps. The first step is to write two empty parentheses to indicate a product.

$$(\qquad)(\qquad)$$

The second step is to write the greatest common factor of the terms in the first parentheses.

$$(2a)(\qquad)$$

The third step is to write the proper terms in the second parentheses so that $2a$ times these terms gives us $2ax + 2ac$.

$$(2a)(x + c)$$

And since the first parentheses are not necessary, the answer can be written as

$$2a(x + c)$$

example 34.2 Factor $a^3x^2m^2 + a^2xm^3 - a^4x^3m^2$.

solution We want to write this sum as a product. We begin by writing two sets of parentheses.

$$(\qquad)(\qquad)$$

In the first parentheses we want to write the greatest common factor of all three terms. To find this GCF, we will write the three terms as products of individual factors,

$$a^3x^2m^2 + a^2xm^3 - a^4x^3m^2$$

$$a \cdot a \cdot a \cdot x \cdot x \cdot m \cdot m + a \cdot a \cdot x \cdot m \cdot m \cdot m - a \cdot a \cdot a \cdot a \cdot x \cdot x \cdot x \cdot m \cdot m$$

Look at the a's. Each term has at least two a's so a^2 is part of the greatest common factor

$$(a^2 \quad)(\qquad)$$

Each term has at least one x, so x is a part of the greatest common factor.

$$(a^2x \quad)(\qquad)$$

Finally each term has at least two m's so m^2 is a part of the greatest common factor. No other factors are common to all three terms.

$$(a^2xm^2)(\qquad)$$

Now, the first entry in the second parentheses must be ax because a^2xm^2ax equals $a^3x^2m^2$, the first term of the original expression.

$$(a^2xm^2)(ax \qquad)$$

The second entry in the second parentheses must be m because $a^2xm^2(m) = a^2xm^3$, the second term of the original expression.

$$(a^2xm^2)(ax + m \qquad)$$

The third entry must be $-a^2x^2$ because $a^2xm^2(-a^2x^2) = -a^4x^3m^2$, the last entry in the original expression. The desired factored expression is $a^2xm^2(ax + m - a^2x^2)$ because

$$a^2xm^2(ax + m - a^2x^2) = a^3x^2m^2 + a^2xm^3 - a^4x^3m^2$$

example 34.3   Factor $4a^3b^4z^3 + 2a^2bz^4$.

solution

$$2 \cdot 2 \cdot a \cdot a \cdot a \cdot b \cdot b \cdot b \cdot b \cdot z \cdot z \cdot z + 2 \cdot a \cdot a \cdot b \cdot z \cdot z \cdot z \cdot z$$

Each term has, at least one 2, two a's, one b, and three z's as factors. Thus the greatest common factor is $2a^2bz^3$, so we write

$$(2a^2bz^3)(\qquad)$$

The first term in the second parentheses is $2ab^3$ because $2a^2bz^3(2ab^3) = 4a^3b^4z^3$, the first term in the original expression.

$$(2a^2bz^3)(2ab^3 \qquad)$$

The second term in the second parentheses is z because $2a^2bz^3(z) = 2a^2bz^4$, the second term in the original expression. Now

$$(2a^2bz^3)(2ab^3 + z)$$

is our answer because

$$(2a^2bz^3)(2ab^3 + z) = 4a^3b^4z^3 + 2a^2bz^4$$

example 34.4   Factor $6a^2x^2 + 2a^3x^3 + 4a^4x^3$.

solution   The greatest common factor is $2a^2x^2$.

$$(2a^2x^2)(\qquad)$$

The entry in the second parentheses is

$$(3 + ax + 2a^2x)$$

because

$$(2a^2x^2)(3 + ax + 2a^2x) = 6a^2x^2 + 2a^3x^3 + 4a^4x^3$$

example 34.5   Factor $3m^3xy^2 + m^2y$.

solution $m^2y(3mxy + 1)$ is the answer because

$$m^2y(3mxy + 1) = 3m^3xy^2 + m^2y$$

problem set 34

1. If the product of 5 and a number is increased by 7, the result is -42. What is the number?

2. If the product of 5 and a number is increased by 7 and this sum multiplied by 3, the result is 11 greater than the opposite of the number. Find the number.

3. $4\frac{1}{3}$ of what number is $3\frac{5}{8}$?

4. What decimal part of .42 is .00504?

5. $3\frac{2}{5}$ of $3\frac{1}{8}$ is what number?

Solve:

6. $3\frac{1}{2}n - \frac{1}{2} = \frac{4}{3}$ **7.** $x - 4 - 2x + 5 = 3(2x - 4)$

8. $.2m + 4.34 - m = 2.3$ **9.** $3(-k - 4) + 6 = k + 7$

10. $a - 3a + 5 = a - 4(2 - a)$

Factor the greatest common factor:

*11. $2ax + 2ac$ ***12.** $a^3x^2m^2 + a^2xm^3 - a^4x^3m^2$

*13. $4a^3b^4z^3 + 2a^2bz^4$ **14.** $3x^4y^2p - 6x^2y^5p^4$

15. $6a^3x^2m^5 + 2a^4x^5m^5 + 4a^2x^2m$

Write as products of prime factors:

16. 250   **17.** 360

Simplify:

18. 3^{-4} **19.** $\dfrac{1}{4^{-3}}$

Expand:

20. $y^0x^{-2}(x^2 - 4x^4y^6)$ **21.** $(p^5y^5 - y^{-5})p^0y^5$

22. $p^0x^2y(x^3y^{-1} - 3x^5y^{-2})$ **23.** $x^2(2x^{-2} - 4x^0p^5y^5)$

Simplify by adding like terms:

24. $4x^2y^{-2}p^4 - 3y^{-2}x^2p^4 + 7yyx^2p^4$

25. $3xxy^2x^{-2} - 2x^0yy + 5y^2 - 6x^2 - 4x^3x^{-1}$

Evaluate:

26. $-p(x - px)$ if $p = -3, x = 4$ **27.** $x^3 - x^2 + 2x$ if $x = -2$

28. $x(y - xy^0)$ if $x = -2, y = 4$

Simplify:

29. $3^2 - 3^3 - 3^0 + |-3^0|$ **30.** $\dfrac{4(-2 - 3) - 4^0}{-2(-4 + 6) - 3}$

LESSON 35 *Canceling*

35.A

canceling   We have been solving equations by using the fact that multiplication and division are inverse operations because they "undo" one another. If we want to solve the equation

$$4x = 20$$

we see that x is multiplied by 4. To undo multiplication by 4, we must divide by 4, and because it is an equation, we divide both sides by 4.

$$\frac{\cancel{4}x}{\cancel{4}} = \frac{20}{4} \quad \longrightarrow \quad x = 5$$

On the left, we say that we have canceled the 4s. Some people prefer to say that 4 over 4 "reduces to 1" instead of saying "canceled" because the use of these words helps to prevent canceling when canceling is not permissible. For instance, the 4s cannot be canceled in the following expression because addition and division are not inverse operations and do not undo one another.

$$\frac{x + \cancel{4}}{\cancel{4}} = x + 1 \qquad \text{incorrect}$$

In this problem nothing "reduces to 1." However, the following expression can be simplified by canceling

$$\frac{\cancel{4}(x + 1)}{\cancel{4}}$$

because multiplication by 4 and division by 4 do undo each other. We can see that 4 over 4 "reduces to 1."

$$\frac{4(x + 1)}{4} = x + 1$$

Cancellation or reduction to 1 is possible when the numerator and the denominator contain one or more common factors.[†]

example 35.1   Simplify:

$$\text{(a)} \quad \frac{4(a - 3)}{4} \qquad\qquad \text{(b)} \quad \frac{3(x - 2)}{x - 2}$$

solution   (a)   Here the common factor is 4, and 4 over 4 equals 1.

$$\frac{4(a - 3)}{4} = a - 3$$

(b)   Here the common factor is $x - 2$, and $x - 2$ over $x - 2$ reduces to 1.

  We will assume in all problems of this type that the denominator does not equal zero.

$$\frac{3(x - 2)}{x - 2} = 3$$

example 35.2   Simplify $\dfrac{3p + 3}{3}$.

solution   We cannot simplify in this form because the numerator is not a product. However, if we factor $3p + 3$, we see that we can cancel because both the numerator and the denominator will have 3 as a factor.

$$\frac{3p + 3}{3} = \frac{3(p + 1)}{3} = p + 1$$

[†] In the expression being discussed we remember that the 4 in the denominator can be written as $4 \cdot 1$. Thus both the numerator and the denominator have 4 as a factor.

example 35.3   Simplify:

$$\text{(a)} \quad \frac{3x - 9x^2}{3x} \qquad \text{(b)} \quad \frac{5x - 25x^2}{5xy}$$

solution   In (a) we can factor out a $3x$ and in (b) a $5x$†

$$\text{(a)} \quad \frac{3x(1 - 3x)}{3x} = \mathbf{1 - 3x} \qquad \text{(b)} \quad \frac{5x(1 - 5x)}{5x(y)} = \frac{\mathbf{1 - 5x}}{\mathbf{y}}$$

problem set 35

1. Jay and Bill found that 4 times the sum of a number and -6 equaled 20. What was the number?

2. If the sum of 4 times a number and 6 is multiplied by 3, the result is 5 greater than the opposite of the number. Find the number.

3. $5\frac{7}{10}$ of what number is $9\frac{1}{2}$?

4. What fraction of $2\frac{1}{4}$ is $7\frac{1}{8}$?

5. 1.05 of $.043$ is what number?

Solve:

6. $-5\frac{1}{2} + 2\frac{2}{5}p = 6\frac{1}{4}$

7. $-n + .4n + 1.8 = -3$

8. $x - (3x - 2) + 5 = 2x + 4$

9. $5(x - 2) - (-x + 3) = 7$

Factor the greatest common factor:

10. $4a^2xy^4p - 6a^2x^4$

11. $3a^2x^4y^6 + 9ax^2y^4 - 6x^4a^2y^5z$

Simplify. Factor if necessary:

12. $\dfrac{4(x - 1)}{4}$

13. $\dfrac{4x + 4}{4}$

*14. $\dfrac{3x - 9x^2}{3x}$

15. $\dfrac{2x + 6}{2}$

16. $\dfrac{mx + mxy}{mx}$

17. $\dfrac{k^2x - k^3x}{k^2xy}$

18. Write 270 as a product of prime factors.

19. Simplify: $\dfrac{1}{4^{-4}}$

Expand:

20. $(x^3y^0 - p^0x^2y^4)x^{-3}$

21. $3x^0y^{-3}(4y^3z - 7x^2)$

22. $3x^4y^2(xy^{-4} - 3x^{-4}y^5)$

23. $2x^0y^{-5}(4xyy^5 - 3x^5y^4)$

Simplify by adding like terms:

24. $3x^2ym^5 - 2xym^5 + 4m^5yx^2 - 6m^5yx$

25. $2x^4y^{-3} - 3x^2x^2y^{-7}y^4 + 6x^3xy^{-1}y^{-3} + xxy^{-3}$

Evaluate:

26. $a^2 - a^0(a - ab)$ if $a = -3, b = 5$

---

†Note the use of the words *factor out*. This phrase is meaningful even though some authorities insist that it is redundant and that the single word *factor* will suffice. However, this slight redundancy is not harmful, especially since it is a natural redundancy.

27. $b - ab(b - a)$ if $a = -3, b = 5$

28. $-k - kp^0 - (-pk^2)$ if $k = -3, p = 2$

Simplify:

29. $2^2 - 2^3 - (-3)^2$

30. $\dfrac{-3^2 + 4^2 + 3^3}{2(-5 + 2) - 3^0}$

LESSON 36 *Multiplying fractions · Minus signs and negative exponents*

Two fractions are multiplied by multiplying the numerators to form the new numerator and multiplying the denominators to form the new denominator. For example,

$$\frac{3}{2} \cdot \frac{5}{7} = \frac{15}{14}$$

Since variables stand for unspecified real numbers, all the rules for real numbers also apply to variables. Thus fractions that contain variables are multiplied by using the same rule.

$$\frac{mx}{4y} \cdot \frac{ax}{2y} = \frac{amx^2}{8y^2}$$

The distributive property of real numbers is also applicable to expressions that contain fractions. Expressions that contain fractions are often called **rational expressions.** We can expand the following rational expression by multiplying x over y by both of the terms inside the parentheses.

$$\frac{x}{y}\left(\frac{a}{y} - b\right) = \frac{xa}{y^2} - \frac{bx}{y}$$

example 36.1 Use the distributive property to expand: $\dfrac{x^2}{y^2}\left[\dfrac{x^2}{y} - \dfrac{3y^3}{m}\right]$.

solution Two multiplications are indicated. We multiply $\dfrac{x^2}{y^2}$ by $\dfrac{x^2}{y}$ and then multiply $\dfrac{x^2}{y^2}$ by $\dfrac{-3y^3}{m}$. This gives us

$$\frac{x^2 x^2}{y^2 y} - \frac{x^2 3y^3}{y^2 m}$$

Lastly, we simplify both expressions and get

$$\frac{x^4}{y^3} - \frac{3yx^2}{m}$$

example 36.2 Expand: $\dfrac{m}{z}\left[\dfrac{axp}{mk} - 2m^4 p^4\right]$.

solution   Again we will use two steps. First, we multiply and then we simplify.

$$\frac{maxp}{zmk} + \frac{m}{z}(-2m^4p^4) = \frac{axp}{zk} - \frac{2m^5p^4}{z}$$

example 36.3   Expand: $\dfrac{ab}{c^2}\left[\dfrac{xab}{c} + 2bx - \dfrac{4}{c^2}\right]$.

solution $\dfrac{ab}{c^2}$ must be multiplied by all three terms inside the parentheses. We do this and get

$$\frac{xa^2b^2}{c^3} + \frac{2ab^2x}{c^2} - \frac{4ab}{c^4}$$

36.B
minus signs and negative exponents

Expressions that contain both minus signs and negative exponents can be troublesome. **A minus sign in front of an expression indicates the opposite of the expression, while a negative exponent has a meaning that is entirely different.**

(a) 4^2 (b) 4^{-2} (c) -4^2
(d) $(-4)^2$ (e) -4^{-2} (f) $(-4)^{-2}$

The notations in (a) and (b) are easy to simplify because in each of these we have a positive number raised to a power.

$$\text{(a)} \quad 4^2 = \mathbf{16} \qquad \text{(b)} \quad 4^{-2} = \frac{1}{4^2} = \frac{1}{\mathbf{16}}$$

The notations in (c) and (d) often give difficulty because of the problem caused by the minus sign in front of the 4. When the minus sign is not enclosed in parentheses as in (c),

$$-4^2$$

it is sometimes helpful to think of moving the minus sign far to the left and the 4^2 far to the right as

$$-_\nwarrow \qquad\qquad 4^2$$
$$\text{sign}$$

and then performing the operation of squaring 4 to get 16,

$$-_\nwarrow \qquad\qquad 16$$
$$\text{sign}$$

and then moving the sign back in front of the 16 to get the final answer of negative 16.

$$-16$$

From this we see that -4^2 is read as *the opposite of* 4 squared and is not read as *the opposite of 4* squared. The notation in (d) which is

$$(-4)^2$$

does mean that -4 is to be multiplied by itself.

$$(-4)^2 = (-4)(-4) = +16$$

The notations in (e) and (f) are similar to the two we have just discussed. They are

$$\text{(e)} \quad -4^{-2} \qquad \text{and} \qquad \text{(f)} \quad (-4)^{-2}$$

In the expression on the left we may again think of moving the minus sign far to the left and the rest of the expression far to the right.

$$-\quad\quad\quad\quad\quad\quad 4^{-2}$$
$$\text{sign}$$

Now we simplify 4^{-2} as

$$-\quad\quad\quad\quad 4^{-2} = \frac{1}{4^2} = \frac{1}{16}$$
$$\text{sign}$$

and lastly we bring back the negative sign for our final answer of negative one-sixteenth.

$$-\frac{1}{16}$$

In (f) the $-$ sign is inside the parentheses and cannot be separated from the rest of the problem. Thus our simplification is

$$(-4)^{-2} = \frac{1}{(-4)^2} = \frac{1}{(-4)(-4)} = \frac{1}{16}$$

We will practice problems like these in many future homework problem sets.

problem set 36

1. If the sum of twice a number and -3 is multiplied by 4, the answer is 28. Find the number.

2. If the product of a number and -3 is reduced by 5, the result is 25 less than twice the opposite of the number. Find the number.

3. $2\frac{5}{8}$ of what number is 14?

4. What fraction of $3\frac{3}{4}$ is $22\frac{1}{2}$?

5. 2.625 of what number is 8.00625?

Solve:

6. $3\frac{1}{4}n - \frac{2}{5} = 3$

7. $x - 3(x - 2) = 7x - (2x + 5)$

8. $-3m - 3 + 5m - 2 = -(2m + 3)$

9. $.2k - 4.21 - .8k = 2(-k + .1)$

Factor the greatest common factor:

10. $12a^2x^5y^7 - 3ax^2y^2$

11. $15a^5x^4y^6 + 3a^4x^3y^7 - 9a^2x^6y$

Simplify. Factor if necessary:

12. $\dfrac{4x^2 - 4x}{4x}$

13. $\dfrac{2 - 6x}{2}$

14. $\dfrac{x^2ym + xym}{xym}$

15. $\dfrac{x^2y - xy}{xym}$

16. Write 750 as a product of prime factors.

Simplify:

17. -3^{-2}

18. $(-3)^{-2}$

19. $\dfrac{1}{-3^{-2}}$

20. $\dfrac{1}{(-3)^2}$

Expand:

21. $\dfrac{x^2}{y}\left[\dfrac{x^2}{y} - \dfrac{3y^3}{m}\right]$

*22. $\dfrac{m}{z}\left[\dfrac{axp}{mk} - 2m^4p^4\right]$

23. $\dfrac{ax}{c^2}\left[\dfrac{b^4}{xk} - 2b^2\right]$

24. $(p^0x^2 - 4p^{-6}xy^5)x^{-2}$

Simplify by adding like terms:

25. $-xym^2 + 6ym^2x - 3x^2ym^2 - 9yx^2m^2$

26. $3x^4x^{-3}y^0 + xy^0y^{-2}y^2 - 7x^4x^{-3}p^0$

Evaluate:

27. $m(a^0 - ma)(-m) + |m^2 - 2|$ if $m = 2, a = -4$

28. $k^3 - k(a)^2$ if $k = -3, a = 2$

29. $-mx(a - x) - a$ if $m = -3, x = 2, a = 2$

30. Simplify: $\dfrac{-3^2 + 4^2 - 5(4 - 2)}{3^0(5 - 2)}$

LESSON 37 *Graphing inequalities*

37.A
inequalities

We use the symbols

(a) $\neq$ (b) $>$ (c) $<$

to designate that quantities are not equal, and we say that these symbols are symbols of **inequality.** They can be read from left to right or from right to left. We read

(d) $4 \neq 5$

from left to right as "4 is not equal to 5" or from right to left as "5 is not equal to 4." The symbols $>$ and $<$ are inequality symbols and are also called greater than/less than symbols. The small or pointed end is read as "less than" and the big or open end is read as "greater than." When we read, we only read one end of the symbol, the end that we come to first. Thus we read

(e) $4 > 2$

from left to right as "4 is greater than 2" or from right to left as "2 is less than 4." If the sign is combined with an equals sign, only one of the conditions must be met. We read

(h) $4 \geq 2 + 2$

from right to left as "2 plus 2 is less than or equal to 4," or from left to right as "4 is greater than or equal to 2 plus 2." This combination symbol is also called an inequality symbol although half of it is an equals sign.

Inequalities can be false inequalities, true inequalities, or conditional inequalities.

(a) $4 + 2 \leq 3$ (b) $x + 2 \geq x$ (c) $x < 4$

Inequality (a) is false, (b) is true, and the truth or falsity of (c) depends on the replacement value used for the variable. If a number that we use as a replacement for the variable makes the inequality a true inequality, we say that the number is a **solution of the inequality** and say that the number **satisfies the inequality.** If the number that we use as a replacement for the variable makes the inequality a false inequality, then the number is not a solution of the inequality and does not satisfy the inequality. Since more than one number will often satisfy a given inequality, there is often more than one solution to the inequality.

> We call the **set of numbers** that will satisfy a given equation or inequality the **solution set** of the equation or inequality.

37.B
greater than and less than

Zero is a real number. Also, any other number that can be used to describe a physical distance is a real number, and the opposite of each of these numbers is also a real number.

We use the number line to help us picture the way real numbers are ordered (are arranged in order) and to help us define what we mean by greater than. On this number line we have graphed 2 and 4.

We remember that one number is greater than another number if its graph is to the right of the graph of the other number. Since the graph of 4 is to the right of the graph of 2, we say that 4 is greater than 2.

Using the same definition, we can say that -1 is greater than -4 because the graph of -1 is to the right of the graph of -4.

In the following section the small arrows will not be drawn on the ends of the number line because these arrows can be confused with the arrows drawn to indicate the solutions to the problems.

37.C
graphical solutions of inequalities

We can use the number line to display the graph or the picture of the solution to many problems.

example 37.1   Graph: $x > 2$.

solution   This problem asks that we graph all numbers that are greater than 2. We draw an arrow to designate these numbers.

The open circle at 2 indicates that 2 is not a part of the solution because 2 is not greater than 2.

example 37.2   Graph the solution of $x \leq 2$.

solution   This inequality is read from left to right as x **is less than or equal to** 2. Thus we are asked to show the location on the number line of all numbers that equal 2 or are less than 2.

The locations of the numbers that satisfy the condition are indicated by the heavy line. The solid circle at 2 indicates that the number 2 is a part of the solution of $x \le 2$.

example 37.3   Graph the solution of $x < -3$.

solution   This time we graph the numbers that are less than -3.

The open circle at -3 indicates that -3 is not a part of the solution.

**problem
set 37**

1. If a number is multiplied by 3 and this product is reduced by 5, the result is 40. What is the number?

2. If a number is multiplied by 7 and this product is increased by 7, the result is one less than 9 times the number. What is the number?

3. $\dfrac{1}{4}$ of what number is $\dfrac{3}{8}$?

4. What decimal part of 41.25 is 2.475?

5. $\dfrac{1}{3}$ of $7\dfrac{1}{6}$ is what number?

Solve:

6. $-\dfrac{1}{5} + 2\dfrac{1}{2}p = 2$

7. $3(x - 2) - (2x + 5) = -2x + 10$

8. $4x - 3(x + 2) = 2x - 5$

9. $.004m - .001m + .002 = -.004$

Factor the greatest common factor:

10. $4a^2x^3y^5 - 8a^4x^2y^4$

11. $6a^2xm^5 + 2ax^4m^6 - 18a^5x^3m^5$

Simplify. Factor if necessary:

12. $\dfrac{3x - 9}{3}$

13. $\dfrac{4px^2 - 8px}{px^2}$

Draw four number lines and graph the solutions to the inequalities:

*14. $x > 2$ *15. $x \le 2$ 16. $x < 2$ 17. $x \ge 2$

Simplify:

18. -3^{-3}

19. $(-3)^{-3}$

20. $\dfrac{1}{-5^{-2}}$

Expand:

21. $\dfrac{p^2}{x}\left[\dfrac{k^2p}{x^2} - \dfrac{p^2}{x}\right]$

22. $\left[\dfrac{a^2}{x} - \dfrac{2x}{a}\right]\dfrac{4x^2}{a}$

23. $\dfrac{ax}{c^2}\left[\dfrac{ax^2}{c} - \dfrac{3c^2a}{x^3}\right]$

24. $x^{-2}(x^2p^0 - 5x^4p^7)$

Simplify by adding like terms:

25. $x^3y^3p + px^3y^3 - 4x^2xyy^2p^2p^{-1}$ **26.** $5a^2x + 7xa^2 + aax^2x^{-1} - 2ax^3x^{-1}$

Evaluate:

27. $m^2 - (m - p)$ if $m = 2, p = -2$

28. $a^2 - y^3(y - a)$ if $a = -2, y = -3$

29. $k(x - ka)$ if $a = -2, k = 3, x = -3$

30. Simplify: $\dfrac{-3^2 - (-3)^3 - 3}{-3(-3)(+3)}$

LESSON 38 *Ratio*

38.A
ratio

When we write the numbers 3 and 4 separated by a fraction line as

$$\frac{3}{4}$$

we say that we have written the fraction three-fourths. Another name for a fraction is **ratio,** and we can also say that we have written the ratio of 3 to 4. All of the following ratios designate the same number and thus are equal ratios.

$$\frac{3}{4} \qquad \frac{6}{8} \qquad \frac{300}{400} \qquad \frac{15}{20} \qquad \frac{27}{36} \qquad \frac{111}{148}$$

An equation or statement in which two ratios are equal is called a proportion. Thus, we can say that

$$\frac{3}{4} = \frac{15}{20}$$

is a proportion. We note that the cross products of equal ratios are equal as shown here.

$$\begin{array}{c} 4 \cdot 15 \\ \dfrac{3}{4} \diagdown \diagup \dfrac{15}{20} \\ 3 \cdot 20 \end{array} \qquad \begin{array}{l} 4 \cdot 15 = 3 \cdot 20 \\ 60 = 60 \quad \text{True} \end{array}$$

We can solve proportions that contain an unknown by setting the cross products equal and then dividing to complete the solution. To solve

$$\frac{7}{5} = \frac{91}{g}$$

we first set the cross products equal

$$7g = 5 \cdot 91$$

and then finish by dividing both sides by 7.

$$\frac{7g}{7} = \frac{5 \cdot 91}{7} \longrightarrow g = \frac{455}{7} \longrightarrow g = 65$$

When we set the cross products equal, we say that we have cross multiplied.

example 38.1   Solve $\dfrac{4}{m} = \dfrac{21}{5}$.

solution   We begin by setting the cross products equal.

$$4 \cdot 5 = 21m$$

Now we finish by dividing both sides by 21

$$\frac{4 \cdot 5}{21} = \frac{21m}{21} \longrightarrow \frac{20}{21} = m$$

We use proportions and cross multiplication to solve ratio word problems. In these problems, we wish to maintain a constant ratio between two things. We will use meaningful variables to represent the things and avoid the meaningless variables x, y, and z.

example 38.2   The ratio of pigs to goats in the barnyard was 7 to 5. If there were 91 pigs, how many goats were there?

solution   We first note that we are comparing pigs and goats. Either one may be on top. If it is on top on one side, it must also be on top on the other side. We will demonstrate this by working the problem two ways.

$$(a) \quad \frac{P}{G} = \frac{P}{G} \qquad\qquad (b) \quad \frac{G}{P} = \frac{G}{P}$$

Now we reread the problem and find that the ratio of pigs to goats was 7 to 5. So on the left side of both equations, we replace P with 7 and G with 5.

$$(a) \quad \frac{7}{5} = \frac{P}{G} \qquad\qquad (b) \quad \frac{5}{7} = \frac{G}{P}$$

Now we read the problem again and find that there were 91 pigs. Thus, we can replace P in both equations with 91.

$$(a) \quad \frac{7}{5} = \frac{91}{G} \qquad\qquad (b) \quad \frac{5}{7} = \frac{G}{91}$$

We solve both of these the same way—by cross multiplying and then dividing by 7.

$$7G = 5 \cdot 91 \longrightarrow 7G = 455 \longrightarrow \frac{7G}{7} = \frac{455}{7} \longrightarrow G = 65$$

example 38.3   In the same barnyard, the ratio of chickens to ducks was 9 to 4, and there were 108 chickens. How many ducks were there?

solution   Either chickens or ducks may go on top. We decide to put the ducks on top, so we write

$$\frac{D}{C} = \frac{D}{C}$$

The ratio of chickens to ducks was 9 to 4, so on the left we replace C with 9 and D with 4.

$$\frac{4}{9} = \frac{D}{C}$$

Finally, we replace C on the right with 108, and finish by cross multiplying and then dividing.

$$\frac{4}{9} = \frac{D}{108} \longrightarrow 9D = 4 \cdot 108 \longrightarrow \frac{9D}{9} = \frac{4 \cdot 108}{9} \longrightarrow D = 48$$

**problem
set 38**

*1. The ratio of pigs to goats in the barnyard was 7 to 5. If there were 91 pigs, how many goats were there?

*2. In the same barnyard the ratio of chickens to ducks was 9 to 4, and there were 108 chickens. How many ducks were there?

3. In a picaresque novel about the Spanish Main the ratio of picaros to picaras was 13 to 5. If there were 600 picaras, how many picaros were in the novel?

4. If the sum of twice a number and -7 is increased by 8, the result is 16 greater than the opposite of the number. What is the number?

5. 2.125 of what number equals .1275?

6. What fraction of $\frac{7}{8}$ is $2\frac{5}{11}$?

Solve:

7. $.06 + .06x = -.042$

8. $3\frac{1}{2}k + \frac{3}{4} = -\frac{7}{8}$

9. $2(5 - x) - (-2)(x - 3) = -(3x - 4)$

10. $3(-2x - 2 - 3) - (-x + 2) = -2(x + 1)$

11. $-x - 2(-x - 3) = -4 - x$

Factor the greatest common factor:

12. $3x^2y^3z^5 - 9xy^6z^6$

13. $4x^2y - 12xy^2 + 24x^3y^3$

Simplify. Factor if necessary:

14. $\dfrac{7x + 7}{7}$

15. $\dfrac{k^4p - 2k^5p^2}{k^4p}$

Draw number lines and graph these inequalities:

16. $x \geq -2$

17. $x > 4$

Simplify:

18. $(-2)^{-2}$

19. -2^{-2}

20. $(-2)^{-3}$

Expand:

21. $\dfrac{p^2}{x}\left[\dfrac{4x}{p^2} - \dfrac{p^4}{p^2x}\right]$

22. $\dfrac{4a^2}{x}\left[\dfrac{ax}{x} - \dfrac{3x}{a}\right]$

23. $\dfrac{mp}{k}\left[\dfrac{mp}{k} - \dfrac{k}{mp}\right]$

24. $x^{-3}(y^{-2}k^0 - 3xk^5)$

Simplify by adding like terms:

25. $3p^3x^{-4}xp^4 - 2pp^2p^3x^{-1}x^{-2} - 4x^2x^{-2}x^{-3}pp^6$

26. $xy - 3yx + 7x^3y^2x^{-2}y^{-1} - 2x^2yy^5y^{-4}y^{-1}x^{-1}$

Evaluate:

27. $(m - x^2)x - (-m)$ if $x = -2, m = -3$

28. $(a^3 - y^2)(a - y)$ if $a = -3, y = 4$

29. $x(x - y)(3 - 2xy)$ if $x = -2, y = 5$

30. Simplify: $-2^2 + (-2)^3 - 2(-2 - 2) - 2$

LESSON 39 *Trichotomy axiom · Negated Inequalities · Advanced ratio problems*

39.A
trichotomy axiom

Johnny wrote a number on a piece of paper. Then he turned the paper over and wrote a number on the other side. There are exactly three possibilities.

1. The second number is the same number as the first number.
2. The second number is less than the first number.
3. The second number is greater than the first number.

While this is seemingly self-evident, it is not trivial. Mathematicians recognize that this property of real numbers reveals that the real numbers are an ordered set. Since this property has three parts, we give it the name **trichotomy.**[†]

> TRICHOTOMY AXIOM
>
> For any two real numbers a and b, exactly one of the following is true:
>
> $$a = b \qquad a > b \qquad a < b$$

39.B
negated inequalities

Symbols of inequality can be negated by drawing a slash through the symbol. We read

(a) $x \not> 10$

from left to right as "x is not greater than 10." There are only three possibilities under the trichotomy axiom, and if x is not greater than 10, then it must be less than or equal to 10. So we can say the same thing by writing

(b) $x \leq 10$

In the same way, if we write

(c) $x \not\geq 6$

that x is not greater than or equal to 6, then x must be less than 6 because that's the only other possibility. Thus, both (c) and (d) make the same statement.

(c) $x \not\geq 6$ (d) $x < 6$

example 39.1   Graph the solution of $x \not< 2$.

solution   If x is not less than 2, then x has to be equal to or greater than 2. Thus, the solution is the graph of $x \geq 2$.

The solid circle at 2 indicates that 2 is a part of the solution of this inequality.

[†] From the Greek *trikha* meaning "in three parts."

example 39.2   Graph the solution of $x \ngeq 3$.

solution   If x is not greater than or equal to 3, then it must be less than 3. Thus, we will graph $x < 3$.

The open circle at 3 indicates that 3 is not a part of the solution of this inequality.

39.C
advanced ratio problems

If we are told that the ratio of red marbles to blue marbles is 5 to 7,

$$\frac{R}{B} = \frac{5}{7}$$

we are also told that the ratio of red marbles to total marbles is 5 to 12 and that the ratio of blue marbles to total marbles is 7 to 12.

$$\frac{R}{T} = \frac{5}{12} \qquad \frac{B}{T} = \frac{7}{12}$$

These last two statements are not always evident; but if we use the following format to write down what is given, we can get the equations by inspection.

$$R = 5$$
$$B = 7$$
$$T = 12$$

The top two give us

$$\frac{R}{B} = \frac{5}{7}$$

The top and bottom give us

$$\frac{R}{T} = \frac{5}{12}$$

and the bottom two give us

$$\frac{B}{T} = \frac{7}{12}$$

Often it is helpful to write all three equations and then reread the problem to see which equation to use.

example 39.3   The ratio of red marbles to blue marbles is 5 to 7. If there are 156 marbles total, how many red marbles are there?

solution   We write down the red and blue and also the total:

$$R = 5$$
$$B = 7$$
$$T = 12$$

Now by looking at what we have written, we can see that the three possible equations are

$$\text{(a)} \quad \frac{R}{B} = \frac{5}{7} \qquad \text{(b)} \quad \frac{R}{T} = \frac{5}{12} \qquad \text{(c)} \quad \frac{B}{T} = \frac{7}{12}$$

The problem asks about red and total, so we will use equation (b) and replace total with 156.

$$\frac{R}{156} = \frac{5}{12} \quad \longrightarrow \quad 12R = 5 \cdot 156 \quad \longrightarrow \quad \frac{12R}{12} = \frac{5 \cdot 156}{12} \quad \longrightarrow \quad R = 65$$

example 39.4   The ratio of fish to crabs in the sea cave was 13 to 4. If there were 119 fish and crabs in the cave, how many were fish?

solution   First we write

$$F = 13$$
$$C = 4$$
$$T = 17$$

From this we can write the three equations

$$\text{(a)} \quad \frac{F}{C} = \frac{13}{4} \qquad \text{(b)} \quad \frac{F}{T} = \frac{13}{17} \qquad \text{(c)} \quad \frac{C}{T} = \frac{4}{17}$$

Since we are given the total and asked for fish, we will use equation (b) and substitute 119 for total.

$$\frac{F}{119} = \frac{13}{17} \quad \longrightarrow \quad 17F = 13 \cdot 119 \quad \longrightarrow \quad \frac{17F}{17} = \frac{13 \cdot 119}{17} \quad \longrightarrow \quad F = 91$$

problem set 39

1. Courtney and Kristofer thought of the same number. They multiplied the number by 3 and then increased the product by 5 for a final result of -55. What number were they thinking of?

2. Twice the opposite of a number is increased by 5, and this sum is multiplied by 2. The result is 22 greater than the number. Find the number.

*3. The ratio of red marbles to blue marbles is 5 to 7. If there are 156 marbles total, how many red marbles are there?

*4. The ratio of fish to crabs in the sea cave was 13 to 4. If there were 119 fish and crabs in the cave, how many were fish?

5. $\frac{2}{5}$ of $23\frac{1}{3}$ is what number?

Solve:

6. $2\frac{1}{3}k - 4 = 7$

7. $p - 3(p - 4) = 2 + (2p + 5)$

8. $5x - 4(2x - 2) = 5 - x$

9. $.02x - 4 - .01x - 2 = -6.3$

Factor the greatest common factor:

10. $12a^2x + 4ax^4$

11. $2x^2a^3y - xay + 4ay^2$

Simplify. Factor first if necessary:

12. $\dfrac{3x - 9}{3}$

13. $\dfrac{5xy + 20xy^2}{5xy}$

Draw four number lines and graph the solutions to these inequalities.

*14. $x \not< 2$ **15.** $x \not\leq 2$ **16.** $y \not> 3$ **17.** $y \geq 3$

Simplify:

18. -2^{-2} **19.** $(-2)^{-2}$ **20.** $-(-2)^{-2}$

Expand:

21. $\left[\dfrac{p}{a} - \dfrac{p}{xa}\right]\dfrac{p^2}{a}$ **22.** $\dfrac{4x^2y}{m}\left[\dfrac{x^3}{m^2} - \dfrac{y}{m}\right]$

23. $\dfrac{mp}{x}\left[\dfrac{a}{x} - \dfrac{1}{x^2}\right]$ **24.** $p^{-2}x^0(p^2 - 3p^5x^5)$

Simplify by adding like terms:

25. $5yxp^2 - p^2yx + 2ppyx - 3p^2y^2y^{-1}x$ **26.** $3ay - 5ya - 6y^{-2}y^3a - 4a^3ay$

Evaluate:

27. $kp^0 - (k - p)$ if $k = -4$, $p = 1$

28. $m^2 - m^3(p)$ if $m = -2$, $p = 3$

29. $a^2b - (a - b)$ if $a = -2$, $b = -1$

30. Simplify: $\dfrac{-4(3 - 5) - 2^0}{(-2)^3 - 2(-2)}$

LESSON 40   *Quotient theorem for exponents*

40.A
quotient theorem for exponents

Let's review our theorems and definitions for exponents.

DEFINITION :	$\underbrace{x \cdot x \cdot x \cdots x}_{n \text{ factors}} = x^n$
DEFINITION :	If x is not zero, $x^0 = 1$
DEFINITION :	$x^1 = x$
PRODUCT THEOREM :	If x is not zero, $x^m \cdot x^n = x^{m+n}$
DEFINITION :	If x is not zero, $x^{-n} = \dfrac{1}{x^n}$

The **quotient theorem for exponents** is really an extension of the last definition above that says $x^{-n} = \dfrac{1}{x^n}$. If we wish to multiply

$$x^5 \qquad \text{times} \qquad \frac{1}{x^2}$$

we can use the definition of x^{-n} to write $\dfrac{1}{x^2}$ as x^{-2} and then multiply by using the product theorem

$$x^5 \cdot \frac{1}{x^2} = x^5 \cdot x^{-2} = x^{5-2} = \boldsymbol{x^3}$$

The quotient theorem permits the same procedure in just one step.

QUOTIENT THEOREM FOR EXPONENTS

If m and n are real numbers and $x \neq 0$,

$$\frac{x^m}{x^n} = x^{m-n} = \frac{1}{x^{n-m}}$$

---

example 40.1   Simplify $\dfrac{x^6}{x^4}$.

solution   We know that we can move the x^4 from the denominator to the numerator if we change the sign of the 4 from plus to minus.

$$\frac{x^6}{x^4} = x^6 x^{-4} = x^{6-4} = \boldsymbol{x^2}$$

If we use the quotient theorem, we can omit the first step and write

$$\frac{x^6}{x^4} = x^{6-4} = \boldsymbol{x^2}$$

example 40.2   Simplify $\dfrac{x^6}{x^4}$, but this time write the exponential in the denominator.

solution
$$\frac{x^6}{x^4} = \frac{1}{x^{4-6}} = \frac{1}{\boldsymbol{x^{-2}}}$$

example 40.3   Simplify $\dfrac{x^6}{x^{-4}}$.

solution   We will work the problem twice. The first time we will use the quotient theorem so that the final exponential is in the numerator.

(a)
$$\frac{x^6}{x^{-4}} = x^{6+4} = \boldsymbol{x^{10}}$$

(b)   This time we will put the exponential in the denominator.

$$\frac{x^6}{x^{-4}} = \frac{1}{x^{-4-6}} = \frac{1}{\boldsymbol{x^{-10}}}$$

example 40.4   Simplify $\dfrac{x^{-a}}{x^b}$.

solution   We will use the quotient theorem to simplify so that the final exponential is in the numerator.

(a)
$$\frac{x^{-a}}{x^b} = x^{-a-b}$$

(b) And this time we will put the exponential in the denominator.
$$\frac{x^{-a}}{x^b} = \frac{1}{x^{b+a}}$$

The two answers shown in (a) and (b) are **equivalent expressions,** which means that they have the same value no matter which nonzero real number is used as a replacement for x. Since they are equivalent expressions, neither expression can be designated as the preferred answer because the preference of one person will not necessarily be the same as the preference of another person.

example 40.5   Simplify $\dfrac{x^{-5}y^6z}{z^{-3}y^2x}$.

solution   We will find four equivalent expressions for this expression.

(a) $x^{-6}y^4z^4$ (b) $\dfrac{1}{x^6y^{-4}z^{-4}}$ (c) $\dfrac{y^4z^4}{x^6}$ (d) $\dfrac{x^{-6}}{y^{-4}z^{-4}}$

Answer (a) is written with all exponentials in the numerator; (b) has all exponentials in the denominator; (c) has all exponents positive; and (d) has all exponents negative. No one of these forms is more correct than another. We will emphasize this by using different forms for the answers in the back of the book. However, many authors and teachers like to have final results written with all exponents positive as in (c) above.

problem set 40

1. Seven times a number is decreased by 4 for a result of -25. What is the number?

2. Three times the opposite of a number is increased by 16, and this result is 26 greater than the number. Find the number.

3. The ratio of gaudy scarves to tawdry scarves was 7 to 11. If there were 2520 scarves in the pile, how many were merely gaudy?

4. What fraction of $2\dfrac{1}{8}$ is 6?   5. 1.205 of 3.2 is what number?

Solve:

6. $\dfrac{2}{3}k + 5 = 12$ 7. $x - 5x + 4(x - 2) = 3x - 8$

8. $2p - 5(p - 4) = 2p + 12$ 9. $.4x - .02x + 1.396 = .598$

Factor the greatest common factor:

10. $3x^2y^5p^6 - 9x^2y^4p^3 + 12x^2yp^4$ 11. $2x^2y^2 - 6y^2x^4 - 12xy^5$

Simplify. Factor first if necessary:

12. $\dfrac{5x^2 - 25x}{5x}$ 13. $\dfrac{4xy + 16x^2y^2}{4xy}$ 14. Graph $x \not\geq -2$.

15. Write 360 as a product of prime factors.

Simplify:

16. -2^{-4} 17. $\dfrac{1}{-(-3)^{-3}}$

Simplify. Write all exponentials in the numerator.

*18. $\dfrac{x^{-5}y^6z}{z^{-3}y^2x}$ **19.** $\dfrac{p^{-4}y^5z}{z^{-5}y^3p^2}$ **20.** $\dfrac{m^5zx^2y^{-5}}{y^4z^2x^{-3}}$

Expand:

21. $\dfrac{p^2}{x}\left[\dfrac{a}{bc}-\dfrac{1}{c}\right]$ **22.** $\left[\dfrac{x}{m^2}-\dfrac{2x}{m}\right]\dfrac{3x^2y}{m}$

23. $p^{-2}k^0(p^2-4p^4k^5)$ **24.** $3x^2y^{-4}(x^4y^{-2}-2x^{-2}y^2)$

Evaluate:

25. $k-k^3p$ if $k=-2$, and $p=4$

26. $xa-a(x-xa)$ if $x=3$, and $a=-1$

27. $m(-a^0-m)$ if $m=-4$, and $a=1$

28. $p-(m-pm)$ if $p=-3$, and $m=4$

29. Simplify: $\dfrac{-2^0(-5-7)(-3)-|-4|}{-2(-(-6))}$

LESSON *41*   *Distributive property and negative exponents*

41.A
distributive property and negative exponents

In the last five problem sets, we have used the distributive property to expand expressions that contain fractions, such as

$$\frac{4x^2}{y^4}(xy+2y^2x^3)$$

In this lesson, we will do the same expansions, but now we will also consider expressions that contain negative exponents.

example 41.1   Expand $\dfrac{4x^{-2}}{y^4}\left[y^4x^2-\dfrac{3x^4}{y^{-2}}\right]$ and write the answer with all exponents positive.

solution   We will use two steps. First we multiply to get

$$\frac{4x^{-2}y^4x^2}{y^4}-\frac{12x^{-2}x^4}{y^4y^{-2}}$$

Now we simplify and write the answer with all exponents positive.

$$4-\frac{12x^2}{y^2}$$

example 41.2   Multiply $x^{-2}y\left(\dfrac{y^4}{x^2}-\dfrac{x^4}{y^2}\right)$ and write the product with all exponentials in the numerator.

solution   We begin by using the distributive property to multiply

$$x^{-2}y\left(\frac{y^4}{x^2}-\frac{x^4}{y^2}\right)=\frac{x^{-2}yy^4}{x^2}-\frac{x^{-2}yx^4}{y^2}=\frac{y^5}{x^4}-\frac{x^2}{y}$$

Now we can write this result with all exponentials in the numerator as

$$x^{-4}y^5 - x^2y^{-1}$$

example 41.3 Multiply $\dfrac{k^2b}{p^{-2}}\left(\dfrac{ab^{-1}}{k^2} - \dfrac{4p^2}{b}\right)$ and write the product with all exponentials in the denominator.

solution Again we begin by using the distributive property to multiply

$$\frac{k^2bab^{-1}}{p^{-2}k^2} - \frac{4p^2k^2b}{p^{-2}b}$$

Now we simplify and write the product with all exponentials in the denominator.

$$\frac{1}{a^{-1}p^{-2}} - \frac{4}{p^{-4}k^{-2}}$$

problem set 41

1. Three more than 5 times a number is -27. What is the number?

2. If 7 times a number is decreased by 5 and this difference is doubled, the result is 14 less than twice the number. What is the number?

3. War Eagle spied 1428 antelope and wildebeests grazing on the savannah. If they were in the ratio of 9 to 5, how many antelope were there?

4. What fraction of 72 is 16?

5. Write 130 as a product of prime factors.

Solve:

6. $\dfrac{3}{5}p + 7 = 22$

7. $3x - (x - 2) + 5 = 4x + 6$

8. $3p - 2(p - 4) = 7p + 6$

9. $.004k - .002 + .002k = 4$

Factor the greatest common factor:

10. $4x^2m^5y - 2x^4m^3y^3$

11. $4m^2x^5 - 2m^2x^2 + 6m^5x^2$

Simplify:

12. $\dfrac{4 - 4x}{4}$

13. $\dfrac{9x - 3x^2}{3x}$

Graph:

14. $x \nleq -5$

15. $x > -2$

Simplify:

16. -3^{-2}

17. $\dfrac{1}{(-3)^2}$

Simplify. Write all exponentials with positive exponents:

18. $\dfrac{x^3y^2}{xy^4}$

19. $\dfrac{x^3y^{-3}z}{z^5x^2y}$

20. $\dfrac{x^{-4}y^{-3}p^2}{x^{-5}yp^4}$

Expand. Write the answers with positive exponents:

*21. $x^{-2}y\left[\dfrac{y^4}{x^2} - \dfrac{x^4}{y^2}\right]$

*22. $\dfrac{k^2b}{p^{-2}}\left[\dfrac{ab^{-1}}{k^2} - \dfrac{4p^2}{b}\right]$

23. $\left[\dfrac{x^{-4}}{a^3} - \dfrac{a^3}{x}\right]\dfrac{a^{-3}}{x}$

24. $\dfrac{m^{-2}}{b}\left[\dfrac{b^2}{m^3} - \dfrac{4am^2}{b^4}\right]$

Simplify by adding like terms:

25. $x^2yp - 4xxyp - 3x^2py$

26. $3x^2ym - 2m^2x^2y - 5x^2my + 4ym^2x^2$

Evaluate:

27. $m - 3m^3$ if $m = -3$ **28.** $a(b^0 - ab)$ if $a = 3, b = -5$

29. $x - (-y) - y^2$ if $x = -2, y = 3$

30. Simplify: $-3 - 2^0|-4 - 3| - 2(-2)$

LESSON 42   *Like terms and negative exponents*

42.A
adding like terms

Sometimes it is difficult to determine if terms in an expression are like terms. For instance, if we look at the expression

$$\frac{bx}{x^3y^{-2}} - 3by^2x^{-2} + \frac{4y^2}{bx^2}$$

it is rather difficult to see that two of the terms are like terms and thus may be added. If each of the terms is written in the same form, however, it is easy to identify like terms. Let's write each of the three terms with all exponents positive.

$$\frac{by^2}{x^2} - \frac{3by^2}{x^2} + \frac{4y^2}{bx^2}$$

We see that the first two terms are like terms and may be added. The last term is different and thus cannot be added.

$$\frac{by^2}{x^2} - \frac{3by^2}{x^2} + \frac{4y^2}{bx^2} = -\frac{2by^2}{x^2} + \frac{4y^2}{bx^2}$$

Now we will work the problem again by first writing the original terms so that all exponents are negative.

$$\frac{x^{-2}}{b^{-1}y^{-2}} - \frac{3x^{-2}}{b^{-1}y^{-2}} + \frac{4b^{-1}x^{-2}}{y^{-2}} = -\frac{2x^{-2}}{b^{-1}y^{-2}} + \frac{4b^{-1}x^{-2}}{y^{-2}}$$

Again we see that the first two terms are like terms and may be added. To see if terms are like terms, we can put them in any form we wish as long as we use the same form for every term. However, most people feel more comfortable with the exponentials written with all positive exponents.

example 42.1   Add like terms:

$$\frac{bx}{x^3y^{-2}} - 3by^2x^{-2} + \frac{4y^2}{b^{-1}x^2}$$

solution   To help us identify like terms, we will rewrite each term so that all exponents are positive.

$$\frac{by^2}{x^2} - 3\frac{by^2}{x^2} + \frac{4by^2}{x^2}$$

We see that all three terms are like terms and can be added by adding the numerical coefficients.

$$\frac{by^2}{x^2} - \frac{3by^2}{x^2} + \frac{4by^2}{x^2} = \frac{2by^2}{x^2}$$

example 42.2 Add like terms:

$$\frac{a^{-3}b}{b^{-3}} + \frac{2b^4}{a^3}$$

solution We begin by writing each term with all exponents positive.

$$\frac{b^4}{a^3} + \frac{2b^4}{a^3}$$

Now we see that the terms are like terms and may be added.

$$\frac{b^4}{a^3} + \frac{2b^4}{a^3} = \frac{3b^4}{a^3}$$

example 42.3 Add like terms:

$$\frac{7a^{-3}b^2}{c^{-1}} - \frac{5b^2}{a^3c^{-1}} + \frac{3b^2}{a^3c}$$

solution We begin by writing each term with all exponents positive.

$$7\frac{cb^2}{a^3} - 5\frac{cb^2}{a^3} + 3\frac{b^2}{a^3c}$$

Now we see that the first two terms are like terms and may be added but that the third term is different. Thus the simplification is

$$\frac{2cb^2}{a^3} + \frac{3b^2}{a^3c}$$

problem set 42

1. Three less than 7 times a number is -31. What is the number?

2. The sum of 7 times a number and -3 is multiplied by 2. This result is 114 greater than the opposite of the number. Find the number.

3. The ratio of poseurs to outright frauds was 14 to 3. If they totaled 2244, how many were poseurs?

4. What decimal part of 7 is 14.14?

5. $2\frac{1}{4}$ of $3\frac{5}{8}$ is what number?

Solve:

6. $\frac{1}{2} + \frac{1}{8}x - 5 = 10\frac{1}{2}$

7. $4(x - 2) - 4x = -(3x + 2)$

8. $-5x + 2 = -2(x - 5)$

9. $.3z - .02z + .2 = 1.18$

Factor the greatest common factor:

10. $6k^5m^2 - 2mk^3 - mk$

11. $x^4y^2m - x^3y^3m^2 + 5x^6y^2m^2$

Simplify:

12. $\frac{3x - 9x^2}{3x}$

13. $\frac{4xy - 4x}{4x^2}$

Graph:

14. $x \not> 5$ **15.** $x \not\leq 2$

16. Write 280 as a product of prime factors.

Simplify. Write all exponentials in the numerator:

17. $\dfrac{x^2y^5}{x^4y^{-3}m^2}$ **18.** $\dfrac{x^5y^5mm^{-2}}{xx^3y^{-3}m^4}$ **19.** $\dfrac{x^2xyp^{-5}}{p^{-3}p^{-4}y^{-4}}$

Expand. Write answers with positive exponents:

20. $\dfrac{x^{-2}}{y}\left[\dfrac{xz}{y} - \dfrac{1}{y^{-4}}\right]$ **21.** $\left[\dfrac{a}{b} - \dfrac{2b}{a}\right]\dfrac{a^{-2}}{b^{-2}}$

Simplify by adding like terms:

*22. $\dfrac{bx}{x^3y^{-2}} - 3by^2x^{-2} + \dfrac{4y^2}{b^{-1}x^2}$ ***23.** $\dfrac{a^{-3}b}{b^{-3}} + \dfrac{2b^4}{a^3}$

*24. $\dfrac{7a^{-3}b^2}{c^{-1}} - \dfrac{5b^2}{a^3c^{-1}} + \dfrac{3b^2}{a^3c}$ **25.** $\dfrac{m^2}{y^2} - \dfrac{3y^{-2}}{m^{-2}}$

Evaluate:

26. $-a(a - a^2x)$ if $a = -2, x = 4$ **27.** $b - (-c^0)$ if $b = -2, c = 4$

28. $k^3 - (k - c)$ if $k = -2, c = 4$

Simplify:

29. $\left|-3^{-3}\right|$ **30.** $\dfrac{-(-2 - 5) - (-3 - 6)}{-2^0(-4)(-2)}$

LESSON 43   *Solving multivariable equations*

43.A
equations
with more than
one variable

When we are asked to solve an equation in one unknown such as

$$12 + 4x - 3 + 4 - 2x = 6x - 3 - 2 + 5x$$

we are asked to simplify both sides and to finally write the equation with x all by itself on one side and a number on the other side. When we do this, we say that we have **isolated** x on one side of the equation. When we have isolated x in this problem we get

$$x = 2$$

If an equation contains more than one variable and we are asked to solve the equation for one of the variables, our task is the same as that described above. **We are asked to rearrange the equation so that the designated variable is the sole member of one side of the equation (either side).** In the following problems, however, the other side of the equation will contain variables as well as numbers.

example 43.1   Solve for y: $6y - x + z = 4$.

solution   We will begin the process of **isolating** y by eliminating $-x$ and $+z$ from the left side of the equation.

$$6y - x + z = 4 \qquad \text{original equation}$$
$$\underline{ + x - z \qquad + x - z} \qquad \text{add } +x - z \text{ to both sides}$$
$$6y \qquad = 4 + x - z$$

Now we complete the isolation by dividing every term by 6.

$$\frac{6y}{6} = \frac{4}{6} + \frac{x}{6} - \frac{z}{6} \longrightarrow y = \frac{2}{3} + \frac{x}{6} - \frac{z}{6}$$

example 43.2   Solve for y: $4x - 2y + 2 = y - 4$.

solution   The first step is to eliminate the y term on one side or the other. We choose to eliminate the $-2y$ so we add $+2y$ to both sides.

$$4x - 2y + 2 = \qquad y - 4 \qquad \text{original equation}$$
$$\underline{ + 2y \qquad \qquad + 2y} \qquad \text{add } +2y \text{ to both sides}$$
$$4x \qquad + 2 = \qquad 3y - 4$$

Now we have all the y's on the right side. To **isolate** y on the right we must eliminate the -4 and the 3 that are on the right side. To eliminate the -4, we add $+4$ to both sides.

$$4x + 2 = 3y - 4$$
$$\underline{ + 4 \qquad + 4} \qquad \text{add } +4 \text{ to both sides}$$
$$4x + 6 = 3y$$

Now we complete the isolation of y by dividing every term by 3.

$$\frac{4x}{3} + \frac{6}{3} = \frac{3y}{3} \qquad \text{divide by 3}$$

$$\frac{4x}{3} + 2 = y \qquad \text{simplified}$$

example 43.3   Solve for p: $4p + 2a - 5 = 6a + p$.

solution   We begin by eliminating the p on the right side of the equation.

$$4p + 2a - 5 = 6a + p \qquad \text{original equation}$$
$$\underline{-p \qquad \qquad \qquad -p} \qquad \text{add } -p \text{ to both sides}$$
$$3p + 2a - 5 = 6a$$

Now we eliminate the $+2a$ and -5 on the left side.

$$3p + 2a - 5 = \qquad 6a$$
$$\underline{ - 2a + 5 \qquad -2a + 5} \qquad \text{add } -2a + 5 \text{ to both sides}$$
$$3p \qquad = \qquad 4a + 5$$

As the final step we divide every term by 3 and get

$$p = \frac{4}{3}a + \frac{5}{3} \qquad \text{divided every term by 3}$$

example 43.4   Solve for x: $5y + x - 2y - 4 + 3x = 0$.

solution   We will begin by adding like terms. Then we will eliminate the $3y$ and the -4 by adding $-3y$ and $+4$ to both sides.

$$3y + 4x - 4 = \qquad 0 \qquad \text{added like terms}$$
$$\underline{-3y \qquad \quad + 4 = -3y + 4} \qquad \text{add } -3y + 4 \text{ to both sides}$$
$$4x \qquad \quad = -3y + 4$$

Now we complete the process of isolating y by dividing every term by 4.

$$\frac{4x}{4} = \frac{-3y}{4} + \frac{4}{4} \qquad \text{divide by 4}$$

$$x = -\frac{3}{4}y + 1 \qquad \text{simplified}$$

example 43.5   Solve for y: $4y + 6x - 4 = 2$.

solution   Since only one term contains a y, we begin by moving all other terms to the right side.

$$
\begin{array}{ll}
4y + 6x - 4 = 2 & \text{original equation} \\
\underline{ - 6x + 4 \phantom{={}} +4 - 6x} & \text{add } +4 - 6x \text{ to both sides} \\
4y = 6 - 6x &
\end{array}
$$

$$\frac{4y}{4} = \frac{6}{4} - \frac{6x}{4} \qquad \text{divide by 4}$$

$$y = \frac{3}{2} - \frac{3}{2}x \qquad \text{simplified}$$

problem set 43

1.   If the sum of twice a number and -10 is multiplied by -4, the result is 61 greater than the opposite of the number. What is the number?

2.   The defense budget was spent on halberds and other armor in the ratio of 2 to 19. If the total budget was 84,000 farthings, how much went for halberds?

3.   What fraction of 30 is 18?

Solve:

4. $2\frac{1}{3}x + 5 = 19$

5. $3(-x - 4) = 2x + 3(x - 5)$

6. $-(5 - 2x) + x = 7(x - 2)$

7. $-(.2 - .4z) - .4 = z - 1.47$

*8.   Solve for y: $4x - 2y + 2 = y - 4$

*9.   Solve for p: $4p + 2a - 5 = 6a + p$

*10.   Solve for x: $5y + x - 2y - 4 + 3x = 0$

*11.   Solve for y: $4y + 6x - 4 = 2$

12.   Solve for y: $3x + 2y = 5 - y$

13.   Solve for y: $-2y + 6y - x - 4 = 0$

Factor the greatest common factor:

14. $6x^4y^2 - 4zx^2y^2$

15. $8x^5y^2z - 16x^2y^2z^2 - xyz$

Simplify:

16. $\dfrac{4x - 8xy}{4x}$

17. $\dfrac{5x^2y^2 - 25x^3y^3}{x^2y^2}$

18.   Graph $x \not\le 2$.

19.   Write 1125 as a product of prime factors.

Simplify. Write all exponentials with positive exponents:

20. $\dfrac{x^5yx^{-7}y^2}{x^4yy^3x^3}$

21. $\dfrac{4x^{-2}y^{-6}m}{x^5y^5m^{-4}}$

22.   Expand: $x^2z^{-2}\left[\dfrac{x^4z^{-4}}{x} - \dfrac{3z^2}{x^2}\right]$

Simplify by adding like terms:

23. $\dfrac{3x^{-2}x^3y}{y^{-4}} - 2xy^5$

24. $\dfrac{3x^{-2}y^2}{m^{-2}} - \dfrac{5m^2y^2}{x^2} + \dfrac{2my^2}{m^{-1}x^2}$

25. $xy^2 - \dfrac{3xy}{y^{-1}} + \dfrac{2x^0y^2}{x^{-1}} - \dfrac{4x^2}{y^2} + 2x^2y^{-2}$

Evaluate:

26. $|x| - x(y)(-x)$ if $x = -2$, and $y = 4$

27. $cx - c^3x$ if $c = -3$, $x = 5$

28. $ab^0(a^2 - b^2)$ if $a = -2$, $b = 3$

Simplify:

29. $-3 - \dfrac{1}{3^{-2}}$

30. $\dfrac{-4[(-2 + 5) - (-3 + 8)]}{-2^0|5 - 1|}$

LESSON 44   *Least common multiple*

If we are given the numbers

$$4, \quad 5, \quad \text{and} \quad 8$$

and are asked to find the **smallest number that is evenly divisible by each of the numbers,** a reasonable guess would be the product of the numbers, which is 160, because we know that each of the numbers will divide 160 evenly.

$$\frac{160}{4} = 40 \qquad \frac{160}{5} = 32 \qquad \frac{160}{8} = 20$$

But 160 is not the smallest number that is evenly divisible by the three numbers. The number 40 is.

$$\frac{40}{4} = 10 \qquad \frac{40}{5} = 8 \qquad \frac{40}{8} = 5$$

We call the smallest number that can be divided evenly by each of a group of specified numbers the least common multiple of the specified numbers.

We can give a two-step procedure for finding the least common multiple.

1. Write each of the given numbers as a product of prime factors.
2. Write the least common multiple as a product of factors in which **each factor of the numbers appears as many times as it is used as a factor in any one of the original numbers.**

Thus to find the least common multiple of 4, 5, and 8, we first write each of the numbers as a product of prime factors.

$$\overset{4}{\overbrace{2 \cdot 2}} \qquad \overset{5}{\overbrace{5}} \qquad \overset{8}{\overbrace{2 \cdot 2 \cdot 2}}$$

Now, 2 is a factor of two of the numbers, and it appears three times as a factor of 8. Thus it must appear three times as a factor of the least common multiple.

$$2 \cdot 2 \cdot 2$$

Every factor of the original numbers must be a factor of the least common multiple so 5 must be factor. Thus the least common multiple of 4, 5, and 8 is 40 because

$$5 \cdot 2 \cdot 2 \cdot 2 = \textbf{40}$$

example 44.1   Find the least common multiple (LCM) of 100, 15, and 8.

solution   We begin by writing each of the numbers as a product of prime numbers.

$$
\begin{array}{ccc}
100 & 15 & 8 \\
2 \cdot 2 \cdot \underbrace{5 \cdot 5} & 3 \cdot 5 & \underbrace{2 \cdot 2 \cdot 2}
\end{array}
$$

We note that 5 is a factor twice, 3 is a factor once, and 2 is a factor three times. Thus the LCM is

$$\underbrace{5 \cdot 5} \cdot \underline{3} \cdot \underbrace{2 \cdot 2 \cdot 2} = \textbf{600}$$

So 600 is the smallest number that is evenly divisible by each of the three numbers, 100, 15, and 8.

example 44.2   Find the LCM of 80, 75, and 30.

solution   We begin by writing each of the numbers as a product of prime numbers.

$$
\begin{array}{ccc}
80 & 75 & 30 \\
\underbrace{2 \cdot 2 \cdot 2 \cdot 2} \cdot 5 & 3 \cdot \underbrace{5 \cdot 5} & 2 \cdot 3 \cdot 5
\end{array}
$$

So the LCM is

$$\underbrace{2 \cdot 2 \cdot 2 \cdot 2} \cdot \underline{3} \cdot \underbrace{5 \cdot 5} = \textbf{1200}$$

problem set 44

1. If twice a number is increased by 5 and this sum is multiplied by -3, the result is -57. What is the number?

2. The village was polyglot. If the ratio of bilingual denizens to trilingual denizens was 14 to 3 and the denizens totaled 3400, how many were trilingual?

3. What fraction of $2\frac{1}{8}$ is $\frac{1}{5}$?

Solve:

4. $-4\frac{3}{4} + 8\frac{1}{3}x = 13\frac{1}{4}$

5. $-2 - |-3| - 2^2 - (3 - x) = -(-3)^3$

6. $-3x - 2(5 - 7x) = 14$

7. $5p - 6(2p + 1) = -4p - 2$

Solve each of the following equations for y:

8. $x + 3y - 4 = 0$

9. $4y - x = 7$

10. $2y + 2k + 4x - 4 = 0$

11. $3y - 2x - 7 = 0$

Factor the greatest common factor:

12. $3a^2b^4c^5 - 6a^2b^6c^6$

13. $8x^2a - 4x^2a^2 + 2xa^2$

Simplify:

14. $\dfrac{3xyz - 3xy}{3xy}$

15. $\dfrac{2xy - 2x}{y - 1}$

16. Graph $x \not> 5$.

Find the least common multiple:

*17.   100, 15, 8   ***18.**   80, 75, 30   **19.**   16, 12, 50

Simplify. Write all exponentials in the numerator:

20. $\dfrac{p^5 p^{-4} z^2}{z^{-5} z p^3}$ **21.** $\dfrac{akp^2 p^4}{a^{-3} a^5 p^5 k^4}$ **22.** $\dfrac{mm^{-4} pp^5}{m^{-3} pp^6}$

23.   Expand: $m^{-2} z^4 \left[m^2 z^{-4} - \dfrac{3m^6 z}{z^4} \right]$

Simplify by adding like terms:

24. $aaxxy^{-3} + \dfrac{2a^2 x^2}{y^3} - \dfrac{4axx}{y^3}$

25. $m^2 xy^{-2} - 3mmxy^{-2} + \dfrac{4m^2 x}{y^2} - 3mmx^{-1} y^2$

Evaluate:

26. $a^2 - b^3 c$ if $a = 3, b = -2, c = -1$

27. $ab(b^0 - bc)$ if $a = 3, b = -1, c = -2$

28. $m - (m - x)$ if $m = -5, x = 3$

Simplify:

29. $-2^4 + \dfrac{1}{-(-2)^3}$ **30.** $\dfrac{-2[(-3 + 2)(-3 + 5^0)]}{-2 - |-3 - 1|}$

LESSON 45   *Least common multiples of algebraic expressions*

45.A

algebraic least common multiple

The least common multiple is most often encountered when it is used as the least common denominator. If we are asked to add the fractions

$$\frac{1}{4} + \frac{5}{8} + \frac{7}{12}$$

we rewrite each of these fractions as a fraction whose denominator is 24, which is the least common multiple of 4, 8, and 12.

$$\frac{6}{24} + \frac{15}{24} + \frac{14}{24} = \frac{35}{24}$$

In Lesson 47 we will discuss the method of adding algebraic fractions such as

$$\frac{b}{15a^2 b} + \frac{c}{10ab^3}$$

To prepare for this lesson we will practice finding the least common multiple of algebraic expressions.

example 45.1   Find the LCM of $15a^2 b$ and $10ab^3$.

solution   We begin by writing the expressions as products of factors whose exponents are 1.

$$15a^2 b \qquad\qquad 10ab^3$$

$$3 \cdot 5 \cdot a \cdot a \cdot b \qquad 2 \cdot 5 \cdot a \cdot b \cdot b \cdot b$$

The LCM is

$$2 \cdot 3 \cdot 5 \cdot a \cdot a \cdot b \cdot b \cdot b = \mathbf{30a^2b^3}$$

example 45.2   Find the LCM of $4x^2m$ and $6x^3m$.

solution   We begin the same way.

$$4x^2m \qquad\qquad\qquad 6x^3m$$

$$\underbrace{2 \cdot 2} \cdot x \cdot x \cdot m \qquad 3 \cdot 2 \cdot \underbrace{x \cdot x \cdot x} \cdot \underbrace{m}$$

The LCM is

$$2 \cdot 2 \cdot 3 \cdot x \cdot x \cdot x \cdot m \qquad \text{or} \qquad \mathbf{12x^3m}$$

example 45.3   Find the LCM of $12x^2am^2$ and $14x^3am^4$.

solution   The LCM of 12 and 14 is 84. The most that x, a, and m are used as factors is x^3am^4. Thus, the LCM is

$$\mathbf{84x^3am^4}$$

problem set 45

1. If the sum of twice a number and -10 is multiplied by 4, the result is 2 less than the number. What is the number?

2. At the Mardi gras ball the guests roistered and rollicked until the wee hours. If the ratio of roisterers to rollickers was 7 to 5 and 1080 were in attendance, how many were rollickers?

3. What decimal part of 2.25 is 1.3995?

Solve:

4. $3\frac{1}{3}x - 4 = 21$

5. $7(x - 3) - 6x + 4 = 2 - (x + 3)$

6. $-2x + 3(-5 - x) = x$

7. $.04x + .2 - .4x = .38$

Solve each equation for y:

8. $2x - 5y + 4 = 0$

9. $4 + 2x + 2y - 3 = 5$

Factor the greatest common factor:

10. $5x^2y^5m^2 - 10x^4y^2m^3$

11. $3x^2yz - 4zyx^2 + 2xyz^2$

Simplify:

12. $\dfrac{2x + 2}{2}$

13. $\dfrac{4xy + 4x}{4x}$

14. Graph $x \nleq 2$.

Find the least common multiple of :

15. 75, 8, and 30

16. 18, 27, and 45

*17. $15a^2b$ and $10ab^3$

*18. $4x^2m$ and $6x^3m$

19. $3x^2y^5z$ and $6y^5z^2$

Simplify. Write all exponentials with positive exponents.

20. $\dfrac{x^{-4}}{y^2p^{-4}}$

21. $\dfrac{k^5m^2}{k^7m^{-5}}$

22. $\dfrac{a^2bc^{-2}c^5}{a^2b^{-3}a^2c^3}$

23. Expand: $\left[\dfrac{m^2}{y^{-1}} + 4m^5y^6 \right] m^{-2}y$

Simplify by adding like terms:

24. $axy^2 + \dfrac{2ax}{y^{-2}} - \dfrac{3ay^2}{x^{-1}} + 5ay^2x^{-1}$

25. $3m^2k^5 - \dfrac{2m^3k^6}{mk} + 4mmk^6k^{-1} - 3mk^5$

Evaluate:

26. $-|-a|(a - x)$ if $a = -2$ and $x = 4$

27. $-xy(y - x^0)$ if $x = 3$ and $y = -2$

28. $p(a - y)$ if $p = 2$, $y = -4$, and $a = -1$

Simplify:

29. -3^{-3}

30. $\dfrac{4(-3 + 2) - |-5 + 2^0|}{3(-2)^2}$

LESSON 46   *Addition of rational expressions*

46.A

addition of rational expressions

If we add one-eleventh to two-elevenths, we get three-elevenths.

$$\frac{1}{11} + \frac{2}{11} = \frac{1 + 2}{11} = \frac{3}{11}$$

This is a demonstration of the rule for adding fractions whose denominators are the same.

> #### RULE FOR ADDING FRACTIONS
>
> Fractions with equal denominators are added by adding the numerators algebraically and recording the sum over a single denominator.

$$\frac{4}{11} - \frac{14}{11} + \frac{2}{11} - \frac{5}{11} = \frac{4 - 14 + 2 - 5}{11} = \frac{-13}{11}$$

This rule also applies if the denominators are algebraic expressions.

$$\frac{5}{a + 6} + \frac{a + b}{a + 6} + \frac{2}{a + 6}$$

We see that the denominators are the same, so we can add the numerators and record the sum over a single denominator.

$$\frac{5}{a + 6} + \frac{a + b}{a + 6} + \frac{2}{a + 6} = \frac{5 + a + b + 2}{a + 6} = \frac{7 + a + b}{a + 6}$$

example 46.1   Add $\dfrac{4}{2x^2 + y} - \dfrac{6ax}{2x^2 + y}$.

solution   The denominators are the same so we can add the numerators.

$$\frac{4}{2x^2 + y} - \frac{6ax}{2x^2 + y} = \frac{4 - 6ax}{2x^2 + y}$$

example 46.2   Add $\dfrac{5}{a^2 + 7y} - \dfrac{3}{a^2 + 7y} + \dfrac{z}{a^2 + 7y}$.

solution   The denominators are the same so we can add the numerators.

$$\frac{5}{a^2 + 7y} - \frac{3}{a^2 + 7y} + \frac{z}{a^2 + 7y} = \frac{5 - 3 + z}{a^2 + 7y} = \frac{2 + z}{a^2 + 7y}$$

example 46.3   Add $\dfrac{5x + 7}{5a^2x} - \dfrac{3x - 2}{5a^2x}$.

solution   The denominators are the same so we add the numerators and get

$$\frac{5x + 7 - 3x + 2}{5a^2x} = \frac{2x + 9}{5a^2x}$$

problem set 46

1. The sum of twice a number and -5 is 35 greater than the opposite of the number. Find the number.

2. When in a defensive posture, the ratio of pikemen to archers was 3 to 10. If the defense totalled 27,989 soldiers, how many were archers?

Solve:

3. $-2x - 4(x - 2) = 3x + 5$

4. $-2(x - 4) + 8 = 4 - (x + 2)$

5. $(-2)^3(-x - 4) - |-2| - 3^2 = -2(x - 4) - x$

Solve for y:

6. $3x + 2y = 5$

7. $x - 3y + 7 = 0$

Add:

*8. $\dfrac{1}{11} + \dfrac{2}{11}$

*9. $\dfrac{5}{a + 6} + \dfrac{a + b}{a + 6} + \dfrac{2}{a + 6}$

*10. $\dfrac{4}{2x^2 + y} - \dfrac{6ax}{2x^2 + y}$

*11. $\dfrac{5}{a^2 + 7y} - \dfrac{3}{a^2 + 7y} + \dfrac{z}{a^2 + 7y}$

12. $\dfrac{a}{b} + \dfrac{c^2 - a}{b} + \dfrac{4}{b}$

13. Graph $x \not> -4$.

Find the least common multiple of:

14. 125, 75, and 45

15. c^3, c^2, and 2

16. $4c^3$, c^2, and $3c^4$

17. b^3, b^2c, and b^2c^2

18. Factor the greatest common factor: $4x^2y^5p^2 - 3x^5y^4p^2$

19. Simplify: $\dfrac{4x^2 - 4x}{4x}$

Simplify. Write all exponentials in the denominator.

20. $\dfrac{x^5y^2}{x^3y^5}$

21. $\dfrac{m^4p^5}{p^{-3}m^6}$

22. $\dfrac{xxx^3y^5y^{-2}}{x^{-3}yy^{-6}}$

23. Expand: $\dfrac{x^{-2}p^0}{y^4}\left(\dfrac{x^2}{p^4} - x^4p^6\right)$

Simplify by adding like terms:

24. $\dfrac{x^2}{y^2} - 3x^2y^{-2} + 4x^{-3}x^5y^{-2} - \dfrac{8y^{-2}}{x^{-2}}$ **25.** $xa^2y - 2a^2xy + 4ya^2x - \dfrac{6ax}{ay}$

Evaluate:

26. $-a - |a - x|$ if $a = -2, x = -3$

27. $xy^0 - (x - y)$ if $x = -3, y = 5$

28. $pn^2 - n(-p)$ if $n = -2, p = -3$

Simplify:

29. $-\dfrac{1}{(-4)^{-2}}$ **30.** $\dfrac{2[(-4 - 6^0)(5 - 2)]}{6 - (-(-2))}$

LESSON 47   *Addition of abstract fractions*

47.A

addition of abstract fractions

There are three rules of algebra that some people believe are more important than all the rest of the rules put together. Two of them are the addition rule for equations and the multiplication rule for equations. We have used these, and they are restated very informally here:

1. **The same quantity can be added to both sides of an equation.**
2. **Every term on both sides of an equation can be multiplied or divided by the same quantity.**[†]

The other important rule is that

3. **The denominator and numerator of a fraction can be multiplied by the same quantity.**[†]

This theorem is usually called the *fundamental theorem of fractions* or the *fundamental theorem of rational expressions*. We will call it the **denominator-numerator same-quantity theorem** because this name is more meaningful.

> DENOMINATOR-NUMERATOR SAME-QUANTITY THEOREM
>
> The denominator and the numerator of a fraction may be multiplied by the same nonzero quantity without changing the value of the fraction.

[†] Except zero

We cannot find the sum of

$$\frac{1}{4} + \frac{1}{2}$$

in this form because the denominators are not the same. But if we use the **denominator-numerator same-quantity theorem** and multiply both the numerator and the denominator of $\frac{1}{2}$ by 2, we get $\frac{2}{4}$, which is an equivalent expression for $\frac{1}{2}$,

$$\frac{1}{4} + \frac{1}{2} = \frac{1}{4} + \frac{1(2)}{2(2)}$$

Now the fractions may be added, for they both have a denominator of 4.

$$\frac{1}{4} + \frac{2}{4} = \frac{3}{4}$$

If the fractions to be added have different denominators, the procedure shown here can be used to rewrite the fractions as equivalent fractions that have the same denominators.

example 47.1 Add $\frac{3}{4} + \frac{2}{b}$.

solution We will use the **denominator-numerator same-quantity theorem** and a three-step procedure to rewrite the fractions as equivalent fractions that have the same denominators.

(a)   As the first step we write the fraction lines with the proper sign between them:

$$- + -$$

(b)   Now we write the least common multiple of the denominators as the new denominators. When the least common multiple is used in this fashion, we call it the **least common denominator.**

$$\frac{}{4b} + \frac{}{4b}$$

(c)   The first two steps were automatic. We used no theorems or rules. Now we use the **denominator-numerator same-quantity theorem.** We have multiplied the denominator, 4, of the first fraction by b to get $4b$, so we must also multiply the numerator, 3, by b and get $3b$.

$$\frac{3b}{4b} + \frac{}{4b}$$

We have multiplied the denominator, b, of the second fraction by 4 to get $4b$, so we must also multiply the numerator, 2, by 4 to get the new numerator of 8.

$$\frac{3b}{4b} + \frac{8}{4b}$$

Now the fractions have the same denominators and can be added.

$$\frac{3b}{4b} + \frac{8}{4b} = \frac{3b + 8}{4b}$$

example 47.2 Add $\frac{4}{b} + \frac{1}{c} + \frac{1}{2}$.

solution   (a) $- + - + -$ write the fraction lines

(b) $\dfrac{}{2bc} + \dfrac{}{2bc} + \dfrac{}{2bc}$ use the LCM as the new denominator of every term

(c) $\dfrac{8c}{2bc} + \dfrac{2b}{2bc} + \dfrac{bc}{2bc} = \dfrac{8c + 2b + bc}{2bc}$ find the new numerators and add

example 47.3   Add $\dfrac{5}{x} + \dfrac{x}{b} + \dfrac{a}{c}$.

solution   (a) $\dfrac{}{} + \dfrac{}{} + \dfrac{}{}$ write the fraction lines

(b) $\dfrac{}{xbc} + \dfrac{}{xbc} + \dfrac{}{xbc}$ use the LCM as the new denominator of every term

(c) $\dfrac{5bc}{xbc} + \dfrac{x^2c}{xbc} + \dfrac{axb}{xbc} = \dfrac{5bc + x^2c + axb}{xbc}$ find the new numerators and add

example 47.4   Add $\dfrac{m}{c^3} + \dfrac{4}{c^2} - \dfrac{1}{2}$.

solution   (a) $\dfrac{}{} + \dfrac{}{} - \dfrac{}{}$ write the fraction lines

(b) $\dfrac{}{2c^3} + \dfrac{}{2c^3} - \dfrac{}{2c^3}$ use the LCM as the new denominator of every term

(c) $\dfrac{2m}{2c^3} + \dfrac{8c}{2c^3} - \dfrac{c^3}{2c^3} = \dfrac{2m + 8c - c^3}{2c^3}$ find the new numerators and add

example 47.5   Add $\dfrac{p}{4} - \dfrac{a}{2} + \dfrac{c}{b}$.

solution   (a) $\dfrac{}{} - \dfrac{}{} + \dfrac{}{}$ write the fraction lines

(b) $\dfrac{}{4b} - \dfrac{}{4b} + \dfrac{}{4b}$ use the LCM as the new denominator of every term

(c) $\dfrac{pb}{4b} - \dfrac{2ab}{4b} + \dfrac{4c}{4b} = \dfrac{pb - 2ab + 4c}{4b}$ find the new numerators and add

example 47.6   Add $\dfrac{a}{c^2} + \dfrac{3}{4c^3} + \dfrac{m}{3c^4}$.

solution   (a) $\dfrac{}{} + \dfrac{}{} + \dfrac{}{}$ write the fraction lines

(b) $\dfrac{}{12c^4} + \dfrac{}{12c^4} + \dfrac{}{12c^4}$ use the LCM as the new denominator of every term

(c) $\dfrac{12ac^2}{12c^4} + \dfrac{9c}{12c^4} + \dfrac{4m}{12c^4} = \dfrac{12ac^2 + 9c + 4m}{12c^4}$ find the new numerators and add

problem set 47

1. The product of 8 and a number is 10 less than 3 times the number. What is the number?

2. When Oberon and Titania assembled the little people, they found that the pixies and leprechauns were in the ratio of 3 to 13. If there were 6816 in all, how many were pixies?

3. What decimal part of .46 is .01058?

Solve:

4. $2\frac{1}{5}x + 5 = -15$ 5. $-4x - 3(x - 3) = x + 2$

6. $-4x + (-2x + 5) = -2x$ 7. $.2p + 2.2 + 2.2p = 4.36$

Solve for y:

8. $5x + 4 = 3y$ 9. $2y - 5 = x$

Add:

*10. $\frac{1}{4} + \frac{1}{2}$ 11. $\frac{3}{7} + \frac{2}{5}$ 12. $\frac{1}{2} + \frac{3}{4} + \frac{6}{7}$

*13. $\frac{4}{b} + \frac{1}{c} + \frac{1}{2}$ *14. $\frac{m}{c^3} + \frac{4}{c^2} - \frac{1}{2}$ *15. $\frac{p}{4} - \frac{a}{2} + \frac{c}{b}$

*16. $\frac{a}{c^2} + \frac{3}{4c^3} + \frac{m}{3c^4}$ 17. $\frac{x}{y} + \frac{b}{4y} + c$ 18. $\frac{4}{a} + \frac{c}{4a} + 5$

19. Graph $x \not\le 2$.

20. Find the least common multiple of: 40, 35, and 18

21. Factor: $x^3y^2m^5 - 3x^2ym^6$ 22. Expand: $\left(\dfrac{a^2}{x^{-1}} - 4a^6x^4\right)\dfrac{a^{-2}}{x}$

Simplify:

23. $\dfrac{4ax - axy}{ax}$ 24. $\dfrac{x^4y^2}{x^{-2}y^{-3}}$ 25. $\dfrac{x^2xyy^{-4}}{x^4y^{-5}}$

26. Simplify by adding like terms: $\dfrac{a^2x^3}{y} - \dfrac{2x^3a^2}{y} + \dfrac{4xx^2a^2}{y} - 3a^2x^3$

Evaluate:

27. $-a - |p|(p - 2)$ if $p = -2, a = 3$

28. $x - y(x^0 - y)$ if $x = -2, y = 3$

Simplify:

29. -3^{-3} 30. $\dfrac{-2[(-4 - 2) - (5^0 - 3)]}{-2 - |2|}$

LESSON 48 *Conjunctions*

48.A

conjunctions   If we wish to designate the numbers that are greater than 5, we could write

$$x > 5$$

If we wish to designate the numbers that are less than 10, we could write

$$x < 10$$

If we wish to designate the numbers that are greater than 5 and that are also less than 10, we could write either of the following.

$$x > 5 \text{ and } x < 10 \qquad \text{or} \qquad 5 < x < 10$$

Both of these notations mean the same thing and designate the numbers that are between 5 and 10. We use the word **conjunction** to describe a statement of two conditions, both of which must be met. Thus, both of the above statements are conjunctions. In the concise notation on the right, we note that the symbols point in the same direction. They always do, as we see when we reverse both the symbols and the numbers to make the same statement another way.

$$10 > x > 5$$

But, we must be careful when we write conjunctions because

$$10 < x < 5$$

designates the numbers that are greater than 10 and are also less than 5. Of course, there are no numbers that fall into this category.

example 48.1   Graph the solution to $5 < x < 10$.

solution   This conjunction designates the numbers between 5 and 10.

example 48.2   Graph the solution to $-2 < x \le 4$.

solution   This conjunction asks for the graph of the numbers between -2 and 4. Note that the $<$ excludes -2 and that ≤ 4 includes 4 in the solution set.

example 48.3   Write the conjunction for which this graph is the solution.

solution   The graph shows the numbers that are less than 0 and greater than or equal to -5. We write this conjunction as

$$-5 \le x < 0$$

example 48.4   Write the conjunction that designates the numbers that are greater than -1 and less than or equal to 5.

solution   The conjunction is $-1 < x \le 5$.

problem set 48

1. If the sum of 4 times a number and 5 is multiplied by 3, the result is 24 less than the opposite of the number.

2. At the prestidigitator's banquet, the ratio of necromancers to pyromancers was 7 to 2. If there were 324 at the banquet, how many were necromancers?

3. $4\frac{1}{5}$ of what number equals 28?

Solve:

4. $3\frac{1}{3}x + 7 = -2$ 5. $5p - 4p - (p - 2) = 3(p + 4)$

6. $(-2)^3(-k - |-3|) - (-2) - 2k = k - 3^2$

Solve for y:

7. $2x + 4y = 6$

8. $3y - 4 = 2x$

Find the least common multiple of:

9. 8, 36, and 75

10. x, c^2x^2, and cdx

Add:

11. $\dfrac{1}{3} + \dfrac{2}{5} + \dfrac{3}{10}$

12. $\dfrac{3}{7} + \dfrac{8}{9} - \dfrac{1}{3}$

13. $\dfrac{a}{x} + \dfrac{b}{c^2x^2} + d$

14. $\dfrac{4}{x^2} + \dfrac{6}{2x^3} - \dfrac{3}{4x^4}$

15. $\dfrac{4}{x^2} + \dfrac{c}{4x^3} + m$

16. $\dfrac{a}{b} + \dfrac{c}{4b^2} + \dfrac{a^2}{8b^3}$

17. $\dfrac{m}{a^5} + \dfrac{k}{2a^4} - \dfrac{3}{4a^3}$

18. $\dfrac{1}{2a^3} + \dfrac{3}{4ab^2} + \dfrac{c}{8a^2b^2}$

Graph:

*19. $5 < x < 10$

*20. $-2 < x \le 4$

21. Factor: $8m^3x^2y^4p - 4m^2xpm$

22. Expand: $\dfrac{x^{-2}}{y^{-3}}\left(\dfrac{x^2}{y^3} - \dfrac{ax^3}{y^{-4}}\right)$

Simplify:

23. $\dfrac{4x^4 - 4}{4}$

24. $\dfrac{mm^2p^3y^{-3}}{m^{-3}m^{-2}p^{-3}y^4}$

25. $\dfrac{xx^{-3}y^5x^0}{x^2y^{-5}xy^2}$

Simplify by adding like terms:

26. $\dfrac{m}{y} - \dfrac{3m^2y}{my^2} - \dfrac{5m^{-3}m^4}{y^{-3}y^4} + \dfrac{2ym}{ym^2}$

Evaluate:

27. $-x - |xa|(x^0 - a)$ if $x = -2, a = -3$

28. $-x^2 - y^2(xy)$ if $x = -2, y = 3$

Simplify:

29. $(-3)^{-2}$

30. $-4[(-3 - 2^0) - (5 - 2) + |3|]$

LESSON 49 *Percents less than 100*

49.A

percents less than 100

We have been working problems about fractional and decimal parts of numbers by using one of the following equations.

(a) $(F) \times (\text{of}) = \text{is}$ or   (b) $(D) \times (\text{of}) = \text{is}$

The percent equation is exactly the same as (a) except that the fraction has a denominator of 100. *Centum* is the Latin word for 100, and thus percent literally means "by the 100." We often use the symbol % to represent the word percent. The percent equation is

(c) $\dfrac{P}{100} \times (\text{of}) = \text{is}$ which can also be written as   (d) $\dfrac{P}{100} = \dfrac{\text{is}}{\text{of}}$

The part identified by the word *of* is often called the **base,** and the part identified by the word *is* is called the **percentage.** If we use these words, we get equation (e). In equation (f) $\frac{P}{100}$ is called the **rate.**

(e) $\frac{P}{100} \times$ base = percentage   (f) rate × base = percentage

All four equations produce the same result. We prefer equation (c) because it is just like equation (a), which we have used for fractional parts of numbers. However, your teacher may prefer one of the other forms. Many teachers like (d) because this form can be explained using the concepts and vocabulary of ratios. Others prefer (e) or (f) because these forms are used almost exclusively in the business world. **The four equations (c), (d), (e), and (f) are not different equations but different forms of the same equation.** Each form has advantages and disadvantages. None is perfect. Some people find form (c) to be the most difficult to solve. Form (d) is seldom used except in mathematics classes. Form (f) does not use percent as such but uses rate which is percent divided by 100. We find that it is best to pick one form and stick with it. If you don't like form (c), use another.

To solve word problems about percent it is necessary to visualize the problem. We will begin to work on understanding this visualization by drawing diagrams of percent problems after we work the problems.

example 49.1 Twenty percent of what number is 15? Work the problem and then draw a diagram of the problem.

solution We will use equation (c) and use 20 for *percent*, *WN* for *what number* and 15 for *is*.

$$\frac{P}{100} \times \text{of} = \text{is} \longrightarrow \frac{20}{100} \cdot WN = 15$$

We will solve by multiplying both sides by $\frac{100}{20}$.

$$\frac{100}{20} \cdot \frac{20}{100} WN = 15 \cdot \frac{100}{20} \longrightarrow WN = \frac{1500}{20} \longrightarrow WN = 75$$

The "before" diagram is 75, which represents 100 percent. The "after" diagram shows that 15 is 20 percent. Thus the other part must be 60, which is 80 percent.

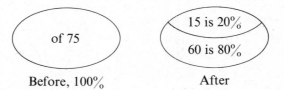

Before, 100%   After

example 49.2 What percent of 140 is 98? Work the problem and then draw a diagram of the problem.

solution We use *WP* for *what percent*, 140 for *of*, and 98 for *is*.

$$\frac{P}{100} \times \text{of} = \text{is} \longrightarrow \frac{WP}{100} \cdot 140 = 98$$

We will solve by multiplying both sides by $\frac{100}{140}$.

$$\frac{100}{140} \cdot \frac{WP}{100} \cdot 140 = 98 \cdot \frac{100}{140} \longrightarrow WP = \frac{9800}{140} \longrightarrow WP = 70\%$$

The diagrams show that 140 or 100 percent was divided into two parts: 98, which is 70 percent, and 42, which is 30 percent.

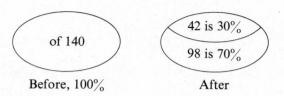

Before, 100%   After

example 49.3   Fifteen percent of 300 is what number? Work the problem and then draw a diagram of the problem.

solution   We use 15 for percent, 300 for *of*, and WN for *is*.

$$\frac{P}{100} \times \text{of} = \text{is} \quad \longrightarrow \quad \frac{15}{100} \cdot (300) = WN$$

We multiply to solve and get

$$\frac{4500}{100} = WN \quad \longrightarrow \quad \mathbf{45 = WN}$$

The diagrams show that 300 is divided into two parts. One part is 45 or 15 percent. Thus the other part must be 255 or 85 percent.

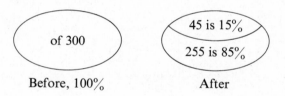

Before, 100%   After

problem   **1.**   The sum of 3 times a number and 25 is 5 greater than the opposite of the number.
set 49   What is the number?

2.   Leonardo and Michelangelo turned out paintings whose areas were in the ratio of 14 to 13. During the period in question, the total area of their paintings was 1080 square units. How many square units were painted by Leonardo?

3.   What fraction of $2\frac{1}{8}$ is $\frac{1}{4}$?

Draw the diagram after completing each problem.

*4.   Twenty percent of what number is 15?

*5.   Fifteen percent of 300 is what number?

6.   Forty percent of what number is 640?

7.   What percent of 8300 is 996?

8.   Eighty percent of what number is 1120?

Solve:

9. $4x - 5(x + 2) = -(2x - 4)$ **10.** $.02 + .02x - .4 - .4x = 3.116$

Solve for y:

11. $3x - 4y = 7$

12. $-2y + 5 + 3x = 0$

Find the least common multiple:

13. 21, 24, and 60

14. $4x^2, yx^2, 8m^3x^2$

Add:

15. $\dfrac{3}{4} + \dfrac{2}{5} - \dfrac{3}{20}$

16. $\dfrac{a}{x} + \dfrac{b}{c} + d$

17. $\dfrac{m}{xc} + \dfrac{d^2}{xkc^3} - \dfrac{3p}{xk^2c^3}$

18. $\dfrac{4}{a^2b^2} - \dfrac{c}{ad} - \dfrac{m}{a^3b}$

Graph:

19. $x \nleq 2$

20. $2 \leq x < 5$

21. Factor: $18x^5y^2m - 9x^3ym^5$

22. Expand: $\dfrac{x^{-1}}{y}\left(\dfrac{y}{x} - \dfrac{3xy^{-5}}{p^6}\right)$

Simplify:

23. $\dfrac{4x^2 - 4x}{4x}$

24. $\dfrac{kp^2k^{-1}p^{-3}p^{-4}}{k^2pp^2k^{-5}}$

25. $\dfrac{m^2xym^3x^{-5}}{yy^{-4}m^{-3}x^2}$

26. Simplify by adding like terms: $m^2x^2 - \dfrac{3m^{-2}x^{-2}}{m^4x^{-4}} + \dfrac{2m^2}{xm} - \dfrac{5x^{-1}}{m^{-1}}$

Evaluate:

27. $-p - (p - x)$ if $p = -3, x = 5$

28. $-p^2 - p^0x^2$ if $p = -3, x = 2$

Simplfy:

29. $\dfrac{1}{-(-3)^{-2}}$

30. $-5[(-2 + 3)(-2 - 4^0) - |-5|]$

LESSON 50 *Polynomials · Addition of polynomials*

50.A
polynomials

Thus far we have encountered expressions such as

(a) $\dfrac{x^{-3} + m}{x}$

(b) $\dfrac{2y}{x^3}$

(c) $4a^{-2}x + m^3$

(d) rt^{n-2}

(e) 4

(f) $-4a^2$

(g) $7x^2 + 2$

(h) $4x^2y$

(i) $7y^3 + 3y + 2$

All of these expressions are called **algebraic expressions** or **mathematical expressions** and are individual terms or indicated sums of terms. The more complicated expressions as shown in (a) through (d) have no special names and are just called **terms** or **expressions.** The simple expressions as shown in (e) through (i) occur so often and are so useful that we give these expressions a special name: **polynomial.**[†]

[†] It is unfortunate that we use such an intimidating word to describe the simplest type of expression.

A polynomial in one variable is one term or a sum of individual terms each of which has the form

$$ax^n$$

where **a is a real number** and **n is a whole number,** such as the following:

(a) $4x^2$ (b) $-x^3$ (c) $-1.414x^{32}$

(d) $2x^4$ (e) $-7x$ (f) -7

Each of the six expressions meets all three of the requirements for being called a polynomial:

1. **Each expression is in the form ax^n.**
2. **The numerical coefficient of each expression, a, is a real number.**
3. **The exponent of the variable, n, is a positive integer or is the number zero.**

The last polynomial shown, -7, can be thought of as being $-7x^0$, which is the same as -7 if x has a value other than zero because **any nonzero quantity raised to the zero power has a value of 1!**
    We refresh our memory with the following examples.

$$(-15)^0 = 1 \qquad \left(\frac{75}{14}\right)^0 = 1 \qquad x^0 = 1 \qquad (x + y)^0 = 1 \qquad (x, x + y \neq 0)$$

Thus, since x^0 equals 1 if x is not zero, we can write

$$-7x^0 = -7(1) = -7$$

A **polynomial** of one term is called a **monomial,** so each of the six expressions in (a) through (f) above can be called a polynomial or it may be described by using the more restrictive name of monomial.
    A **polynomial of two terms** is also called a **binomial,** and a **polynomial of three terms** is also called a **trinomial.** Thus each of the following

(a) $4 + x$ (b) $p^{15} - 4p$ (c) $-y^{10} + 3y$

can be described by using either word polynomial or the word binomial. The following expressions

(a) $x^2 + 2x + 4$ (b) $y^{14} - 1.6y^2 + 4$ (c) $m^4 + 2m - 1.6$

can be called either polynomials or trinomials. Indicated sums of more than three monomial terms have no special names and are just called polynomials. Thus we can call any of the following expressions a polynomial.

(a) -14.2 (b) $\dfrac{7}{2}$ (c) $4x^2$ (d) $6x^2 + 4$

(e) $x^4 - 3x^2 + 2x$ (f) $-7x^{15} + 2x^3 - 5x^2 + 6x + 4$

Expressions (a) and (b) are polynomials because the exponents of the understood variable is the number zero.

$$(a) \quad -14.2m^0 = -14.2 \qquad (b) \quad \frac{7}{2}y^0 = \frac{7}{2}$$

Although expressions (c), (d), and (e) are all polynomials they can also be called a monomial, a binomial, and a trinomial in that order. The last expression, (f), is a polynomial because each term in the indicated sum is a monomial. This expression does not have a more restrictive name.

> **DEFINITION OF A POLYNOMIAL IN ONE UNKNOWN**
>
> A polynomial in one unknown is an algebraic expression of the form
>
> $$4x^{15} + 2x^{14} - 3x^{10} + \cdots + 2x + 2$$
>
> where the coefficients are real numbers, x is a variable, and the exponents are whole numbers.

Thus none of the following expressions is a polynomial.

(a) $4x^{-3}$ (b) $\dfrac{-6x + y}{z}$ (c) $-15y^{-5}$

The expressions (a) and (c) have real number coefficients but the exponent of the variable is not a whole number. Expression (b) is not a polynomial because it is not in the required form of ax^n.

50.B
polynomials, general

> **DEFINITION OF A POLYNOMIAL**
>
> A polynomial in one or more unknowns is an algebraic expression having only terms of the form $ax^n y^m z^p \cdots$ where the coefficient a is a real number and the exponents $n, m, p, \ldots$ are whole numbers.

Thus the general definition of a polynomial has the same restrictions that are given by the definition of a polynomial in one unknown, namely:

1. The numerical coefficients of the individual terms must be **real numbers.**
2. The exponents of the variables must be **whole numbers.**

The following can therefore be called polynomials in more than one variable.

(a) $xyz^2 m$ (b) $4x^{15}ym^3 + pq^5$ (c) $-11x^2p^4 + 2$

50.C
degree

The **degree of a term** of a polynomial is the sum of the exponents of the variables in the term. Thus

$$4x^3, \; 6xym, \; 2x^2y \qquad \text{are third-degree terms}$$

$$4x^2m^3, \; 3y^5, \; 2xypmz \qquad \text{are fifth-degree terms}$$

The **degree of a polynomial** is the same as the degree of its highest-degree term.

$3x^2 + xyz + m$ is a third-degree polynomial because the degree of its highest degree term (xyz) is 3.

$4x^5 + yx^3 + 2x^2 + 2$ is a fifth-degree polynomial because the degree of its highest degree term ($4x^5$) is 5.

Polynomials are usually written in descending powers of one of the variables. The polynomials

$$x^5 - 3x^4 + 2x^2 - x + 5$$

$$x^4m + x^3m^2 - 2xm^5 - 6$$

$$-2xm^5 + x^3m^2 + x^4m - 6$$

are written in descending powers of a particular variable. The first two polynomials are written in descending powers of x. The third polynomial is the same polynomial as the second but is written in descending powers of m instead of descending powers of x.

50.D
addition of polynomials

Since polynomials are composed of individual terms, the rule for adding polynomials is the same rule that we use for adding terms—**like terms may be added.**

example 50.1 Add $(x^3 + 3x^2 + 2) + (2x^3 + 4)$.

solution We remember that we can discard parentheses preceded by a plus sign without changing the signs of the terms therein.

$$(x^3 + 3x^2 + 2) + (2x^3 + 4) = x^3 + 3x^2 + 2 + 2x^3 + 4$$
$$= 3x^3 + 3x^2 + 6$$

example 50.2 Add $(3x^4 - 2x^2 + 3) - (x^4 - 2x^3 + x^2)$.

solution When we discard the parentheses in this problem, we remember to **change the sign of every term in the second parentheses** because the second parentheses are preceded by a minus sign.

$$3x^4 - 2x^2 + 3 - x^4 + 2x^3 - x^2 = 2x^4 + 2x^3 - 3x^2 + 3$$

example 50.3 Add $(3x^3 + 2x^2 - x + 4) - (x^2 - 7x - 5)$.

solution The first parentheses can be removed with no change to the terms inside, but because the second parentheses are preceded by a minus sign, when we remove these parentheses, we must **change the sign of all terms therein.**

$$(3x^3 + 2x^2 - x + 4) - (x^2 - 7x - 5) = 3x^3 + 2x^2 - x + 4 - x^2 + 7x + 5$$
$$= 3x^3 + x^2 + 6x + 9$$

problem set 50

1. Glicken found that 4 times the sum of twice a number and -5 was 92 less than the opposite of the number. What was the number?

2. In the remuda, the ratio of piebalds to mottled mares was 7 to 11. If there were 756 remounts in the remuda, how many were piebalds?

3. What fraction of $\frac{1}{3}$ is $\frac{2}{27}$?

Draw the diagrams after working these problems.

4. Harding and Jack found that 8 percent of their number was 72. What was their number?

5. What percent of 860 is 43?

6. Sixteen percent of 4200 is what number?

7. Forty-three percent of what number is 2150?

8. What percent of 5400 is 108?

Solve:

9. $3x - 5(-2x - 8) + 4 = 2 - (x - 4)$

10. $-(-3)^3 - |-2| - 2^2 - (-k - 3) = -4^2 - (3k - 4)$

11.   Solve for y: $3x - 4y + 7 = 0$

12.   Find the least common multiple of 250, 75, and 20.

Add:

13. $\dfrac{1}{7} + \dfrac{5}{21} + \dfrac{3}{5}$

14. $\dfrac{m}{x} + \dfrac{b}{xy} + \dfrac{ac}{x^2ym}$

15. $\dfrac{3}{cd} + \dfrac{5}{4c^2d} + \dfrac{7}{8cd^2}$

16. $\dfrac{p}{xa} + \dfrac{5}{xam} + x$

Add. Write the answers in descending order of the variable:

*17. $(3x^3 + 2x^2 - x + 4) - (x^2 - 7x - 5)$

18. $-(3x^4 - 2x^2 + 4x + 2) + (3x^3 - x^2 - 2x + 5)$

19. $(5x - 2x^5 + 6x^2 - 5) - (x^4 + 3x^2 + 2x + 10)$

20. $(-x^3 - 2x - 3x^2 + 5) - 2(x^3 - x + 2x^2 - 3)$

21.   Graph $-4 < x \le 1$.

22.   Factor $x^2ym - 4x^2ym^3 + 2x^4y^3m^6$

Simplify:

23. $\dfrac{4x^2 - 4x^4}{4x^2}$

24. $\dfrac{x^2y^{-2}m^{-5}y^0}{xxy^2y^{-5}x^{-3}}$

25.   Expand: $\left(\dfrac{x^2}{yp^{-4}} - \dfrac{3x^2y}{p^{-4}}\right)\dfrac{x^{-2}}{y^4p}$

26.   Simplify by adding like terms: $x^2y^{-2}p + \dfrac{3xxp}{y^2} - \dfrac{4x}{y^{-2}} + 6xy^2$

Evaluate:

27. $-xy(x - y)$ if $x = -3, y = 5$

28. $-p^2 - p^3(xp^0)$ if $p = -2, x = -4$

Simplify:

29. -3^{-3}

30. $-2[(-4 - 3^0)(5 - 2) - (-6)]$

LESSON 51   *Multiplication of polynomials*

51.A
multiplication of polynomials

We remember that we use the distributive property

$$a(b + c) = ab + ac$$

when we multiply a monomial by a binomial. The expression on the outside is multiplied by both terms inside.

example 51.1   Multiply $4x(x^2 - 2)$.

solution   We must multiply $4x$ by x^2 and by -2. Then we sum the products.

$$4x(x^2 - 2) = 4x(x^2) + 4x(-2) = \mathbf{4x^3 - 8x}$$

To develop a procedure for multiplying a binomial by a binomial, we will use the distributive property to multiply

$$(a + b)(c + d)$$

The notation $(a + b)(c + d)$ tells us that $(a + b)$ is to be multiplied by c and that $(a + b)$ is also to be multiplied by d and that the two products are to be summed.

$$(a + b)(c + d) = (a + b)c + (a + b)d$$

Now we use the distributive property again to multiply $(a + b)$ by c

$$(a + b)c = ac + bc$$

and to multiply $(a + b)$ by d

$$(a + b)d = ad + bd$$

and the products are summed

$$\mathbf{ac + bc + ad + bd}$$

This has been a rather involved development to demonstrate the following rule.

RULE FOR MULTIPLYING BINOMIALS

To multiply one binomial by a second binomial, each term of the first binomial is multiplied by each term of the second binomial and then the products are summed.

---

example 51.2   Multiply $(4x + 5)(3x - 2)$.

solution   The notation indicates that $4x$ is to be multiplied by both $3x$ and -2

$$4x(3x - 2) = \underline{12x^2 - 8x}$$

and that $+5$ is to be multiplied by both $3x$ and -2

$$+5(3x - 2) = \underline{15x - 10}$$

and that the products are to be added algebraically.

$$(4x + 5)(3x - 2) = \underline{12x^2 - 8x} + \underline{15x - 10} = \mathbf{12x^2 + 7x - 10}$$

We can also do the multiplications if the binomials are written one above the other. Either one may be on top.

example 51.3   Multiply $(4x + 2)(x - 5)$.

solution   We begin writing the binomials one above the other.

$$\begin{array}{r} 4x + 2 \\ \underline{x - 5} \end{array}$$

Now, the x of $x - 5$ is multiplied by both terms of $4x + 2$ and the products are recorded.

$$\begin{array}{r} 4x \ + 2 \\ \underline{x \ - 5} \\ 4x^2 + 2x \end{array} \qquad \text{product of } x \text{ and } 4x + 2$$

Now the -5 of $x - 5$ is multiplied by both terms of $4x + 2$, and the products are recorded so that like terms (if any) are recorded below like terms to facilitate addition.

$$
\begin{array}{l}
4x \;+\; 2 \\
\underline{\;x \;-\; 5\;} \\
4x^2 \;+\; 2x \qquad\qquad \text{product of } x \text{ and } 4x + 2 \\
\underline{\qquad\; -\; 20x \;-\; 10} \qquad \text{product of } -5 \text{ and } 4x + 2 \\
\mathbf{4x^2 \;-\; 18x \;-\; 10} \qquad \text{sum of the products}
\end{array}
$$

The product $-20x$ was recorded below the term $+2x$. There was no constant in the first product for -10 to be recorded below, so -10 was written out to the right.

example 51.4   Multiply $(4x + 2)(3x - 5)$.

solution   We will use the vertical format to multiply.

$$
\begin{array}{l}
4x \;+\; 2 \\
\underline{\;3x \;-\; 5\;} \\
12x^2 \;+\; 6x \qquad\qquad \text{product of } 3x \text{ and } 4x + 2 \\
\underline{\qquad\; -\; 20x \;-\; 10} \qquad \text{product of } -5 \text{ and } 4x + 2 \\
\mathbf{12x^2 \;-\; 14x \;-\; 10} \qquad \text{sum of the products}
\end{array}
$$

example 51.5   Expand $(3x + 2)^2$.

solution   When we write x^2 we mean x times x. Thus, when we write $(3x + 2)^2$, we mean $3x + 2$ times $3x + 2$. We will use a vertical format.

$$
\begin{array}{l}
3x \;+\; 2 \\
\underline{\;3x \;+\; 2\;} \\
9x^2 \;+\; 6x \qquad\qquad \text{product of } 3x \text{ and } 3x + 2 \\
\underline{\qquad\; +\; 6x \;+\; 4} \qquad \text{product of } 2 \text{ and } 3x + 2 \\
\mathbf{9x^2 \;+\; 12x \;+\; 4} \qquad \text{sum of the products}
\end{array}
$$

This same procedure can be used if one or both expressions have three or more terms. Each term in one expression is multiplied by every term in the other expression and the products are then summed algebraically. The next example illustrates the procedure for multiplying a binomial by a trinomial.

example 51.6   Multiply $(4x - 2)(x^2 + x + 4)$.

solution   We will multiply both $4x$ and -2 by all three terms in the second parentheses. This will give us six products. Then we simplify by adding the like terms.

$$4x^3 + 4x^2 + 16x - 2x^2 - 2x - 8 = \mathbf{4x^3 + 2x^2 + 14x - 8}$$

problem set 51

1. Jo Ellen found that if the product of a number and 5 is increased by 20, the result is 28 less than the opposite of the number. What is the number?

2. The ratio of expressions frowned on to expressions proscribed was 3 to 17. If 1620 expressions were considered, how many were proscribed?

3. What fraction of $7\frac{1}{8}$ is $3\frac{2}{7}$?

Draw the diagrams after working the next three problems.

4. Seventeen percent of what number is 952?

5. What percent of 300 is 60?

6. Thirty-eight percent of 700 is what number?

Solve:

7. $-4x - x - 3(x - 2) = 4 - (x - 2)$

8. $1.591 + .003k - .002 + .002k = -(.003 - k)$

Add. Write the answer in descending order of the variable:

9. $(4x^2 - 2x + 7x^5) - 2(x - 4 + 2x^2 - 3x^4)$

Multiply:

*10. $(4x + 5)(3x - 2)$ *11. $(4x + 2)(x - 5)$

*12. $(4x + 2)(3x - 5)$ *13. $(3x + 2)^2$

*14. $(4x - 2)(x^2 + x + 4)$ 15. $(2x - 1)^2$

16. Find the least common multiple of 15, 175, and 225.

Add:

17. $\dfrac{1}{2a} + \dfrac{k}{4a^2} + \dfrac{x}{8m^2a}$ 18. $\dfrac{m}{x^2y} + \dfrac{4}{yx^2} - \dfrac{3y}{x^4}$

19. $\dfrac{1}{c} + \dfrac{x}{mc^2} + d$ 20. $\dfrac{x}{ya} + \dfrac{b}{xa^2} - k$

21. Graph $x \nleq 5$.   22. Factor: $xy - 4x^2y^2m - 3x^2y$

Simplify:

23. $\dfrac{3x^4 - 3x^2}{3x^2}$ 24. $\dfrac{x^{-4}yy^{-3}x^0x^2}{x^{-3}y^3y^2x^{-4}}$

25. Expand: $\dfrac{x}{y^{-1}}\left(\dfrac{x}{y} - \dfrac{3x^2}{xy}\right)$

26. Simplify by adding like terms: $\dfrac{x^2y}{p^{-3}} - \dfrac{4x^2p^3}{y^{-1}} - \dfrac{2xp}{y^{-1}p^2} - \dfrac{5y}{p^{-3}x^{-2}}$

Evaluate:

27. $-x - x^2 - xy$ if $x = -2, y = 5$

28. $-x(x - y^0)|y|$ if $x = -1, y = -4$

Simplify:

29. $\dfrac{1}{-4^{-2}}$ 30. $-3[(-2^0 + 5) - (-3 + 7) - |-2|]$

LESSON 52   *Percents greater than 100*

52.A

percents greater than 100

When a problem discusses a quantity that increases, the final quantity is greater than the initial quantity. If we let the initial quantity represent 100 percent, the final percent will be greater than 100. This means that the "after" diagram representing the final quantity will be larger than the "before" diagram. To demonstrate, we will work three problems of this type. We will finish each problem by drawing diagrams that give a visual representation of the problem.

example 52.1 What number is 160 percent of 60? Find the number and then draw diagrams that depict the problem.

solution We substitute *WN* for *is*, 160 for *p*, and 60 for *of*.

$$\frac{P}{100} \times of = is \longrightarrow \frac{160}{100} \times 60 = WN \longrightarrow \frac{9600}{100} = WN \longrightarrow 96 = WN$$

Thus, our diagrams show a before of 60 and an after with 96, which is 160 percent.

Before, 100%   After

example 52.2 If 75 is increased by 150 percent, what is the result? Work the problem and then draw diagrams that depict the problem.

solution **We must be careful here. Seventy-five is the original number and is 100 percent. If we increase the percentage by 150 percent, the final percentage will be 250 percent.** We can restate the problem as: 250 percent of 75 is what number? We will use 250 for percent, 75 for *of*, and *WN* for *is*.

$$\frac{250}{100} \times 75 = WN \longrightarrow 2.5 \times 75 = WN \longrightarrow 187.5 = WN$$

Thus, our diagrams show a before of 75 and an after with 187.5, which is 250 percent.

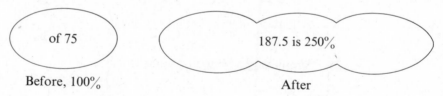

Before, 100%   After

example 52.3 What percent of 90 is 306? Work the problem and then draw diagrams that depict the problem.

We use *WP* for *P*, 90 for *of*, and 306 for *is*.

$$\frac{P}{100} \times of = is \longrightarrow \frac{WP}{100} \times 90 = 306 \longrightarrow \frac{100}{90} \cdot \frac{WP}{100} \times 90 = 306 \cdot \frac{100}{90}$$

$$\longrightarrow WP = 340\%$$

Thus the diagram of the problem should show a before of 90 and an after with 306, which is 340 percent of 90.

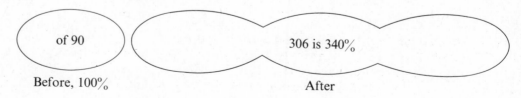

Before, 100%   After

1. Max and Coon found that the sum of twice a number and 5 is 13 less than the opposite of the number. Find the number.

2. The ratio of quotidian chores to mundane chores was 7 to 2. If 3780 chores were considered, how many were quotidian?

3. What decimal part of 1.07 is 2.1721?

Work each problem, and then draw the diagram.

*4. What number is 160 percent of 60?

*5. If 75 is increased by 150 percent, what is the result?

6. What percent of 50 is 700?

Solve:

7. $5 + 3\frac{1}{2}x = 2\frac{1}{4}$
8. $3x - 5(x - 4) = 2x + 7$

9. $-2[(-k - 3)(-2) - 3] = (-3 - 3k)(-2)^3 - 3^2$

Add. Write the answer in descending order of the variable:

10. $4(x^2 - 3x + 5) - 2(x^3 + 2x^2 - 4) - (2x^4 - 3x^3 + x^2 + 3)$

Multiply:

11. $(2x + 4)(5x - 3)$
12. $(x + 3)^2$

13. $(x + 3)(3x - 4)$
14. $(2x + 7)(2x - 7)$

15. Find the least common multiple of 24, 60, and 450.

Add:

16. $\dfrac{a}{x} + \dfrac{b}{cx^2} + d$
17. $\dfrac{m}{p^2k} - \dfrac{4a}{3pk} + \dfrac{6}{5pk^2}$

18. $a + \dfrac{bc}{m} - \dfrac{4mc}{x^2}$
19. $\dfrac{x}{mc} + \dfrac{b}{c} - \dfrac{4}{2kc^2}$

20. Graph $4 \le x < 9$ on a number line.

21. Factor: $4x^2ym - 6xym + 2x^2y^2m^2$

Simplify:

22. $\dfrac{5x - 5}{5}$
23. $\dfrac{x^3y^{-4}p^0y^4p^2}{x^4xx^{-7}y^2p^4}$

24. Expand: $\left[\dfrac{ax^{-5}}{y^{-2}} + \dfrac{4x^3}{ay^2}\right]\dfrac{x^5}{ay^2}$

25. Simplify by adding like terms: $\dfrac{m^2x}{y^{-1}} - \dfrac{3m^2y}{x^{-1}} + 5mmyx - \dfrac{4x^2ym^2}{x}$

Evaluate:

26. $-a(-a^0 + ax)$ if $a = -3, x = 7$

27. $(p - a)(a - 2pa)$ if $a = -2, p = 5$

Simplify:

28. -3^{-4}
29. $\dfrac{1}{-2^{-4}}$

30. $4 - [5(6 - 5^0) - (-5 - 2) - |-7|]$

LESSON 53   *Rectangular coordinates*

53.A
first-degree equations in two unknowns

In Lesson 50, it was said that the **degree of a term** of a polynomial is the sum of the exponents of the variables of the term. Thus

$$2x^3 \quad \text{is a third-degree term}$$
$$x \quad \text{is a first-degree term}$$
$$xyz \quad \text{is a third-degree term}$$

Also we said that the **degree of a polynomial** is the same as the degree of its highest-degree term. Thus

$$x^3 + xy + m \quad \text{is a third-degree polynomial}$$
$$2x + 4y \quad \text{is a first-degree polynomial}$$
$$2x \quad \text{is a first-degree polynomial}$$

If two polynomial expressions are connected by an equals sign, we call the equation a **polynomial equation. The degree of a polynomial equation is the same as the degree of its highest-degree term.** Thus

$$x^4 - 3x^2 + 2 = 0 \quad \text{is a fourth-degree polynomial equation}$$
$$xyz + y = 4 \quad \text{is a third-degree polynomial equation}$$
$$y = 2x + 4 \quad \text{is a first-degree polynomial equation}$$

There is an infinite number of pairs of values of x and y that are solutions to any first-degree polynomial equation in two variables. We will use the equation

$$y = 2x + 4$$

to investigate. If we assign a value to x, the equation will then indicate the value of y that is paired with the assigned value of x. For instance, if we assign to x a value of 2, then

$$y = 2(2) + 4 \quad \longrightarrow \quad y = 8$$

the paired value of y is 8. If we give x a value of 2 and y a value of 8 in the original equation, we find that these values of x and y satisfy the equation and are solutions to the equation because the replacement of the variables by these numbers makes the equation a true statement.

$$y = 2x + 4$$
$$8 = 2(2) + 4$$
$$8 = 4 + 4$$
$$8 = 8 \quad \text{True}$$

If, in the original equation, we give x a value of -5, then

$$y = 2(-5) + 4 \quad \longrightarrow \quad y = -10 + 4 \quad \longrightarrow \quad \mathbf{y = -6}$$

we find that the paired value of y is -6. Thus the pair of values $x = -5$ and $y = -6$

will also satisfy the equation, for the use of both of these numbers in the original equation in place of x and y will cause the equation to become a true equation, as shown here.

$$y = 2x + 4$$
$$-6 = 2(-5) + 4$$
$$-6 = -10 + 4$$
$$-6 = -6 \qquad \text{True}$$

We can replace x with any real number and use the equation to find the value of y that the equation pairs with this value of x.

In both of the foregoing examples we **assigned** a value to the variable x. The variable x, to which we assign a value, is called the **independent variable.** We see that, in each case, the value of y **depended** on the value that we assigned to x. Therefore, in our examples, we call the variable y the **dependent variable.** We could have assigned a value to y and then used the equation to find the corresponding value of x, in which case y would be the independent variable and x would be the dependent variable. **It is customary, however, to avoid confusion, to use the letter x to designate the independent variable and to use the letter y to designate the dependent variable. We will follow this custom in this book.**

53.B
ordered pairs

On the last page, we found that, given the equation

$$y = 2x + 4$$

if we let $x = 2$ and $y = 8$, this pair of values of x and y will make the equation a true equation. Also the pair of values $x = -5$ and $y = -6$ will make the equation a true equation. Since writing $x = 2$ and $y = 8$ and writing $x = -5$ and $y = -6$ is rather cumbersome, it is customary to write just $(2, 8)$ and $(-5, -6)$, **with the x value always designated by the first number in the parentheses and the y value always designated by the second number.** Since the numbers are written in order with x first and y second, we designate this notation as an **ordered pair** of x and y. The general form of an ordered pair of x and y is (x, y). If two other variables are used instead of x and y, it is necessary to designate which variable will be represented by each of the entries in the parentheses. If the variables m and p are to be used and we wish to write the m value first, we could designate this at the outset by making a statement about the ordered pair (m, p).

It is important to remember that in ordered pairs of x and y, the first number will always designate the value of x and the second number will always designate the value of y.

53.C
cartesian coordinate system

In Lesson 37 we learned that the solution of an equation or inequality in one variable can be presented in graphical form on a single number line as we show here.

$x = 4$

$x \geq -2$

The graphical solution to equations or inequalities that contain two variables cannot be displayed on a single number line. We must have one number line for one of the variables and another number line for the other variable. It is customary to use the variables x and y and draw the x number line horizontally, and the y number line vertically, and to let the number lines intersect at the origin of both number lines. The positive values of x are located to the right of the origin on the horizontal or x number line, and the positive values of y are located above the origin on the vertical or y number line. The x number line is called the **x axis** or **horizontal axis** and the y number line is called the **y axis** or **vertical axis.**

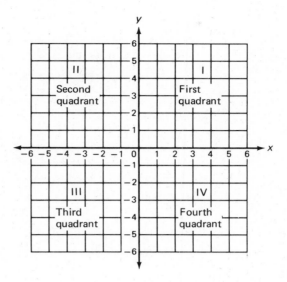

The figure shows that the two number lines divide the plane into four *quarters*, or *quadrants*. The quadrants are named the **first quadrant, second quadrant, third quadrant,** and **fourth quadrant** as shown. The figure in its entirety is called a system of **Cartesian coordinates** or a **Cartesian coordinate system** after the famous seventeenth-century French philosopher and mathematician René Descartes. It is also called a **rectangular coordinate system** and is sometimes called a **coordinate plane.**

When we use a single number line to graph the solution to an equation in one variable such as $x = 2$ (below), we call the mark we make on the number line the **graph** of the point, and conversely we call the number the **coordinate** of the point on the line designated by the graph. The point is defined to be without size, and thus the graph is not the point but denotes the location of the point.

Since a rectangular coordinate system has two number lines, **it is necessary to associate with every point on a rectangular coordinate system two numbers or coordinates.** These numbers designate the location of the point. The following figure shows the graphs of four points. Written by each point is the ordered pair of values of x and y that we associate with the point; these numbers are the x and y coordinates of the point. **The number written first is always the x coordinate of the point and is called the abscissa of the point.** This number denotes the measure of the distance of the point to the right ($+$) or left ($-$) of the vertical axis. **The number written second is always the y coordinate of the point and is called the ordinate of the point.** This number denotes the measure of the distance of the point above ($+$) or below ($-$) the horizontal axis.

example 53.1   Graph the points whose coordinates are: (a) $(3, 3)$, (b) $(-4, 0)$, (c) $(0, -3)$, (d) $(4, -2)$.

solution   In the figure below we place four dots, one to mark the location of each point.

In the first quadrant we show the graph of the point (3, 3), which is **3 units to the right** of the *y* axis and thus has an ***x* coordinate of +3.** The point is also **3 units above** the *x* axis and thus has a ***y* coordinate of +3.** The next point, (−4, 0), is located 4 units to the left of the *y* axis and thus has an *x* coordinate of −4. The point is no units above or below the *x* axis (it is on the *x* axis) and thus has a *y* coordinate of 0. The exact positions of the other two points shown are designated by the ordered pairs (0, 3) and (4, −2). Thus we see that the location of every point on the coordinate plane can be designated by starting the *x* and *y* coordinates of the point.

We have seen that a point on a number line is without size and that the graph of a point is not the point but only denotes the location of the point. In the same way, a point in a rectangular coordinate system is without size, and the graph of the point is not the point but designates the location of the point.

problem   **1.**   Gwen and Norma found that the sum of 4 times the opposite of a number and −3
set 53   is 27 larger than the number. What is the number?

2.   Pragmatism was on the rise and thus the ratio of pragmatists to quixotics was 7 to 4. If 14,740 delegates attended the convention, how many were pragmatists?

3.   What fraction of $2\frac{1}{8}$ is $\frac{1}{4}$?

Work each problem and then draw the diagrams.

4.   One hundred and thirty percent of what number is 78?

5.   Fifteen percent of what number is 10.5?

6.   What percent of 450 is 288?

Solve:

7. $3\frac{2}{5}x - 3 = 4\frac{1}{8}$

8. $.4m - 2 - .2m = 1.4 + m$

Multiply:

9. $(3x - 2)(x + 4)$

10. $(5x - 3)^2$

11. $(2x - 5)(3x - 2)$

12. $(5x - 7)(6x - 1)$

Graph the ordered pairs of x and y on a rectangular coordinate system:

*13. $(3, 3)$ ***14.** $(4, -2)$ **15.** $(-3, 4)$ **16.** $(-1, -3)$

Add:

17. $4 + \dfrac{2}{x^2} - \dfrac{5}{a^2 x}$

18. $\dfrac{p}{a^2 m} - \dfrac{4}{a} - k$

19. $\dfrac{3ax}{m} + \dfrac{4x}{am^2} + \dfrac{2}{mx}$

20. $\dfrac{2a}{x} - \dfrac{5}{p^2 x} - 3m$

21. Graph $x \nleq 4$ on a number line.

Simplify:

22. $\dfrac{3ax - 3a}{3a^2}$

23. $\dfrac{myy^{-3}m^{-4}y^{-2}}{x^0 y^2 y^{-4} y^2 m^{-7}}$

24. Expand: $\dfrac{3x^{-4}}{y^4}\left[\dfrac{2x^{-4}}{y^4} - \dfrac{3x^2}{y^2 a}\right]$

25. Simplify by adding like terms: $\dfrac{3x^2 y^{-2}}{m^5} - \dfrac{3x^2 y^2}{m^5} - \dfrac{4xx^3 m^{-5}}{x^2 y^2} + \dfrac{6m^{-5}}{x^{-2} y^{-2}}$

Evaluate:

26. $-xa(a - xa^0)$ if $a = -3$, $x = 7$

27. $(xa - a)(-ax)$ if $a = -3$, $x = 7$

Simplify:

28. $-(-2)^{-2}$

29. $\dfrac{1}{(-2)^{-3}}$

30. $4[(6 - 2^0) - (-3 + 5)2]$

LESSON 54   *Graphs of linear equations*

54.A
graphs of linear equations

If we use a rectangular coordinate system to graph the ordered pairs that satisfy a first-degree polynomial equation in two variables, we find that the graph is the graph of a straight line. For this reason, we call a first-degree polynomial equation in one or more variables a **linear equation.** The following equations are examples of linear equations in two variables,

$$y - 2x + 1 = 0 \qquad x - 4 = -2y \qquad x + y = 0$$

for the graph of each of the equations is a straight line.

To graph a linear equation in two variables, we need to know only two ordered pairs of x and y that satisfy the equation since two points adequately determine the graph of a line. But since the topic is a new one, we will learn how to find three or more ordered pairs of x and y that lie on the line.

example 54.1 Graph $y = 2x - 1$.

solution We begin by making a table and choosing convenient values for x. The values chosen for x should not be too close together. Also they should not be so large that they will graph off of our coordinate system. Numbers such as 0, 3, -3, 2, and -2 usually work well.

x	0	3	-3	2	-2
y					

Now we will use the numbers 0, 3, -3, 2, and -2 one at a time as replacement values for x in the equation $y = 2x - 1$ to find the paired values of y.

WHEN $x = 0$	WHEN $x = 3$	WHEN $x = -3$	WHEN $x = 2$	WHEN $x = -2$
$y = 2(0) - 1$	$y = 2(3) - 1$	$y = 2(-3) - 1$	$y = 2(2) - 1$	$y = 2(-2) - 1$
$y = -1$	$y = 5$	$y = -7$	$y = 3$	$y = -5$

Next we complete the table by entering the values of y that the equation has paired with the chosen values of x.

x	0	3	-3	2	-2
y	-1	5	-7	3	-5

Thus we have found five ordered pairs of x and y that satisfy the equation and therefore lie on the graph of the equation. These ordered pairs are $(0, -1)$, $(3, 5)$, $(-3, -7)$, $(2, 3)$, and $(-2, -5)$. In the next figure we have graphed the points designated by these ordered pairs and have connected them with a straight line. The point $(-3, -7)$ was not graphed since this point fell outside the borders of our coordinate system.

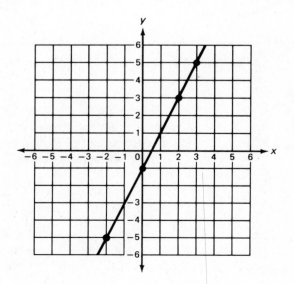

Since a line (straight line) in mathematics is defined to be infinite in length, we have graphed only a segment of the line, or a **line segment.** Also, we say that a mathematical line has no width and since the line we have drawn has a width that can be measured, it is not a mathematical line but is a **graph of a mathematical line** and indicates the location of the mathematical line in question.

example 54.2   Graph the equation: $y = -\dfrac{1}{2}x + 2$.

solution   We begin by making a table and choosing convenient values for x.

x	0	2	-2
y			

Now we find the values of y that the equation $y = -\dfrac{1}{2}x + 2$ pairs with the chosen values of x.

WHEN $x = 0$ WHEN $x = 2$ WHEN $x = -2$

$y = -\dfrac{1}{2}(0) + 2$ $y = -\dfrac{1}{2}(2) + 2$ $y = -\dfrac{1}{2}(-2) + 2$

$y = 2$ $y = 1$ $y = 3$

We insert 2, 1, and 3 to complete the table.

x	0	2	-2
y	2	1	3

We finish by graphing the points and drawing the line through the points.

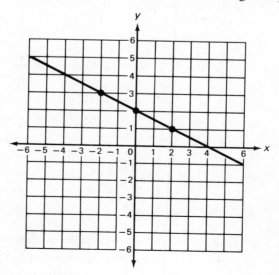

example 54.3   Graph the equation $y = -x$.

solution   (a)   Make a table.

x	0	3	−3
y			

(b)   Find the value of *y* for each value of *x*.

WHEN $x = 0$ WHEN $x = 3$ WHEN $x = -3$

$y = -(0)$ $y = -(3)$ $y = -(-3)$

$y = 0$ $y = -3$ $y = 3$

(c)   Complete the table.

x	0	3	−3
y	0	−3	3

(d)   Graph the points and draw the line.

problem   **1.**   If the product of 5 and a number is decreased by 10 and this difference is multiplied
set 54   by 3, the result is 22 less than the opposite of the number. What is the number?

2.   The mourners wailed and ululated. If the ratio of the former to the latter was 2
to 21 and 805 mourners were in attendance, how many merely wailed?

3.   What fraction of $3\frac{1}{5}$ is $2\frac{3}{4}$?

Work the problems and then draw the diagrams:

4.   Three hundred seventy-five percent of 1300 is what number?

5.   Sixty-five percent of what number is 260?

6.   What percent of 18 is 27?

Solve:

7. $-1\frac{2}{9} + 2\frac{1}{5}p = -\frac{1}{3}$

8. $3x - [-(-2)]x + (-3)(x + 2) = 5x + (-7)$

Multiply:

9. $(5x - 4)(2x + 3)$ **10.** $(-5x - 2)(-x + 4)$ **11.** $(7x - 5)^2$

Graph on a rectangular coordinate system:

*12. $y = 2x - 1$ ***13.** $y = -\frac{1}{2}x + 2$ ***14.** $y = -x$

15. $y = -2x + 3$ **16.** $y = x - 3$

Add:

17. $4 + \dfrac{x}{y^2} - \dfrac{2a}{xy^3}$ **18.** $\dfrac{a}{x^2c} - \dfrac{3a}{x^3c^2} - \dfrac{4}{xc} + \dfrac{3a + 2}{x^3c}$

19. $\dfrac{b}{m^2p} - \dfrac{1}{c^2m^3} + 4$ **20.** $x + \dfrac{4}{2x^2p^5} + \dfrac{xy}{4xp}$

21. Graph $-2 < x \le 4$ on a number line.

Simplify:

22. $\dfrac{4xy^2 - xy^2}{xy}$ **23.** $\dfrac{m^2p^4x^{-2}x^2x^0p^6}{m^2p^{-4}x^0p^0x^2}$ **24.** $\dfrac{x^3y^0x^{-1}p^2y^{-2}}{ppx^{-4}x^6y^{-2}}$

25. Expand: $\dfrac{2x^{-4}}{y^2}\left[\dfrac{x^4}{2y^{-2}} - \dfrac{6x^4}{y^7p}\right]$

26. Simplify by adding like terms: $\dfrac{x}{y} - \dfrac{3x^2x^{-1}y^2}{y^3} + \dfrac{2x^2}{xy^2} - \dfrac{4xxy^{-1}}{xy}$

Evaluate:

27. $p - x(p^2 - x^0)$ if $p = -2, x = 5$ **28.** $x - ax(x - ax)$ if $x = -4, a = 1$

Simplify:

29. $\dfrac{1}{-3^{-2}} - [3 - (-3)^3]$ **30.** $-3[-2(-2^0 - 5) - (3 - 2) - |3|]$

LESSON 55 *Vertical and horizontal lines*

55.A

vertical and horizontal lines

In Lesson 54 we graphed the following equations.

$$y = 2x - 1 \qquad y = -\frac{1}{2}x + 2 \qquad y = -x$$

We note that each of these equations has both an x term and a y term. We also note that none of the graphs was a vertical line or a horizontal line. **Some equations of a straight line contain either an x term or a y term but not both an x term and a y term, and the graph of these equations is either a vertical line or a horizontal line.**

example 55.1 Graph $y = 4$.

solution This equation tells us that the y coordinate of every point on the line is 4. On the line, we indicate three points. Note that each of these points has a y-coordinate of 4.

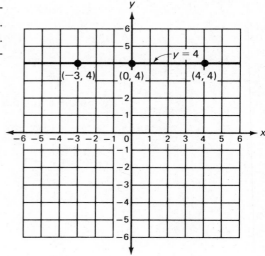

example 55.2 Graph $x = -2$.

solution This equation places no restriction on the y coordinates of the points on the line. It tells us that the x coordinate of every point on the line is -2 regardless of the y coordinate of the point. Note that each point graphed has an x coordinate of -2.

We will remember that equations of the form

$$y = \pm k \quad \text{or} \quad x = \pm k$$

such as

$$y = 2, \quad y = -4, \quad x = 5, \quad x = -4$$

are special forms of the equation of a straight line. Equations of the form

$$y = \pm k$$

are horizontal lines, while equations of the form

$$x = \pm k$$

are vertical lines.

example 55.3   Graph $y = 2x$ and $y = 2$ on the same coordinate system.

solution   These equations are often confused. The equation $y = 2$ is the equation of a horizontal line, while the equation $y = 2x$ is not.

$$y = 2x$$

x	0	2	-2
y	0	4	-4

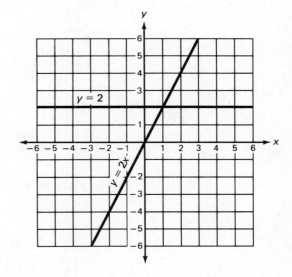

problem   **1.**   If the product of a number and -5 is increased by 6, the result is 2 less than 3
set 55   times the opposite of the number. Find the number.

2.   Unfortunately, the ratio of the erudite to the unlettered was 2 to 7. If there were 3717 under consideration, how many were erudite?

3.   What fraction of $\dfrac{1}{5}$ is $2\dfrac{7}{8}$?

Work the problems and then draw the diagrams:

4.   Four hundred-sixty percent of 700 is what number?

5.   Ninety-three percent of what number is 651,000?

6.   What percent of 2000 is 10?

Solve:

7. $3\dfrac{1}{6}k + \dfrac{3}{4} = 2\dfrac{1}{5}$

8. $-(-x) - 3(-2x) + 3(-2 + x) = -(x - 5)$

Multiply:

9. $(3x + 5)(2x + 7)$ **10.** $(4x - 3)^2$

11.   Find the least common multiple of 1575, 25, and 14.

Graph these lines on a rectangular coordinate system:

*12. $y = 4$ ***13.** $x = -2$ **14.** $y = -3$

*15. $y = 2x$ **16.** $y = \dfrac{1}{3}x - 3$

Add:

17. $x + \dfrac{2x}{m^2} - \dfrac{3}{m^2 x^3}$

18. $\dfrac{x}{k^2 p} - \dfrac{3x}{k^2 p} + 7$

19. $-3x + \dfrac{2}{xp^2} - \dfrac{5x}{x^3 p}$

20. $\dfrac{k}{4m^2} + \dfrac{k^2}{8} - 3$

21. Graph $-3 \not\leq x$ on a number line.

22. Factor: $4k^2 pz - 6k^3 p^2 z^5 - 2k^2 p^2 z^2 - 4kp$

Simplify:

23. $\dfrac{6xay - 24xay^2}{6xay}$

24. $\dfrac{k^2 m^{-2} pk^{-2} k^5 p^0}{p^5 k p^{-5} k^{-4} m^4}$

25. Expand: $\left[\dfrac{x^2 m^2}{3} - \dfrac{5x^5 p^0}{m^{-4}} \right] \dfrac{3x^{-2} y^0}{m^2}$

26. Simplify by adding like terms: $\quad a^2 k^2 y^{-1} - \dfrac{4k^2}{a^{-2} y} + \dfrac{2k^2 a}{a^{-1} y} - \dfrac{6k^{-4}}{k^2 y}$

Evaluate:

27. $mx - m(x^2) - m^0 \qquad$ if $m = -2, x = -5$

28. $p - pm(m - pm) \qquad$ if $m = -2, p = 4$

Simplify:

29. $27(-3)^{-3} - 5^2$

30. $-2(-1) - 3[(2 - 4^0) - 2(-3 - 5)] + |2|$

LESSON 56   *Addition of rational expressions*

56.A
**addition
of rational
expressions**

If the denominator of a term in an addition problem is in the form of a sum, this sum must be a factor of the least common multiple of the denominators.

example 56.1   Add $\dfrac{a}{x} + \dfrac{b}{x + y}$.

solution   The least common multiple of the denominators is $x(x + y)$, and we will use this expression as the new denominator. We will use the **denominator-numerator same-quantity theorem** to find the new numerators.

$$\dfrac{}{x(x + y)} + \dfrac{}{x(x + y)}$$

The original denominator of the first term was x, and it has been multiplied by $(x + y)$, so the original numerator, a, must also be multiplied by $x + y$.

$$\dfrac{a(x + y)}{x(x + y)} + \dfrac{}{x(x + y)}$$

The original denominator of the second term was $x + y$, and it has been multiplied by x, so the original numerator, b, must also be multiplied by x. Then we add the numerators.

$$\frac{a(x + y)}{x(x + y)} + \frac{xb}{x(x + y)} = \frac{a(x + y) + xb}{x(x + y)}$$

There are many equally correct forms of the answer. We may multiply out in either the numerator or denominator or both. Thus all three of the following forms are correct.

$$\frac{ax + ay + xb}{x(x + y)}, \qquad \frac{a(x + y) + xb}{x^2 + xy}, \qquad \frac{ax + ay + xb}{x^2 + xy}$$

example 56.2   Add $\dfrac{4x + a}{a + b} + \dfrac{c}{x}$.

solution   We begin by writing the fraction lines and the new denominators. Then we find the new numerators and add.

$$\frac{}{x(a + b)} + \frac{}{x(a + b)}$$

$$\frac{x(4x + a)}{x(a + b)} + \frac{c(a + b)}{x(a + b)} = \frac{4x^2 + ax + ca + cb}{x(a + b)}$$

We could multiply $x(a + b)$ in the denominator, but we decide to leave it in the more concise form.

example 56.3   Add $\dfrac{a + b}{x} + \dfrac{c}{m} + d$.

solution   We begin by writing the fraction lines with the new denominators and then we find the new numerators and add.

$$\frac{}{xm} + \frac{}{xm} + \frac{}{xm}$$

$$\frac{m(a + b)}{mx} + \frac{cx}{mx} + \frac{dxm}{mx} = \frac{m(a + b) + cx + dxm}{xm}$$

We could multiply $m(a + b)$ in the numerator, but we decide to leave it as it is.

example 56.4   Add $\dfrac{a}{b + c} - x + \dfrac{d + m}{k}$.

solution   The first step is to write the new denominators. Then we find the new numerators and add.

$$\frac{}{k(b + c)} - \frac{}{k(b + c)} + \frac{}{k(b + c)}$$

$$\frac{ka}{k(b + c)} - \frac{xk(b + c)}{k(b + c)} + \frac{(d + m)(b + c)}{k(b + c)} = \frac{ka - xk(b + c) + (d + m)(b + c)}{k(b + c)}$$

Again we choose not to use the distributive property, and we leave the answer in the more concise form.

problem set 56

1. Virginia and Weir found that if the product of 3 and a number was increased by 13, the result was 12 less than twice the opposite of the number. What was their number?

2. The apes were either prognathous or platyrrhine. If the ratio of the former to the latter was 2 to 23 and 3500 were observed, how many were prognathous?

3. 2.14 times what number is .00642?

Solve and then draw the diagrams:

4. Three hundred forty-seven percent of what number equals 2429?

5. Eighty-seven percent of 300 is what number?

6. What percent of 460 is 1150?

Solve:

7. $3\frac{1}{8}p + 2\frac{1}{4} = \frac{1}{6}$

8. $-[-|-3|(-2 - m) - 4] = -2[(-3 - m) - 2m]$

Multiply:

9. $(4x - 2)(3x + 5)$ 10. $(5x - 1)^2$ 11. $(5x - 7)(7x - 5)$

Graph on a rectangular coordinate system:

12. $y = -3$ 13. $x = 4$

14. $y = -2x + 1$ 15. $y = 3x - 4$

Add:

16. $\dfrac{3}{2x^2y} - \dfrac{ab}{4x^3y} - c$ *17. $\dfrac{a}{x} + \dfrac{b}{x + y}$ *18. $\dfrac{4x + a}{a + b} + \dfrac{c}{x}$

*19. $\dfrac{a + b}{x} + \dfrac{c}{m} + d$ 20. $\dfrac{a}{b + c} - x + \dfrac{4}{b^2}$

21. Graph $4 \le x < 7$ on a number line.

22. Factor: $9k^2bm^4 - 3kb^4m^2 + 12kb^3m^3$

Simplify:

23. $\dfrac{3ap^2m - 6ap^2m^2}{3ap^2m}$ 24. $\dfrac{m^{-2}mmm^{-3}xx^{-4}}{xm^3x^{-3}m^{-3}x^4}$

25. Expand: $\dfrac{x^2y^0p}{m^{-2}}\left[\dfrac{p^{-3}m^2}{k} - \dfrac{p^0pm^2}{x^2}\right].$

26. Simplify by adding like terms: $mx - \dfrac{3}{m^{-1}x^{-1}} + \dfrac{4mx}{m^2x^2} + \dfrac{5m^2x^{-1}}{mx^{-2}}$

Evaluate:

27. $k^2 - kp - p(-k)$ if $k = -3, p = 5$

28. $x - xy(y - x^0)$ if $x = -3, y = -1$

Simplify:

29. $\dfrac{1}{(-2)^{-3}} + 2$

30. $-\{[(-2) - 3^0] - [(2 - 3)(-2) + 3]\}$

LESSON 57 *Power theorem for exponents*

57.A
power theorem for exponents

The last eight problem sets have contained problems whose solutions have required the use of one or more of the following definitions and theorems.

DEFINITION:	$x \cdot x \cdot x \cdot x \cdots x = x^m$	(m is a whole number)
DEFINITION:	$x^0 = 1$	($x \neq 0$)
DEFINITION:	$x^1 = x$	
DEFINITION:	$x^{-n} = \dfrac{1}{x^n}$	($x \neq 0$)
PRODUCT THEOREM:	$x^m \cdot x^n = x^{m+n}$	($x \neq 0$)
QUOTIENT THEOREM:	$\dfrac{x^m}{x^n} = x^{m-n} = \dfrac{1}{x^{n-m}}$	($x \neq 0$)

To complete the list of rules for exponents, we will introduce the **power theorem for exponents,** which is a logical extension of the first definition listed above. By using this definition, we can show that x is to be used as a factor three times by writing x^3

$$x \cdot x \cdot x = x^3$$

If we wish to show that x^5 should be used as a factor three times, we could write $(x^5)^3$:

$$x^5 \cdot x^5 \cdot x^5 = (x^5)^3$$

Since $x^5 \cdot x^5 \cdot x^5$ equals x^{15}, then $(x^5)^3$ must also equal x^{15}.

$$(x^5)^3 = x^{15}$$

Thus, when an exponential has one base and two exponents as shown here, we simplify by multiplying the exponents. We call this rule the power theorem for exponents.

> POWER THEOREM FOR EXPONENTS
>
> If m and n and x are real numbers and $x \neq 0$
>
> $$(x^m)^n = x^{mn}$$

EXAMPLES

(a) $(a^5)^2 = a^{10}$ (b) $(x^{-4})^{-2} = x^8$ (c) $(m^{-2})^4 = m^{-8}$

(d) $(p^{-7})^{-7} = p^{49}$ (e) $(x^3)^5 = x^{15}$ (f) $(m^k)^2 = m^{2k}$

The use of the power theorem in expressions such as the above usually gives little trouble, but the use of this theorem to simplify expressions such as $(2x^5y^2z)^3$ is more difficult. We know that the notation indicates that $2x^5y^2z$ is to be used as a factor three times because 3 is the exponent of the whole expression.

$$(2x^5y^2z)(2x^5y^2z)(2x^5y^2z)$$

Since the order of multiplication of factors of a product does not affect the value of the product, we will rearrange the order of multiplication to get

$$2 \cdot 2 \cdot 2 \cdot x^5 \cdot x^5 \cdot x^5 \cdot y^2 \cdot y^2 \cdot y^2 \cdot z \cdot z \cdot z$$

which could be written by use of the product theorem as

$$2^3 x^{15} y^6 z^3 \quad \text{which equals} \quad \mathbf{8x^{15}y^6z^3}$$

To obtain the same result by using the power theorem

$$(2x^5y^2z)^3$$

we would have to multiply the exponent of each factor of the given term by 3 or raise each factor of the given term to the third power.

$$(2x^5y^2z)^3 = (2)^3(x^5)^3(y^2)^3(z)^3 = 8x^{15}y^6z^3$$

This is clear to the mathematician from the notation that $(x^m)^n = x^{mn}$, but it is sometimes not clear to the student at this level. Therefore, we will state it as follows:

To raise a term that contains no indicated additions to a given power, the exponent indicating the power is multiplied by the exponent of every factor of the numerator and the denominator (if there is one) **of the term.** For example,

(a) $\left(\dfrac{3x^{-2}y^5}{z^4}\right)^{-2} = \dfrac{x^4y^{-10}}{9z^{-8}}$

(b) $(2a^{-2}b^2z^{-10})^{-5} = \dfrac{a^{10}b^{-10}z^{50}}{32}$

(c) $\left(\dfrac{4xy}{m^{-2}}\right)^2 = \dfrac{4^2x^2y^2}{m^{-4}} = \dfrac{16x^2y^2}{m^{-4}}$

(d) $\left(\dfrac{3xy^{-2}}{p^5}\right)^{-2} = \dfrac{3^{-2}x^{-2}y^4}{p^{-10}} = \dfrac{x^{-2}y^4}{9p^{-10}}$

(e) $\left(\dfrac{3x^0y}{k^4}\right)^3 = \dfrac{27y^3}{k^{12}}$

(f) $\left(\dfrac{2^0x^{-2}y^4}{z}\right)^{15} = \dfrac{x^{-30}y^{60}}{z^{15}}$

problem set 57

1. If the sum of a number and -5 is multiplied by -3, the result is 2 greater than 9 times the opposite of the number. What is the number?

2. The proboscidians led the parade of pachyderms. If the ratio of proboscidians to nonproboscidians was 13 to 5 and there were 756 pachyderms in the parade, how many were proboscidians?

3. What fraction of $3\dfrac{7}{8}$ is $\dfrac{1}{4}$?

Solve and then draw the diagrams:

4. Three hundred-twenty percent of what number equals 192?

5. What percent of 98 is 3.92?

6. If 72 is increased by 130 percent, what is the result?

Solve:

7. $\dfrac{1}{4} + 2\dfrac{1}{5}k + 3\dfrac{2}{9} = 0$

8. $-[-(-k)] - (-2)(-2 + k) = -k - (4k + 3)$

Multiply:

9. $(2x - 4)(x - 3)$

10. $(3x + 5)^2$

Graph on a rectangular coordinate system:

11. $x = -1\dfrac{1}{2}$

12. $y = 2x + 2$

13. $y = -\dfrac{1}{3}x - 2$

Add:

14. $\dfrac{x + y}{x^2y} + \dfrac{y}{x^4}$

15. $\dfrac{4}{x - y} - \dfrac{3}{y}$

16. $\dfrac{9 + 2b}{x} - 3 + \dfrac{6}{x^2y}$

17.   Graph $4 \nleq x$ on a number line.

18.   Factor: $12x^4yp^3 - 4x^3y^2pz - 8x^2p^2y^2$

Simplify:

19. $\dfrac{mx - 5mx}{mx}$ *20. $\left(\dfrac{3x^{-2}y^5}{z^4}\right)^{-2}$ *21. $(2a^{-2}b^2z^{-10})^{-5}$

22. $(2x^2y^{-2}z)^{-2}(xy)^4$ 23. $(5x^{-3}y^2)^2(x^0y)^5xy$ 24. $(xy)x(x^{-4}y)^2(x)^3$

25.   Expand: $\dfrac{3x^{-2}}{y^{-4}}\left(\dfrac{ax^2}{3y^4} - \dfrac{3x^5}{y^{-2}}\right)$

26.   Simplify by adding like terms: $2xy - 3yx + x^2y - y^2x + 4y^2xy^{-1} - \dfrac{2x^4x^{-2}}{y^{-1}}$

Evaluate:

27. $xy - y^2(x - y)$ if $x = -2, y = 3$

28. $x^3 - x^2(-x)$ if $x = -2$

Simplify:

29. $\dfrac{1}{-2^{-3}} - 2$

30. $-2\{[2 - 4] - [3 - 6^0][-2] - 2[(-3)(-2 - 1)]\}$

LESSON 58   *Substitution axiom*

58.A
substitution axiom

Since Lesson 13 we have been finding the value of a particular expression that contains variables by assigning a value to each variable. Thus the value of

$$\frac{x^2y}{p} \qquad \text{when } x = 2, y = 4, \text{ and } p = -1 \text{ is} \qquad \frac{(2)^2(4)}{(-1)} = \frac{16}{-1} = -16$$

To do this we have been using what is usually called the **substitution axiom.**

The substitution axiom is stated in different ways by different authors. Three statements of this axiom that are frequently used follow:

SUBSTITUTION AXIOM

1. Changing the numeral by which a number is named in an expression does not change the value of the expression.
2. For any numbers a and b, if $a = b$, then a and b may be substituted for each other.
3. If $a = $ b, then a may replace b or b replace a in any statement without changing the truth or falsity of the statement.

Definition 1 seems to apply only to individual expressions. Definition 2 is general enough but not sufficiently specific. Definition 3 seems to apply only to statements and not to individual expressions. **We will use the definition below to state formally and exactly the thought that if two expressions have equal value, it is permissible to use either expression.**

> **SUBSTITUTION AXIOM**
>
> If two expressions a and b are of equal value, $a = b$, then a may replace b or b may replace a in any expression without changing the value of the expression. Also, a may replace b or b may replace a in any statement without changing the truth or falsity of the statement. Also, a may replace b or b replace a in any equation or inequality without changing the solution set of the equation or inequality.

Thus the substitution axiom applies to expressions, equations, and inequalities. We have already been using this axiom to evaluate expressions as shown in the problem worked out at the beginning of this section. Now we will use the axiom to solve a system of first-degree linear equations in two unknowns.

58.B
simultaneous equations

If we consider the two equations

$$\text{(a)} \quad 2x - y = 1 \quad \text{and} \quad \text{(b)} \quad x = -3y + 11$$

we see that $(3, 5)$ is a solution to equation (a) and that $(5, 2)$ is a solution to equation (b).

(a) $\qquad 2x - y = 1$ $\qquad\qquad\qquad$ (b) $\quad x = -3y + 11$

$\qquad 2(3) - (5) = 1$ $\qquad\qquad\qquad\qquad\quad 5 = -3(2) + 11$

$\qquad\qquad 6 - 5 = 1$ $\qquad\qquad\qquad\qquad\qquad\quad 5 = -6 + 11$

$\qquad\qquad\qquad 1 = 1 \quad$ True $\qquad\qquad\qquad\qquad 5 = 5 \quad$ True

But neither $(3, 5)$ nor $(5, 2)$ is a solution to both equations at the same time or simultaneously. If we need one solution that will satisfy two or more equations, we call the equations a **system of simultaneous equations.** We can designate that two or more equations form a system of equations by so stating in words or by using a brace as shown here.

$$\begin{cases} 2x - y = 1 \\ x = -3y + 11 \end{cases}$$

58.C
solution of simultaneous equations by substitution

We can find the common solution to a system of two first-degree simultaneous equations in two unknowns by using the substitution axiom.

example 58.1 Solve: (a) $\begin{cases} 2x - y = 1 \\ x = -3y + 11 \end{cases}$
(b)

solution **First we must assume that an ordered pair of real numbers exists which will satisfy both of these equations and that x and y in the equations represent these real numbers.** If our assumption is correct, we can find the value of the members of this ordered pair. If our assumption is incorrect, the attempted solution will degenerate into an expression involving real numbers, such as $1 = 2$ or $4 + 2 = 6$ or $0 = 0$.[†]

[†] To be discussed in a later lesson.

Now, since both equations are assumed to be true equations and also since x in both equations stands for the number that will satisfy both equations, we can **replace the variable x in the equation (a) with the equivalent expression for x given by equation (b).**

$2x - y = 1$	equation (a)
$2(-3y + 11) - y = 1$	replaced x with $-3y + 11$
$-6y + 22 - y = 1$	multiplied
$-7y + 22 = 1$	added like terms
$-7y = -21$	added -22 to both sides
$y = 3$	divided both sides by -7

We have found that the y value of the desired ordered pair is 3. Now we may find the value of x by substituting the number 3 for y in either of the original equations.

IN EQUATION (a)	IN EQUATION (b)
$2x - y = 1$	$x = -3y + 11$
$2x - (3) = 1$	$x = -3(3) + 11$
$2x - 3 = 1$	$x = -9 + 11$
$2x = 4$	$x = 2$
$x = 2$	

Thus the ordered pair of x and y that will satisfy both equations is **(2, 3)**.

example 58.2   Solve $\begin{cases} 2x + 3y = -13 \\ y = x - 6 \end{cases}$

solution   The bottom equation states that y equals $x - 6$. Therefore, in the top equation we will replace y with $x - 6$ and solve for x.

$2x + 3y = -13$	top equation
$2x + 3(x - 6) = -13$	substituted $x - 6$ for y
$2x + 3x - 18 = -13$	multiplied
$5x = 5$	added like terms
$x = 1$	divided both sides by 5

Now the paired value for y may be found by substituting 1 for x in either of the original equations.

TOP EQUATION	BOTTOM EQUATION
$2x + 3y = -13$	$y = x - 6$
$2(1) + 3y = -13$	$y = (1) - 6$
$2 + 3y = -13$	$y = 1 - 6$
$3y = -15$	$y = -5$
$y = -5$	

Thus the solution is the ordered pair **(1, −5)**.

example 58.3   Solve $\begin{cases} 3x + 2y = 3 \\ x = 3y - 10 \end{cases}$

solution   We will replace x in the top equation with its equivalent $(3y - 10)$ from the bottom equation.

$$3x + 2y = 3 \qquad \text{top equation}$$
$$3(3y - 10) + 2y = 3 \qquad \text{replaced } x \text{ with } (3y - 10)$$
$$9y - 30 + 2y = 3 \qquad \text{multiplied}$$
$$11y = 33 \qquad \text{simplified}$$
$$\boldsymbol{y = 3} \qquad \text{divided by 11}$$

Now the paired value for x may be found by replacing y with 3 in either of the original equations.

TOP EQUATION	BOTTOM EQUATION
$3x + 2(3) = 3$	$x = 3(3) - 10$
$3x + 6 = 3$	$x = 9 - 10$
$3x = -3$	$\boldsymbol{x = -1}$
$\boldsymbol{x = -1}$	

Thus the solution is the ordered pair $\boldsymbol{(-1, 3)}$.

example 58.4   Solve $\begin{cases} -x - 2y = 4 \\ y = -3x + 8 \end{cases}$

solution   We will replace y in the top equation by its equivalent $(-3x + 8)$ from the bottom equation.

$$-x - 2y = 4 \qquad \text{top equation}$$
$$-x - 2(-3x + 8) = 4 \qquad \text{replaced } y \text{ with } (-3x + 8)$$
$$-x + 6x - 16 = 4 \qquad \text{multiplied}$$
$$5x - 16 = 4 \qquad \text{simplified}$$
$$5x = 20 \qquad \text{added 16 to both sides}$$
$$x = 4 \qquad \text{divided by 5}$$

Now we can replace x with 4 in either of the original equations to find the value of y.

TOP EQUATION	BOTTOM EQUATION
$-x - 2y = 4$	$y = -3x + 8$
$-(4) - 2y = 4$	$y = -3(4) + 8$
$-4 - 2y = 4$	$y = -12 + 8$
$-2y = 8$	$\boldsymbol{y = -4}$
$\boldsymbol{y = -4}$	

Thus the solution is the ordered pair $\boldsymbol{(4, -4)}$.

example 58.5   Solve $\begin{cases} 2x + 3y = 5 \\ x = y \end{cases}$

solution   In the top equation we replace x with its equivalent (y) from the bottom equation.

$$2x + 3y = 5 \qquad \text{original equation}$$
$$2(y) + 3y = 5 \qquad \text{substituted } y \text{ for } x$$
$$5y = 5 \qquad \text{added}$$
$$y = 1 \qquad \text{divided by 5}$$
$$\text{and since } x = y, \; \boldsymbol{x = 1}$$

Thus the solution is the ordered pair **(1, 1)**.

**problem
set 58**   Solve and then draw the diagrams:

1. Forty-seven percent of what number equals 188?

2. What percent of 68 is 95.2?

3. If 80 is increased by 210 percent, what is the result?

4. Solve: $-2[(-3 - 2) - 2(-2 + m)] = -3m - 4^2 - |-2|$

Use substitution to solve for x and y:

*5. $\begin{cases} 2x - y = 1 \\ x = -3y + 11 \end{cases}$
   *6. $\begin{cases} 2x + 3y = -13 \\ y = x - 6 \end{cases}$

*7. $\begin{cases} 3x + 2y = 3 \\ x = 3y - 10 \end{cases}$
   *8. $\begin{cases} -x - 2y = 4 \\ y = -3x + 8 \end{cases}$

*9. $\begin{cases} 2x + 3y = 5 \\ x = y \end{cases}$

10. Find the least common multiple of 175, 147, and 45.

11. Expand: $(3 - x)^2$

Graph on a rectangular coordinate system:

12. $y = -3\dfrac{1}{2}$
   13. $y = 2x - 5$
   14. $y = -\dfrac{1}{2}x + 4$

Add:

15. $\dfrac{-x}{a^2 b} + \dfrac{a - b}{b}$
   16. $\dfrac{a + b}{a} - \dfrac{3}{ax}$
   17. $4 - \dfrac{7}{a} + \dfrac{a + b}{a^2}$

18. Factor: $25x^2 y^5 p^2 - 15x^5 y^4 p^4 + 10x^4 y^4 p^4 z$

Simplify:

19. $\dfrac{5pq - 5p^2 q^2}{5pq}$
   20. $x(x^2)(x^3)^{-3} x^{-2} y^2$
   21. $(x^2 y^0)(x^3 y^2) xy$

22. $xx(x^2)^{-2}(x^2)(x^{-3} y^0)^5$
   23. $\left(\dfrac{3x^{-2} y^5}{p^{-3}} \right)^{-2} \left(\dfrac{2x^{-2}}{y} \right)^2$

24. $\dfrac{x(x^2 y^0)^{-2} xy^{-2}}{(xy^{-3})^2 xy^{-3}}$
   25.   Expand: $\left(\dfrac{x^2}{2y^{-1}} - \dfrac{4x^{-2}}{y} \right) \dfrac{2x^{-2}}{y}$

26. Simplify by adding like terms: $x^2 (p^5)^2 y - \dfrac{3x^2 p^{10}}{y^{-1}} + \dfrac{4x^3 py}{x^2 p^{-2}} - 2xp^{10} y$

Evaluate:

27. $x^2 - xy - (-y)$ if $x = 3$ and $y = -3$

28. $p^2x - xp^0(x - p)$ if $x = -3$ and $p = 3$

Simplify:

29. $\dfrac{1}{-2^{-3}} + 8$

30. $-3\{(-4 - 2) - (-2 - 1^0) - [(-2) - (-2 - 1)]\}$

LESSON 59 *Dividing fractions*

59.A
complex fractions

To review the **denominator-numerator same-quantity theorem** we will begin with the number 5. Then if we multiply 5 by 2 and divide by 2 the answer is 5

$$5 = \frac{5 \cdot 2}{2} = \frac{10}{2}$$

because $\frac{10}{2}$ is another way to write 5. We have changed the numeral, but the number it represents is unchanged because 2 over 2 has a value of 1. This is why we can multiply the denominator and the numerator of a fraction by any nonzero number without changing the value of the fraction.

Fractions of fractions are called **complex fractions.** We will simplify complex fractions by multiplying the numerator and the denominator by the same quantity.

$$\frac{\dfrac{a}{b}}{\dfrac{c}{d}} \qquad\qquad \begin{pmatrix} b \neq 0 \\ c \neq 0 \\ d \neq 0 \end{pmatrix}^{\dagger}$$

If we multiply the denominator of this complex fraction by the reciprocal of the denominator, we obtain a product of 1 because we remember that the product of a number and the reciprocal of the same number is the number 1.

$$\frac{c}{d} \cdot \frac{d}{c} = \frac{cd}{dc} = 1$$

But if we wish to multiply the denominator of the fraction by its reciprocal, which is $\dfrac{d}{c}$, we must also multiply the numerator by $\dfrac{d}{c}$ so that the value of the original expression will not be changed.

$$\frac{\dfrac{a}{b}}{\dfrac{c}{d}} = \frac{\dfrac{a}{b} \cdot \dfrac{d}{c}}{\dfrac{c}{d} \cdot \dfrac{d}{c}} = \frac{\dfrac{ad}{bc}}{\dfrac{cd}{dc}} = \frac{\dfrac{ad}{bc}}{1} = \frac{ad}{bc}$$

We have simplified the original fraction by multiplying both the denominator and the numerator by the **reciprocal of the denominator.**

example 59.1   Simplify $\dfrac{\dfrac{a}{b}}{c}$ $(b, c \neq 0)$.

---

† To ensure that we are not implying that division by zero is permissible.

solution   We multiply both the denominator and the numerator by $\dfrac{1}{c}$, which is the reciprocal of the denominator c.

$$\frac{\dfrac{a}{b}}{c} = \frac{\dfrac{a}{b} \cdot \dfrac{1}{c}}{\dfrac{c}{1} \cdot \dfrac{1}{c}} = \frac{\dfrac{a}{bc}}{\dfrac{c}{c}} = \frac{\dfrac{a}{bc}}{1} = \frac{a}{bc}$$

example 59.2   Simplify $\dfrac{\dfrac{1}{c}}{\dfrac{1}{b}}$ $(b, c \neq 0)$.

solution   We multiply both the denominator and the numerator by b, which is the reciprocal of $\dfrac{1}{b}$.

$$\frac{\dfrac{1}{c}}{\dfrac{1}{b}} = \frac{\dfrac{1}{c} \cdot \dfrac{b}{1}}{\dfrac{1}{b} \cdot \dfrac{b}{1}} = \frac{\dfrac{b}{c}}{1} = \frac{b}{c}$$

example 59.3   Simplify $\dfrac{a}{\dfrac{1}{c}}$ $(c \neq 0)$.

solution   We multiply both the numerator and the denominator by c, which is the reciprocal of $\dfrac{1}{c}$.

$$\frac{a}{\dfrac{1}{c}} = \frac{\dfrac{a}{1} \cdot \dfrac{c}{1}}{\dfrac{1}{c} \cdot \dfrac{c}{1}} = \frac{\dfrac{ac}{1}}{1} = ac$$

example 59.4   Simplify $\dfrac{\dfrac{a}{x}}{b}$ $(x, b \neq 0)$.

solution   We multiply both the denominator and the numerator by $\dfrac{1}{b}$, which is the reciprocal of b.

$$\frac{\dfrac{a}{x}}{b} = \frac{\dfrac{a}{x} \cdot \dfrac{1}{b}}{\dfrac{b}{1} \cdot \dfrac{1}{b}} = \frac{a}{xb}$$

The rule for dividing fractions is sometimes stated as follows: **To divide one fraction by another fraction, invert the fraction in the denominator and multiply.** If we use this rule to simplify

$$\frac{\dfrac{m}{n}}{\dfrac{x}{y}} \qquad (x, n, y \neq 0)$$

the solution is $\dfrac{m}{n} \cdot \dfrac{y}{x} = \dfrac{my}{nx}$

If we use the **denominator-numerator same-quantity theorem** to perform the same simplification, we obtain

$$\frac{\dfrac{m}{n}}{\dfrac{x}{y}} = \frac{\dfrac{m}{n}\cdot\dfrac{y}{x}}{\dfrac{x}{y}\cdot\dfrac{y}{x}} = \frac{my}{nx}$$

This procedure yields the same result as that obtained by using the rule, but hopefully we have some understanding of what we did and can justify our procedure.

Many algebraic manipulations can be justified by one of the following:

1. **The denominator-numerator same-quantity theorem.**
2. **The multiplicative property of equality.**
3. **The additive property of equality.**

We will remember to justify our manipulations by one of these three rules whenever possible.

problem set 59

1. If the product of a number and 2 is increased by 7, the result is 2 less than the opposite of the number. What is the number?

2. The prognosticators outscored the retrospectors in the ratio of 5 to 2. If 98 total points were scored, how many belonged to the prognosticators?

3. What fraction of $\dfrac{1}{4}$ is $3\dfrac{7}{8}$?

Solve and then draw the diagrams:

4. Seventy-two percent of what number equals 216?

5. One hundred thirty-five percent of what number equals 405?

6. If 48 is increased by 250 percent, what is the result?

Solve:

7. $2\dfrac{1}{4} + 3\dfrac{1}{5}k + \dfrac{1}{8} = 0$

8. $-[-(-p)] - (-4)(-2 - p) = -(4 - 2p)$

Use substitution to solve for both x and y:

9. $\begin{cases} 3x + 2y = 7 \\ x = 7 - 3y \end{cases}$ 10. $\begin{cases} x + 2y = -6 \\ y = 3x + 4 \end{cases}$ 11. $\begin{cases} x + y = 6 \\ x = 9 - 2y \end{cases}$

Multiply:

12. $(3x - 4)(2 - x)$ 13. $(x + 1)(x^2 + 2x + 2)$

Graph on a rectangular coordinate system:

14. $y = -3$ 15. $y = -3x - 2$ 16. $y = \dfrac{1}{2}x - 2$

Simplify:

*17. $\dfrac{\dfrac{a}{b}}{\dfrac{c}{d}}$ *18. $\dfrac{\dfrac{a}{b}}{c}$ *19. $\dfrac{\dfrac{1}{c}}{\dfrac{1}{b}}$ 20. $\dfrac{a}{\dfrac{1}{a + x}}$

Add:

21. $\dfrac{4}{x + y} - \dfrac{3}{y^2}$ 22. $\dfrac{a}{x^2y} + 4a - \dfrac{m}{x + y}$ 23. $\dfrac{x + y}{a^2} + \dfrac{y^2}{a}$

Simplify:

24. $\dfrac{(x^2)^{-3}(yx)^2x^0}{x^2y^{-2}(xy^{-2})^3}$

25. $\left(\dfrac{x^{-2}}{y^4}\right)^2\left(\dfrac{x}{y}\right)$

26. $\dfrac{p^9(3x)^2y}{x^0(x^{-1}y^{-2})^3}$

27. $\dfrac{4x^2-4}{4}$

28. Expand: $\dfrac{x^{-4}y}{p^2}\left(\dfrac{y^{-1}}{x^4}+\dfrac{2x^{-4}p^{-2}}{y^{-1}}\right)$

29. Simplify by adding like terms: $\dfrac{xy}{z}-\dfrac{3x^2y^2}{xyz}+\dfrac{2x^3x^{-2}yz^{-2}}{z^{-1}}+\dfrac{5xy}{zy^{-2}}$

30. Simplify: $-\{3(-3\cdot2^0)[-(3-2)(-2)]-|-4|\}$

LESSON 60   *Set notation · Rearranging before graphing*

60.A
finite and infinite sets

The words **finite** and **infinite** are basic words and are difficult to define. The word *finite* implies the thought of bounded or limited, while the word *infinite* implies the thought of unbounded or without limit. Thus when we say that a set has a finite number of members, we are describing a set such as

$$\{6,7,8,9,10\}$$

in which the listing of the members has an end. **A set with a finite number of members is called a finite set.**

When we say that a set has an infinite number of members, we are describing a set such as

$$\{1,2,3,4,5,\ldots\}$$

in which the listing of the members of the set continues without end. **Sets that have an infinite number of members are called infinite sets.**

Some authors define a finite set as a set whose members could be counted if we could live the number of lifetimes necessary to do the counting. Thus the set that has

$$63,072,000,000,000,000$$

members is a finite set because we could count the members of this set if we counted for 1 billion years at two counts per second (not considering leap years).

Using the same definition the set

$$\{1,2,3,4,\ldots\}$$

is an infinite set because we could never count the number of members of this set since the listing has no end.

60.B
membership in a set

If we have the numbers

$$0,0,0,0,0,0,1,1,1,1,2,2,2,2$$

and wish to designate these numbers as members of a set B, we would write

$$B=\{0,1,2\}$$

which is read "B is the set whose members are the numbers 0, 1, and 2." **Note that each of the numbers is listed only once.** Thus, if we have the set

$$D=\{0,1,2,3,4\}$$

we have said that the set consists of 0, or any number of 0s; also the set consists of 1s, 2s, 3s, and 4s, but not necessarily just one 1 and one 2 and one 3 and one 4.

example 60.1 Represent the following numbers as being members of set K: 1, 0, 2, 1, 0, 5, 7, 4, 5, 7.

solution We list each number only once. The order in which the numbers are listed is unimportant.

$$K = \{0, 1, 2, 4, 5, 7\}$$

example 60.2 Represent the following numbers as constituting set L: 7, 15, 0, 1, 15, 0, 8, -13, 42.

solution We list each number only one time. The order in which the numbers are listed is unimportant.

$$L = \{7, 15, 0, 1, 8, -13, 42\}$$

We use the symbols

$$\in \quad \text{and} \quad \notin$$

to designate that a particular symbol or number is a member of a given set or is not a member of the given set.

$$0 \in L \qquad 23 \notin L$$

We would read the above as "zero is a member of set L" and "23 is not a member of set L."

example 60.3 Given the sets $A = \{0, 1, 3, 5\}$, $B = \{0, 4, 6, 7\}$, $C = \{1, 2, 3, 5, 7\}$, are the following statements true or false? (a) $5 \in A$; (b) $4 \in C$; (c) $5 \notin B$.

solution (a) $5 \in A$ True, because 5 is a member of set A.
(b) $4 \in C$ False, because 4 is not a member of set C.
(c) $5 \notin B$ True, because 5 is not a member of set B.

example 60.4 Given the sets $M = \{0, 1, 2, 3\}$, $L = \{5, 6, 7\}$, $N = \{0, 1\}$, are the following statements true or false? (a) $6 \in M$; (b) $0 \in N$.

solution (a) False, because 6 is not a member of set M.
(b) True, because 0 is a member of set N.

60.C
rearranging before graphing

Often we encounter linear equations that have not been solved for y. Graphing these equations is easier if we first rearrange them by solving for y.

example 60.5 Graph $3x + 2y = 4$.

solution We first solve for y by adding $-3x$ to both sides and then dividing by 2.

$$3x + 2y = 4 \qquad \text{original equation}$$
$$\underline{-3x \qquad\qquad -3x} \qquad \text{add } -3x \text{ to both sides}$$
$$2y = 4 - 3x$$

$$y = 2 - \frac{3}{2}x \qquad \text{divided each term by 2}$$

$$y = -\frac{3}{2}x + 2 \qquad \text{order of terms changed}$$

Now we make a table and choose convenient values for x.

x	0	2	-2
y			

Next we find the values of y that the equation $y = -\frac{3}{2}x + 2$ pairs with 0, 2, and -2.

WHEN $x = 0$ WHEN $x = 2$ WHEN $x = -2$

$$y = -\frac{3}{2}(0) + 2 \qquad y = -\frac{3}{2}(2) + 2 \qquad y = -\frac{3}{2}(-2) + 2$$

$$y = 2 \qquad\qquad y = -1 \qquad\qquad y = 5$$

Now we complete the table, graph the points, and draw the line.

x	0	2	-2
y	2	-1	5

example 60.6   Graph $y - x = 0$.

solution   We (a) solve for y, (b) complete the table, and then (c) draw the graph.

(a) $\quad y - x = 0$

$$\underline{ + x \quad + x}$$

$$\quad\quad y = x$$

(c)

(b)

x	0	3	-3
y	0	3	-3

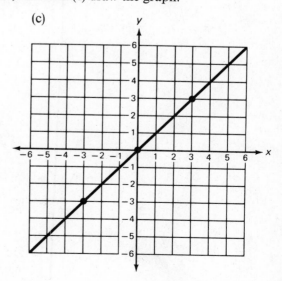

problem set 60

Work the problems and then draw the diagrams:

1. Lisa and Dennis found that 190 percent of their number equaled 76. What was their number?

2. What percent of 180 is 36?

3. Ninety percent of .4 is what number?

4. Tricia and Heather increased 170 by 60 percent. What number did they get?

*5. Represent 1, 0, 2, 1, 0, 5, 7, 4, 5, 7 as making up set K.

6. Given the sets $A = \{0, 1, 3, 5\}$, $B = \{0, 4, 6, 7\}$, $C = \{1, 2, 3, 5, 7\}$, which of the following is true?

   *(a) $5 \in A$ *(b) $4 \in C$ *(c) $5 \notin B$ (d) $3 \in B$

Use substitution to solve for x and y:

7. $\begin{cases} 2x - 2y = 18 \\ x = 6 - 2y \end{cases}$ 8. $\begin{cases} 3x - y = 4 \\ y = 6 - 2x \end{cases}$ 9. $\begin{cases} 5x - 3y = 6 \\ y = 2x + 3 \end{cases}$

Multiply:

10. $(x + 12)^2$ 11. $(2x - 3)(2x^2 - 3x + 4)$

Graph these lines on a rectangular coordinate system:

*12. $3x + 2y = 4$ *13. $y - x = 0$

Simplify:

14. $\dfrac{\frac{1}{a}}{x}$ 15. $\dfrac{\frac{b}{c}}{\frac{1}{a+b}}$ 16. $\dfrac{\frac{x}{c}}{x+y}$ 17. $\dfrac{m}{\frac{a}{mc^2}}$

Add:

18. $\dfrac{4}{2x^2} - \dfrac{3}{4x^2y} + \dfrac{2a}{8x^3p}$ 19. $\dfrac{m}{b(b+c)} + \dfrac{k}{b}$ 20. $\dfrac{3p}{xm^2} - \dfrac{a}{2m^3} + \dfrac{4k}{m^4a}$

21. Factor: $4k^2ax - 8ka^2x^2 + 12k^3a^4x^2$

Simplify:

22. $\dfrac{x^2y - x^2yz}{xyz}$ 23. $(4x^0y^2m)^{-2}(2y^{-4}m^0x)^4$

24. $\dfrac{(k^3p^0)^{-2}k^2p^5}{p^{-5}p^0k^{-1}}$ 25. $\left(\dfrac{x^2y^{-2}}{p^4k^0}\right)^{-2}\left(\dfrac{y^{-2}}{x}\right)$

26. Expand: $\dfrac{x^{-2}}{4m^2}\left(\dfrac{4x^2}{m^{-2}} - \dfrac{8m^{-2}k}{x^{-2}}\right)$

27. Simplify by adding like terms: $3x^2y^2m - \dfrac{m}{x^{-2}y^{-2}} + \dfrac{4x^2y^2}{m^{-1}} - \dfrac{3x^4y^4}{xy}$

28. Evaluate: $a^3 - a(x - a)$ if $a = -2$, $x = 4$

29. Solve: $-3(-2 - x) - 3^2 - |-2| = -(-2x - 3)$

30. Simplify: $-2\{-[(3^0 - 5) - (2 - 4)] - |-3| + 2\}$

LESSON 61   *Percent word problems*

61.A
percent
word problems

It is absolutely necessary to be able to visualize percent word problems in order to work them effectively. There is no shortcut that can be used as a substitute for understanding the problem. We will use the statement of the problem to draw "before" and "after" diagrams to help us write the percent equation that will give us the missing parts.

example 61.1   Kathy, John, and Susie have only 20 chickens left. If they began with 80 chickens, what percent of the original flock remains?

solution   **A diagram should always be drawn as the first step.**

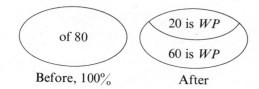

The original flock of 80 is on the left. It is divided into the 20 that remain and the 60 that are missing. We see that we can write two statements that can be used to solve the problem.

   (a)   20 is what percent of 80?   or   (b)   60 is what percent of 80?

We will write an equation to solve question (a).

$$\frac{WP}{100} \times 80 = 20 \longrightarrow \frac{\cancel{100}}{80} \cdot \frac{WP}{\cancel{100}} \times \cancel{80} = 20 \cdot \frac{100}{80} \longrightarrow WP = \frac{2000}{80} \longrightarrow WP = 25\%$$

Thus the other percent is 75 percent. We could also have found 75 percent by solving question (b). Now we can draw the diagram with all numbers in place.

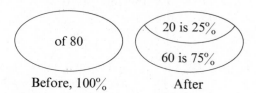

example 61.2   Meme and Jim have 75 thingamabobs. They want to give Hal 20 percent of them. How many thingamabobs do they give Hal?

solution   **A diagram should always be drawn as the first step.**

We can write two statements from this diagram.

   (a)   What number is 20% of 75?
   (b)   What number is 80% of 75?

We don't have to solve both equations. We will solve equation (a) and subtract this answer from 75 to get the other number.

$$\frac{P}{100} \times \text{of} = \text{is} \longrightarrow \frac{20}{100} \times 75 = WN \longrightarrow \frac{1500}{100} = WN \longrightarrow \mathbf{15 = WN}$$

Thus, they give Hal 15 and save $75 - 15 = 60$ for themselves. We finish by drawing the diagram with all numbers in place.

Before, 100%   After

example 61.3   Beau and Christy hide 600 raisins. This is 60 percent more than they hid last month. How many raisins did they hide last month?

solution   **A diagram should always be drawn as the first step.**

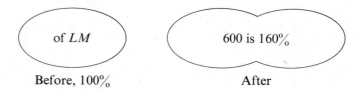

Before, 100%   After

A 60 percent increase gives us 160 percent. The single statement is

600 is 160 percent of last month.

$$\frac{160}{100} \times LM = 600 \longrightarrow \frac{\cancel{100}}{160} \cdot \frac{\cancel{160}}{\cancel{100}} \times LM = 600 \cdot \frac{100}{160} \longrightarrow LM = \frac{60000}{160} \longrightarrow \mathbf{LM = 375}$$

Thus, last month they hid 375 raisins. We finish with diagrams that have all numbers in place.

Before, 100%   After

problem set 61

1. Six times a number is increased by 7. Then this sum is multiplied by 4 and the result is 10 larger than 30 times the opposite of the number. What is the number?

Draw the diagrams and then solve the problems:

*2. Kathy, John, and Susie have only 20 chickens left. If they began with 80 chickens, what percent of the original flock remains?

*3. Meme and Jim have 75 thingamabobs. They want to give Hal 20 percent of them. How many thingamabobs do they give Hal?

*4. Beau and Christy hide 600 raisins. This is 60 percent more than they hid last month. How many raisins did they hide last month?

5. $5\frac{1}{8}$ of what number is $\frac{3}{16}$?   6.   What decimal part of .004 is .00008?

Solve:

7. $3\frac{2}{5}p + 1\frac{1}{2} = -2\frac{3}{8}$

8. $-(-k) - (-2)(2k - 5) + 7 = -(2k + 4)$

9. Graph $2 < x \le 4$ on a number line.

10. The numbers $0, 1, 5, 0, 7, 5, 2, 0, 7$ make up set K. Designate set K using set notation.

11. Given: $A = \{1, 2, 3\}$, $B = \{2, 3, 4, 0\}$. Which of the following statements are true and which are false?

   (a) $7 \in B$ (b) $3 \notin A$ (c) $0 \in B$

Use substitution to solve for x and y.

12. $\begin{cases} x = -19 - 6y \\ 2x + 3y = -11 \end{cases}$

13. $\begin{cases} 2x - 3y = 5 \\ x = -2y - 8 \end{cases}$

14. $\begin{cases} y = 4x + 9 \\ 3x + y = -12 \end{cases}$

Multiply:

15. $(5 + 3x)(8 - 2x)$

16. $(4x + 2)^2$

Graph on a rectangular coordinate system:

17. $y = -3$

18. $3y + x = -9$

Simplify:

19. $\dfrac{x}{\dfrac{1}{a + b}}$

20. $\dfrac{\dfrac{1}{a + b}}{x}$

21. $\dfrac{\dfrac{a}{b}}{\dfrac{1}{x}}$

22. $\dfrac{\dfrac{a + b}{x}}{\dfrac{1}{x}}$

Add:

23. $\dfrac{x}{y} + \dfrac{1}{y + 1}$

24. $1 + \dfrac{x}{y}$

25. $y - \dfrac{1}{y}$

26. Factor: $10x^2y^5z - 5x^5y^2z^5 - 10x^4y^4z^4$

Simplify:

27. $\dfrac{(x^2y^0m)(m^{-2}y)}{m^2(my^{-2})}$

28. $\dfrac{(x^0y^2)^{-2}y^5x}{x^2x^{-5}yy^{-3}}$

29. $\left(\dfrac{xy^{-2}}{m^2}\right)^{-3}\dfrac{(y^{-2})^0}{(2x^2)^{-3}}$

30. Simplify: $-2\{[(3 - 5) - (2^0 - 6) - 2] - [(4 - 3) - 2(-3)]\}$

LESSON 62 *Rearranging before substitution*

62.A
rearranging before substitution

In every substitution problem encountered thus far, one of the equations has expressed x in terms of y as in the bottom equation in (a), or y in terms of x as in the top equation in (b).

(a) $\begin{cases} 2x + 3y = 5 \\ x = 2y + 3 \end{cases}$ (b) $\begin{cases} y = 2x + 4 \\ 2x - y = 7 \end{cases}$

If neither of the equations is in one of these forms, we begin by rearranging one of the equations.

example 62.1   Use substitution to solve the system (a) $\begin{cases} x - 2y = -1 \\ 2x - 3y = 4 \end{cases}$
(b)

solution   To use substitution to solve this system, it is necessary to rearrange one of the equations. We choose to solve for x in equation (a) because the x term in this equation has a co-efficient of 1 and thus we can solve this equation for x in just one step.

$$\begin{array}{ll} x - 2y = -1 & \text{equation (a)} \\ \underline{\quad 2y \quad\quad 2y} & \text{add } 2y \text{ to both sides} \\ x \quad\quad = 2y - 1 & \end{array}$$

Now we can substitute the expression $2y - 1$ for x in equation (b) and complete the solution.

$$\begin{array}{ll} 2x - 3y = 4 & \text{equation (b)} \\ 2(2y - 1) - 3y = 4 & \text{substituted } 2y - 1 \text{ for } x \\ 4y - 2 - 3y = 4 & \text{multiplied} \\ y - 2 = 4 & \text{added like terms} \\ \boldsymbol{y = 6} & \text{added 2 to both sides} \end{array}$$

We can find the value of x by replacing the variable y with the number 6 in either of the original equations. We will use both of the original equations to demonstrate that either one can be used to find x.

$$\begin{array}{cc} \text{USING EQUATION (a)} & \text{USING EQUATION (b)} \\ x - 2y = -1 & 2x - 3y = 4 \\ x - 2(6) = -1 & 2x - 3(6) = 4 \\ x - 12 = -1 & 2x - 18 = 4 \\ \boldsymbol{x = 11} & \boldsymbol{x = 11} \end{array}$$

Thus the ordered pair of x and y that satisfies both equations is **(11, 6)**.

example 62.2   Solve (a) $\begin{cases} 2x - y = 10 \\ 4x - 3y = 16 \end{cases}$
(b)

solution   We will first solve equation (a) for y and then substitute the resulting expression for y in equation (b).

$$\begin{array}{ll} 2x - y = \quad 10 & \text{equation (a)} \\ \underline{-2x \quad\quad\quad -2x} & \text{add } -2x \text{ to both sides} \\ -y = \quad 10 - 2x & \\ y = -10 + 2x & \text{multiplied both sides by } -1 \end{array}$$

Now we substitute $-10 + 2x$ for y in equation (b).

$$\begin{array}{ll} 4x - 3y = 16 & \text{equation (b)} \\ 4x - 3(-10 + 2x) = 16 & \text{substituted } -10 + 2x \text{ for } y \\ 4x + 30 - 6x = 16 & \text{multiplied} \\ -2x + 30 = 16 & \text{added like terms} \\ -2x = -14 & \text{added } -30 \text{ to both sides} \\ \boldsymbol{x = 7} & \text{divided by } -2 \end{array}$$

To finish the solution we can use either of the original equations to solve for y.

USING EQUATION (a)

$$2x - y = 10$$
$$2(7) - y = 10$$
$$14 - y = 10$$
$$-y = -4$$
$$y = 4$$

USING EQUATION (b)

$$4x - 3y = 16$$
$$4(7) - 3y = 16$$
$$28 - 3y = 16$$
$$-3y = -12$$
$$y = 4$$

The solution is the ordered pair **(7, 4)**.

example 62.3   Solve (a) $\begin{cases} 4x - 2y = 38 \\ 2x + y = 25 \end{cases}$
(b)

solution   We will first solve equation (b) for y and then substitute the resulting expression for y in equation (a).

$$
\begin{array}{ll}
2x + y = 25 & \text{equation (b)} \\
\underline{-2x \qquad\quad -2x} & \text{add } -2x \text{ to both sides} \\
y = 25 - 2x &
\end{array}
$$

Now we substitute $25 - 2x$ for y in equation (a).

$$
\begin{array}{ll}
4x - 2y = 38 & \text{equation (a)} \\
4x - 2(25 - 2x) = 38 & \text{substituted} \\
4x - 50 + 4x = 38 & \text{multiplied} \\
8x = 88 & \text{added like terms} \\
x = 11 & \text{divided both sides by 8}
\end{array}
$$

Now we can use either of the original equations to solve for y.

USING EQUATION (a)

$$4x - 2y = 38$$
$$4(11) - 2y = 38$$
$$44 - 2y = 38$$
$$-2y = -6$$
$$y = 3$$

USING EQUATION (b)

$$2x + y = 25$$
$$2(11) + y = 25$$
$$22 + y = 25$$
$$y = 3$$

The solution is the ordered pair **(11, 3)**.

problem set 62   Use "before" and "after" diagrams as aids in working the following problems. After working the problem, fill in the missing parts in the diagrams.

1.   When the votes were counted, we found that 15 percent of the people in Brenham had voted. If 2100 had voted, how many people lived in Brenham?

2.   T-Willy got 128 pounds of honey from his hive. If this was a 60 percent increase from last year, how much honey did he get last year?

3. The store owner gave a 35 percent discount, yet Joe and Carol still had to pay $247 for the camera. What was the original price of the camera?

4. The weight of the elephant was 1040 percent of the weight of the bear. If the elephant weighed 20,800 pounds, what did the bear weigh?

5. The gallimaufry contained things large and small in the ratio of 7 to 2. If the total was 1098 items, how many were large?

6. Given: $A = \{1, 5, 7, 0\}$ and $B = \{0, 1, 5\}$; which of the following are true and which are false?

   (a) $1 \notin A$ (b) $7 \notin B$ (c) $0 \in A$

Use substitution and solve for x and y:

*7. $\begin{cases} x - 2y = -1 \\ 2x - 3y = 4 \end{cases}$

*8. $\begin{cases} 2x - y = 10 \\ 4x - 3y = 16 \end{cases}$

*9. $\begin{cases} 4x - 2y = 38 \\ 2x + y = 25 \end{cases}$

10. $\begin{cases} 3x - 5y = 36 \\ x + 3y = -16 \end{cases}$

Multiply:

11. $(2x - 4)(x - 4)$

12. $(4 - 2x)(x^2 + 3x + 2)$

Graph on a rectangular coordinate system:

13. $y = -2$

14. $y = -2x$

15. $y + 2x + 2 = 0$

Simplify:

16. $\dfrac{\dfrac{a}{b}}{a + b}$

17. $\dfrac{\dfrac{a}{c}}{d + c}$

18. Find the least common multiple of 35, 45, and 80.

Add:

19. $\dfrac{x}{x + y} + y$

20. $\dfrac{a}{x^2 c} + c^3$

21. $\dfrac{m}{c^2} - \dfrac{1}{c} + \dfrac{b}{c + b}$

22. Factor: $3x^3 yp^6 - 3x^2 y^2 p^2 + 9x^2 yp^5$

Simplify:

23. $\dfrac{x^2 ym + 3x^2 ymz}{x^2 ym}$

24. $\dfrac{(x^{-2})^0 (x^2 y^{-2})^{-2}}{p^0 x^{-4} (y^2)^{-2}}$

25. $\left(\dfrac{x^2 p^2}{4y^{-2}}\right)^{-2} \dfrac{(y^{-2})^2}{2y^0 p^5}$

26. $\dfrac{(m^2 xy)mxy}{(m^2 x^2 y)^{-2}(xy)}$

27. Expand: $\dfrac{-x^{-2}}{y^2}\left(-x^2 y^2 + \dfrac{3x^4}{xy^2}\right)$

28. Simplify by adding like terms: $xym - \dfrac{3x^2 y^2 m}{xy} + \dfrac{4x^{-2} y^2}{x^{-3} m^{-1} y} - \dfrac{3xy}{m^{-1}}$

29. Evaluate: $x - xy(y^0 - x)$ if $x = -2$ and $y = -5$

30. Solve: $-2(-3 - |-2|)x - 2 - 3x = -2 - (-2)^3$

LESSON 63 *Subsets*

63.A
subsets

If all the members of one set are also members of a second set, the first set is said to be a subset of the second set. If we have the two sets

$$B = \{1, 2\} \qquad A = \{1, 2, 3\}$$

then we can say that set B is a subset of set A because all the members of set B are also members of set A. We use the symbol $\subset$ to mean **is a subset of**. Therefore, we can write

$$B \subset A$$

which is read as "set B is a subset of set A" or as "B is a subset of A." If we also consider set C where

$$C = \{3, 2, 1\}$$

we would say that $C \subset A$, read "set C is a subset of set A" because all the members of set C are also members of set A. Sets A and C are said to be **equal sets** because they have the **same members.** Set C is said to be an **improper subset** of set A and set A is said to be an **improper subset** of set C because **equal sets are defined to be improper subsets of each other.** Conversely, set B is a **proper subset** of set A because every member of set B is also a member of set A and the sets are not equal sets.

The **set that has no members** is defined to be a **proper subset** of every set that has members and to be an **improper subset** of itself. This set is called the **empty set** or the **null set** and can be designated by using either of the symbols shown here.

$$\{ \quad \} \text{ is the empty set} \qquad \varnothing \text{ is the null set}$$

Thus we can say that $\{ \quad \} \subset A$ or $\varnothing \subset A$, read "the empty set is a subset of set A" or "the null set is a subset of set A" because **this set is considered to be a subset of every set.** The slash can be used to negate the symbol $\subset$, as

$$A \not\subset B$$

read "set A is not a subset of set B" because all the members of set A are not members of set B.

example 63.1 Given the sets $D = \{0, 1, 2\}, E = \{1, 2, 3, \ldots\}, G = \{1, 3, 5\}$, tell which of the following assertions are true and which are false and why.

solution
(a) $E \subset G$ False   All members of set E are not members of set G
(b) $G \subset E$ True   All members of set G are members of set E
(c) $D \subset E$ False   0 is not a member of set E

63.B
subsets of the set of real numbers

The set of **real numbers** is said to be an **infinite set** because there is an infinite number of members of this set. An infinite set has an infinite number of subsets, but we normally restrict our attention to five subsets of the set of real numbers. These are the sets of **natural numbers, whole numbers, integers, rational numbers,** and **irrational numbers.** It would seem reasonable to assume that all authorities use the same definitions for these sets of numbers, but unfortunately this is not the case. In this book we will use the definitions used by almost all authors of recent algebra textbooks. We define the set of natural numbers (counting numbers) as follows:

$$\text{Natural numbers} \qquad N = \{1, 2, 3, \ldots\}$$

If we list the number zero in addition to the set of natural numbers, we have designated the set of whole numbers.

$$\text{Whole numbers} \quad W = \{0, 1, 2, 3, \ldots\}$$

Many students want to argue that numbers such as -3 and -15 should also be called whole numbers because they are not fractions and they are not decimal numbers. Maybe they should be called whole numbers, but if we use the definitions given above we must call these numbers **integers** and we cannot call them whole numbers. We designate the set of integers by listing the set of whole numbers and including the opposite of every natural number.

$$\text{Integers} \quad J = \{\ldots, -3, -2, -1, 0, 1, 2, 3, \ldots\}$$

Every real number is either a rational number or an irrational number. A rational number is a number that **can be** represented as a quotient[†] (fraction) of integers. Thus each of the following numbers is a rational number.

$$\text{(a)} \quad \frac{1}{4} \qquad \text{(b)} \quad \frac{-3}{17} \qquad \text{(c)} \quad -10 \qquad \text{(d)} \quad .013$$

The numbers $\frac{1}{4}$ and $\frac{-3}{17}$ are already expressed as fractions of integers. The number -10 can be written as a quotient of integers in many ways such as

$$\frac{-100}{10} \quad \text{and} \quad \frac{3000}{-300} \quad \text{and} \quad \frac{270}{-27}$$

and the number $.013$ can be written as a fraction of integers in many ways, two of which are

$$\frac{13}{1000} \quad \text{and} \quad \frac{39}{3000}$$

There is an infinite number of numbers, such as the square root of 2, that cannot be written as fractions of integers. These numbers constitute the set of **irrational numbers.**

Real Numbers
(The coordinates of all points on the number line)

$Q =$ **Rational numbers**	$P =$ **Irrational numbers**
(All real numbers that **can** be expressed as fractions of integers)	(All real numbers that **cannot** be expressed as fractions of integers)

In this book when we ask to which sets a particular number belongs, we restrict the possible answers to the sets of natural numbers, whole numbers, integers, rational numbers, irrational numbers, and real numbers. Problems such as the following will give practice in distinguishing between these sets.

example 63.2 $\frac{1}{2} \in \{\text{What sets of numbers?}\}$

solution   This asks that we identify the sets of which the number $\frac{1}{2}$ is a member. If we restrict our reply to the sets discussed above, the answer is the sets of **rationals** and **reals.**

example 63.3 $5 \in \{\text{What sets of numbers?}\}$

[†] A fraction of integers whose denominator is not zero.

solution The **naturals, wholes, integers, rationals,** and **reals.**

example 63.4 Tell whether the following statements are true or false and tell why.

solution (a) {Reals} ⊂ {Integers}   False   The reals are not a subset of the integers. The integers are a subset of the reals.

(b) {Irrationals} ⊂ {Reals}   True   All irrational numbers are also real numbers.

(c) {Irrationals} ⊂ {Rationals}   False   The irrationals are not a subset of the rationals. In fact, no irrational number is also a rational number and vice versa.

(d) {Wholes} ⊂ {Naturals}   False   It's the other way around.

(e) {Integers} ⊂ {Reals}   True   All five sets just discussed are subsets of the real numbers.

problem set 63

1. The patricians and plutocrats controlled the society. If they were in the ratio of 2 to 13 and a total of 315 belonged, how many were plutocrats?

Use diagrams as an aid in solving Problems 2 and 3:

2. The girls' weightlifting team outlifted the boys' team by 140 percent. If the boys' team lifted 1400 pounds, how many pounds did the girls' team lift?

3. Harry and Jack raised roses and petunias in their garden. If 15 percent of the flowers were roses and there were 120 roses, how many flowers were there in all?

4. $7\frac{2}{5}$ of what number is $1\frac{7}{10}$?   5. What fraction of $14\frac{1}{4}$ is $\frac{3}{8}$?

Solve:

6. $7\frac{2}{5}x + 5\frac{1}{3} = \frac{1}{15}$

7. $-[-(-2p)] - 3(-3p + 15) = -(-4)(p - 12)$

8. Graph $x \not\geq -2$ on a number line.   *9. $\frac{1}{2} \in$ {What sets of numbers}?

10. $3\sqrt{2} \in$ {What sets of numbers}?

11. Are the following statements true or false?

   *(a) {Reals} ⊂ {Integers}   *(b) {Irrationals} ⊂ {Reals}
   *(c) {Wholes} ⊂ {Naturals}   *(d) {Integers} ⊂ {Reals}

Use substitution and solve for x and y:

12. $\begin{cases} 4x + y = -5 \\ 2x - y = -1 \end{cases}$ 13. $\begin{cases} x - 3y = -7 \\ 3x + y = -1 \end{cases}$ 14. $\begin{cases} 4x - y = -7 \\ 2x + 2y = 4 \end{cases}$

Multiply:

15. $(4x - 3)(x + 2)$ 16. $(4x + 3)^2$

Graph on a rectangular coordinate system:

17. $y = -3x$ 18. $y = 3x$ 19. $4 + 3x - y = 0$

Simplify:

20. $\dfrac{\dfrac{a}{b}}{c + x}$

21. $\dfrac{\dfrac{a}{b}}{c + x}$

22. $\dfrac{\dfrac{1}{1}}{a + b}$

Add:

23. $\dfrac{4}{xyc} - \dfrac{5m}{xy(c + 1)} - \dfrac{3k}{xy^2}$

24. $k + \dfrac{1}{k}$

25. $my + \dfrac{p}{y}$

26. Factor: $20x^2m^5k^6 - 10xm^4k^4 + 30x^5m^4k^6$

Simplify:

27. $\dfrac{(x^{-2}y^0)^2 y^0 k^2}{(2x^2k^5)^{-4}y}$

28. $\dfrac{a^0 x^2 x^0}{m^2 y^0 m^{-2}}$

29. $\left(\dfrac{p^2 x}{y}\right)^2 \left(\dfrac{x^{-2}y}{p^2}\right)^3$

30. $-2\{[(-2 - 2) - 3^0(-2 - 1)] - [-2(-3 + 5) - 2]\} - |-2|$

LESSON 64 *Square roots*

64.A
roots of
a positive
real number

The operations of addition and subtraction are inverse operations because subtraction can undo addition and addition can undo subtraction. For example, if we begin with a number, say 7, and add 2 and then subtract 2, the result is 7, the number with which we began.

$$7 + 2 - 2 = 7$$

The operations of multiplication and division are also inverse operations because multiplication can undo division and division can undo multiplication. If again we begin with the number 7 and then multiply and divide by 2, the result is 7, the number with which we began.

$$\frac{7 \cdot 2}{2} = 7$$

By the definition of exponential notation,

$$2^2 = 2 \cdot 2 = 4$$

This operation is called raising to a power. **The inverse operation of raising to a power is called extracting a root or taking the root of.** Since

$$2^2 = (2)(2) = 4$$

we say that 2 is a square root of 4, and since

$$(-2)^2 = (-2)(-2) = 4$$

we also say that -2 is a square root of 4. To indicate the square root of 4, we use the expression $\sqrt{4}$. This expression is called a **radical expression.** The symbol $\sqrt{}$ is called a **radical sign** and the number or expression under the radical sign is called the **radicand.**

We note that $2^2 = 4$ and also $(-2)^2 = 4$ so the number 4 has two square roots. In a like manner every positive number has two square roots, one positive and one negative. The number zero has only one square root, which is zero itself. **We reserve the symbol $\sqrt{x}$ to indicate the positive square root or principal square root of the positive[†] number x. Thus when we write $\sqrt{4}$, we are indicating the principal square root of 4, which is $+2$.** To consolidate the above, we write

$$2^2 = 4 \quad \text{and} \quad (-2)^2 = 4 \quad \text{but} \quad \sqrt{4} = 2 \text{ only}$$

If we wish to indicate the negative square root of 4, we must write

$$-\sqrt{4} = -2$$

DEFINITION OF SQUARE ROOT

If x is greater than zero, then $\sqrt{x}$ is the unique **positive** real number such that

$$(\sqrt{x})^2 = x$$

This definition says that the square root of a given positive number is that positive number which multiplied by itself equals the given positive number.

(a) $\sqrt{2}\sqrt{2} = 2$ (b) $\sqrt{a}\sqrt{a} = a$ $(a > 0)$
(c) $\sqrt{2.42}\sqrt{2.42} = 2.42$ (d) $\sqrt{x}\sqrt{x} = x$ $(x > 0)$

Unfortunately the square root of every counting number is not another counting number. **If the square root of a counting number is not another counting number, then the square root cannot be represented exactly by using a decimal numeral that has a finite number of digits.** The **decimal approximation** often used for the square root of 2 is 1.414.

$$\sqrt{2} \cong 1.414 \qquad (\cong \text{ is read "approximately equal to"})$$

A more exact approximation is given by a particular pocket calculator as

$$\sqrt{2} \cong 1.414213562$$

and a table of square roots gives an approximation to 19 decimal places as

$$\sqrt{2} \cong 1.4142135623730950488$$

But we cannot write the square root of 2 exactly by using decimal numerals because an exact representation would require that we use an infinite number of digits.

In Lesson 63 we said that an irrational number is a number that cannot be written as a fraction of integers. **Another way to define an irrational number is to say that an irrational number is a number whose decimal representation is a nonrepeating decimal numeral of infinite length. Thus the $\sqrt{2}$ is an irrational number. The square roots of the following counting numbers are not counting numbers so all of these are irrational numbers.**

$$\sqrt{2}, \quad \sqrt{3}, \quad \sqrt{5}, \quad \sqrt{6}, \quad \sqrt{7}, \quad \sqrt{10}, \quad \sqrt{11}, \quad \sqrt{12}, \quad \sqrt{13}$$

The easiest way to find an approximation for the square root of a positive real number is to use a table of square roots or a pocket calculator. If these are not available, some variation of a *cut-and-try* approach will give a good approximation.[‡] Let's cut and try to find an approximation for the square root of 5.

[†] In this course, square roots of negative numbers will not be discussed. This is a topic for a more advanced course in algebra.

[‡] There are several rote methods that can be used to find the square root of a number. Instead of using one of these we use cut and try because this method emphasizes what a square root is.

Now, $2 \times 2 = 4$ and $3 \times 3 = 9$, so $\sqrt{5}$ lies between 2 and 3 and is closer to 2.

Try 2.1	$2.1 \times 2.1 = 4.41$	not large enough
Try 2.2	$2.2 \times 2.2 = 4.84$	still not large enough
Try 2.3	$2.3 \times 2.3 = 5.29$	too large
Try 2.25	$2.25 \times 2.25 = 5.0625$	a little too large
Try 2.24	$2.24 \times 2.24 = 5.0176$	still too large
Try 2.23	$2.23 \times 2.23 = 4.9729$	now too small

Thus we see that $\sqrt{5}$ is a number somewhere between 2.23 and 2.24. We could continue this cut-and-try procedure to find an approximation of $\sqrt{5}$ that is accurate to as many decimal places as we desire.

Although the decimal representation of $\sqrt{2}$ is a nonrepeating decimal numeral of infinite length, the $\sqrt{2}$ is a real number and can be associated with a specific point on the number line. Here we graph $\sqrt{2}$ and also graph several other irrational numbers.

problem set 64

Use "before" and "after" diagrams to help with Problems 1, 2 and 3.

1. The doctor increased the dosage to 128 percent of the original dosage. If the new dosage was 3840 units, what was the original dosage?

2. Odessa could afford the coat because it was sold for 28 percent less than the original price. If the sale price was $324, what was the original price?

3. When the Huns debouched from the Alpine passes, Attila found that 18 percent of the spearpoints were dull. If 720 spearpoints were dull, how many spears did the Huns bring with them?

4. $5\frac{1}{2} \in$ {What sets of numbers}?   5. $-2 \in$ {What sets of numbers}?

*6. Use the cut-and-try method to find $\sqrt{5}$ that is accurate to one decimal place.

7. Use the cut-and-try method to find $\sqrt{8}$ that is accurate to one decimal place.

Use substitution and solve for x and y:

8. $\begin{cases} 3x - 2y = -1 \\ y = x - 1 \end{cases}$ 9. $\begin{cases} 5x - 3y = 1 \\ 7x - y = -5 \end{cases}$ 10. $\begin{cases} 5x + 2y = -21 \\ -2x + y = 3 \end{cases}$

Multiply:

11. $(2x - 5)(2x + 5)$ 12. $(3x - 2)^2$

Graph on a rectangular coordinate system:

13. $y = -3$ 14. $y = -3x$ 15. $3y + 9x + 9 = 0$

Simplify:

16. $\dfrac{\dfrac{m}{x+y}}{\dfrac{a}{x+y}}$ 17. $\dfrac{\dfrac{m}{x+y}}{\dfrac{m}{x}}$ 18. $\dfrac{m}{\dfrac{1}{m}}$

Add:

19. $\dfrac{a}{x^2y} + \dfrac{m+c}{x+y}$ **20.** $1 + \dfrac{y}{x}$ **21.** $y + \dfrac{y}{x}$

22. Factor: $\quad 9x^3ym^5 + 6m^2y^4p^4 - 3y^3m^3$

Simplify:

23. $\dfrac{4kp + 4kpx}{4kp}$ **24.** $mx(x^0y)m^2x^2(y^2)$

25. $\dfrac{a^{-2}p^2a(a^0)^2}{(a^{-3})^2(p^{-2})^{-2}}$ **26.** $\dfrac{(mx)(mx^0)}{3^{-2}xyyx^{-3}m}$

27. Expand: $\quad \dfrac{-x^{-3}}{y}\left(\dfrac{x^3}{y^{-1}} - \dfrac{3x^{-3}}{y^2}\right)$

28. Simplify by adding like terms: $\quad -2^0x^2y^2x^{-2} + \dfrac{3y^2}{x^2} - \dfrac{4x^{-2}}{y^{-2}} + 5y^2$

29. Evaluate: $\quad -x^0 - x(x^0 - y) \quad$ if $x = -3$ and $y = -2$

30. Solve: $\quad -k(-2-3) - (-2)(-k-5) = -2 - (-2k+4) - 3^2$

LESSON 65 *Product of square roots theorem*

65.A
product of square roots theorem

Square roots of many numbers such as $\sqrt{50}$, $\sqrt{200}$, and $\sqrt{147}$ can be written in a simplified form. To write one of these numbers in a simplified form, we use the following theorem.

> PRODUCT OF SQUARE ROOTS THEOREM
>
> If m and n are nonnegative real numbers, then
> $$\sqrt{m}\sqrt{n} = \sqrt{mn} \qquad \text{and} \qquad \sqrt{mn} = \sqrt{m}\sqrt{n}$$

This theorem can be generalized to the product of any number of factors, and we say that the square root of any product may be written as the product of the square roots of the factors of the product. For example,

$$\sqrt{2 \cdot 5 \cdot 5} \qquad \text{can be written as} \qquad \sqrt{2}\sqrt{5}\sqrt{5}$$

and $\quad \sqrt{3 \cdot 3 \cdot 3 \cdot 5} \qquad$ can be written as $\qquad \sqrt{3}\sqrt{3}\sqrt{3}\sqrt{5}$

We will use this theorem in the following problems to help us simplify the expressions $\sqrt{50}$, $\sqrt{200}$, $\sqrt{147}$, and $\sqrt{108}$.

example 65.1 Simplify $\sqrt{50}$.

solution We will first write 50 as a product of prime factors.

$$\sqrt{5 \cdot 5 \cdot 2}$$

Now we use the product of square roots theorem to write the square root of the product as a product of square roots.

$$\sqrt{5}\sqrt{5}\sqrt{2}$$

Now, by definition $\sqrt{5}\sqrt{5} = 5$, so we have

$$\sqrt{5}\sqrt{5}\sqrt{2} = \mathbf{5\sqrt{2}}$$

example 65.2   Simplify $\sqrt{200}$.

solution   First we write 200 as a product of prime factors

$$\sqrt{200} = \sqrt{2 \cdot 2 \cdot 2 \cdot 5 \cdot 5}$$

Now the square root of the product is written as the product of square roots.

$$\sqrt{2 \cdot 2 \cdot 2 \cdot 5 \cdot 5} = \sqrt{2}\sqrt{2}\sqrt{2}\sqrt{5}\sqrt{5}$$

Since by definition $\sqrt{2}\sqrt{2}$ equals 2 and $\sqrt{5}\sqrt{5} = 5$, we can simplify as

$$(\sqrt{2}\sqrt{2})\sqrt{2}(\sqrt{5}\sqrt{5}) = (2)\sqrt{2}(5) = \mathbf{10\sqrt{2}}$$

example 65.3   Simplify $\sqrt{147}$.

solution   First we write 147 as a product of prime factors.

(a) $\sqrt{147} = \sqrt{3 \cdot 7 \cdot 7}$ write as product of prime factors
(b) $\sqrt{3 \cdot 7 \cdot 7} = \sqrt{3}\sqrt{7}\sqrt{7}$ root of product equals product of roots
(c) $\sqrt{3}(\sqrt{7}\sqrt{7}) = \mathbf{7\sqrt{3}}$ definition of square root

example 65.4   Simplify $\sqrt{108}$.

solution   First we write 108 as a product of prime factors.

(a) $\sqrt{108} = \sqrt{2 \cdot 2 \cdot 3 \cdot 3 \cdot 3}$ write as product of prime factors
(b) $\sqrt{2 \cdot 2 \cdot 3 \cdot 3 \cdot 3} = \sqrt{2}\sqrt{2}\sqrt{3}\sqrt{3}\sqrt{3}$ root of product equals product of roots
(c) $(\sqrt{2}\sqrt{2})(\sqrt{3}\sqrt{3})\sqrt{3} = 2 \cdot 3\sqrt{3} = \mathbf{6\sqrt{3}}$ definition of square root

problem set 65

1. If the product of -3 and the opposite of a number is decreased by 7, the result is one greater than the number. What is the number?

Draw diagrams for the following percent problems and work the problems.

2. Between Karnak and Edfu, the Pharaoh kept 1020 white goats. If these goats represented 17 percent of the total flock, how many goats did the Pharaoh have?

3. After the temple was destroyed, Amenhotep found 1200 precious stones in the ruins. If 3 percent of these stones were rubies, how many rubies did Amenhotep find?

4. 1.05 of what number is 4.221?

5. What fraction of $3\frac{1}{8}$ is $\frac{7}{4}$?

Solve:

6. $-3\frac{1}{2} + 1\frac{2}{5}p - 4\frac{2}{3} = 0$

7. $.3k - .2 + .2k - .05 = -2(k - 3)$

8. Graph $4 < x \leq 7$ on a number line.

9. $\dfrac{\sqrt{3}}{2} \in \{$What sets of numbers$\}$? 10. $4 \in \{$What sets of numbers$\}$?

Use substitution and solve for x and y:

11. $\begin{cases} 5x - 4y = -6 \\ x - 2y = -6 \end{cases}$ 12. $\begin{cases} x - 2y = 7 \\ 2x - 3y = -4 \end{cases}$ 13. $\begin{cases} 4x + y = 14 \\ 2x - 2y = 22 \end{cases}$

Simplify:

*14. $\sqrt{50}$ *15. $\sqrt{200}$ *16. $\sqrt{147}$ *17. $\sqrt{108}$

18. Multiply: $(3x - 2)(5 - 3x)$

Graph on a rectangular coordinate system:

19. $y = 2$ 20. $y = 2x$ 21. $y - 2x = 2$

Simplify:

22. $\dfrac{a}{\dfrac{1}{a + b}}$ 23. $\dfrac{x}{\dfrac{x + y}{x}}$

Add:

24. $\dfrac{1}{x} + \dfrac{3}{x + y}$ 25. $x + \dfrac{1}{y}$ 26. $1 + \dfrac{1}{y}$

27. Factor: $15m^2x^5k^4 - 5m^6x^6k^6 + 20m^4xk^5$

Simplify:

28. $\dfrac{3x^2m^5(2x^4m^2)}{3x^2m^5m^{-4}}$ 29. $\dfrac{4(p^{-2})^0(p^5)}{4^{-1}p^6x^{-5}}$

30. $\left(\dfrac{3x^{-2}}{y^{-3}}\right)^{-2}\left(\dfrac{x^4}{y^6}\right)^2$ 31. $-2^0 - 2^2(2^0) - (2^0)^{-3}$

LESSON 66 *Domain*

66.A

the domain In problems in mathematics (and in physics, chemistry, and other mathematically based disciplines) the numbers that may be used as replacements for the variables are often restricted by the nature of the problem or by a restriction stated in the problem. For instance, if a girl goes to the store with 25 cents to buy eggs for her mother, and eggs cost 10 cents each, the total amount of money that she can spend on eggs can be represented by the equation

$$\text{Total cost} = 10N_E \quad \text{cents}$$

where N_E represents the number of eggs she buys. She may buy no eggs or one egg or two eggs. Thus the total cost of her eggs is as shown in (a), (b), or (c).

(a) $\text{Cost} = 10N_E$ (b) $\text{Cost} = 10N_E$ (c) $\text{Cost} = 10N_E$

   $\text{Cost} = 10(0)$ $\text{Cost} = 10(1)$ $\text{Cost} = 10(2)$

   $\text{Cost} = 0 \text{ cents}$ $\text{Cost} = 10 \text{ cents}$ $\text{Cost} = 20 \text{ cents}$

In (a) we use 0 as the replacement for the variable, and in this case the cost is 0 cents. In (b), we use 1 as the replacement for the variable and find the cost to be 10 cents. In (c) we use 2 as the replacement for the variable and find that in this case the cost is 20 cents. We cannot use 3 as a replacement for the variable because she has only 25 cents to spend. We cannot use $2\frac{1}{2}$ as a replacement for the variable because only whole eggs are sold at the market. Neither could we use -4 as a replacement for the variable because buying -4 eggs makes no sense. We are restricted by the statement of the problem to using only the whole numbers

$$\{0, 1, 2\}$$

as replacements for the variable in the equation.

The set of numbers that constitutes the set of permissible replacement values for the variables in a particular equation or inequality is called the domain for that equation or inequality.

Since **every equation and every inequality have a domain**[†] and since this is an important concept, it is customary in courses in algebra to include problems in which the domain is specified. These problems should help with the concept of domain. The domains for the problems in this book were chosen by the author, sometimes with a purpose and sometimes just arbitrarily so that the problems would have specified domains. We will use the capital letter D as the symbol for the domains in this book and will indicate the domains as sets by enclosing them with braces. For instance,

(a) $D = \{0, 1, 2\}$ (b) $D = \{\text{Reals}\}$ (c) $D = \{\text{Positive integers}\}$

The domains specified here are (a) the numbers 0, 1, and 2; (b) the set of real numbers, and (c) the set of positive integers.

example 66.1   Graph $x < 3$, $D = \{\text{Integers}\}$.

solution   We are asked to indicate the **integers** that are less than 3.

On the number line we have indicated the integers whose values are less than 3. Note that it was not necessary to place an open circle at 3. The arrow on the left indicates an infinite continuation.

example 66.2   Graph $x \geq -1$, $D = \{\text{Reals}\}$.

solution   We are asked to indicate all **real numbers** that are greater than or equal to -1.

The graph indicates all real numbers that are greater than or equal to -1. The solid circle at -1 indicates that -1 is a member of the solution set.

example 66.3   Graph $x < -1$, $D = \{\text{Positive integers}\}$.

[†] If the domain is not stated, it is implied.

solution The solution is $\varnothing$, the null set, or $\{\ \ \}$, the empty set, because there are *no* positive integers that are less than -1.

example 66.4 Graph $x \geq -5$, $D = \{\text{Positive integers}\}$.

solution The graph indicates the numbers that are greater than or equal to -5 and that are also members of the set of positive integers.

1. The number of bacteria increased by 280 percent overnight. If there were 30,000 bacteria yesterday, how many bacteria were present this morning?

2. When Charles inspected the troops that survived, he found that 3600 were still alive. If 40 percent died in the fight, how many troops did he begin the day with?

3. Edna and Mabel climbed 40 percent of the mountains in the whole country. If they climbed 184 mountains, how many mountains were in the country?

Simplify:

4. $\sqrt{72}$

5. $3\sqrt{75}$

6. $4\sqrt{324}$

7. Multiply: $(4x - 3)(12x + 2)$

Graph on a rectangular coordinate system:

8. $y = -2$

9. $y = -2x - 5$

10. $y = -2x + 5$

Graph on a number line:

11. $x \geq -1$: $D = \{\text{Integers}\}$

*12. $x \geq -5$, $D = \{\text{Positive integers}\}$

Use substitution and solve for x and y:

13. $\begin{cases} 3x - 2y = 15 \\ 5x + y = 12 \end{cases}$

14. $\begin{cases} y + 2x = 12 \\ x + 2y = 12 \end{cases}$

15. Use the cut-and-try method to find $\sqrt{27}$ to one decimal place.

Simplify:

16. $\dfrac{\frac{x}{y}}{\frac{1}{y}}$

17. $\dfrac{x}{\frac{x}{y}}$

18. $\dfrac{1}{\frac{1}{y}}$

Add:

19. $\dfrac{4}{a^2x} + \dfrac{7}{x(x + a)}$

20. $2 + \dfrac{3}{y}$

21. $1 + \dfrac{x}{y}$

22. Factor: $40x^4ym^7z - 20x^5y^5m^2z + 20xy^2m$

Simplify:

23. $\dfrac{4x + 4x^2}{4x}$

24. $\dfrac{kp^{-2}k(p^0)^2}{kp(k)(p^{-2})^2}$

25. $\left(\dfrac{3m^2}{y^{-4}}\right)^2\left(\dfrac{m}{y}\right)$

26. $\dfrac{2p^2x^{-4}(x)(x^2)}{y^{-4}(p^2)^{-2}x}$

27. $-|-3^0| - 3^0(-2)(-3)(-2 - 3)$

28. Simplify by adding like terms: $-\dfrac{3x^2 y^{-2}}{x^{-2} y^{-2}} - 2x^4 yy^{-1} + 4x^3 xyy^{-1} - \dfrac{2x^2}{x^{-2}}$

29. Expand: $-\dfrac{x^{-2}}{y^4}\left(x^2 y^4 - \dfrac{3x^{-2}}{y^4}\right)$

30. Evaluate: $x - (x^2)^0(x - y) - |x - y|$ if $x = -2$ and $y = -3$

LESSON 67   *Additive property of inequality*

67.A
additive property of inequality

We restate the additive property of equality here. Note that we write $a + c = b + c$ and also write $c + a = c + b$. We do this to emphasize that the order of the addends does not affect the result.

ADDITIVE PROPERTY OF EQUALITY

If a, b, and c are any real numbers such that

$$a = b$$

then $a + c = b + c$ and $c + a = c + b$

We have learned that we can use the additive property of equality to help us solve some equations. For example, we can solve $x + 4 = 8$ by adding -4 to both sides of the equation.

$$
\begin{array}{rcr}
x + 4 = & & 8 \\
-4 & & -4 \\
\hline
x & = & 4
\end{array}
$$

The **additive property of inequality** is stated in the same way as the additive property of equality except that we use the $>$ symbol rather than the $=$ sign.

ADDITIVE PROPERTY OF INEQUALITY

If a, b, and c are any real numbers such that

$$a > b$$

then $a + c > b + c$ and also $c + a > c + b$

This statement can be used to prove that the same quantity can be added to both sides of an inequality without changing the solution set of the inequality.

example 67.1   Graph $x + 2 < 0$, $D = \{\text{Reals}\}$.

solution   We isolate x by adding -2 to both sides of the inequality, as permitted by the additive property of inequality.

$$x + 2 < \quad 0 \qquad \text{given}$$
$$\underline{\quad -2 \quad -2 \qquad} \text{add } -2 \text{ to both sides}$$
$$x \qquad < -2$$

Now we graph the inequality $x < -2$.

The open circle at -2 indicates that -2 is not a member of the solution set.

example 67.2   Graph $x - 3 \geq -5$, $D = \{\text{Integers}\}$.

solution   We isolate x by adding $+3$ to both sides of the inequality, as permitted by the additive property of inequality.

$$x - 3 \geq -5 \qquad \text{given}$$
$$\underline{\quad +3 \quad +3 \qquad} \text{add } +3 \text{ to both sides}$$
$$x \qquad \geq -2$$

Now we graph the solution $x \geq -2$ and remember that the domain is the set of integers.

Note that -2 is a solution to the inequality $x \geq -2$.

problem set 67

1. The postprandial exercises were situps and pushups in the ratio of 7 to 2. If Hominoid did 9180 exercises, how many were pushups?

2. We estimate that the giant pyramid of Cheops near Cairo contains 2,300,000 blocks of stone. If the builders only used 80 percent of the available blocks, how many blocks were available?

3. When Hannibal increased his army by 17 percent, he found that he had 5850 soldiers. How many soldiers did he have before the increase?

4. $2\frac{1}{8}$ of what number is $\frac{1}{16}$?   5. What fraction of $2\frac{1}{7}$ is $\frac{3}{14}$?

Solve:

6. $3\frac{1}{5}x + \frac{2}{3} = 5\frac{1}{2}$

7. $-|-2| - 2^2 - (-3 - k) = -2(k - |-3|)$

*8. Graph $x - 3 \geq -5$; $D = \{\text{Integers}\}$

9. Graph $x + 5 \nleq 2$; $D = \{\text{Positive integers}\}$

Use substitution and solve for x and y:

10. $\begin{cases} 3x - y = 1 \\ x + 2y = 5 \end{cases}$ 11. $\begin{cases} 3x + y = -12 \\ 2x - 3y = 3 \end{cases}$ 12. $\begin{cases} 5x - 4y = 12 \\ 9x + y = 38 \end{cases}$

Simplify:

13. $4\sqrt{50}$ **14.** $6\sqrt{45}$ **15.** $2\sqrt{12}$ **16.** $5\sqrt{32}$

17. Multiply: $(2x - 5)^2$

Graph on a rectangular coordinate system:

18. $y = -4$ **19.** $y = 3x - 2$ **20.** $2y + 3x = 4$

Simplify:

21. $\dfrac{x}{\dfrac{1}{xy + b}}$ **22.** $\dfrac{\dfrac{x}{a}}{\dfrac{1}{b}}$ **23.** $-3^0 - 3^2 - 3(2^0 - 1)$

Add:

24. $\dfrac{1}{x^2} + \dfrac{m}{x^3 y} + \dfrac{c}{y}$ **25.** $y + \dfrac{1}{x}$ **26.** $1 + \dfrac{1}{xy}$

27. Factor: $30a^2 b^3 c^4 - 15ab^4 c^5 + 45ab^4 c^4$

Simplify:

28. $\dfrac{xx(x^{-2})^2 y(y)}{x^{-2} x^0 xx^{-3}}$ **29.** $\dfrac{4(2x^2 y^4 p^{-2})}{4x(y^2)^{-2} p^0}$ **30.** $\dfrac{(x^{-2})^{-2}(2y^2)^2(y)}{x^{-2} x^{-5} x^0 xx^{-2}}$

LESSON 68   *Addition of radical expressions*

68.A

addition of radical expressions

In Lesson 16, we found that like terms may be added by adding the numerical coefficients of the terms. Radical expressions that have the same index and the same radicand designate the same number and are like terms. Thus the rule for adding like terms can be used to add like radical expressions. We add like radical expressions by adding the numerical coefficients.

example 68.1   Add $4\sqrt{2} - 5\sqrt{2} + 12\sqrt{2}$.

solution   We add these like terms by adding their numerical coefficients.

$$4\sqrt{2} - 5\sqrt{2} + 12\sqrt{2} = (4 - 5 + 12)\sqrt{2} = \mathbf{11\sqrt{2}}$$

example 68.2   Add $4\sqrt{3} + 3\sqrt{5} - 6\sqrt{3}$.

solution   Only like radical terms may be added.

$$4\sqrt{3} - 6\sqrt{3} + 3\sqrt{5} = (4 - 6)\sqrt{3} + 3\sqrt{5} = \mathbf{-2\sqrt{3} + 3\sqrt{5}}$$

example 68.3   Add $-3\sqrt{2} + 5\sqrt{3} - 2\sqrt{2} + 8\sqrt{3}$.

solution   We omit the intermediate step and write the answer directly by simply adding the coefficients of like radical terms.

$$\mathbf{-5\sqrt{2} + 13\sqrt{3}}$$

example 68.4   Add $4\sqrt{3} - 2\sqrt{2} + 6\sqrt{5}$.

solution   No two of these radical terms are like radical terms so no addition is possible.

problem set 68

1. The troll became incensed when he saw the billy goats prancing across the bridge. Finally, he tore the bridge down—but not before 18 percent of the goats had crossed. If 45 goats had crossed, how many goats were there?

2. Hansel ate 34 percent of the gingerbread. If he ate 170 ounces, how much gingerbread was available?

3. A 130 percent increase in the doll population resulted in a total of 1610 dolls. How many dolls were present before the population increased?

Simplify:

4. $5\sqrt{80}$

5. $3\sqrt{120}$

*6. $4\sqrt{2} - 5\sqrt{2} + 12\sqrt{2}$

7. $7\sqrt{5} - \sqrt{5} + 5\sqrt{3} - 3\sqrt{3}$

8. Multiply: $(3p - 4)(2p + 5)$

Graph on a rectangular coordinate system:

9. $x = -\dfrac{1}{2}$

10. $y = -\dfrac{1}{2}x$

11. $2y = x - 8$

12. Graph on a number line: $x + 3 > -7$, $D = \{\text{Positive integers}\}$

Use substitution and solve for x and y:

13. $\begin{cases} x + y = 10 \\ -x + y = 0 \end{cases}$

14. $\begin{cases} 3x - 3y = 3 \\ x - 5y = -3 \end{cases}$

15. $\begin{cases} 3x - y = 8 \\ x - 3y = -8 \end{cases}$

Simplify:

16. $\dfrac{\dfrac{a}{x}}{\dfrac{1}{a^2}}$

17. $\dfrac{\dfrac{a}{a+b}}{a}$

18. $\dfrac{\dfrac{x}{y}}{\dfrac{1}{y}}$

Add:

19. $\dfrac{a}{x + y} + \dfrac{5}{x^2}$

20. $1 + \dfrac{a}{b}$

21. $x + \dfrac{1}{x}$

22. Factor: $4x^2y^2z - 8x^2y^2z^5$

23. Expand: $-3x^{-2}y^2\left(\dfrac{y^{-2}}{x^{-2}} + 4x^2y\right)$

Simplify:

24. $\dfrac{4kx - 4kx^2}{4kx}$

25. $\dfrac{m^0(p^{-2})^2x^2y^4}{(y^{-2})^2yy^0x^{-2}}$

26. $\left(\dfrac{2x^{-2}y}{p}\right)^2\left(\dfrac{p^2x}{2}\right)^{-2}$

27. $\dfrac{x^2x^{-2}x^0y^2}{y^2(x^{-4})^2}$

28. Simplify by adding like terms: $\dfrac{3a^2x}{m} + \dfrac{5xm^{-1}}{a^{-2}} - \dfrac{4aax^{-1}}{x^{-2}m}$

29. Evaluate: $-x^0 - x^2(x - m)$ if $x = -2$ and $m = 3$

30. Simplify: $-3^0 - 3(-2 - 2^0)(-8^0 - 5)$

LESSON 69   Simplification of radical expressions

69.A
simplification
of radical
expressions

In Lesson 65 we learned to simplify expressions such as $\sqrt{50}$ by using the product of square roots theorem.

$$\sqrt{50} = \sqrt{5 \cdot 5 \cdot 2} = \sqrt{5}\sqrt{5}\sqrt{2} = 5\sqrt{2}$$

We can use the same procedure to simplify expressions such as

$$\sqrt{8} - \sqrt{50} + \sqrt{98}$$

if we first simplify each expression and then add the like radical terms. We begin by writing each radicand as a product of prime factors.

$$\sqrt{2 \cdot 2 \cdot 2} - \sqrt{5 \cdot 5 \cdot 2} + \sqrt{7 \cdot 7 \cdot 2}$$

Now we write the roots of products as products of roots.

$$\sqrt{2}\sqrt{2}\sqrt{2} - \sqrt{5}\sqrt{5}\sqrt{2} + \sqrt{7}\sqrt{7}\sqrt{2}$$

And to finish we simplify and add like radical terms

$$2\sqrt{2} - 5\sqrt{2} + 7\sqrt{2} = \mathbf{4\sqrt{2}}$$

example 69.1   Simplify $\sqrt{18} + \sqrt{8}$.

solution

$\sqrt{18} + \sqrt{8}$	given
$\sqrt{2 \cdot 3 \cdot 3} + \sqrt{2 \cdot 2 \cdot 2}$	write each radicand as a product of prime factors
$\sqrt{2}\sqrt{3}\sqrt{3} + \sqrt{2}\sqrt{2}\sqrt{2}$	write roots of products as products of roots
$3\sqrt{2} + 2\sqrt{2} = \mathbf{5\sqrt{2}}$	simplify and add like radical terms

example 69.2   Simplify $8\sqrt{27} - 3\sqrt{75}$.

solution   The radicals in this problem have coefficients, so we will use parentheses to help us prevent errors. We can simplify $\sqrt{27}$ and $\sqrt{75}$ as

$$\sqrt{27} = \sqrt{3 \cdot 3 \cdot 3} = \sqrt{3}\sqrt{3}\sqrt{3} = \mathbf{3\sqrt{3}}$$

and

$$\sqrt{75} = \sqrt{5 \cdot 5 \cdot 3} = \sqrt{5}\sqrt{5}\sqrt{3} = \mathbf{5\sqrt{3}}$$

and now we replace $\sqrt{27}$ with $3\sqrt{3}$ and replace $\sqrt{75}$ with $5\sqrt{3}$.

$$8\sqrt{27} - 3\sqrt{75} = 8(3\sqrt{3}) - 3(5\sqrt{3}) = 24\sqrt{3} - 15\sqrt{3} = \mathbf{9\sqrt{3}}$$

example 69.3   Simplify $\sqrt{27} - 3\sqrt{18} - 6\sqrt{45}$.

solution

$\sqrt{27} - 3\sqrt{18} - 6\sqrt{45}$	given
$\sqrt{3 \cdot 3 \cdot 3} - 3\sqrt{3 \cdot 3 \cdot 2} - 6\sqrt{3 \cdot 3 \cdot 5}$	products of prime factors
$\sqrt{3}\sqrt{3}\sqrt{3} - 3\sqrt{3}\sqrt{3}\sqrt{2} - 6\sqrt{3}\sqrt{3}\sqrt{5}$	roots of products as products of roots
$3\sqrt{3} - 3 \cdot 3\sqrt{2} - 6 \cdot 3\sqrt{5}$	definition of square root
$\mathbf{3\sqrt{3} - 9\sqrt{2} - 18\sqrt{5}}$	simplify

No further simplification is possible since no two of the radical terms are like radical terms.

1. The opposite of a number is tripled and then decreased by 7. The result is 3 greater than twice the number. What is the number?

2. Rubella found 60 escargots in the dell. This was only 80 percent of her largest find. What was the size of her largest find?

3. When the moot assembled, the village leader found that only 37 percent of those who attended had oil for their lamps. If 300 people attended the moot, how many had oil for their lamps?

4. $4\frac{2}{7}$ of what number is $20\frac{1}{2}$?

5. What decimal part of 20.2 is 1.01?

Solve:

6. $2\frac{1}{8}x - \frac{1}{5} = 3\frac{1}{8}$

7. $.003k + .188 - .001k = .2k - .01$

8. Graph on a number line: $x - 3 \not< -5$, $D = \{\text{Positive integers}\}$

Use substitution and solve for x and y:

9. $\begin{cases} 5x - y = 18 \\ 4x - 3y = 10 \end{cases}$

10. $\begin{cases} x + 2y = 0 \\ 3x + y = -10 \end{cases}$

11. $\begin{cases} 5x + 4y = -28 \\ x - y = -2 \end{cases}$

Simplify:

*12. $\sqrt{18} + \sqrt{8}$

*13. $8\sqrt{27} - 3\sqrt{75}$

14. $5\sqrt{20} - 6\sqrt{32}$

15. $2\sqrt{45} - 3\sqrt{20}$

16. Expand: $(4x + 5)^2$

Graph on a rectangular coordinate system:

17. $y = 2x + 2$

18. $y = 2x - 2$

19. $3\sqrt{2} \in \{\text{What sets of numbers}\}$?

Simplify:

20. $\dfrac{a}{\frac{1}{x}}$

21. $\dfrac{\frac{a}{b}}{\frac{1}{c}}$

22. $\dfrac{\frac{a}{b}}{c}$

Add:

23. $\dfrac{a}{x^2y} + \dfrac{b}{x + y}$

24. $k + \dfrac{1}{y^2}$

25. $m + \dfrac{1}{m^2}$

26. Factor: $12x^2y^3p^4 - 4x^3y^2p^6 + 16x^4y^4p^4$

Simplify:

27. $(3x^2y^5m^2)^2(x^2y)^{-2}$

28. $\dfrac{(2xy)^3(xy)^{-2}}{x^2y^0y^{-4}}$

29. $\dfrac{(4x^{-2})^2(x^2y^0)^{-3}}{x^2yy^{-2}y^4}$

30. $[(-3 - 4^0) - (-3 - 2)] - 5$

LESSON 70   *Elimination*

We have said that equivalent equations are equations that have the same solutions. Thus, the solution sets for equivalent equations must be equal sets. The number 2 is a solution to $x + 4 = 6$ and is also a solution to $x^2 - 4 = 0$.

$$\begin{array}{ll}
\text{(a)} \quad x + 4 = 6 & \text{(b)} \quad x^2 - 4 = 0 \\
(2) + 4 = 6 & (2)^2 - 4 = 0 \\
\qquad 6 = 6 \quad \text{True} & \qquad 4 - 4 = 0 \quad \text{True}
\end{array}$$

But these equations are not equivalent equations because equation (b) has another solution which is not a solution to equation (a). The other solution of equation (b) is -2.

$$\begin{array}{ll}
\text{(a)} \quad x + 4 = 6 & \text{(b)} \quad x^2 - 4 = 0 \\
(-2) + 4 = 6 & (-2)^2 - 4 = 0 \\
\qquad 2 = 6 \quad \text{False} & \qquad 4 - 4 = 0 \quad \text{True}
\end{array}$$

We remember that if every term of a particular equation is multiplied by the same nonzero quantity, the resulting equation is an equivalent equation to the original equation. On the left below we write the equation $x + y = 6$. On the right we write the equation $2x + 2y = 12$, which is the original equation with every term having been multiplied by 2. The ordered pair $(4, 2)$ is a solution to both equations.

$$\begin{array}{ll}
x + y = 6 & 2x + 2y = 12 \\
(4) + (2) = 6 & 2(4) + 2(2) = 12 \\
\qquad 6 = 6 \quad \text{True} & \qquad 8 + 4 = 12 \quad \text{True}
\end{array}$$

Of course, there is an infinite number of ordered pairs of x and y that will satisfy either of these equations, but it can be shown that any ordered pair that satisfies either one of the equations will satisfy the other equation, and thus we say that these equations are equivalent equations!

Thus far we have been using the substitution method to solve systems of linear equations in two unknowns. Now we will see that these equations can also be solved by using another method. **This new method is called the elimination method and is sometimes called more meaningfully the addition method.** To solve the following system of equations by using elimination

$$\begin{array}{ll}
\text{(a)} & \{ x + 2y = 8 \\
\text{(b)} & \{ 5x - 2y = 4
\end{array}$$

we first assume that values of x and y exist that will make both of these equations true equations and that x and y in the equations represent these numbers.[†] Thus $x + 2y$ equals the number 8 and $5x - 2y$ equals the number 4. The additive property of equality permits the addition of equal quantities to both sides of an equation. Thus

---

[†] If values of x and y do not exist that will simultaneously make both equations true statements, the attempted solution will degenerate into an equation that contains only numbers such as $2 = 4$. $4 + 2 = 6, 0 = 0$, or $0 = 5$. The reasons for results like this will be discussed in Lesson 115.

we can add $5x - 2y$ to the left side of equation (a), and add 4 to the right side of equation (a).

$$
\begin{array}{rl}
\text{(a)} & x + 2y = 8 \\
\text{(b)} & \underline{5x - 2y = 4} \\
& 6x = 12
\end{array}
$$

By doing this we have eliminated the variable y. Now we can solve the equation $6x = 12$ for x, find that $x = 2$, and use this value for x in *either* of the original equations to find that $y = 3$.

<table>
<tr><td align="center">IN EQUATION (a)</td><td align="center">IN EQUATION (b)</td></tr>
<tr><td align="center">$x + 2y = 8$</td><td align="center">$5x - 2y = 4$</td></tr>
<tr><td align="center">$(2) + 2y = 8$</td><td align="center">$5(2) - 2y = 4$</td></tr>
<tr><td align="center">$2y = 6$</td><td align="center">$-2y = -6$</td></tr>
<tr><td align="center">$y = 3$</td><td align="center">$y = 3$</td></tr>
</table>

example 70.1   Solve by using the elimination method:
$$
\begin{array}{l}
\text{(a)} \\
\text{(b)}
\end{array}
\left\{
\begin{array}{l}
2x - y = 13 \\
3x + 4y = 3
\end{array}
\right.
$$

solution   If we add the equations in their present form

$$
\begin{array}{rl}
\text{(a)} & 2x - y = 13 \\
\text{(b)} & \underline{3x + 4y = 3} \\
& 5x + 3y = 16
\end{array}
$$

we find that we have accomplished nothing because we have not eliminated one of the variables. But by proper use of the multiplicative property of equality we can change the equations into equivalent equations that, when added, will result in the elimination of one of the variables. We choose to eliminate the variable y and thus we will multiply every term in equation (a) by 4 and every term in equation (b) by 1.[†] Now the equations are added and we find that we have eliminated the variable y.

$$
\begin{array}{llllll}
\text{(a)} & 2x - y = 13 & \longrightarrow & (4) & \longrightarrow & 8x - 4y = 52 & \text{multiplied by 4} \\
\text{(b)} & 3x + 4y = 3 & \longrightarrow & (1) & \longrightarrow^{‡} & \underline{3x + 4y = 3} & \text{multiplied by 1} \\
& & & & & 11x = 55 & \text{added} \\
& & & & & x = 5 & \text{divided by 11}
\end{array}
$$

The number 5 can now be used to replace x in either of the original equations or either of the equivalent equations to find the corresponding value of y. We will demonstrate this by replacing x with 5 in both equation (a) and equation (b).

<table>
<tr><td align="center">EQUATION (a)</td><td align="center">EQUATION (b)</td></tr>
<tr><td align="center">$2(5) - y = 13$</td><td align="center">$3(5) + 4y = 3$</td></tr>
<tr><td align="center">$10 - y = 13$</td><td align="center">$15 + 4y = 3$</td></tr>
<tr><td align="center">$-y = 3$</td><td align="center">$4y = -12$</td></tr>
<tr><td align="center">$y = -3$</td><td align="center">$y = -3$</td></tr>
</table>

[†] This will leave equation (b) unchanged. We say that we multiply by 1 to establish a general procedure for this type of problem.

[‡] The notations $\rightarrow (4) \rightarrow$ and $\rightarrow (1) \rightarrow$ are just bookkeeping notations and have no mathematical meaning. We find them convenient to help us remember the number that we have used as a multiplier.

example 70.2   Solve by using the elimination method: $\begin{matrix} \text{(a)} \\ \text{(b)} \end{matrix} \begin{cases} 2x - 3y = 5 \\ 3x + 4y = -18 \end{cases}$

solution   There are many ways that the multiplicative property of equality can be used to form equivalent equations that when added will result in one of the variables being eliminated. We will show one way here and then repeat the problem and show another way. Look at the x terms in both equations. If we multiply the x term in the top equation by -3, the product will be $-6x$. If we multiply the x term in the bottom equation by 2, the product will be $+6x$.[†] Now if we add the equations we can eliminate x since the sum of $+6x$ and $-6x$ is zero.

$$\begin{array}{lllll} \text{(a)} & 2x - 3y = & 5 & \longrightarrow \quad (-3) \quad \longrightarrow & -6x + 9y = -15 & \text{multiplied by } -3 \\ \text{(b)} & 3x + 4y = & -18 & \longrightarrow \quad (2) \quad \longrightarrow & \underline{6x + 8y = -36} & \text{multiplied by } 2 \\ & & & & 17y = -51 \quad \longrightarrow \quad \boldsymbol{y = -3} \end{array}$$

Now we will use -3 for y in the original equation (a) to find the corresponding value for x.

$$\begin{array}{l} \text{(a)} \quad 2x - 3y = 5 \quad \longrightarrow \quad 2x - 3(-3) = 5 \quad \longrightarrow \quad 2x + 9 = 5 \\ \qquad\qquad \longrightarrow \quad 2x = -4 \quad \longrightarrow \quad \boldsymbol{x = -2} \end{array}$$

Thus the solution is the ordered pair $\boldsymbol{(-2, -3)}$.

example 70.3   Solve by using the elimination method but this time eliminate y.

$$\begin{matrix} \text{(a)} \\ \text{(b)} \end{matrix} \begin{cases} 2x - 3y = 5 \\ 3x + 4y = -18 \end{cases}$$

solution   Look at the y terms in both equations. One of them already has a minus sign. If we multiply the y term in equation (a) by $+4$, the product will be $-12y$; and if we multiply the y term in equation (b) by $+3$, the product will be $+12y$; and, of course, the sum of $-12y$ and $+12y$ is zero.

$$\begin{array}{lllll} \text{(a)} & 2x - 3y = & 5 & \longrightarrow \quad (4) \quad \longrightarrow & 8x - 12y = & 20 & \text{multiplied by } 4 \\ \text{(b)} & 3x + 4y = & -18 & \longrightarrow \quad (3) \quad \longrightarrow & \underline{9x + 12y = -54} & & \text{multiplied by } 3 \\ & & & & 17x \qquad\quad = -34 \quad \longrightarrow \quad \boldsymbol{x = -2} \end{array}$$

Now we could use $x = -2$ in either of the original equations or either of the equivalent equations to find that the corresponding value of y is -3. Again we find that the solution is the ordered pair $\boldsymbol{(-2, -3)}$.

example 70.4   Use elimination to solve the system: $\begin{matrix} \text{(a)} \\ \text{(b)} \end{matrix} \begin{cases} 2x + 5y = -7 \\ 3x - 4y = 1 \end{cases}$

solution   Since one of the y terms already has a minus sign, we can find satisfactory equivalent equations by multiplying both equations by positive numbers.

$$\begin{array}{lllll} \text{(a)} & 2x + 5y = & -7 & \longrightarrow \quad (4) \quad \longrightarrow & 8x + 20y = -28 & \text{multiplied by } 4 \\ \text{(b)} & 3x - 4y = & 1 & \longrightarrow \quad (5) \quad \longrightarrow & \underline{15x - 20y = 5} & \text{multiplied by } 5 \\ & & & & 23x \qquad\quad = -23 \quad \longrightarrow \quad \boldsymbol{x = -1} \end{array}$$

[†] Of course we must multiply every term in the equations by -3 and by 2 as required by the multiplicative property of equality.

Now we will use -1 for x in equation (a) and find the corresponding value of y.

(a) $2x + 5y = -7 \longrightarrow 2(-1) + 5y = -7 \longrightarrow -2 + 5y = -7$

$\longrightarrow 5y = -5 \longrightarrow y = -1$

Thus the solution is the ordered pair $(-1, -1)$.

problem set 70

1. For some strange reason Jim's new diet caused him to gain weight rather than lose weight. If his weight increased 35 percent to 297 pounds, what did he weigh before he began his diet?

2. Only 13 percent of the tribe did not want Sleeping Bear to be chief. If there were 3000 members of the tribe, how many wanted Sleeping Bear?

3. The potato bugs decimated the potato crop and the harvest was down 22 percent from last year. If 3900 tons were harvested, what was the harvest last year?

Simplify:

4. $6\sqrt{45} + \sqrt{180}$

5. $2\sqrt{8} - 3\sqrt{32}$

6. $2\sqrt{12} - 3\sqrt{18}$

Graph on a rectangular coordinate system:

7. $y = -\dfrac{1}{2}x + 3$

8. $x = -2\dfrac{1}{2}$

9. Use substitution and solve for x and y: $\begin{cases} 4x + y = 25 \\ x - 3y = -10 \end{cases}$

10. Graph on a number line: $x - 3 \not> 1, D = \{\text{Reals}\}$

Use elimination and solve for x and y:

*11. $\begin{cases} 2x - y = 13 \\ 3x + 4y = 3 \end{cases}$

*12. $\begin{cases} 2x - 3y = 5 \\ 3x + 4y = -18 \end{cases}$

*13. $\begin{cases} 2x + 5y = -7 \\ 3x - 4y = 1 \end{cases}$

14. $\begin{cases} 2x - 4y = -4 \\ 3x + 2y = 18 \end{cases}$

15. $\begin{cases} 3x - y = 7 \\ 2x + 2y = 10 \end{cases}$

Simplify:

16. $\dfrac{a}{\dfrac{1}{x^2 a}}$

17. $\dfrac{a}{\dfrac{a^2}{a + b}}$

Add:

18. $\dfrac{m}{x^2 a} + \dfrac{3}{a(a + x)}$

19. $4x + \dfrac{1}{y}$

20. $1 + \dfrac{x}{y}$

Simplify:

21. $\dfrac{12mx + 12mxy}{12mx}$

22. $\dfrac{x^2(y^{-2})^2 x x^4 (p^0)^2}{(x^{-3}y^{-2})^2 y}$

23. $(4x^2 y^3 p^4)^3$

24. Expand: $\left(\dfrac{y^{-5}}{x^2} - \dfrac{3y^5 x^{-2}}{p}\right)\dfrac{x^{-2}}{y^5}$

25. Evaluate: $-x^0 - x^2 - a(x - a) - |x^3|$ if $x = -2$ and $a = 4$

26. Simplify: $-2\{[(-2 - 3) - (-2^0 - 2) - 2] - 2\}$

LESSON 71   *More about complex fractions*

71.A
complex
fractions

In Lesson 63 we defined a **rational number** to be a number that **can** be expressed as a fraction of integers. Thus the following are all rational numbers.

$$13, \quad \frac{4}{7}, \quad \frac{-5}{14}, \quad \frac{6}{-13}, \quad 2\frac{1}{3}$$

The number 13 is a rational number because the number 13 can be expressed as a fraction of integers by writing, say, $\frac{52}{4}$. Of course, the mixed number $2\frac{1}{3}$ can be expressed as a fraction of integers as $\frac{7}{3}$.

An algebraic expression containing variables that is written in fractional form is called a **rational expression** because it has the same form as a rational number that is written as a fraction of integers. Thus all the following are **rational expressions.**

$$\frac{x+y}{4}, \quad \frac{a}{-b+c}, \quad 5, \quad \frac{-7}{x}, \quad \frac{a+b}{14-x}$$

The number 5 is a rational number and is also considered to be a rational expression because rational expression is a general term that describes both rational numbers and rational expressions that include variables and/or numbers. Of course, the denominators of none of these expressions can equal zero. We recognize this fact and in this section we will omit the restrictive notations that are normally used to emphasize that division by zero is not permissible.

We have used the **denominator-numerator same-quantity theorem** to help us change denominators as required so that rational expressions may be added in three steps as shown here by adding *a* over *b* to *x* over *y*.

$$\frac{a}{b} + \frac{x}{y} \longrightarrow \quad (1) \quad \frac{}{by} + \frac{}{by} \qquad \text{LCM used as new denominator}$$

$$(2) \quad \frac{ay}{by} + \frac{bx}{by} \qquad \text{new numerators determined}$$

$$(3) \quad \frac{ay + bx}{by} \qquad \text{added}$$

We have also used this same theorem to help us simplify fractions of rational expressions (complex fractions).

$$\frac{\dfrac{x}{y}}{\dfrac{b}{c}} = \frac{\dfrac{x}{y} \cdot \dfrac{c}{b}}{\dfrac{b}{c} \cdot \dfrac{c}{b}} = \frac{\dfrac{xc}{yb}}{\dfrac{bc}{bc}} = \frac{\dfrac{xc}{yb}}{1} = \frac{xc}{yb}$$

We will use both of these procedures when we simplify expressions that are fractions of sums of rational expressions. These expressions are also called **complex fractions.**

example 71.1   Simplify $\dfrac{\dfrac{x}{y} + \dfrac{1}{y}}{\dfrac{x}{y} - \dfrac{1}{y}}$

solution   The simplification is performed in two steps. The first step is to add the two expressions in the numerator and add the two expressions in the denominator.

$$\frac{\dfrac{x}{y}+\dfrac{1}{y}}{\dfrac{x}{y}-\dfrac{1}{y}}=\frac{\dfrac{x+1}{y}}{\dfrac{x-1}{y}}$$

Now use the **denominator-numerator same-quantity theorem** by multiplying both the numerator and denominator by $\dfrac{y}{x-1}$, which is the reciprocal of the denominator $\dfrac{x-1}{y}$

$$\frac{\dfrac{x+1}{\cancel{y}}\cdot\dfrac{\cancel{y}}{x-1}}{\dfrac{\cancel{x-1}}{\cancel{y}}\cdot\dfrac{\cancel{y}}{\cancel{x-1}}}=\frac{x+1}{x-1}$$

example 71.2 Simplify $\dfrac{1+\dfrac{1}{x}}{7}$

solution First we add in the numerator.

$$\frac{1+\dfrac{1}{x}}{7}=\frac{\dfrac{x+1}{x}}{7}$$

Now we finish by multiplying the denominator and the numerator by the reciprocal of 7, which is $\frac{1}{7}$.

$$=\frac{\dfrac{x+1}{x}\cdot\dfrac{1}{7}}{\dfrac{\cancel{7}}{1}\cdot\dfrac{1}{\cancel{7}}}=\frac{x+1}{7x}$$

example 71.3 Simplify $\dfrac{\dfrac{1}{x}}{1-\dfrac{1}{x}}$

solution First we add in the denominator.

$$\frac{\dfrac{1}{x}}{1-\dfrac{1}{x}}=\frac{\dfrac{1}{x}}{\dfrac{x-1}{x}}$$

Now we finish by multiplying the denominator and the numerator by $\dfrac{x}{x-1}$, which is the reciprocal of $\dfrac{x-1}{x}$

$$=\frac{\dfrac{1}{\cancel{x}}\cdot\dfrac{\cancel{x}}{x-1}}{\dfrac{\cancel{x-1}}{\cancel{x}}\cdot\dfrac{\cancel{x}}{\cancel{x-1}}}=\frac{1}{x-1}$$

example 71.4   Simplify $\dfrac{\dfrac{a}{b} + 1}{\dfrac{x}{b} + 4}$

solution   First we add in the numerator and add in the denominator.

$$\frac{\dfrac{a}{b} + 1}{\dfrac{x}{b} + 4} = \frac{\dfrac{a + b}{b}}{\dfrac{x + 4b}{b}}$$

Now we finish by multiplying both the denominator and the numerator by the reciprocal of the denominator.

$$= \frac{\dfrac{a + b}{\cancel{b}} \cdot \dfrac{\cancel{b}}{x + 4b}}{\dfrac{\cancel{x + 4b}}{\cancel{b}} \cdot \dfrac{\cancel{b}}{\cancel{x + 4b}}} = \frac{a + b}{x + 4b}$$

problem set 71

1. If a number is multiplied by 7 and this product is increased by 42, the result is 87 greater than twice the opposite of the number. Find the number.

2. When Batman's entourage joined the motorcade, the total number of vehicles increased 130 percent. If the final count was 345, how many vehicles were present in the beginning?

3. The new drug saved lives but had side effects on 37 percent of the people who took it. If 1110 people had side effects, how many took the new drug?

4. $24\dfrac{1}{2}$ of what number is 120?   5.   What fraction of 105 is $5\dfrac{1}{3}$?

Solve:

6. $20\dfrac{1}{4}x + 5\dfrac{1}{2} = 7\dfrac{1}{16}$ 7. $-(-3)^3 - 2^2 = -2(-3k - 4)$

8. Graph on a number line: $x - 3 \not> -2$, $D = \{\text{Positive integers}\}$

9. True or false: $\{\text{Integers}\} \subset \{\text{Reals}\}$

Use substitution and solve for x and y:

10. $\begin{cases} x + 2y = 15 \\ 3x - y = 10 \end{cases}$ 11. $\begin{cases} 4x - 3y = 14 \\ x + 3y = -4 \end{cases}$

Graph on a rectangular coordinate system:

12. $x = 3\dfrac{1}{2}$ 13. $y = -2x + 4$

Use elimination and solve for x and y:

14. $\begin{cases} 5x - 2y = 10 \\ 7x - 3y = 13 \end{cases}$ 15. $\begin{cases} 5x + 3y = 1 \\ 7x + 3y = 5 \end{cases}$ 16. $\begin{cases} 14x - 2y = 12 \\ x + 2y = 3 \end{cases}$

Simplify:

17. $\dfrac{\dfrac{a}{b}}{\dfrac{x}{y}}$ *18. $\dfrac{\dfrac{x}{y} + \dfrac{1}{y}}{\dfrac{x}{y} - \dfrac{1}{y}}$ *19. $\dfrac{1 + \dfrac{1}{x}}{7}$ *20. $\dfrac{\dfrac{1}{x}}{1 - \dfrac{1}{x}}$

*21. $\dfrac{\dfrac{a}{b} + 1}{\dfrac{x}{b} + 4}$

22. $\dfrac{3 - \dfrac{a}{b}}{\dfrac{1}{b} + b}$

23. $4\sqrt{8} - 3\sqrt{12}$

24. $2\sqrt{75} - 4\sqrt{243}$

25. Simplify by adding like terms: $-x^2y + 3yx^2 - \dfrac{4y^3x}{y^2x^{-1}} - \dfrac{7x^{-2}}{x^{-4}y^{-1}}$

Simplify:

26. $(4x^{-2}y^2m)^{-2}y$

27. $\dfrac{(x^{-2}y^2p)^2(x^2yp)^{-4}}{(xyp^2)^2}$

28. $\left(\dfrac{x^{-1}}{y^{-1}}\right)^{-2}\left(\dfrac{y^2}{x^2}\right)^{-4}$

29. $\dfrac{x^{-2}y^{-2}(p^0)^2}{(x^2y^{-2}p^3)^{-2}}$

30. $-3^2 - (-3)^3 + (-3)^0$

LESSON 72 *Factoring trinomials*

72.A
factoring
trinomials

To begin a quick review of the nomenclature of polynomials in one variable, we say that **a monomial is a single expression of the form ax^n where a is any real number and n is any whole number.** Thus the following expressions are monomials.

$$4, \qquad 6x^2, \qquad -2x^{15}, \qquad 4.163x^4$$

The number 4 can be classified as a monomial because it can be thought of as $4x^0$, and if x is any nonzero real number, then x^0 equals 1, so $4x^0 = 4 \cdot 1 = 4$.

A binomial is the indicated algebraic sum of two monomials and a trinomial is the indicated algebraic sum of three monomials. We use the word polynomial as the general descriptive term to describe monomials, binomials, trinomials, and algebraic expressions that are the indicated sum of four or more monomials.

We are familiar with the vertical format for multiplying binomials as shown here.

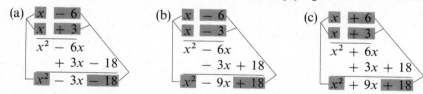

In each of these three examples the product is a trinomial. We call these trinomials **quadratic trinomials in x** or more simply **quadratic trinomials.** The word *quadratic* tells us that the highest power of the variable is 2.

To reverse the process and factor the trinomials into a product of binomials, we must observe the pattern that developed when we did the multiplications. Note that:

1. The first term of the trinomial is the product of the first terms of the binomials.
2. The last term of the trinomial is the product of the last terms of the binomials.
3. The *coefficient* of the middle term of the trinomial is the *sum* of the last terms of the binomials.
4. If all signs in the trinomial are positive, all signs in both binomials are positive. If a negative sign appears in the trinomial, at least one of the terms of the binomials is negative.

We use these observations to help us factor trinomials. To factor the trinomial

$$x^2 - 3x - 18$$

we first write down two sets of parentheses to form an indicated product.

$$(\quad)(\quad)$$

Since the first term in the trinomial is the product of the first terms of the binomials, we enter x as the first term of each binomial.

$$(x\quad)(x\quad)$$

Now, the product of the last terms of the binomials must equal -18, their sum must equal -3, and at least one of them must be negative. There are six pairs of integral[†] factors of -18

$$(-18)(1) = -18 \qquad (2)(-9) = -18 \qquad (3)(-6) = -18$$
$$(18)(-1) = -18 \qquad (-2)(9) = -18 \qquad (-3)(6) = -18$$

Their sums are

$$(-18) + (1) = -17 \qquad (2) + (-9) = -7 \qquad (3) + (-6) = -3$$
$$(18) + (-1) = 17 \qquad (-2) + (9) = 7 \qquad (-3) + (6) = 3$$

Note that while all six pairs have a product of -18, only one pair, $3, -6$, sums to -3. Therefore, the last terms of the binomials are 3 and -6, and so $(x + 3)(x - 6)$ are the factors of $x^2 - 3x - 18$ because

$$(x + 3)(x - 6) = x^2 - 3x - 18$$

Thus, the general approach to factoring a quadratic trinomial that has a leading coefficient of 1 is to determine the pairs of integral factors of the last term of the trinomial whose sum equals the coefficient of the middle term. To factor $x^2 - 8x + 16$, we list the factors of $+16$ and see which pair, if any, sums to -8. If no pair of integral factors has a sum of -8, the trinomial cannot be factored over the integers.

PRODUCT	SUM
$(16)(1) = 16$	$(16) + (1) = 17$
$(-16)(-1) = 16$	$(-16) + (-1) = -17$
$(2)(8) = 16$	$(2) + (8) = 10$
$(-2)(-8) = 16$	$(-2) + (-8) = -10$
$(4)(4) = 16$	$(4) + (4) = 8$
$(-4)(-4) = 16$	$(-4) + (-4) = -8$

Thus we find that the factors of $x^2 - 8x + 16$ are $(x - 4)$ and $(x - 4)$ because the product of the first terms is x^2, the product of the last terms is $+16$, and the sum of the last terms is -8. This may seem to be a complicated procedure, but there is no shortcut until one becomes sufficiently familiar with the process to perform some of the calculations mentally. We will check our solution by multiplying the factors.

[†] Since $(2\sqrt{3})(3\sqrt{3}) = 18$, both $2\sqrt{3}$ and $3\sqrt{3}$ are factors of 18. We will not consider nonintegral factors such as these and will concentrate on factors that are integers. The process of factoring a polynomial into expressions all of whose coefficients are integers is defined as factoring over the set of integers.

$$\begin{array}{r} x \quad - 4 \\ x \quad - 4 \\ \hline x^2 - 4x \\ - 4x + 16 \\ \hline x^2 - 8x + 16 \end{array} \quad \text{Check}$$

example 72.1 Factor $x^2 - 14x - 15$.

solution The last term of the trinomial is -15, so the products of the last terms in the binomial must be -15. Four pairs of integral factors have a product of -15,

$$(3)(-5) = -15, \quad (-3)(5) = -15, \quad (-15)(1) = -15, \quad (15)(-1) = -15,$$

but only one pair sums to -14,

$$(3) + (-5) = -2, \quad (-3) + (5) = 2, \quad \mathbf{(-15) + (1) \ = \ -14}, \quad (15) + (-1) = 14$$

Thus the constant terms of the binomials are -15 and 1 because these are the only two factors whose product is -15 and whose sum is -14. So $x^2 - 14x - 15$ in factored form is $(\mathbf{x - 15})(\mathbf{x + 1})$. We will check by multiplying the two factors:

$$\begin{array}{r} x \quad - 15 \\ x \quad + 1 \\ \hline x^2 - 15x \\ + \quad x - 15 \\ \hline x^2 - 14x - 15 \end{array} \quad \text{Check}$$

example 72.2 Factor $x^2 + 3x - 10$.

solution The constant term is -10. Negative 10 has four pairs of integral factors. They are 1 and -10, 10 and -1, -5 and 2, and lastly 5 and -2. The only pair whose sum is $+3$ is the pair 5 and -2, so

$$x^2 + 3x - 10 = (\mathbf{x + 5})(\mathbf{x - 2})$$

We note that the trinomial in this problem was written as $x^2 + 3x - 10$ with the powers of the variable x in descending order. **If the trinomial is not in this form, the first step in factoring is to write the trinomial in descending powers of the variable.**

example 72.3 Factor $-5x + x^2 + 6$.

solution We begin by writing the trinomial in descending powers of the variable as

$$x^2 - 5x + 6$$

The minus sign in the middle term indicates that at least one of the constant terms is a negative number. The last term, $+6$, is a positive number and is a product of the constant terms, so both of the constants must be negative since their product is positive. Two pairs of negative integers have a product of $+6$,

$$(-3)(-2) = 6 \quad \text{and} \quad (-1)(-6) = 6$$

but only the first pair sums to -5:

$$(-3) + (-2) = -5$$

Thus $\qquad\qquad x^2 - 5x + 6 = (\mathbf{x - 3})(\mathbf{x - 2})$

example 72.4 Factor $x^2 + 5 + 6x$.

solution We begin by writing the trinomial in descending powers of the variable

$$x^2 + 6x + 5$$

There are no minus signs in the trinomial, so all constants in the binomial factors will be positive. The constants therefore are positive integers whose product is $+5$ and whose sum is $+6$. The constants are $+5$ and $+1$ because

$$(5)(1) = 5 \quad \text{and} \quad 5 + 1 = 6$$

Thus

$$x^2 + 6x + 5 = (x + 5)(x + 1)$$

problem set 72

Factor. Remember that the product of the constant terms must equal the constant in the trinomial and the sum of the constant terms must equal the coefficient of the x term.

*1. $x^2 - 3x - 18$ ***2.** $x^2 - 14x - 15$ **3.** $x^2 - 6x - 16$

4. $x^2 + 6x - 16$ **5.** $x^2 - 6x + 9$ **6.** $x^2 - 6x - 27$

7. $p^2 - p - 20$ **8.** $x^2 - 2x - 15$ **9.** $p^2 - 4p - 21$

10. $p^2 + p - 20$ **11.** $k^2 - 3k - 40$ **12.** $m^2 + 9m + 20$

First rearrange in descending order of the variable. Then factor.

13. $x^2 + 33 + 14x$ **14.** $-13p + p^2 + 36$ **15.** $-30 + m^2 - m$

16. $11n + n^2 + 18$ **17.** $x^2 + 27 + 12x$ **18.** $x^2 + 90 - 19x$

19. $x^2 + x - 132$ **20.** $a^2 + 90 - 47a$ **21.** $10m + m^2 + 16$

Use substitution and solve for x and y:

22. $\begin{cases} 3x + y = 9 \\ x - 4y = -10 \end{cases}$ **23.** $\begin{cases} 2x + 5y = 7 \\ x + 3y = 4 \end{cases}$

Use elimination and solve for x and y:

24. $\begin{cases} 3x + 4y = -7 \\ 3x - 3y = 21 \end{cases}$ **25.** $\begin{cases} 2x - 2y = -2 \\ 4x - 5y = -9 \end{cases}$

Simplify:

26. $7\sqrt{20} - 5\sqrt{32}$ **27.** $2\sqrt{18} - 5\sqrt{8} + 4\sqrt{50}$

Simplify:

28. $\dfrac{1 + \dfrac{1}{y}}{\dfrac{1}{y}}$ **29.** $\dfrac{\dfrac{a}{b} - 4}{\dfrac{x}{b} - b}$ **30.** $\dfrac{\dfrac{a}{x} - a}{x + \dfrac{y}{x}}$

LESSON 73   *Trinomials with common factors.*
Subscripted variables

73.A
trinomials with common factors

As the first step in factoring we always check the terms to see if they have a common factor. If they do, we begin by factoring out this common factor. Then we finish by factoring one or both of the resulting expressions.

example 73.1   Factor $x^3 + 6x^2 + 5x$.

solution   If we first factor out the greatest common factor x, we find

$$x^3 + 6x^2 + 5x = x(x^2 + 6x + 5)$$

and now the trinomial can be factored as in the last lesson to get

$$x(x + 5)(x + 1)$$

example 73.2   Factor $4bx^3 - 4bx^2 - 80bx$.

solution   Here we see that the greatest common factor of all three terms is $4bx$, and if we factor out $4bx$, we find

$$4bx^3 - 4bx^2 - 80bx = 4bx(x^2 - x - 20)$$

Now the trinomial can be factored, and the final result is

$$4bx(x - 5)(x + 4)$$

example 73.3   Factor $-x^2 + x + 20$.

solution   **To factor trinomials in which the coefficient of the second-degree term is negative, it is helpful first to factor out a negative quantity.** Here we will factor out (-1).

$$-x^2 + x + 20 = (-1)(x^2 - x - 20)$$

and now we factor the trinomial to get

$$(-1)(x - 5)(x + 4)$$

Finally, the (-1) can be multiplied by either of the binomials to yield two possible final results,

$$(-x + 5)(x + 4) \quad \text{or} \quad (x - 5)(-x - 4)$$

either of which is correct.

example 73.4   Factor $-3x^3 - 6x^2 + 72x$.

solution   First we factor out the greatest common factor $-3x$, and then we factor the trinomial.

$$-3x(x^2 + 2x - 24) = -3x(x + 6)(x - 4)$$

Thus we again find that the original trinomial has three factors.

73.B
subscripted variables

We have used the letter N to represent an unknown number but have always used x and y as variables in systems of two equations such as

$$\text{(a)} \quad \begin{cases} 5x + 10y = 125 \\ x + y = 16 \end{cases} \qquad \text{(b)} \quad \begin{cases} 5x + 25y = 290 \\ x = y + 2 \end{cases}$$

We have solved these systems by using either the substitution method or the elimination method. In Lesson 84, we will look at word problems about coins: nickels, dimes, and quarters. In the equations, we will use N_N for the number of nickels, N_D for the number of dimes, and N_Q for the number of quarters. We will solve the equations by using either the substitution method or the elimination method.

example 73.5   Use elimination to solve: $\begin{cases} 5N_N + 10N_D = 125 \\ N_N + N_D = 16 \end{cases}$

solution   We will multiply the bottom equation by -5 and add it to the top equation.

$$5N_N + 10N_D = 125 \longrightarrow \quad (1) \quad \longrightarrow \quad 5N_N + 10N_D = 125$$
$$N_N + \quad N_D = 16 \longrightarrow (-5) \longrightarrow \quad \underline{-5N_N - \quad 5N_D = -80}$$
$$5N_D = 45$$
$$N_D = 9$$

and since $N_N + N_D = 16$, $N_N = 7$.

example 73.6   Use substitution to solve: $\begin{cases} 5N_N + 25N_Q = 290 \\ N_Q = N_N + 2 \end{cases}$

solution   We will replace N_Q in the top equation with $N_N + 2$ and then solve.

$$5N_N + 25(N_N + 2) = 290 \qquad \text{replaced } N_Q \text{ with } N_N + 2$$
$$5N_N + 25N_N + 50 = 290 \qquad \text{multiplied}$$
$$30N_N + 50 = 290 \qquad \text{simplified}$$
$$30N_N = 240 \qquad \text{added } -50 \text{ to both sides}$$
$$N_N = 8 \qquad \text{divided by 30}$$

Now, since $N_Q = N_N + 2$, $N_Q = 10$.

problem set 73   Factor. First rearrange in descending order of the variable if necessary.

1. $x^2 - 3x - 10$ 2. $x^2 + 12 + 7x$ 3. $-30 - x + x^2$

4. $x^2 + 10 + 7x$ 5. $x^2 + 12 + 8x$ 6. $4 - 4x + x^2$

7. $x^2 + 14 + 9x$ 8. $x^2 - 14 - 5x$ 9. $-3x - 18 + x^2$

10. $6x + 8 + x^2$ 11. $-8 + 2x + x^2$ 12. $-8 - 2x + x^2$

Each of the following has a common factor. Factor out the common factor first.

13. $2x^2 + 10x + 12$ 14. $5x^2 + 30x + 40$

*15. $4bx^3 - 4bx^2 - 80bx$ 16. $ax^2 + 6ax + 9a$

17. $abx^2 - 6ab + abx$ 18. $x^3 + 20x + 9x^2$

Factor a negative common factor from each of these first.

*19. $-x^2 + x + 20$ 20. $-x^3 - x^2 + 20x$ *21. $-3x^3 + 6x^2 + 72x$

22. $-2p^2 + 110 + 12p$ 23. $-3m^2 - 30m - 48$ 24. $-b^3 + 5b^2 + 24b$

Use substitution:

*25. $\begin{cases} 5N_N + 25N_Q = 290 \\ N_Q = N_N + 2 \end{cases}$

Use elimination:

*26. $\begin{cases} 5N_N + 10N_D = 125 \\ N_N + N_D = 16 \end{cases}$

Simplify:

27. $5\sqrt{8} - 14\sqrt{50}$

28. $\dfrac{\dfrac{a}{x} + x}{\dfrac{1}{x} - 1}$

LESSON 74   *Factors that are sums*

74.A
factors
that are sums

Sometimes a trinomial has a common factor that is a sum as we see in the following example.

example 74.1   Factor $(a + b)x^2 - (a + b)x - 6(a + b)$.

solution   Each of the terms has the sum $(a + b)$ as a factor. If we factor out $(a + b)$, we get

$$(a + b)(x^2 - x - 6)$$

Now we finish by factoring the trinomial:

$$(a + b)(x - 3)(x + 2)$$

example 74.2   Factor $(x + y)x^2 + 9x(x + y) + 20(x + y)$.

solution   First we factor out $(x + y)$.

$$(x + y)(x^2 + 9x + 20)$$

Now we finish by factoring the trinomial:

$$(x + y)(x + 4)(x + 5)$$

example 74.3   Factor $m(x - 1)x^2 + 7mx(x - 1) + 10m(x - 1)$.

solution   The greatest common factor of each term of the trinomial is $m(x - 1)$. If we begin by factoring this term, we get

$$m(x - 1)(x^2 + 7x + 10)$$

Now we complete the solution by factoring the trinomial $x^2 + 7x + 10$, and the result is

$$m(x - 1)(x + 2)(x + 5)$$

problem
set 74

Factor. Rearrange in descending order of the variable if necessary.

1. $m^2 + 10m + 16$ 2. $-48 - 8n + n^2$ 3. $y^2 + 56 - 15y$

4. $p^2 - 55 - 6p$ 5. $12t + 35 + t^2$ 6. $y^2 + 50 + 51y$

7. $77 - 18r + r^2$ 8. $m^2 + 21m + 90$ 9. $55 + v^2 + 16v$

10. $-63 - 2h + h^2$ 11. $-30 - 13x + x^2$ 12. $w^2 + 22 - 13w$

Begin by factoring out -1.

13. $-x^2 - 12 + 7x$ 14. $-s^2 - 15 - 8s$ 15. $-a^2 + 40 - 3a$

Begin by factoring the greatest common factor.

16. $2x^3 + 30x + 16x^2$ 17. $4a^2 - 160 + 12a$ 18. $abx^2 - 24ab - 5abx$

The GCFs here are a little different. Factor the GCFs first.

19. $(x - 1)x^2 + 7x(x - 1) + 10(x - 1)$

*20. $m(x - 1)x^2 + 7mx(x - 1) + 10m(x - 1)$

21. $(a + b)x^2 + 8x(a + b) + 15(a + b)$

Use substitution and solve for x and y:

22. $\begin{cases} x + y = 10 \\ x + 2y = 15 \end{cases}$

23. $\begin{cases} 10N_D + 25N_Q = 495 \\ N_Q = N_D + 10 \end{cases}$

Use elimination and solve for x and y:

24. $\begin{cases} 5x - 2y = 9 \\ 3x - y = 6 \end{cases}$

25. $\begin{cases} 2x - 2y = 2 \\ 3x + y = 7 \end{cases}$

Simplify:

26. $3\sqrt{98} - 4\sqrt{50}$

27. $2\sqrt{45} - 2\sqrt{180}$

28. $\dfrac{a + \dfrac{b}{a}}{\dfrac{1}{a} - 4}$

29. $\dfrac{\dfrac{m}{p} + p}{\dfrac{1}{p} - x}$

30. Solve: $-.003k - .03k - .3k - 666 = 0$

LESSON 75 *Difference of two squares*

75.A
difference
of
two squares

Since each of the terms in the following binomials is a perfect square,

$$x^2 - y^2 \qquad 4p^2 - 25 \qquad m^2 - 16$$

these binomials are sometimes called the **difference of two squares.** They can be generated by multiplying the sum and difference of two monomials.

$$
\begin{array}{lll}
\begin{array}{r} x + y \\ x - y \\ \hline x^2 + xy \\ - xy - y^2 \\ \hline x^2 - y^2 \end{array}
&
\begin{array}{r} 2p + 5 \\ 2p - 5 \\ \hline 4p^2 + 10p \\ - 10p - 25 \\ \hline 4p^2 - 25 \end{array}
&
\begin{array}{r} m + 4 \\ m - 4 \\ \hline m^2 + 4m \\ - 4m - 16 \\ \hline m^2 - 16 \end{array}
\end{array}
$$

We note in each case that the middle term is eliminated because the numerical coefficients of the addends that would form the middle term have the same absolute value but are opposite in sign.

If we are asked to factor a binomial that is the difference of two squares such as

$$9m^2 - 49$$

the problem is a problem in recognition. There is no procedure to follow. We recognize that each term of the binomial is a perfect square and that the binomial can be written as

$$(3m)^2 - (7)^2$$

Now from the pattern developed above, we can write

$$9m^2 - 49 = (3m + 7)(3m - 7)$$

example 75.1   Factor $-4 + x^2$.

solution   **We recognize that both of the terms are perfect squares.** We begin by reversing the order of the terms and writing the squared terms as

$$x^2 - 4 = (x)^2 - (2)^2$$

and now we can write the factored form as

$$(x + 2)(x - 2)$$

example 75.2   Factor $49m^2 - a^2$.

solution   **We recognize that each of the terms is a perfect square** and that the binomial can be written as

$$(7m)^2 - (a)^2$$

and the factored form of this binomial is

$$(7m + a)(7m - a)$$

example 75.3   Factor $-36a^2 + 25y^2$.

solution   **We recognize that both of the terms are perfect squares.** We begin by reversing the order of the terms and writing the squared terms as

$$25y^2 - 36a^2 = (5y)^2 - (6a)^2$$

Now we write the factored form as

$$(5y + 6a)(5y - 6a)$$

example 75.4   Factor $-36x^6y^4 + 49a^2$

solution   **Again we recognize that both of the terms are perfect squares.** We write down the answer by inspection as

$$(7a + 6x^3y^2)(7a - 6x^3y^2)$$

**problem
set 75**   Factor:

*1. $x^2 - y^2$ *2. $4p^2 - 25$ *3. $m^2 - 16$

*4. $-4 + x^2$ *5. $49m^2 - a^2$ *6. $-36a^2 + 25y^2$

7. $4p^2x^2 - k^2$ 8. $-4m^2 + 25p^2x^2$ 9. $-9x^2 + 4y^2$

10. $9k^2a^2 - 49$ 11. $p^2 - 4k^2$ 12. $36a^2x^2 - k^2$

Factor. It may be necessary to factor the common factor first.

13. $x^2 - x - 20$ 14. $4x^2 - 4x - 80$ 15. $2b^2 - 48 - 10b$

16. $-90 - 39x + 3x^2$ 17. $(a + b)x^2 + 7(a + b)x + 10(a + b)$

18. $pm^2 + 9pm + 20p$ 19. $5k^2 + 30 + 25k$ 20. $-x^2 - 8x - 7$

21. $5m^2 + 5 - 10m$

Use substitution and solve for x and y:

22. $\begin{cases} x + 2y = 12 \\ 3x + y = 16 \end{cases}$ 23. $\begin{cases} 2x - y = 9 \\ 3x + y = 6 \end{cases}$

Use elimination and solve for x and y:

24. $\begin{cases} 5x - 2y = 3 \\ 2x - 3y = -1 \end{cases}$ 25. $\begin{cases} N_P + N_N = 175 \\ N_P + 5N_N = 475 \end{cases}$

Simplify:

26. $3\sqrt{125} + 2\sqrt{45}$ 27. $5\sqrt{12} - 2\sqrt{27}$

28. $\dfrac{\dfrac{1}{x} + 1}{\dfrac{y}{x} + x}$

29. Add: $\dfrac{x}{x(x + y)} + \dfrac{1}{x}$

30. Solve: $-[2(-3 - k)] = -4(-3) - |-3|k$

LESSON 76   *Scientific notation*

76.A

scientific notation

In science courses, it is sometimes necessary to use extremely large numbers and/or extremely small numbers. For example, to calculate the number of molecules in 1000 liters of a gas, it would be necessary to multiply 1000 times 1000 times a very large number such as 26,890,000,000,000,000,000, which represents the number of molecules in a gas in a cubic centimeter. Besides requiring large pieces of paper, multiplying these numbers in their present form is cumbersome and often leads to errors since it is easy to miscount the number of zeros. If we use a type of mathematical shorthand called **scientific notation,** however, computations such as the above can be performed easily and accurately.

To write a number in scientific notation, the numerator and the denominator are multiplied by the required power of 10 that will place the decimal point immediately to the right of the first nonzero digit in the number **(denominator-numerator same-quantity theorem).** For example, if we wish to write the number

$$.0000416$$

in scientific notation, we would like to place the decimal point between the 4 and the 1.

$$4.16$$

To accomplish this we multiply the number by 10^5, and we must also divide by 10^5 to keep from changing the value of the expression.

$$.0000416 = \frac{.0000416}{1} \frac{10^5}{10^5} = \frac{4.16}{10^5}$$

Now, if we remember that $\dfrac{1}{10^5}$ can be written as 10^{-5}, we can write

$$.0000416 = \frac{4.16}{10^5} = 4.16 \times 10^{-5}$$

We have described the algebraically correct procedure, but since scientific notation is used so often, we prefer to use another thought process. This thought process is much easier to use, but it is not quite so rigorous.

When we look at numbers written in scientific notation such as

$$4.16 \times 10^{+b} \quad \text{and}^{\dagger} \quad 4.16 \times 10^{-b}$$

we think of 10^{+b} as a decimal point indicator that tells us that the true location of the decimal point is really b places to the right of where it is written and 10^{-b} as a decimal

---

† The replacements for b are restricted to positive integers.

point indicator that tells us that the true location of the decimal point is really b **places to the left of where it is written.** If we use this thought process, the 10^{-7} in the notation

$$4.165 \times 10^{-7}$$

tells us that the **true location** of the decimal point **is really** seven places **to the left** of where it is written, giving

$$.0000004165$$

as the number being designated. In a like manner, the exponential 10^7 in the notation

$$4.165 \times 10^7$$

tells us that the **true location** of the decimal point **is really** seven places **to the right** of where it is written, giving

$$41,650,000$$

which is the number being designated.

It is helpful to use a two-step procedure to write a number in scientific notation. The first step is to place the decimal point immediately to the right of the first nonzero digit in the number. Then we follow this notation with the power of 10 that designates the true location of the decimal point. If we use this procedure to write

$$714,600,000$$

in scientific notation, we begin by placing the decimal point immediately to the right of first nonzero digit (which is 7) and dropping the terminal zeros.

$$7.146$$

Now we follow this with $\times 10^8$ to indicate that the **true location of the decimal point is really eight places to the right of where we have written it.**

$$\mathbf{7.146 \times 10^8}$$

example 76.1   Write .000316 in scientific notation.

solution   **We always begin by writing the decimal point immediately after the first nonzero digit.**

$$3.16 \times 10^?$$

Now we must choose an exponent for 10 that tells us what the true location of the decimal point **really is.** Since it **really is** four places to the left of where we have written it, the proper exponent is -4. Thus

$$.000316 \qquad \text{equals} \qquad \mathbf{3.16 \times 10^{-4}}$$

example 76.2   Write $.000316 \times 10^{-7}$ in scientific notation.

solution   We begin by writing .000316 in scientific notation

$$.000316 = 3.16 \times 10^{-4}$$

Thus we can rewrite the original expression as

$$3.16 \times 10^{-4} \times 10^{-7}$$

and this simplifies to

$$\mathbf{3.16 \times 10^{-11}}$$

example 76.3   Write $.000316 \times 10^7$ in scientific notation.

solution   We write .000316 as 3.16×10^{-4} and simplify

$$.000316 \times 10^7 = 3.16 \times 10^{-4} \times 10^7 = \mathbf{3.16 \times 10^3}$$

example 76.4   Write the following numbers in scientific notation.

$$\text{(a)} \quad 47{,}800 \qquad \text{(b)} \quad 47{,}800 \times 10^{-7} \qquad \text{(c)} \quad 47{,}800 \times 10^{7}$$

solution   (a) 47,800   equals   **4.78×10^{4}**
              (b) $47{,}800 \times 10^{-7}$ equals   **4.78×10^{-3}**
              (c) $47{,}800 \times 10^{7}$ equals   **4.78×10^{11}**

A careful study of the four preceding examples is recommended. Often the student searches for a simple rule or shortcut that can be used to solve the problem. No such shortcut exists for writing a number in scientific notation, as the examples will verify.

problem set 76

1. The ratio of withs to withouts was 3 to 11. If 5600 were huddled in the forest, how many were with?

2. The cost of building a house increased 20 percent every year. If it cost $74,000 to build a house one year, what would it cost the next year to build the same house?

3. The fine for sedition was reduced 30 percent. If the new fine was $4900, what was the amount of the fine before the reduction?

4. The ratio of bivalves to other crustaceans was 9 to 1. If there were a total of 130 crustaceans on the table, how many were bivalves?

5. Graph on a number line: $x - 3 \not> 4$, $D = \{\text{Reals}\}$

6. Multiply: $(4 + x)(x^2 + 2x + 3)$

7. Add: $\dfrac{1}{xc^2} + \dfrac{b}{x(c + x)} + \dfrac{5}{x^2c^2}$

8. Simplify: $\dfrac{x + \dfrac{a}{b}}{1 - \dfrac{1}{b}}$

9. Solve: $.4x - 4 - .4 = -.2(4 - x)$

Write these numbers in scientific notation:

*10. 47,800

*11. $47{,}800 \times 10^{-7}$

*12. $47{,}800 \times 10^{7}$

13. .000478

14. Graph $y = -2x - 3\dfrac{1}{2}$ on a rectangular coordinate system.

Use substitution to solve:

15. $\begin{cases} 7x + y = -18 \\ 4x - 2y = 0 \end{cases}$

Use elimination to solve:

16. $\begin{cases} N_D + N_Q = 40 \\ 10N_D + 25N_Q = 475 \end{cases}$

17. Expand: $\dfrac{2x^2}{y^2}\left(\dfrac{-x^2}{2y^{-2}} + \dfrac{x^2a^4}{a^{-2}4^{-2}}\right)$

18. Simplify: $\dfrac{x^{-3}xy^2(y)^{-2}x^{-4}}{(x^0yy^{-2})^2(x^2y^{-3})^{-2}}$

Simplify:

19. $4\sqrt{60} - 7\sqrt{135}$

20. $4\sqrt{80} + 8\sqrt{45}$

Factor the trinomials. If there is a common factor, factor it as the first step.

21. $x^2 + 9x + 20$

22. $x^2 + 15 + 8x$

23. $x^2 + 28 + 11x$

24. $x^3 + 10x^2 + 24x$

25. $ax^2 - 2ax - 15a$

26. $5x^2 - 140 + 15x$

Now factor these binomials. If there is a common factor, factor it as the first step.

27. $5x^2 - 5y^2$

28. $45x^2 - 20m^2$

29. $4a^2 - 9b^2$

30. $49a^2p^2 - k^2$

> The appendix contains four enrichment lessons. These lessons are complete with homework problem sets so that they can be taught, if desired, after Lessons 76, 94 (two enrichment lessons), and 99. Review problems on enrichment topics will be designated at the end of future problem sets. The first enrichment lesson is on closure and should be included at this point if the enrichment lessons are to be included.

LESSON 77   *Consecutive integers*

77.A
integer word problems

If we use the letter N to designate an unspecified integer and then look at the number line

we see that the integer to the right of any given integer is one greater than the given integer. Thus we may use

$$N + 1$$

to designate the next greater integer and $N + 2$ to designate the next greater integer, etc. Integers that are 1 unit apart are called **consecutive integers.**

example 77.1   Find three consecutive integers such that the sum of the first and third is 146.

solution   We will designate the consecutive integers as

$$N \qquad N + 1 \qquad \text{and} \qquad N + 2$$

The sum of the first integer and the third integer is 146, so we write

$$\underset{\sim}{N} + \underbrace{N + 2} = 146 \qquad \text{equation}$$

and now we solve for N:

$$2N + 2 = 146 \qquad \text{added}$$
$$2N = 144 \qquad \text{simplified}$$
$$N = 72 \qquad \text{divided}$$

Thus $N + 1 = 73$, and $N + 2 = 74$. The desired integers are **72, 73,** and **74.**

Check:

$$72 + 74 = 146$$
$$146 = 146 \qquad \text{Check}$$

example 77.2   Find three consecutive integers such that twice the sum of the first two is 2 less than 3 times the third.

solution   We designate the integers as

$$N \qquad N + 1 \qquad \text{and} \qquad N + 2$$

We write the equation as

$$2(N + N + 1) + 2 = 3(N + 2)$$

Note that we added 2 to the sum of the first two integers because this sum was 2 less than. Now we solve

$$2(2N + 1) + 2 = 3(N + 2) \quad \text{simplified}$$
$$4N + 2 + 2 = 3N + 6 \quad \text{multiplied}$$
$$4N + 4 = 3N + 6 \quad \text{simplified}$$
$$N = 2 \quad \text{solved}$$

Thus, the integers are **2**, **3**, and **4**.

Check: $2(2 + 3) + 2 = 3(4) \longrightarrow 12 = 12$ Check

example 77.3 Find four consecutive integers such that 6 times the sum of the first and fourth is 26 less than 10 times the third.

solution We will use N, $N + 1$, $N + 2$, and $N + 3$ to designate the four integers. When we write the equation, we add 26 because 6 times the sum is 26 less than. We want it to be equal to.

$$6(N + N + 3) + 26 = 10(N + 2) \quad \text{equation}$$
$$6(2N + 3) + 26 = 10(N + 2) \quad \text{simplified}$$
$$12N + 18 + 26 = 10N + 20 \quad \text{multiplied}$$
$$12N + 44 = 10N + 20 \quad \text{simplified}$$
$$2N = -24 \quad \text{simplified}$$
$$N = -12 \quad \text{divided}$$

Thus the integers are **−12**, **−11**, **−10**, and **−9**.

Check: $6(-12 - 9) + 26 = 10(-10) \longrightarrow -100 = -100$ Check

problem set 77

*1. Find three consecutive integers such that the sum of the first and the third is 146.

2. Find three consecutive integers such that 2 times the sum of the first two is 6 less than 3 times the third.

3. Find four consecutive integers such that twice the sum of the first and third is 11 greater than 3 times the second.

*4. Find four consecutive integers such that 6 times the sum of the first and the fourth is 26 less than 10 times the third.

Solve:

5. $31\frac{1}{5}x - 2\frac{3}{5} = 14$

6. $-x + 4 - (-2)(-x - 5) = -(-2x + |4|)$

7. True or false? $\{\text{Rationals}\} \subset \{\text{Integers}\}$

8. Graph $y = -\frac{1}{3}x + 2$ on a rectangular coordinate system.

9. Simplify: $\dfrac{1}{(-3)^{-2}} + 3^0 + 2^0$

Write the following numbers in scientific notation:

10. $430,000 \times 10^{-20}$

11. 4300×10^7

12. Seventy-eight percent of the more successful students tended to be serendipitous. If there were 18,400 more successful students in the district, how many tended to be serendipitous?

Add:

13. $\dfrac{a}{x^2} + \dfrac{2}{ax^2} + \dfrac{b}{cx^3}$

14. $\dfrac{a}{x} + \dfrac{b}{x + 6}$

15. Simplify: $\dfrac{aa(a^{-3})^0 a^{-2}}{a^4(x^2)^{-2}}$

16. Simplify by adding like terms: $\dfrac{a^2 x^5}{y} - 3aax^6 x^{-1} y^{-1} + \dfrac{4a^2 y^{-1}}{x^{-5}} - \dfrac{3aax^3 y^{-2}}{y}$

Use substitution:

17. $\begin{cases} N_N = N_Q + 15 \\ 5N_N + 25N_Q = 525 \end{cases}$

Use elimination:

18. $\begin{cases} 2x + 2y = 14 \\ 3x - 2y = -4 \end{cases}$

19. Expand: $\dfrac{3x^2 y^{-2}}{ax^{-1}}\left(\dfrac{x^3}{y^2} - \dfrac{2x^{-2} a}{y^2}\right)$

20. Simplify: $3\sqrt{20} - 2\sqrt{80} + 2\sqrt{125}$

Factor these trinomials:

21. $x^2 - 5x - 14$

22. $-x^3 + 4x^2 + 12x$

23. $ax^2 + 7xa + 10a$

24. $24x + 2x^2 + 70$

25. $24 + 27x + 3x^2$

26. $-px + px^2 - 2p$

Now factor these binomials. If there is a common factor, factor it as the first step.

27. $-9x^2 + m^2$

28. $4x^2 - 9m^2$

29. $125m^2 - 5x^2$

30. $-72k^2 + 2x^2$

ER 77-1, 77-2

LESSON 78 *Consecutive odd and consecutive even integers • Fraction and decimal word problems*

78.A
consecutive odd and consecutive even integers

In the last lesson, we noted that if we use N to represent some unknown integer, then the next larger integer is $N + 1$, the next is $N + 2$, etc. This is because consecutive integers are 1 unit apart on the number line. **Consecutive even integers are different because consecutive even integers are 2 units apart.** If we look at the number line,

we see that -4 and -2 are consecutive even integers and that they are 2 units apart. **Consecutive odd integers are also 2 units apart.** The numbers -3 and -1 are consecutive odd integers and on the number line, and we see that they are 2 units apart. Thus, if we use N to designate an unspecified odd/even integer, the next greater odd/even integer is $N + 2$; the next is $N + 4$; etc.

example 78.1   Find three consecutive even integers such that the sum of the first and second equals the sum of the third and -10.

solution   We will represent the unknown even integers as

$$N \qquad N + 2 \qquad N + 4$$

Thus,

$$N + N + 2 = N + 4 + (-10)$$

now we solve and get

$$2N + 2 = N - 6$$
$$N = -8$$

so the integers are -8, -6, and -4.

Check: $(-8) + (-6) = (-4) + (-10)$
$$-14 = -14 \qquad \text{Check}$$

example 78.2   Find three consecutive odd integers such that the sum of the first and third is 7 greater than the second decreased by 18.

solution   We also use N, $N + 2$, $N + 4$, etc. to represent consecutive odd integers. Thus, we can write the problem as

$$N + N + 4 - 7 = N + 2 - 18$$

and solve

$$2N - 3 = N - 16$$
$$N = -13$$

So $N + 2 = -11$ and $N + 4 = -9$.

Check: $(-13) + (-9) - 7 = (-11) - 18$
$$-29 = -29 \qquad \text{Check}$$

example 78.3   Find four consecutive odd integers such that the sum of the first and fourth is 25 greater than the opposite of the third.

solution   We will use N, $N + 2$, $N + 4$, and $N + 6$ to represent the unknown integers. Thus, we can write

$$N + N + 6 - 25 = -(N + 4)$$

and solve
$$2N - 19 = -N - 4$$
$$3N = 15$$
$$N = 5$$

So $N + 2 = 7$, $N + 4 = 9$, and $N + 6 = 11$.

Check: $5 + 11 - 25 = -9$
$$-9 = -9 \qquad \text{Check}$$

78.B

**fraction
and decimal
word problems**

We have drawn diagrams to help us visualize percent word problems. Since problems involving fractional and decimal parts are essentially the same as percent problems, similar diagrams can be used to help with these. The "before" diagram for these problems will represent 1 instead of 100 percent. The equations we will use are the equations for fractional and decimal parts of a number.

$$(F) \times (\text{of}) = (\text{is}) \quad \text{or} \quad (D) \times (\text{of}) = (\text{is})$$

example 78.4   Lopez used a 5-iron, but the ball covered only $\frac{4}{5}$ of the required distance. If she hit the ball 112 yards, what was the required distance?

solution   The before diagram represents 1 instead of 100 percent.

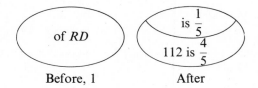

Before, 1   After

From this, we see that $\frac{4}{5}$ of the required distance is 112.

$$(F) \times (\text{of}) = (\text{is}) \quad \rightarrow \quad \frac{4}{5}RD = 112 \quad \longrightarrow \quad \textbf{RD = 140 yards}$$

example 78.5   McAbee guessed that the total was 30.24, but this was 7.2 times the total. What was the total?

solution   The following diagram shows that 30.24 is 7.2 of the total.

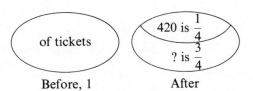

We use this to solve.

$$(D)(\text{of}) = (\text{is}) \quad \longrightarrow \quad (7.2)(T) = 30.24 \quad \longrightarrow \quad T = \frac{30.24}{7.2} \quad \longrightarrow \quad \textbf{T = 4.2}$$

example 78.6   Three-fourths of the tickets had been sold and there were 420 tickets left. How many tickets were printed?

solution

If $\frac{3}{4}$ have been sold, then $\frac{1}{4}$ are left, so we say that $\frac{1}{4}$ of the tickets printed equals 420. Now we write the equation and solve.

$$\frac{1}{4}T = 420 \longrightarrow \frac{\frac{1}{4}T}{\frac{1}{4}} = \frac{420}{\frac{1}{4}} \longrightarrow T = 1680$$

problem set 78

*1. Find three consecutive even integers such that the sum of the first and second equals the sum of the third and -10.

*2. Find three consecutive odd integers such that the sum of the first and third is 7 greater than the second decreased by 18.

*3. Find four consecutive odd integers such that the sum of the first and fourth is 25 greater than the opposite of the third.

*4. Lopez used a 5-iron, but the ball covered only $\frac{4}{5}$ the required distance. If she hit the ball 112 yards, what was the required distance?

*5. McAbee guessed that the total was 30.24, but this was 7.2 times the total. What was the total?

6. Graph on a number line: $4 \le x < 10$; $D = \{\text{Integers}\}$

7. Add: $\dfrac{1}{xc^2} + \dfrac{a}{xc} + \dfrac{m}{c(x + c)}$

8. Simplify: $\dfrac{x + \dfrac{1}{y}}{\dfrac{x^2}{y} - 5}$

9. Solve: $-2|-2| - 2^2 - 3(-2 - x) = -2(x - 3 - 2)$

Write these numbers in scientific notation:

10. 7000×10^{-15}

11. $.000007 \times 10^{-15}$

*12. Three-fourths of the tickets had been sold and there were 420 tickets left. How many tickets were printed?

Graph on a rectangular coordinate system:

13. $y = 2x - 2$

14. $y = -4$

Use substitution to solve:

15. $\begin{cases} x + 3y = 16 \\ 2x - y = 4 \end{cases}$

Use elimination to solve:

16. $\begin{cases} N_N + N_D = 500 \\ 5N_N + 10N_D = 3000 \end{cases}$

17. Expand: $\dfrac{3a^2x}{y^2}\left(\dfrac{y^{-2}a^{-2}}{x^{-1}} - \dfrac{4ax}{y}\right)$

18. Simplify: $\dfrac{(2x^0 x^{-3})^{-2}(y^{-5})^{-2}x}{(x^2 y)(xy^2)}$

Simplify:

19. $2\sqrt{60} - 2\sqrt{135}$

20. $2\sqrt{75} - 6\sqrt{27}$

Factor these trinomials. If there is a common factor, factor it as the first step.

21. $x^2 - 6x + 9$

22. $2x^2 - 8x + 8$

23. $2x^2 + 8x + 8$

24. $2x^2 + 20x + 50$

25. $3x^2 - 30x + 75$

26. $ax^2 - 12ax + 36a$

Now factor these binomials. If there is a common factor, factor it as the first step.

27. $4x^2 - 49$

28. $k^2 - 9x^2y^2$

29. $3p^2 - 12k^2$

30. $-4m^2 + k^2$

ER 78-3, 78-4

LESSON 79   *Rational equations*

79.A
solution
of rational
equations

Often we need to solve an equation in which some of the terms of the equation have denominators that are numbers other than the number 1. The equations

$$\frac{y}{2} + \frac{1}{4} = \frac{y}{6} \quad \text{and} \quad \frac{3y}{2} + \frac{8 - 4y}{7} = 3$$

are equations of this type. Because the terms of the equation are all rational expressions, we call these equations **rational equations. There are many ways that these equations can be solved, but the most straightforward method of attack is to eliminate the denominators first by a judicious application of the multiplicative property of equality.** Since Lesson 47 we have been adding rational expressions by using the least common multiple of the original denominators as the new denominators. To eliminate the denominators in these equations, we will again use the least common multiple, but we will use it in a different way. As permitted by the multiplicative property of equality, we will **multiply the numerator of every term in the equation by the least common multiple of all the denominators of the terms of the equation.** Since every denominator is guaranteed to be a factor of the least common multiple, we are able to eliminate the denominators in one step. The remainder of the solution is straightforward.

example 79.1   Solve $\dfrac{y}{2} + \dfrac{1}{4} = \dfrac{y}{6}$.

solution   **The least common multiple of the denominators is 12. We will multiply the numerator of every term by 12 and cancel the denominators.**

$$\frac{12 \cdot y}{2} + \frac{12 \cdot 1}{4} = \frac{12 \cdot y}{6}$$

Now divide and find

$$6y + 3 = 2y$$

Now solve for y:

$$4y = -3$$

$$y = -\frac{3}{4}$$

example 79.2   Solve $\dfrac{2x}{7} - \dfrac{3x}{2} = \dfrac{1}{3}$.

solution   We begin by multiplying each numerator by 42, which is the **least common multiple of the denominators.**

$$\frac{2x(42)}{7} - \frac{3x(42)}{2} = \frac{1(42)}{3}$$

Now we cancel the denominators and solve.

$$\frac{2x(\overset{6}{\cancel{42}})}{\cancel{7}} - \frac{3x(\overset{21}{\cancel{42}})}{\cancel{2}} = \frac{1(\overset{14}{\cancel{42}})}{\cancel{3}} \longrightarrow 12x - 63x = 14$$

$$\longrightarrow -51x = 14 \longrightarrow x = -\frac{14}{51}$$

example 79.3   Solve $\dfrac{3y}{2} + \dfrac{8 - 4y}{7} = 3$.

solution   The beginner often makes mistakes when trying to eliminate denominators when one or more of the terms has a binomial expression as the numerator. A simple ploy that will prevent the most common mistake is to enclose the binomials in the numerators in parentheses first.

$$\frac{3y}{2} + \frac{(8 - 4y)}{7} = 3$$

Now multiply every numerator by the **least common multiple of the denominators,** which is 14.

$$\frac{14 \cdot 3y}{2} + \frac{14 \cdot (8 - 4y)}{7} = 14 \cdot 3$$

Now divide, simplify, and complete the solution.

$$21y + 2(8 - 4y) = 42 \longrightarrow 21y + 16 - 8y = 42$$
$$\longrightarrow 13y = 26 \longrightarrow y = 2$$

example 79.4   Solve $\dfrac{x + 1}{4} - \dfrac{3}{2} = \dfrac{2x - 9}{10}$.

solution   First we enclose the binomials in parentheses.

$$\frac{(x + 1)}{4} - \frac{3}{2} = \frac{(2x - 9)}{10}$$

Now we multiply every numerator by the **least common multiple of the denominators,** which is 20.

$$\frac{20(x + 1)}{4} - \frac{20 \cdot 3}{2} = \frac{20(2x - 9)}{10}$$

Now we divide, simplify, and complete the solution.

$$5(x + 1) - 30 = 2(2x - 9) \longrightarrow 5x + 5 - 30 = 4x - 18$$
$$\longrightarrow 5x - 25 = 4x - 18 \longrightarrow x = 7$$

problem set 79

1. Find three consecutive integers such that 3 times the sum of the first two is 49 less than the third.

2. Find four consecutive odd integers such that 4 times the sum of the first and third is 4 larger than 4 times the fourth.

3. When the pilgrims counted noses, they got a count that was 128 percent too high. If they counted 9120 noses, what was the correct count?

4. Four thousand two hundred carnations were reserved for use on the float. If this was only $\frac{7}{10}$ of the flowers needed, how many flowers would be on the float?

5. What fraction of 36 is 9?   6. $-2\sqrt{3} \in \{$What sets of numbers$\}$?

7. Add: $\dfrac{a}{x^2y} + \dfrac{b}{x^2y^2} + \dfrac{c}{x^2y^3}$ 8. Simplify: $\dfrac{x^0 x^2 y^{-2} (x^0)^{-2}}{(x^2)^{-3} (y^{-2})^3 y^0}$

Simplify:

9. -3^{-3}

10. $3^0 - (2^0 - 3)(-3 - 2) - (-2)$

11.   Simplify by adding like terms: $\dfrac{m^2xy}{x} - \dfrac{4m^3y}{m} + \dfrac{my}{m^{-1}} - \dfrac{3x^0y}{m^{-2}}$

Write in scientific notation:

12. $.0003 \times 10^{-15}$

13. 4000×10^{-40}

14.   Graph $y = -3x + 4$ on a rectangular coordinate system.

Use substitution to solve:

15. $\begin{cases} N_Q = N_D + 300 \\ 10N_D + 25N_Q = 8200 \end{cases}$

Use elimination to solve:

16. $\begin{cases} 4x - 3y = -3 \\ 2x + 4y = -18 \end{cases}$

17.   Expand: $\left(\dfrac{ax^2}{y^2} - \dfrac{3x}{xy^2}\right)\dfrac{2x^{-3}}{y^2}$

18.   Simplify: $3\sqrt{28} - 5\sqrt{56} + 2\sqrt{63}$

Solve:

*19. $\dfrac{y}{2} + \dfrac{1}{4} = \dfrac{y}{6}$

*20. $\dfrac{3y}{2} + \dfrac{8 - 4y}{7} = 3$

*21. $\dfrac{x + 1}{4} - \dfrac{3}{2} = \dfrac{2x - 9}{10}$

22. $\dfrac{y}{7} + \dfrac{y + 1}{4} = 6$

Factor. Always look for a common factor:

23. $x^2 - 9x + 20$

24. $2ax^2 - 20ax + 42a$

25. $13mx + 42m + mx^2$

26. $16x^2 - 9a^2$

27. $25m^2 - 4$

28. $-36k^2 + 9m^2y^2$

ER 79-5

LESSON 80   *Systems of equations with subscripted variables*

80.A
systems of equations

In Lesson 94 we will introduce uniform motion word problems. These problems will require the use of variables that represent rate or speed and variables that represent time. Rather than use the usual variables x, y, and z, we will use variables that are easy to associate with the words in the problem. If we need a variable to represent the rate of Mike, we will use R_M, which can be read as "the rate of Mike" or "R sub M." If we need a variable to represent Joanie's rate, we will use R_J, which can be read as "the rate of Joanie" *or* "R sub J." In the same way, Bud's time and Sadie's time will be represented by the variables T_B and T_S, which can be read as "the time of Bud" and "the time of Sadie" or as "T sub B" and "T sub S." The first problem we will solve involves the rates and times of Anne and Pat and uses the variables R_A, T_A and R_P, T_P.

example 80.1   Solve the following system of equations for R_A.

$$R_AT_A + R_PT_P = 320, \qquad R_P = 50, \qquad T_P = 4, \qquad T_A = 3$$

solution   In the left equation we will substitute 50, 4, and 3 for R_p, T_p, and T_A, respectively

$$R_A(3) + (50)(4) = 320$$

and solve

$$3R_A + 200 = 320 \longrightarrow 3R_A = 120 \longrightarrow \boldsymbol{R_A = 40}$$

The equations for the next example come from a story problem about a turtle and a rabbit. Thus R_R and T_R stand for the rate of the rabbit and the time of the rabbit and R_T and T_T stand for the rate of the turtle and the time of the turtle.

example 80.2 Solve the following system of equations for T_R and T_T.

$$R_T T_T + 120 = R_R T_R, \qquad R_T = 2, \qquad R_R = 10, \qquad T_T = T_R$$

solution We begin by substituting 2 and 10 for R_T and R_R in the first equation.

$$(a) \quad 2T_T + 120 = 10T_R$$

We have used the first three given equations thus far. The remaining given equation is $T_T = T_R$. We can use this equation to change T_T to T_R in equation (a) or to change T_R to T_T. We choose to change T_T to T_R. We do this and then complete the solution.

$$
\begin{array}{lll}
2T_R + 120 = & 10T_R & \text{substituted } T_R \text{ for } T_T \\
-2T_R & -2T_R & \text{add } -2T_R \text{ to both sides} \\
\hline
120 = & 8T_R &
\end{array}
$$

$$
\frac{120}{8} = \frac{8T_R}{8} \qquad \text{divide both sides by 8}
$$

$$
\mathbf{15 = T_R} \qquad \text{and since } T_T = T_R, \qquad \mathbf{T_T = 15}
$$

Now we will solve the equations from a problem in which Little Brother and Sis take a trip. We will use R_L and T_L for the rate and time of Little Brother and R_S and T_S for the rate and time of Sis.

example 80.3 Solve the following system of equations for T_L and T_S.

$$R_L T_L = R_S T_S, \qquad R_L = 40, \qquad R_S = 80, \qquad T_S = T_L - 5$$

solution We begin with the first equation by replacing R_L with 40 and R_S with 80.

$$40T_L = 80T_S$$

Now we will replace T_S with $T_L - 5$ and multiply out using the distributive property.

$$40T_L = 80(T_L - 5) \longrightarrow 40T_L = 80T_L - 400$$

Now we complete the solution.

$$
\begin{array}{lll}
40T_L = & 80T_L - 400 & \\
-40T_L & -40T_L & \text{add } -40T_L \text{ to both sides} \\
\hline
0 = & 40T_L - 400 & \\
+400 & +400 & \text{add 400 to both sides} \\
\hline
400 = & 40T_L &
\end{array}
$$

$$
\frac{400}{40} = \frac{40T_L}{40} \qquad \text{divide both sides by 40}
$$

so $\mathbf{T_L = 10}$ and since $T_S = T_L - 5$, $T_S = 10 - 5 \longrightarrow \mathbf{T_S = 5}$.

The following equations are from a problem about a freight train and an express train. Thus the rate and time of the freight are symbolized by R_F and T_F and those for the express by R_E and T_E.

example 80.4 Solve the following system of equations for R_F and R_E.

$$R_F T_F = R_E T_E, \qquad T_F = 16, \qquad T_E = 12, \qquad R_E = R_F + 15$$

solution In the left equation we will substitute 16, 12, and $R_F + 15$ for T_F, T_E, and R_E, respectively.

$$R_F T_F = R_E T_E \quad \longrightarrow \quad R_F(16) = (R_F + 15)(12) \quad \longrightarrow \quad 16R_F = 12R_F + 180$$

$$\longrightarrow \quad 4R_F = 180 \quad \longrightarrow \quad \mathbf{R_F = 45}$$

and since $R_E = R_F + 15$, $R_E = 45 + 15 \quad \longrightarrow \quad \mathbf{R_E = 60.}$

problem set 80

1. Find four consecutive even integers such that 4 times the sum of the first and fourth is 8 greater than 12 times the third.

2. Galileo tried for a reasonable result but got $4\frac{3}{5}$. If this was $2\frac{3}{10}$ of his goal, what number was he trying for?

3. In the eighth grade Paul wrestled at 125 pounds. If his weight increased 16 percent in 1 year, at what weight did he wrestle in the ninth grade?

4. With the new tractor Carolyn could plow 67 percent of the farm in 2 weeks. If she plowed 268 acres, how large was the farm?

5. Seven-eighths of the workers in London's fish market used scurrilous language. If 400 did not use scurrilous language, how many worked in the fish market?

6. $2 \in \{\text{What sets of numbers}\}$?

Solve:

7. $3\frac{1}{8}x - 4\frac{2}{5} = 7\frac{1}{2}$

8. $-[-2(x-4) - |-3|] = -2x - 8$

Write the following numbers in scientific notation:

9. $.000135 \times 10^{-17}$

10. $135,000 \times 10^{-17}$

11. Add: $\dfrac{a}{xy} + \dfrac{4}{x(x+y)}$

12. Simplify: $\dfrac{a^0(2x)^{-2}}{a^2(4a^0)^2}$

13. Simplify by adding like terms: $3x^2y - \dfrac{4x^{-2}y^{-2}}{x^{-4}y^{-3}} + \dfrac{5xx}{y^{-1}} - \dfrac{3x^2y^2}{y^{-2}}$

Use substitution:

14. $\begin{cases} x + 3y = 16 \\ 2x - 3y = -4 \end{cases}$

Use elimination:

15. $\begin{cases} N_N + N_D = 22 \\ 5N_N + 10N_D = 135 \end{cases}$

16. Expand: $\dfrac{x^{-2}y^{-2}}{m^2}\left(x^2y^2m^2 - \dfrac{3x^4y^{-4}}{m^{-2}}\right)$

17. Simplify: $5\sqrt{45} - 3\sqrt{180} + 2\sqrt{20}$

Solve:

18. $\dfrac{x}{4} - \dfrac{x+2}{3} = 12$

19. $\dfrac{2y}{4} - \dfrac{y}{7} = \dfrac{y-3}{2}$

20. $\dfrac{p}{6} - \dfrac{2p}{5} = \dfrac{4p-5}{15}$

*21. $R_A T_A + R_P T_P = 320$, $R_P = 50$, $T_P = 4$, $T_A = 3$. Find R_A.

*22. $R_L T_L = R_S T_S$, $R_L = 40$, $R_S = 80$, $T_S = T_L - 5$. Find T_L and T_S.

*23. $R_F T_F = R_E T_E$, $T_F = 16$, $T_E = 12$, $R_E = R_F + 15$. Find R_E and R_F.

*24. $R_T T_T + 120 = R_R T_R$, $R_T = 2$, $R_R = 10$, $T_T = T_R$. Find T_R and T_T.

Factor. Always begin by factoring the common monomial factor if there is one.

25. $p^2 - 55 - 6p$ 26. $-30 - 13x + x^2$ 27. $2m^2 - 24m + 70$

28. $-x^3 + 14x^2 - 40x$ 29. $4m^2 - 49x^2 p^2$ 30. $-16am^2 + 25a^3$

ER 80-6

LESSON 81 *Operations with scientific notation*

81.A

scientific notation

In Lesson 76 we introduced the topic of scientific notation and discussed the method of writing a number in scientific notation. Scientific notation is particularly useful in problems that require the multiplication and division of very large or very small numbers. We will discuss multiplication first.

multiplication

We begin by multiplying the numbers 4,000,000 and 20,000,000 by using scientific notation. As the first step we write both numbers in scientific notation.

$$4,000,000 = 4 \times 10^6 \qquad 20,000,000 = 2 \times 10^7$$

Then we note that the numbers are to be multiplied by writing

$$(4 \times 10^6)(2 \times 10^7)$$

Since the order of multiplication of real numbers does not affect the value of the product, we may rearrange the order of the multiplication and place the exponentials last.

$$(4 \cdot 2)(10^6 \cdot 10^7)$$

Now we multiply 4 by 2 and get 8 and multiply the exponentials by using the product theorem for exponents to get 10^{13}. Thus our answer is the number

$$\mathbf{8 \times 10^{13}}$$

example 81.1   Write the numbers $.003 \times 10^{-4}$ and 2×10^{20} in scientific notation and then multiply.

solution   First we write the numbers in scientific notation.

$$(.003 \times 10^{-4})(2 \times 10^{20}) = (3 \times 10^{-7})(2 \times 10^{20})$$

Next we rearrange the order of the factors and then we multiply:

$$(3 \cdot 2)(10^{-7} \cdot 10^{20}) = \mathbf{6 \times 10^{13}}$$

example 81.2   Multiply $(.00004 \times 10^{-5})(700,000)$.

solution   We write the numbers in scientific notation, rearrange the order of factors, and then multiply.

$$(4 \times 10^{-10})(7 \times 10^5) = (4 \cdot 7)(10^{-10} \cdot 10^5) = 28 \times 10^{-5}$$

$$= \mathbf{2.8 \times 10^{-4}}$$

division   A similar procedure is used to divide numbers written in scientific notation. The exponentials are handled separately from the other numbers. To divide 20,000,000 by 4,000,000, we first write both numbers in scientific notation.

$$\frac{20,000,000}{4,000,000} = \frac{2 \times 10^7}{4 \times 10^6}$$

we can think of this expression as a product of fractions, which we simplify as follows.

$$\left(\frac{2}{4}\right) \cdot \left(\frac{10^7}{10^6}\right) = .5 \times 10^1 = 5$$

example 81.3   Divide .0016 by 400,000.

solution   We write both numbers in scientific notation as the numerator and denominator of a fraction,

$$\frac{1.6 \times 10^{-3}}{4 \times 10^5}$$

which we think of as a product of fractions. Then we simplify both fractions.

$$\left(\frac{1.6}{4}\right)\left(\frac{10^{-3}}{10^5}\right) = .4 \times 10^{-8}$$

Now to finish we write $.4 \times 10^{-8}$ in scientific notation as

$$= 4 \times 10^{-9}$$

multiplication and division   The procedure for simplifying a problem such as

$$\frac{(.06 \times 10^5)(300,000)}{(1000)(.00009)}$$

is first to simplify both the numerator and denominator by using scientific notation. Next we rearrange the expression into a product of fractions and then simplify each fraction.

$$\frac{(6 \times 10^3)(3 \times 10^5)}{(1 \times 10^3)(9 \times 10^{-5})} = \frac{6 \cdot 3}{1 \cdot 9} \times \frac{10^3 \cdot 10^5}{10^3 \cdot 10^{-5}} = \frac{18}{9} \times \frac{10^8}{10^{-2}} = 2 \times 10^{10}$$

example 81.4   Simplify $\dfrac{(.0007 \times 10^{-23})(4000 \times 10^6)}{(.00004)(7,000,000)}$.

solution   We will begin by writing every number in scientific notation. Next we rearrange the expression into a product of two fractions and then simplify each fraction.

$$\frac{(7 \times 10^{-27})(4 \times 10^9)}{(4 \times 10^{-5})(7 \times 10^6)} = \frac{7 \cdot 4}{4 \cdot 7} \times \frac{10^{-27} \cdot 10^9}{10^{-5} \cdot 10^6} = \frac{28}{28} \times \frac{10^{-18}}{10^1} = 1 \times 10^{-19}$$

example 81.5   Simplify $\dfrac{(20 \times 10^{-45})(400 \times 10^{20})}{(100,000)(.0008 \times 10^{-15})}$.

solution   First we write all numbers in scientific notation.

$$\frac{(2 \times 10^{-44})(4 \times 10^{22})}{(1 \times 10^5)(8 \times 10^{-19})}$$

Now we group the exponentials and the other numbers and simplify.

$$\frac{2 \cdot 4}{1 \cdot 8} \times \frac{10^{-44} \cdot 10^{22}}{10^5 \cdot 10^{-19}} = \frac{8}{8} \times \frac{10^{-22}}{10^{-14}} = 1 \times 10^{-8}$$

problem set 81

1. Cindy had three consecutive even integers. Frank and Mark found that the product of their sum and 3 was 20 greater than 8 times the third integer. What were Cindy's integers?

2. Find four consecutive odd integers such that 6 times the sum of the first and the third is 3 more than 5 times the opposite of the fourth.

3. The fairies outnumbered the hamadryads by 130 percent. If there were 460 fairies in the clearing, how many hamadryads were present?

4. Lancelot paid for halberds only 86 percent of what he paid for cuirasses. If he paid 43 farthings for halberds, how much did the cuirasses cost?

5. The sinecure required no work but Querulous was not satisfied because it did not pay enough. It paid 4125 pounds but this was only five-thirteenths of what he expected. How much did Querulous expect?

6. Graph on a number line: $x \nleq 4$; $D = \{\text{Positive integers}\}$

7. Add: $\dfrac{3}{a} + \dfrac{4}{a^2} + \dfrac{7}{a^2(a+x)}$

8. Solve: $-.013 - .013x + .026 = .039$

9. Simplify: $\dfrac{4 + \dfrac{1}{y^2}}{\dfrac{x}{y} + \dfrac{m}{y^2}}$

Simplify:

*10. $\dfrac{(.06 \times 10^5)(300,000)}{(1000)(.00009)}$

*11. $\dfrac{(.0007 \times 10^{-23})(4000 \times 10^6)}{(.00004)(7,000,000)}$

12. Only $\dfrac{2}{17}$ of the teachers were sciolists. If 3000 were not sciolists, how many teachers were there in all?

13. Graph $y = -4\dfrac{1}{2}$ on a rectangular coordinate system.

Use substitution to solve:

14. $\begin{cases} N_N = N_D + 12 \\ 5N_N + 10N_D = 510 \end{cases}$

Use elimination to solve:

15. $\begin{cases} 7x - 4y = 29 \\ 3x + 5y = -1 \end{cases}$

16. Expand: $\dfrac{x^{-2}}{y^2 a}\left(\dfrac{y^2 a^{-3}}{x^{-2}} + \dfrac{3x^{-4}}{y^{-2}a^{-4}}\right)$

17. Simplify: $\dfrac{[x^2(y^5)^{-2}]^{-3}}{(x^0 y^2)y^{-2}}$

18. Simplify: $3\sqrt{8} - 5\sqrt{18} + 6\sqrt{72} - 3\sqrt{50}$

Solve:

19. $\dfrac{3x}{2} - \dfrac{5-x}{3} = 7$

20. $\dfrac{2x-3}{5} - \dfrac{2x}{10} = \dfrac{1}{2}$

21. $R_F T_F = R_S T_S$, $T_S = 6$, $T_F = 5$, $R_F - 16 = R_S$. Find R_S and R_F.

22. $R_M T_M = R_R T_R$, $R_M = 8$, $R_R = 2$, $T_R = 5 - T_M$. Find T_M and T_R.

23. $R_G T_G + R_B T_B = 100$, $R_G = 4$, $R_B = 10$, $T_B = T_G + 3$. Find T_G and T_B.

Factor. Always begin by factoring the common factor if there is one.

24. $x^2 - 14x + 48$

25. $80 - 18x + x^2$

26. $x^3 + 9x^2 + 8x$

27. $-ax^2 + 48a - 13xa$

28. $bcx^2 - a^2 cb$

29. $-m^3 + k^2 m$

ER 81-7

LESSON 82 *Graphical solutions*

graphical solutions of systems of linear equations

In Lessons 50 and 53, we noted that the degree of a term of a polynomial is the sum of the exponents of the variables in the term. Thus $2xy^5$ is a sixth-degree term, $4x$ is a first-degree term, and xy is a second-degree term.

Also, we remember that the degree of a polynomial is the same as the degree of its highest-degree term and that the degree of a polynomial equation is the same as the degree of the highest-degree term in the equation. Thus

$$x^4y + 4y \qquad \text{is a fifth-degree polynomial}$$

$$x^4y + 4y = x \qquad \text{is a fifth-degree polynomial equation}$$

$$x + 2y = 4 \qquad \text{is a first-degree polynomial equation}$$

In Lesson 89 we will learn to solve second degree polynomial equations by factoring, but until then, we will continue to concentrate on first degree equations. We have learned that the graph of a first-degree polynomial equation in two unknowns is a straight line and that we call this kind of equation a linear equation. We have learned to find the solution to a system of two linear equations by using the **substitution method** and the **elimination method.** Here we will see that we can find the solution to a system of two linear equations by **graphing** each of the equations and visually estimating the coordinates of point where the two lines cross. The coordinates of this point will satisfy both equations. The shortcoming of this method is that it is inexact because the coordinates of the crossing point must be estimated.

example 82.1   Solve by graphing: $\begin{cases} y = x + 1 \\ y = -2x + 4 \end{cases}$

solution   We choose values for x and use the equations to find the paired values of y.

$y = x + 1$

x	0	2	-3
y	1	3	-2

$y = -2x + 4$

x	0	2	-2	4
y	4	0	8	-4

It appears that the lines cross at $x = 1$ and $y = 2$, so $(1, 2)$ is our solution.

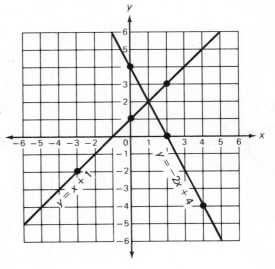

example 82.2   Solve by graphing: $\begin{cases} y = 2 \\ y = x \end{cases}$

solution   We graph the lines as the first step.

$y = x$

x	0	4
y	0	4

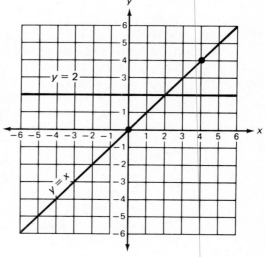

It appears that the lines cross at $x = 2$, $y = 2$, so our graphical solution is **(2, 2)**.

example 82.3   Solve by graphing: $\begin{cases} y = x + 2 \\ y = x - 1 \end{cases}$

solution   It appears from the graph that the lines are parallel and thus never intersect. If this is so, no point lies on both of the lines and no ordered pair will satisfy both equations. An attempt at solving this system by substitution or elimination will degenerate into a false numerical statement. We will look into this in detail in Lesson 115.

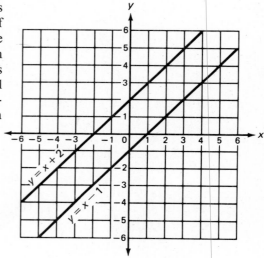

problem set 82

1. Find three consecutive even integers such that 4 times the sum of the first and third is 16 greater than 7 times the second.

2. Find four consecutive integers such that if the sum of the first and third is increased by 10, the result is 6 greater than 4 times the fourth.

3. The new hog food supplement increased the weight gain by $\frac{2}{5}$. If the weight gain used to be 300 pounds, what was the new weight gain?

4. When the leprechauns ran into the forest, the number of little people present was decreased by 35 percent. If 105 ran into the forest, how many were left?

5. What fraction of $7\frac{2}{5}$ is $49\frac{1}{3}$?

6. $-\frac{\sqrt{3}}{2} \in \{\text{What sets of numbers}\}$?

7. Simplify: $\dfrac{\dfrac{x}{y} - 1}{\dfrac{x}{y} + m}$

8. Simplify: $\dfrac{x^2(2y^{-2})^{-3}}{(4x^2)^{-2}}$

9. Simplify: $\dfrac{1}{-3^{-2}} + (-3^0)(-3 - 5)$

10. Simplify by adding like terms: $\dfrac{ax^{-4}}{(x^{-2})^2} + \dfrac{3a^{-2}a^3}{a^0} - \dfrac{6a^5}{(a^{-2})^{-2}} + 3a$

Simplify:

11. $\dfrac{(.003 \times 10^7)(700{,}000)}{(5000)(.0021 \times 10^{-6})}$

12. $\dfrac{(.0007 \times 10^{-10})(4000 \times 10^5)}{(.0004)(7000)}$

13. Four-fifths of the delegates crowded into the convention hall. If 140 could not get in, how many attended the convention?

Solve by graphing:

*14. $\begin{cases} y = x + 1 \\ y = -2x + 4 \end{cases}$

*15. $\begin{cases} y = x \\ y = 2 \end{cases}$

Use substitution to solve:

16. $\begin{cases} x + 5y = 17 \\ 2x - 4y = -8 \end{cases}$

Use elimination to solve:

17. $\begin{cases} N_N + N_D = 30 \\ 5N_N + 10N_D = 250 \end{cases}$

Solve:

18. $\dfrac{x}{3} + \dfrac{5x + 3}{2} = 5$

19. $\dfrac{y + 3}{2} - \dfrac{4y}{3} = \dfrac{1}{6}$

20. Expand: $\dfrac{x^{-2}}{a^2y^{-2}}\left(\dfrac{x^4a^5}{y^4} - \dfrac{3x^{-4}}{a^{-4}y^2}\right)$

21. Simplify: $4\sqrt{28} - 3\sqrt{63} + \sqrt{175}$

22. $R_G T_G + R_B T_B = 120$, $R_G = 4$, $R_B = 10$, $T_G = T_B + 2$. Find T_G and T_B.

23. $R_K T_K = R_N T_N$, $R_K = 6$, $R_N = 3$, $T_K = T_N - 8$. Find T_K and T_N.

Factor. Always look for a common factor.

24. $x^3 - 6x^2 + 5x$

25. $2x^2 + 8 - 10x$

26. $ax^2 + 6a - 7ax$

27. $-mx^2 - 8m - 6mx$

28. $mx^2 - 9ma^2$

29. $-k^2 + 4m^2x^2$

ER 82-8

LESSON 83   *Writing the equation of a line*

83.A
writing
the equation
of a line

We remember that the graph of a vertical line is everywhere equidistant from the y axis. In the left figure on the next page every point on line A is 3 units to the left of the y axis, and the equation of this vertical line is

$$x = -3$$

Every point on line B is 5 units to the right of the y axis, and the equation of this line is

$$x = 5$$

The equation of every vertical line has this form, and if we use k to represent the number, we can say that the general form of a vertical line is

$$x = \pm k$$

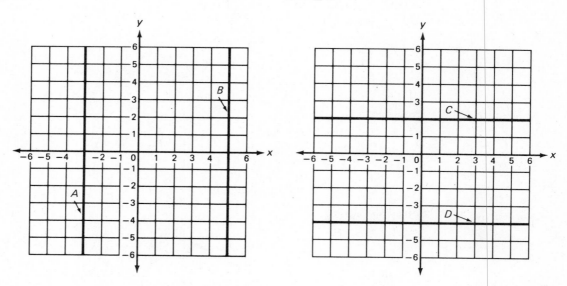

The graph of a horizontal line is everywhere equidistant from the x axis. Every point on line C is 2 units above the x axis, and the equation of this line is

$$y = +2$$

Every point on line D is 4 units below the x axis, and thus the equation of this line is

$$y = -4$$

If we use k to represent the number, we can say that the general form of a horizontal line is

$$y = \pm k$$

Thus, we see that the equations of vertical and horizontal lines can be determined by inspection. These equations contain an x and one number or a y and one number.

$$x = +5 \qquad x = -3 \qquad y = +2 \qquad y = -4$$

The equation of a line that is neither vertical nor horizontal cannot be determined by inspection. However, the equations of these lines can be written in what we call the **slope-intercept form.** The following equations are equations of three different lines written in slope-intercept form.

(a) $y = -6x + 2$ (b) $y = \dfrac{2}{3}x - 5$ (c) $y = .007x + 3$

We note that each equation contains an equals sign, a y, an x, and two numbers. **The only difference in the equations is that the numbers are different.**

We use the letters m and b when we write this equation without specifying the two numbers.

$$y = mx + b$$

Since the equation of any line that is not a vertical line or a horizontal line can be written in this form, the problem of finding the equation of a given line is reduced to the problem of finding the two numbers which will be the values of m and b in the equation.

83.B
intercept

In the slope-intercept form of the equation, $y = mx + b$, we call the constant b the **intercept** of the equation because b represents the value of y when x has a value of zero.

Thus *b* **is the** *y* **coordinate of the line at the point where the line intercepts the** *y* **axis.** The figure shows the graphs of two lines. The *y* coordinate of the upper line at the point where this line intercepts the *y* axis is +4, so the intercept *b* in the equation of this line has a value of 4. The *y* coordinate of the lower line at the point where this line intercepts the *y* axis is −3, so the intercept *b* in the equation of this line has a value of −3.

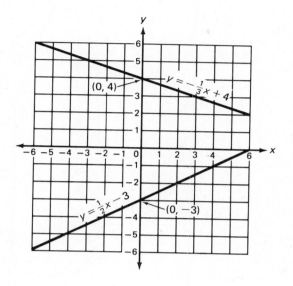

83.C

slope

In the slope-intercept form, $y = mx + b$, we call the constant m the slope of the line. Thus in the equation $y = -2x + 6$, we say that the slope of this line is −2 because the coefficient of x is −2. **We note that the slope has both a sign and a magnitude.**[†] A line represented by a line segment that points toward the upper right part of the coordinate plane has a positive slope. A line represented by a line segment that points toward the lower right part of the coordinate plane has a negative slope. As a mnemonic to help us remember this, we will use the little man and his car. **He always comes from the left side as shown here.**

[†] Absolute value.

He sees the first set of lines as uphill with positive slopes and the second set of lines as downhill with negative slopes.

The magnitude or absolute value of the slope is defined to be the ratio of the absolute value of the change in the y coordinate to the absolute value of the change in the x coordinate as we move from one point on the line to another point on the line.

$$|m| = \frac{|\text{change in } y|}{|\text{change in } x|}$$

 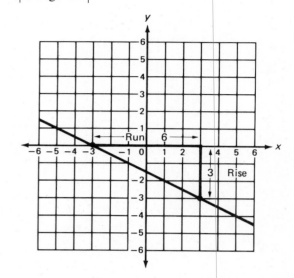

The figure on the left shows the graph of a line that has a negative slope. To find the magnitude of the slope of this line, we arbitrarily choose two points on the line and connect the two points with lines drawn parallel to the coordinate axes. This has been done in the right figure.

The length of the horizontal line is 6 and is the difference of the x coordinate of the two points. The length of the vertical line is 3 and this is the difference of the y coordinates of the two points. Since the magnitude of the slope is the ratio of the absolute value of the change in the y coordinate to the absolute value of the change in the x coordinate, we see that the magnitude or absolute value of the slope of this line is $\frac{1}{2}$.

$$|m| = \frac{|\text{change in } y|}{|\text{change in } x|} \longrightarrow |m| = \frac{3}{6} \longrightarrow |m| = \frac{1}{2}$$

We have labeled the change in x as the **run** and the change in y as the **rise**. Using these words, the magnitude of the slope can be defined as the **absolute value of the rise over the absolute value of the run.**

$$|\text{slope}| = \frac{|\text{rise}|}{|\text{run}|} \qquad \text{or} \qquad |m| = \frac{|\text{rise}|}{|\text{run}|}$$

The general form of the equation of a line is $y = mx + b$, and to write the equation of this line, we need to know (1) the value of the intercept b, (2) the sign of the slope, and (3) the magnitude or absolute value of the slope. We see that

1. The y value of the coordinate of the point where the line intercepts the y axis is approximately -1.4, so $b = -1.4$.
2. The line points to the lower right and thus the *sign* of the slope is negative.
3. The magnitude of the slope is $\frac{3}{6}$, which is equivalent to $\frac{1}{2}$.

So the equation of this line is

$$y = -\frac{1}{2}x - 1.4$$

Not everyone will read the intercept on the graph as -1.4. Some will say -1.5 or some other number close to -1.4. Also, not everyone will compute the magnitude of the slope to be exactly $\frac{1}{2}$. However, the values found for these numbers should be close to -1.4 and $-\frac{1}{2}$.

example 83.1 Find the equations of the lines graphed in the accompanying figures.

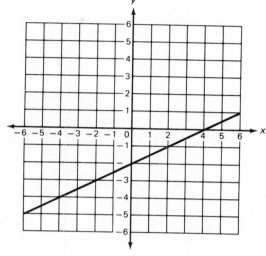

solution The desired equation is $y = mx + b$ and we need to find m and b.

By inspection, $b = +3$.
By inspection, the sign of m is $-$.

The desired equation is $y = mx + b$ and we need to find m and b.

By inspection, $b = -2$.
By inspection, the sign of m is $+$.

Now we need to find the magnitudes or absolute values of the slopes. We will arbitrarily choose two points on each of the lines, draw the triangles, and compute the slopes.

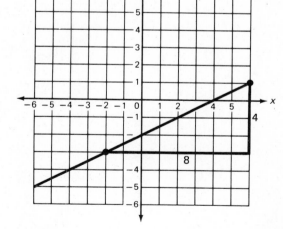

$$|m| = \frac{9}{6} = \frac{3}{2}$$

So $b = +3$ and $m = -\frac{3}{2}$
Using these values in $y = mx + b$ yields

$$y = -\frac{3}{2}x + 3$$

$$|m| = \frac{4}{8} = \frac{1}{2}$$

So $b = -2$ and $m = +\frac{1}{2}$
Using these values in $y = mx + b$ yields

$$y = \frac{1}{2}x - 2$$

The following graphs are provided for classroom use as required for this lesson. Assume that each small square measures 1 unit on a side.

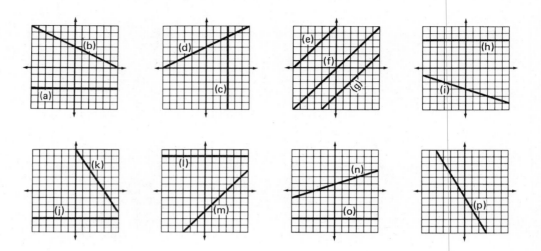

problem set 83

1. Leona found three consecutive integers such that the product of 5 and the sum of the first two was 7 greater than the opposite of the third. What were her integers?

2. Robin and Andrea found four consecutive even integers such that the product of 3 and the sum of the first and second was 18 greater than 5 times the third. What were the integers?

3. The salesman reduced the price 14 percent so he could sell the car for $3440. What was the original price of the car? How much did he reduce the price?

4. When the next rise occurred in the flood tide, the amount of inundated land increased 270 percent. If the former inundation was 1400 acres, how many acres were under water after the rise?

5. Ezekiel had 7452 jeroboams on display. If this was 1.62 times the total number of jeroboams in the next country, how many jeroboams were in the next country?

6. $\dfrac{7\sqrt{2}}{3} \in \{\text{What sets of numbers}\}$?

7. Simplify: $-2(4 - 1)(-1 - 2^0) + |-3 + 5|$

8. What fraction of $2\dfrac{1}{4}$ is $7\dfrac{3}{8}$?

9. Solve: $2.2x - .1x + .02x = -2 - .332$

Find the equations of these lines. Each small square measures 1 unit.

10.   11.   12.   13.

Simplify:

14. $\dfrac{(.0056 \times 10^{-5})(100,000 \times 10^{-14})}{8,000 \times 10^{15}}$

15. $\dfrac{\dfrac{x^2}{y} + y}{a - \dfrac{x}{y}}$

16. Add: $\dfrac{4}{x + y} + \dfrac{6}{x} - \dfrac{4}{ax}$

17. Expand: $\dfrac{x^{-2}a}{y^2}\left(\dfrac{a^4y^{-2}}{x} - \dfrac{3x^2a}{y^2}\right)$

18. Simplify by adding like terms: $a^2xy - \dfrac{3a^2x}{y^{-1}} + \dfrac{4x}{y^{-1}a^2} + \dfrac{5x^{-1}y}{a^2}$

Solve by graphing:

19. $\begin{cases} y = x + 4 \\ y = -x + 2 \end{cases}$

Use substitution:

20. $\begin{cases} 5N_N + 10N_D = 450 \\ N_D = N_N + 30 \end{cases}$

Use elimination:

21. $\begin{cases} 5x - 2y = 7 \\ 4x + y = 3 \end{cases}$

Solve:

22. $\dfrac{p}{4} - \dfrac{p + 2}{6} = -4$

23. $\dfrac{k + 2}{5} - \dfrac{k}{10} = \dfrac{3}{20}$

24. Simplify: $3\sqrt{72} - 14\sqrt{18} + 6\sqrt{8}$

25. $R_T T_T = R_B T_B$, $R_T = 80$, $R_B = 20$, $T_B = T_T + 18$. Find T_B and T_T.

26. $R_L T_L = R_S T_S$, $R_L = 120$, $R_S = 280$, $T_S = 20 - T_L$. Find T_S and T_L.

Factor. Always factor the common factor first.

27. $4x^2 + 40x + 100$

28. $-x^3 - 30x - 11x^2$

29. $ax^2 - 35a + 2ax$

30. $-16x^2y^2 + 4k^2$

ER 83-9

LESSON 84   *Coin problems*

When we solve a word problem, we read the problem and search for statements about quantities that are equal. Each time we find a statement of equality, we write it as an algebraic equation. When we have the same number of equations as we have variables, we solve the equations by using substitution or elimination or by graphing.

The word problems that we have been working thus far have contained only one statement about quantities that are equal. We have worked these problems by using one equation and one variable. Now we will begin solving word problems that contain two statements about quantities that are equal. We will turn each of the statements into an equation that has two variables. Then we will solve the equations by using either substitution or elimination. The first problems of this kind are called **coin problems.**

In coin problems one statement will be about the number of coins. This statement will be like one of the following:

(a)   The number of nickels plus the number of dimes equals 40.

$$N_N + N_D = 40$$

(b)   There were 6 more nickels than dimes.

$$N_N = N_D + 6$$

The other statement will be about the value of the coins. Two examples are:

(c) The value of the nickels plus the value of the dimes equals $4.65.

$$5N_N + 10N_D = 465$$

(d) The value of the dimes and quarters equaled $25.10.

$$10N_D + 25N_Q = 2510$$

To avoid decimal numbers, we will often write all values in cents as in (c) and (d).

example 84.1 Jack and Betty have 28 coins that are nickels and dimes. If the value of the coins is $1.95, how many coins of each type do they have?

solution This is a typical problem about coins. It says that the number of nickels plus the number of dimes equals 28 and that the value of the nickels plus the value of the dimes equals 195 cents.

(a) $N_N + N_D = 28$

(b) $5N_N + 10N_D = 195$

The values of N_N and N_D that will simultaneously satisfy both equations may be found using either the substitution method or the elimination method.

SUBSTITUTION

$$5(28 - N_D) + 10N_D = 195$$
$$140 - 5N_D + 10N_D = 195$$
$$140 + 5N_D = 195$$
$$5N_D = 55$$
$$N_D = 11$$

and since $N_N + N_D = 28$

$$N_N = 17 \quad \text{by inspection.}$$

ELIMINATION

$$-5N_N - 5N_D = -140$$
$$\underline{5N_N + 10N_D = 195}$$
$$5N_D = 55$$

$$N_D = 11$$

and since $N_N + N_D = 28$

$$N_N = 17 \quad \text{by inspection.}$$

example 84.2 Toy has $4.45 in quarters and dimes. She has 8 more quarters than dimes. How many coins of each type does she have?

solution This problem has two statements about things that are equal. The first is that the value of the quarters plus the value of the dimes equals 445 pennies. We write this as

(a) $25N_Q + 10N_D = 445$

Since she has 8 more quarters, we add 8 to the number of dimes so we can write an equation.

(b) $N_D + 8 = N_Q$

To solve, we will substitute equation (b) into equation (a).

$$25(N_D + 8) + 10N_D = 445 \qquad \text{substituted}$$
$$25N_D + 200 + 10N_D = 445 \qquad \text{multiplied}$$
$$35N_D + 200 = 445 \qquad \text{simplified}$$
$$35N_D = 245 \qquad \text{added } -200 \text{ to both sides}$$
$$N_D = 7$$

Thus, $N_Q = (7) + 8 = 15$

example 84.3 Orlando had a hoard of 22 nickels and dimes whose value was $1.35. How many coins of each type did he have?

solution The two statements of equality are: (a) the number of nickels plus the number of dimes equaled 22.

$$\text{(a)} \quad N_N + N_D = 22$$

(b) The value of the nickels plus the value of the dimes equaled 135 pennies.

$$\text{(b)} \quad 5N_N + 10N_D = 135$$

This time we will use elimination. We multiply each term in (a) by -5 to get (a′) which we add to (b).

$$
\begin{array}{rl}
\text{(a′)} & -5N_N - 5N_D = -110 \\
\text{(b)} & \underline{5N_N + 10N_D = 135} \\
& 5N_D = 25 \\
\end{array}
$$

$$N_D = 5 \qquad \text{so } N_N = 17$$

problem set 84

*1. Jack and Betty have 28 coins that are nickels and dimes. If the value of the coins is $1.95, how many coins of each type do they have?

*2. Toy has $4.45 in quarters and dimes. She has 8 more quarters than dimes. How many coins of each type does she have?

*3. Orlando had a hoard of 22 nickels and dimes whose value was $1.35. How many coins of each type did he have?

4. A paroxysm of laughter escaped a few. If the ratio of the laughers to the stolid was 2 to 17 and 7600 were in the throng, how many did not laugh?

Simplify:

5. $\dfrac{(.0016 \times 10^{-7})(3000 \times 10^5)}{1{,}200{,}000}$

6. $\dfrac{(.003 \times 10^{-5})(700 \times 10^{14})}{21{,}000{,}000}$

7. Find the equation of line 7.

8. Find the equation of line 8.

Solve by graphing:

9. $\begin{cases} y = x + 2 \\ y = -x \end{cases}$

Use elimination:

10. $\begin{cases} 3x + 2y = 11 \\ 2x - 3y = 16 \end{cases}$

11. Simplify: $4\sqrt{8} - 5\sqrt{32} + 6\sqrt{18}$

12. $3\dfrac{1}{4}$ of what number is $15\dfrac{1}{2}$?

13. Graph on a number line: $-3 \le x < 2$; $D = \{\text{Positive integers}\}$

14. $R_M T_M + R_S T_S = 170$, $R_M = 20$, $R_S = 30$, $T_M = T_S + 1$. Find T_M and T_S.

15. $R_P T_P = R_M T_M$, $R_P = 45$, $R_M = 15$, $T_P = T_M + 8$. Find T_P and T_M.

16. $R_G T_G + 10 = R_P T_P$, $T_G = 4$, $T_P = 2$, $R_P = R_G + 45$. Find R_P and R_G.

Solve:

17. $\dfrac{x}{3} - 2 = \dfrac{4 - x}{5}$

18. $\dfrac{x}{4} - \dfrac{1}{2} = \dfrac{2 - x}{8}$

19. $-\dfrac{(x + 2)}{3} - \dfrac{2x + 8}{7} = 4$

20. $\dfrac{x}{4} - \dfrac{2x + 5}{2} = 7$

Add:

21. $\dfrac{1}{a^2} + \dfrac{2b}{a^3} - \dfrac{3b}{4a^3}$

22. $\dfrac{4}{a + b} + \dfrac{6}{a^2}$

23. Evaluate $x(x^0 - y) - x^2$ if $x = -2$ and $y = 3$

Simplify:

24. $\dfrac{x + \dfrac{x}{y}}{\dfrac{ax}{y} + 1}$

25. $\dfrac{\dfrac{ab}{c} - \dfrac{1}{c^2}}{4 - \dfrac{a}{c^2}}$

26. -4^{-2}

Factor:

27. $28x + 11x^2 + x^3$

28. $-xy^2 + 4a^2x$

29. Expand: $\dfrac{x^{-2}}{y^2}\left(x^2 y^2 - \dfrac{3a^0 x^2}{y^2}\right)$

Simplify:

30. $\dfrac{(x^2)^{-3} y^2 p^0 x^4}{[(x^2)^{-3} y]^{-2} x^{-4}}$

31. $\dfrac{x^2 y y y^3}{(x^2 y)^0}$

ER 84-10

LESSON 85 *Multiplication of radicals*

85.A

multiplication of radical expressions

We have used the product of square roots theorem to help us simplify radical expressions such as the square root of 50.

$$\sqrt{50} = \sqrt{5 \cdot 5 \cdot 2} = \sqrt{5}\sqrt{5}\sqrt{2} = 5\sqrt{2}$$

The theorem is restated here.

THE PRODUCT OF SQUARE ROOTS THEOREM

If m and n are nonnegative real numbers, then

$$\sqrt{n}\sqrt{m} = \sqrt{mn} \quad \text{and} \quad \sqrt{mn} = \sqrt{m}\sqrt{n}$$

We can use the theorem the other way to multiply radical expressions as shown here.

$$\sqrt{2}\sqrt{3} = \sqrt{6}$$

example 85.1 Simplify $4\sqrt{3} \cdot 3\sqrt{2}$.

solution Since the order of the factors does not affect the product in the multiplication of real numbers, we will rearrange the factors as

$$4 \cdot 3\sqrt{3}\sqrt{2}$$

and 4 is multiplied by 3 and $\sqrt{3}$ is multiplied by $\sqrt{2}$ to yield

$$\mathbf{12\sqrt{6}}$$

example 85.2 Simplify $4\sqrt{3} \cdot 6\sqrt{6}$.

solution Again we will rearrange the order of the indicated multiplication to get

$$4 \cdot 6\sqrt{3}\sqrt{6}$$

and multiply 4 by 6 and $\sqrt{3}$ by $\sqrt{6}$ to get

$$24\sqrt{18}$$

and since $\sqrt{18}$ can be written as $3\sqrt{2}$, we can write

$$24\sqrt{18} = 24(3\sqrt{2}) = \mathbf{72\sqrt{2}}$$

example 85.3 Simplify $4\sqrt{3}(5\sqrt{2} + 6\sqrt{3})$.

solution The notation indicates that $4\sqrt{3}$ is to be multiplied by both terms within the parentheses:

$$4\sqrt{3}(5\sqrt{2} + 6\sqrt{3}) = 4\sqrt{3} \cdot 5\sqrt{2} + 4\sqrt{3} \cdot 6\sqrt{3}$$

Now we rearrange the order of the factors in each term to get

$$4 \cdot 5\sqrt{3}\sqrt{2} + 4 \cdot 6\sqrt{3}\sqrt{3}$$

and lastly we perform the indicated multiplications.

$$\mathbf{20\sqrt{6} + 72}$$

example 85.4 Simplify $4\sqrt{2}(3\sqrt{2} + 5)$.

solution First we multiply as indicated

$$4\sqrt{2} \cdot 3\sqrt{2} + 4\sqrt{2} \cdot 5$$

and now simplify

$$\mathbf{24 + 20\sqrt{2}}$$

problem set 85

1. When the piggy bank was opened, it yielded $4.75 in nickels and pennies. If there were 175 coins in all, how many were nickels and how many were pennies?

2. The nickels and dimes all fell on the floor. There were 12 more nickels than dimes and the total value of the coins was $5.10. How many nickels and how many dimes were on the floor?

3. The room was a mess. In 1 hour Gretchen picked up 80 percent of the toys on the floor. If she picked up 128 toys, how many were on the floor to begin with?

4. Ramses cogitated. He thought of three consecutive even integers and found that 3 times the sum of the first two was 58 less than 14 times the opposite of the third. What were his integers?

5. Simplify: $\dfrac{(.00032 \times 10^{-5})(4000 \times 10^7)}{(160,000)(.00002)}$

6. Find the equation of line 6.

7. Find the equation of line 7.

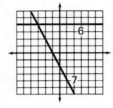

Simplify:

*8. $4\sqrt{3} \cdot 3\sqrt{2}$

*9. $4\sqrt{3}(5\sqrt{2} + 6\sqrt{3})$

10. $5\sqrt{2}(2\sqrt{3} + \sqrt{6})$

11. $5\sqrt{8} - 4\sqrt{32}$

Solve by graphing:

12. $\begin{cases} y = -2x - 2 \\ y = -4 \end{cases}$

Solve by elimination:

13. $\begin{cases} 4x + 3y = -14 \\ 3x + 2y = -10 \end{cases}$

14. What fraction of $3\dfrac{1}{8}$ is $1\dfrac{1}{8}$?

15. $-4 \in \{$What sets of numbers$\}$?

Solve:

16. $\dfrac{3 + x}{4} - \dfrac{x}{3} = 5$

17. $\dfrac{1}{2} - \dfrac{2x}{5} = 7$

18. $R_M T_M = R_K T_K$, $R_M = 30$, $R_K = 10$, $T_M = 16 - T_K$. Find T_M and T_K.

Add:

19. $\dfrac{1}{x^2} - \dfrac{3a}{x - a} + \dfrac{2}{x}$

20. $-\dfrac{a}{x} + \dfrac{b}{x^2 c} - \dfrac{d}{x^3 c^2}$

Simplify:

21. $\dfrac{a + \dfrac{1}{a}}{\dfrac{3}{a^2} - b}$

22. $\dfrac{\dfrac{abx}{c} - 4}{\dfrac{1}{c} + a}$

23. Evaluate $x^2 - y^2 - x(x - y) - |x - y^2|$ if $x = -3$ and $y = 4$

Simplify:

24. $-2^0[(3^0 - 5) - 2^2(2 - 3) + 5]$

25. $-3[(-2 - 2) - 2^0 - (4 - 3)(-2)]$

26. Solve: $-.2 - .02 - .02x = .4(1 - x) - .012$

27. Expand: $\dfrac{3x^{-2}}{a^2 y^5}\left(\dfrac{x^2 a^2}{y^5} - \dfrac{x^{-3} y}{a}\right)$

Simplify:

28. $\dfrac{(a^{-3})^0 (a^2)^0 (a^{-2})^{-2}}{a^4 (x^{-5})^{-2} x x^2}$

29. $\dfrac{(x^2 y) x y^{-2}}{x^2 y y}$

LESSON 86   *Division of polynomials*

86.A

division of polynomials

In the problem sets since Lesson 51 we have been finding the product of two polynomials. Now we investigate the inverse process, the division of polynomials.

The simplest type of polynomial division is the division of a polynomial by a monomial. The desired result can be obtained by dividing each term of the polynomial by the monomial as shown in the following examples.

example 86.1   Divide $3x^3 + 7x^2 - x$ by x.

solution   We will divide each of the three terms by x.

$$\frac{3x^3}{x} + \frac{7x^2}{x} - \frac{x}{x} = 3x^2 + 7x - 1$$

example 86.2   Divide $12x^{12} - 8x^8 + 4x^6$ by $4x^4$.

solution   We will divide every term by $4x^4$.

$$\frac{12x^{12}}{4x^4} - \frac{8x^8}{4x^4} + \frac{4x^6}{4x^4} = 3x^8 - 2x^4 + x^2$$

Before considering the method of dividing a polynomial by a binomial, we will complete a long division problem involving the natural numbers, an algorithm with which we are familiar. The method of dividing a polynomial by a binomial is very similar.

$$
\begin{array}{r}
4 \\
12 \overline{\smash{)}49} \\
48 \\
\hline
1
\end{array}
\qquad \text{so } 49 \div 12 = 4\frac{1}{12}
$$

We multiplied 4 by 12 and recorded the product of 48 below the 49. Next we **mentally changed the sign** of the $+48$ to -48 and then added algebraically to find the remainder of 1.

example 86.3   Divide $-2x^2 + 3x^3 + 5x + 50$ by $-3 + x$.

solution   The first step is to write both polynomials in descending powers of the variable and use the same format for division as we used above.

$$x - 3 \overline{\smash{)}3x^3 - 2x^2 + 5x + 50}$$

To determine the first term of the quotient, divide the first term of the divisor into the first term of the dividend; in this case, divide x into $3x^3$ and get $3x^2$. Record as indicated.

$$
\begin{array}{r}
3x^2 \\
x - 3 \overline{\smash{)}3x^3 - 2x^2 + 5x + 50}
\end{array}
$$

Now multiply the term $3x^2$ by $x - 3$ and record as shown below:

$$
\begin{array}{r}
3x^2 \\
x - 3 \overline{\big)\; 3x^3 - 2x^2 + 5x + 50} \\
3x^3 - 9x^2
\end{array}
$$

Now **mentally change the sign** of both $3x^3$ and $-9x^2$ and add algebraically:

$$
\begin{array}{r}
3x^2 \\
x - 3 \overline{\big)\; 3x^3 - 2x^2 + 5x + 50} \\
3x^3 - 9x^2 \\
\hline
+7x^2
\end{array}
$$

Now bring down the $+5x$:

$$
\begin{array}{r}
3x^2 \\
x - 3 \overline{\big)\; 3x^3 - 2x^2 + 5x + 50} \\
3x^3 - 9x^2 \\
\hline
+7x^2 + 5x
\end{array}
$$

Now divide the x of $x - 3$ into $7x^2$, get $7x$, and record as shown.

$$
\begin{array}{r}
3x^2 + 7x \\
x - 3 \overline{\big)\; 3x^3 - 2x^2 + 5x + 50} \\
3x^3 - 9x^2 \\
\hline
7x^2 + 5x
\end{array}
$$

Multiply $7x$ by $x - 3$ and record:

$$
\begin{array}{r}
3x^2 + 7x \\
x - 3 \overline{\big)\; 3x^3 - 2x^2 + 5x + 50} \\
3x^3 - 9x^2 \\
\hline
7x^2 + 5x \\
7x^2 - 21x
\end{array}
$$

Now **mentally change the sign** of both $7x^2$ and $-21x$ and add algebraically. Repeat the procedure until the remainder of 128 is obtained.

$$
\begin{array}{r}
3x^2 + 7x + 26 \\
x - 3 \overline{\big)\; 3x^3 - 2x^2 + 5x + 50} \\
3x^3 - 9x^2 \\
\hline
7x^2 + 5x \\
7x^2 - 21x \\
\hline
26x + 50 \\
26x - 78 \\
\hline
128
\end{array}
$$

Thus we find that

$$
\frac{3x^3 - 2x^2 + 5x + 50}{x - 3} = 3x^2 + 7x + 26 + \frac{128}{x - 3}
$$

example 86.4   Divide $2x^3 - x - x^2 + 4$ by $-2 + x$.

solution   As the first step, we rearrange both expressions in descending powers of the variable and write

$$
x - 2 \overline{\big)\; 2x^3 - x^2 - x + 4}
$$

Now we divide using the same procedure we used in the last example.

$$
\begin{array}{r}
2x^2 + 3x + 5 \\
x - 2 \overline{\smash{\big)}\ 2x^3 - x^2 - x + 4} \\
\underline{2x^3 - 4x^2} \\
3x^2 - x \\
\underline{3x^2 - 6x} \\
5x + 4 \\
\underline{5x - 10} \\
14
\end{array}
$$

Thus,

$$(2x^3 - x - x^2 + 4) \div (-2 + x) = 2x^2 + 3x + 5 + \frac{14}{x - 2}$$

problem set 86

1. There were 40 dimes and quarters in the drawer. Peggy counted them and found that their total value was $4.75. How many coins of each type were there?

2. Jimmy had $5.25 in nickels and quarters. If he had 15 more nickels than quarters, how many coins of each type did he have?

3. Find four consecutive odd integers such that the product of -3 and the sum of the first and fourth is 30 less than 10 times the opposite of the third.

4. When the tractorcade approached the town square, the driver of the lead vehicle was told that only 64 percent of the tractors had enough fuel. If 224 tractors had enough fuel, how many tractors were in the tractorcade?

5. Simplify: $\dfrac{(.0003 \times 10^{-8})(8000 \times 10^6)}{.004 \times 10^5}$

6. Find the equation of line 6.

7. Find the equation of line 7.

Simplify:

8. $4\sqrt{3} \cdot 6\sqrt{6}$

9. $3\sqrt{2}(7\sqrt{2} - \sqrt{6})$

10. $5\sqrt{45} - \sqrt{180}$

Divide:

*11. $(3x^3 + 7x^2 - x) \div x$

*12. $(-2x^2 + 3x^3 + 5x + 50) \div (-3 + x)$

*13. $(2x^3 - x - x^2 + 4) \div (-2 + x)$

14. Solve: $x^0 + (3x)^0 - 2 - x = -4(x + 2)$

15. Add: $\dfrac{5}{x^2 + y} + \dfrac{3}{x^2} + \dfrac{2}{x}$

16. Simplify: $\dfrac{\dfrac{x}{y^2} + \dfrac{4}{x}}{\dfrac{1}{y^2} + \dfrac{2}{xy}}$

17. Expand: $\dfrac{a^{-2}}{x^4 y}\left(\dfrac{x^4 a^2}{y^{-1}} - \dfrac{3a^{-2}y}{x^4}\right)$

18. Simplify by adding like terms: $\dfrac{4x^2y}{xy} - \dfrac{3x^{-3}y^0}{x^{-4}} + \dfrac{2}{x^{-1}} - \dfrac{4x}{y}$

Solve by graphing: Solve by substitution: Solve by elimination:

19. $\begin{cases} y = 3x - 2 \\ y = -x + 2 \end{cases}$ 20. $\begin{cases} 4x - 5y = -26 \\ x - y = -6 \end{cases}$ 21. $\begin{cases} 4x - 5y = 45 \\ 2x - 3y = 25 \end{cases}$

Solve:

22. $\dfrac{x}{2} - \dfrac{3+x}{4} = \dfrac{1}{6}$ 23. $\dfrac{3x}{2} - \dfrac{4x+1}{5} = 4$

24. $R_M T_M + 6 = R_D T_D$, $R_M = 3$, $R_D = 12$, $T_M = 4 + T_D$. Find T_M and T_D.

25. Simplify: $\dfrac{xx^{-3}x^4(y^2)}{x^2(2x)^{-3}}$

Factor. Always factor the common factor first.

26. $ax^2 + 4ax + 4a$ 27. $-10 - 3x + x^2$

28. $-4ax^2 + 9a$ 29. $20x + 12x^2 + x^3$

ER 86-11

LESSON 87 *More on systems of equations*

87.A
more on systems of equations

When we solve systems of equations using the substitution method, sometimes it is necessary to rearrange one of the equations before the necessary substitutions can be accomplished. In the example problems shown here, we will have to rearrange the equation that relates the time variables before we can substitute.

example 87.1 $R_E T_E = R_W T_W$, $R_E = 200$, $R_W = 250$, $T_E + T_W = 9$. Find T_E and T_W.

solution We begin by replacing R_E with 200 and R_W with 250 to find

$$200T_E = 250T_W$$

Now we must use the equation $T_E + T_W = 9$ to substitute for T_E or for T_W. The equation cannot be used in its present form and must be solved for T_E by adding $-T_W$ to both sides or be solved for T_W by adding $-T_E$ to both sides. We will work the problem both ways to show that the final results will be the same.

$$\begin{array}{rcl} T_E + T_W &=& 9 \\ -T_W & & -T_W \\ \hline T_E &=& 9 - T_W \end{array}$$

$$\begin{array}{rcl} T_E + T_W &=& 9 \\ -T_E & & -T_E \\ \hline T_W &=& 9 - T_E \end{array}$$

Now we will substitute $9 - T_W$ for T_E and solve the resulting equation for T_W:

Now we will substitute $9 - T_E$ for T_W and solve the resulting equation for T_E:

$$\begin{array}{rcl} 200(9 - T_W) &=& 250T_W \\ 1800 - 200T_W &=& 250T_W \\ +200T_W & & +200T_W \\ \hline 1800 &=& 450T_W \end{array}$$

$$\begin{array}{rcl} 200T_E &=& 250(9 - T_E) \\ 200T_E &=& 2250 - 250T_E \\ 250T_E & & +250T_E \\ \hline 450T_E &=& 2250 \end{array}$$

so $T_W = \dfrac{1800}{450}$ or $\mathbf{T_W = 4}$.

so $T_E = \dfrac{2250}{450}$ or $\mathbf{T_E = 5}$.

Since $T_E + T_W = 9$, we can solve for T_E by replacing T_W with 4:

$$T_E + 4 = 9$$
$$\underline{-4 \quad\quad -4}$$
$$T_E \quad = 5$$

Since $T_E + T_W = 9$, we can solve for T_W by replacing T_E with 5.

$$5 + T_W = 9$$
$$\underline{-5 \quad\quad -5}$$
$$T_W = 4$$

Thus we see that while one procedure leads to our solving for T_E first and the other leads to our solving for T_W first, both procedures finally yield the same answers.

example 87.2 $R_1 T_1 + R_2 T_2 = 360$, $R_1 = 30$, $R_2 = 40$, $T_1 + T_2 = 10$. Find T_1 and T_2.

solution   The time equation cannot be used in its present form. We decide to solve the time equation for T_1.

$$T_1 + T_2 = 10$$
$$\underline{\quad - T_2 \quad\quad\quad - T_2}$$
$$T_1 \quad\quad = 10 - T_2$$

Now we replace R_1 with 30, R_2 with 40, and T_1 with $10 - T_2$ and then solve.

$$30(10 - T_2) + 40T_2 = 360 \qquad \text{substituted}$$
$$300 - 30T_2 + 40T_2 = 360 \qquad \text{multiplied}$$
$$300 + 10T_2 = 360 \qquad \text{simplified}$$
$$10T_2 = 60 \qquad \text{added } -300 \text{ to both sides}$$
$$T_2 = 6$$

Thus,

$$T_1 = 4$$

problem set 87

1. The big sack contained \$30 in nickels and dimes. If there were 500 coins in the sack, how many were nickels and how many were dimes?

2. When the box broke open, \$82 in quarters and dimes fell out. If there were 300 more quarters than dimes, how many dimes and how many quarters were there?

3. Find four consecutive integers such that 3 times the sum of the first and third is 84 greater than the opposite of the second.

4. Ninety percent of the people who voted voted for Sammy because he was 7 feet tall. If 1930 people voted, how many voted for Sammy?

5. Simplify: $\dfrac{(.0072 \times 10^{-4})(100{,}000)}{6000 \times 10^{-24}}$

6. Find the equation of line 6.

7. Find the equation of line 7.

Simplify:

8. $3\sqrt{2}(4\sqrt{2} + 6\sqrt{6})$ **9.** $5\sqrt{3}(2\sqrt{3} - 6\sqrt{12})$

Divide:

10. $x^3 - 2x^2 + 4x$ by $x - 2$ **11.** $2x^3 - 3x^2 + 2x - 4$ by $x + 3$

12. .05 of what number is .0009?

Add:

13. $\dfrac{x}{24a} + \dfrac{y}{70a^2}$ **14.** $\dfrac{k}{42} - \dfrac{3x}{18}$

15. Simplify: $\dfrac{\dfrac{x}{y} - 1}{\dfrac{a}{y} + b}$

16. Simplify by adding like terms: $\dfrac{x^3y}{xy^{-1}} + 3xxy^2 - \dfrac{2x^4x}{x^2xy^{-2}} - \dfrac{5x^2}{xy}$

17. Solve: $(3x)^0(-2 - 3x) - x = -3(-2 - 3)$

Solve by graphing: Solve by substitution: Solve by elimination:

18. $\begin{cases} y = -2 \\ y = 2x - 2 \end{cases}$ **19.** $\begin{cases} x + y = 2 \\ 2x - 3y = -1 \end{cases}$ **20.** $\begin{cases} 3x - y = 8 \\ x + 2y = 12 \end{cases}$

21. Expand: $\left(x^2y - \dfrac{3x^{-4}a}{y}\right)\dfrac{x^{-2}}{y}$ **22.** Solve: $\dfrac{4 + x}{2} - \dfrac{1}{3} = \dfrac{1}{6}$

*23. $R_ET_E = R_WT_W$, $R_E = 200$, $R_W = 250$, $T_E + T_W = 9$. Find T_E and T_W.

*24. $R_1T_1 + R_2T_2 = 360$, $R_1 = 30$, $R_2 = 40$, $T_1 + T_2 = 10$. Find T_1 and T_2.

25. Simplify: $\dfrac{(x^2)^{-2}(y^0)^2yy^3}{(y^{-2})^3yy^4y^{-1}x}$

Factor. Always factor the greatest common factor first.

26. $max^2 + 9xma + 14ma$ **27.** $-x^3 - 35x - 12x^2$

28. $a^6m^6 - y^2$ **29.** $4a^3x^2 - a^3y^2$

ER 87-12

LESSON 88   *More on division of polynomials*

88.A
polynomial dividends with missing terms

We remember that the first step in dividing polynomials is to rearrange the terms so that both expressions are written in descending order of the variable. Division can be accomplished without this step, but it is much more difficult. In this lesson, we discuss another helpful procedure that can be used in polynomial division.

example 88.1   Divide $-2x + 5 + 3x^3$ by $-3 + x$.

solution We begin by writing both polynomials in descending order of the variable and using the format for long division.

$$x - 3 \overline{)\, 3x^3 - 2x + 5}$$

We note that the dividend has an x^3 term and an x term but no x^2 term. A good ploy to avoid confusion is to insert an x^2 term with zero as its coefficient as shown below. Of course, zero multiplied by x^2 equals zero so the polynomial is really unchanged. Now we perform the division using the same procedure we learned in Lesson 86.

$$
\begin{array}{r}
3x^2 + 9x\ \ + 25 \\
x - 3 \overline{)\, 3x^3 + 0x^2 -\ \ 2x +\ \ 5} \\
\underline{3x^3 - 9x^2} \\
9x^2 -\ \ 2x \\
\underline{9x^2 - 27x} \\
25x +\ \ 5 \\
\underline{25x - 75} \\
80
\end{array}
$$

Thus,

$$\frac{3x^3 - 2x + 5}{x - 3} = 3x^2 + 9x + 25 + \frac{80}{x - 3}$$

example 88.2 Divide $-4 + x^3$ by $-3 + x$.

solution Again we begin by rearranging each polynomial and using the long division format.

$$x - 3 \overline{)\, x^3 - 4}$$

This time, we see that the dividend does not have an x^2 term or an x term. We will insert these terms and give each of them a coefficient of zero. Then we will divide.

$$
\begin{array}{r}
x^2 + 3x\ \ + 9 \\
x - 3 \overline{)\, x^3 + 0x^2 + 0x -\ \ 4} \\
\underline{x^3 - 3x^2} \\
3x^2 + 0x \\
\underline{3x^2 - 9x} \\
9x -\ \ 4 \\
\underline{9x - 27} \\
23
\end{array}
$$

Thus,

$$(-4 + x^3) \div (-3 + x) = x^2 + 3x + 9 + \frac{23}{x - 3}$$

problem set 88

1. There were 143 more Susan B. Anthony dollars than there were quarters. If the total value of the coins was $153, how many dollars were there?

2. The collection had 60 coins that were nickels and dimes. If the total value of the coins was $5, how many nickels were in the collection?

3. Forty percent of the crop was destroyed by the thunderstorm. If the farm consisted of 570 acres, how many acres were affected by the thunderstorm?

4. Find three consecutive integers such that -4 times the sum of the first and third is 13 less than 7 times the opposite of the second.

5. Simplify: $\dfrac{(.00035 \times 10^{-8})(2000 \times 10^{-3})}{(.0007 \times 10^{6})(2,000,000)}$

6. Find the equation of line 6.

7. Find the equation of line 7.

Simplify:

8. $5\sqrt{2}(3\sqrt{2} - 2\sqrt{12})$

9. $3\sqrt{5}(2\sqrt{5} - 6\sqrt{15})$

Divide:

*10. $(-2x + 5 + 3x^3) \div (-3 + x)$

11. $(3x^3 - x - 7)$ by $(x - 5)$

Solve by graphing:

12. $\begin{cases} x = 2 \\ y = 2x - 4 \end{cases}$

Solve by substitution:

13. $\begin{cases} 2x + y = 1 \\ 5x - 2y = 7 \end{cases}$

Solve by elimination

14. $\begin{cases} 7x + 2y = 3 \\ 3x - 3y = 9 \end{cases}$

Simplify:

15. $3\sqrt{2}(4\sqrt{2} - \sqrt{12})$

16. $2\sqrt{2}(5\sqrt{2} - 3\sqrt{20})$

17. $2\sqrt{3}(\sqrt{3} - 2\sqrt{15})$

18. Solve: $2\dfrac{1}{4}p - 3\dfrac{1}{8} = \dfrac{1}{5}$

19. Graph: $x \not< 3$; $D = \{\text{Positive integers}\}$

Solve:

20. $\dfrac{x}{4} - \dfrac{x - 2}{3} = 8$

21. $\dfrac{x}{3} - \dfrac{x + 2}{4} = 5$

Add:

22. $\dfrac{3}{a + b} - \dfrac{4}{b} + \dfrac{6}{b^2}$

23. $\dfrac{3}{a^2 x} + \dfrac{2b}{a(x + a)}$

24. Simplify: $\dfrac{x - \dfrac{4}{y}}{\dfrac{a}{y} + 3}$

25. Evaluate $x^2 - x^0 - (x^0)^2 + ax(x - a)$ if $x = -2$ and $a = 4$

26. Solve: $(-2 - 3)x^0 - 2(-2)x = 3(x^0 - 2)$

27. Factor: $15ax + 56a + ax^2$

28. $R_P T_P = R_D T_D - 90$, $R_P = 30$, $R_D = 30$, $T_P + T_D = 9$. Find T_P and T_D.

29. Expand: $\dfrac{a^2 y^{-2}}{x^4}\left(\dfrac{a^{-4}y^{-2}}{x} - \dfrac{3a^{-2}x}{y}\right)$

30. Simplify: $\dfrac{xxx^{-2}(y^0)(3y^{-2})^{-1}}{(2x^{-4})^2 y}$

LESSON 89 *Solution of quadratic equations by factoring*

89.A
quadratic equations

Quadratic equations are second-degree polynomial equations that can be written in one of two forms shown here.

$$ax^2 + bx + c = 0 \qquad ax^2 + bx + c = y$$

In this book we will restrict our investigation of quadratic equations to equations in the form of the equation on the left. This equation is a second-degree polynomial equation in one unknown. If the equation is written in descending powers of the variable with all nonzero members on the left side of the equation, we say that the equation is written in **standard form.** Thus the equation

$$x^2 - 3x - 10 = 0$$

is a **quadratic equation in standard form.** To designate a general quadratic equation, we use the letter a to represent the coefficient of x^2, the letter b to represent the coefficient of x, and the letter c to represent the constant term. Using these letters to represent the constants in the equation, we can write **a general quadratic equation in standard form** as

$$ax^2 + bx + c = 0$$

Thus if $a = 1$, $b = -3$, and $c = -10$, we have the equation

$$x^2 - 3x - 10 = 0$$

If we substitute either 5 or -2 for the variable x in the quadratic equation $x^2 - 3x - 10 = 0$, the equation will be transformed into a true equation as shown here.

IF $x = 5$	IF $x = -2$
$(5)^2 - 3(5) - 10 = 0$	$(-2)^2 - 3(-2) - 10 = 0$
$25 - 15 - 10 = 0$	$4 + 6 - 10 = 0$
$0 = 0$	$0 = 0$

The numbers 5 and -2 are the only numbers that will satisfy the equation above. **Every quadratic equation has at most[†] two numbers that will make the equation a true statement.** For that matter, every third-degree polynomial equation in one variable has at most three numbers that will satisfy the equation; every fourth-degree polynomial equation in one variable has at most four numbers that will satisfy the equation. To generalize, we can say that every nth-degree polynomial equation in one variable (n is a natural number) has at most n roots.

89.B
solution of quadratic equations by factoring

Some quadratic equations can be solved by using the zero factor theorem.

> **ZERO FACTOR THEOREM**
>
> If p and q are any real numbers and if $p \cdot q = 0$, then either $p = 0$ or $q = 0$, or both.

This says that **if the product of two real numbers is zero, one or both of the numbers is zero.** Thus if we indicate the product of 4 and an unspecified number by writing

$$4(\quad) = 0$$

[†] We say *at most* because some quadratic equations have only one root. For instance, the only root of the equation $x^2 - 4x + 4 = 0$ is the number $+2$.

the only number that we can place in the parentheses that will make the equation a true equation is the number zero.

In the same way, if we indicate the product of two unspecified numbers by writing

$$(\quad)(\quad) = 0$$

the quantity in the first parentheses or the quantity in the second parentheses must equal zero or the product will not equal zero.

Now let's look at the equation

$$(x - 3)(x + 5) = 0$$

Here we have two quantities multiplied and the product is equal to zero. From the **zero factor theorem,** we know that at least one of the quantities must equal zero if the product is to equal zero, so either

$$x - 3 = 0 \qquad\qquad \text{or} \qquad\qquad x + 5 = 0$$

But if $x - 3 = 0, x = 3$ and if $x + 5 = 0, x = -5$

Thus the two values of x that satisfy the condition stated are 3 and -5.

We can use the zero factor theorem to help us solve quadratic equations that can be factored. We do this by first writing the equation in standard form and factoring the polynomial. Then we set each of the factors equal to zero and solve for the values of the variable.

example 89.1   Find the roots of $x^2 - 18 = 3x$ by the factor method.

solution   First we write the equation in standard form and then factor.

$$x^2 - 3x - 18 = 0 \quad\longrightarrow\quad (x + 3)(x - 6) = 0$$

Since the product $(x + 3)(x - 6)$ equals zero, by the zero factor theorem, we know that one of these factors must equal zero.

If $x + 3 = 0, x = -3$ If $x - 6 = 0, x = +6$

To check, we will use -3 and $+6$ as values for x in the original equation

$$\text{IF } x = -3 \qquad\qquad\qquad \text{IF } x = 6$$
$$(-3)^2 - 18 = 3(-3) \qquad\qquad (6)^2 - 18 = 3(6)$$
$$9 - 18 = -9 \qquad\qquad\qquad 36 - 18 = 18$$
$$-9 = -9 \quad \text{Check} \qquad\qquad 18 = 18 \quad \text{Check}$$

example 89.2   Find the roots of $-25 = -4x^2$.

solution   First we write the equation in standard form, $4x^2 - 25 = 0$, and then we factor to get $(2x - 5)(2x + 5) = 0$. For this to be true, either $2x - 5$ equals zero or $2x + 5$ equals zero.

$$\text{IF } 2x - 5 = 0 \qquad\qquad \text{IF } 2x + 5 = 0$$
$$2x = 5 \qquad\qquad\qquad 2x = -5$$
$$x = \frac{5}{2} \qquad\qquad\qquad x = -\frac{5}{2}$$

To check, we will use $\frac{5}{2}$ and $-\frac{5}{2}$ as values for x in the original equation

$$\text{IF } x = \frac{5}{2}$$

$$-25 = -4\left(\frac{5}{2}\right)^2$$

$$-25 = -4\left(\frac{25}{4}\right)$$

$$-25 = -25 \quad \text{Check}$$

$$\text{IF } x = -\frac{5}{2}$$

$$-25 = -4\left(-\frac{5}{2}\right)^2$$

$$-25 = -4\left(\frac{25}{4}\right)$$

$$-25 = -25 \quad \text{Check}$$

example 89.3   Find the values of x that satisfy $x - 56 = -x^2$.

solution   First we rewrite the equation in standard form.

$$x^2 + x - 56 = 0$$

Now we factor:

$$(x + 8)(x - 7) = 0$$

For this to be true, either $x + 8$ equals zero or $x - 7$ equals zero.

$$\text{If } x + 8 = 0, x = -8 \qquad \text{If } x - 7 = 0, x = 7$$

To check, we will use -8 and $+7$ as replacements for x in the original equation.

$$\text{IF } x = -8$$

$$(-8) - 56 = -(-8)^2$$

$$-8 - 56 = -64$$

$$-64 = -64 \quad \text{Check}$$

$$\text{IF } x = 7$$

$$(7) - 56 = -(7)^2$$

$$7 - 56 = -49$$

$$-49 = -49 \quad \text{Check}$$

example 89.4   Solve $x^2 - x = 42$.

solution   First we write the equation in standard form

$$x^2 - x - 42 = 0$$

and now we factor

$$(x - 7)(x + 6) = 0$$

The zero factor theorem says that if a product of two factors equals zero, either one factor or the other factor must equal zero.

$$\text{If } x - 7 = 0, x = 7 \qquad \text{If } x + 6 = 0, x = -6$$

and check.

$$(7)^2 - (7) = 42$$

$$49 - 7 - 42 = 0$$

$$0 = 0 \quad \text{Check}$$

$$(-6)^2 - (-6) = 42$$

$$36 + 6 - 42 = 0$$

$$0 = 0 \quad \text{Check}$$

problem set 89

1. The bowl contained 150 coins. If they were all pennies and nickels and their total value was \$2.70, how many coins of each type were there?

2. The second bowl also contained \$2.70 in pennies and nickels. If there were 54 more pennies than nickels in this bowl, how many were pennies and how many were nickels?

3. When the home team scored, 78 percent of the crowd stood and cheered, and the rest were dejected. If 8800 were dejected, how many stood and cheered?

4. Find four consecutive odd integers such that 4 times the sum of the first and fourth is 3 greater than 7 times the third.

5. Simplify: $\dfrac{(.016 \times 10^{-5})(300 \times 10^{6})}{(20{,}000 \times 10^{4})(400 \times 10^{-8})}$

6. Find the equation of line 6.

7. Find the equation of line 7.

Simplify:

8. $5\sqrt{5}(2\sqrt{10} - 3\sqrt{5})$

9. $4\sqrt{7}(2\sqrt{7} - 3\sqrt{14})$

Divide:

10. $(2x^3 - 5x + 4) \div (x + 2)$

11. $(3x^3 - 4) \div (x - 5)$

Solve by factoring:

*12. $x^2 - 18 = 3x$

*13. $x^2 - x = 42$

*14. $-25 = -4x^2$

*15. $x - 56 = -x^2$

16. $9x^2 - 16 = 0$

17. $28 = x^2 - 3x$

18. $x^2 = 25$

19. $x^2 - 6 = x$

20. $-x^2 - 8x = 16$

Solve by graphing:

21. $\begin{cases} y = 2x - 3 \\ x = -4 \end{cases}$

Use elimination:

22. $\begin{cases} 3x + 5y = 16 \\ 4x - 3y = 2 \end{cases}$

23. Graph on a number line: $x + 2 \nleq 4$; $D = \{\text{Positive integers}\}$

24. Solve: $4\dfrac{3}{5}x - \dfrac{1}{4} = \dfrac{1}{10}$

25. Simplify: $-(-4 - 2^0) - |-2| + \dfrac{1}{-2^{-2}}$

26. $R_P T_P + R_K T_K = 170$, $T_P = 2$, $T_K = 3$, $R_P = R_K + 10$. Find R_K and R_P.

27. Solve: $\dfrac{2x}{3} - \dfrac{2x - 4}{5} = 7$

28. Add: $\dfrac{1}{a} + \dfrac{2}{a^2} + \dfrac{3}{a + x}$

29. Solve: $(-2x^0 - 3)2 - 3x = -2(x^0 - 2)$

30. Simplify: $\dfrac{axx^{-12}y^{-2}(a^4)^{-2}}{a^{-4}(a^2)a(a^{-4}x^2)}$

ER 89-13

LESSON 90 *Value problems*

90.A
value problems

We remember that when we read word problems, we look for word statements about quantities that are equal. Then we transform each of these word statements into an algebraic equation which makes the same statement of equality. We use as many variables as are necessary. When we have written as many equations as we have variables, we solve the equations by using the substitution method or the elimination method. When we write the equations, instead of x and y, we use meaningful variables so that we can remember what these variables represent. We have used two equations in two variables to solve coin problems. These problems are of a genre called **value problems.** We will look at other types of value problems in this lesson. They are very similar to coin problems.

example 90.1 Airline fares for a flight from Tifton to Adel are $30 for first class and $25 for tourist class. If a flight had 52 passengers who paid a total of $1360, how many first-class passengers were on the trip?

solution There are two statements of equality here. The number of first class plus the number of tourist class equals 52.

$$\text{(a)} \quad N_F + N_T = 52$$

The cost of the first-class tickets plus the cost of the tourist tickets equals 1360.

$$\text{(b)} \quad 30N_F + 25N_T = 1360$$

We will solve these equations by using elimination. We will multiply (a) by -30 to get (a′), which we then add to (b):

$$\begin{array}{rl} \text{(a′)} & -30N_F - 30N_T = -1560 \\ \text{(b)} & \underline{30N_F + 25N_T = 1360} \\ & -5N_T = -200 \end{array} \rightarrow \frac{-5N_T}{-5} = \frac{-200}{-5} \rightarrow N_T = 40$$

So N_T equals **40** and this leaves **12** for N_F because there were 52 in all.

example 90.2 Wataksha's dress shop sold cheaper dresses for $20 each and more expensive ones for $45 each. They took in $1375 and sold 20 more of the cheaper dresses than the expensive dresses. How many of each kind did they sell?

solution Again we have statements of equality. The first is that the value of the cheaper dresses plus the value of the expensive dresses equals $1375.

$$\text{(a)} \quad 20N_C + 45N_E = 1375$$

The other is that they sold 20 more of the cheaper dresses. Thus, we add 20 to the number of expensive dresses to get a statement of equality.

$$\text{(b)} \quad N_C = N_E + 20$$

We will use substitution to solve.

$20(N_E + 20) + 45N_E = 1375$	substituted
$20N_E + 400 + 45N_E = 1375$	multiplied
$65N_E + 400 = 1375$	simplified
$65N_E = 975$	added -400 to both sides
$N_E = 15$	divided
$N_C = 35$	20 more than N_E

problem set 90

*1. Airline fares for a flight from Tifton to Adel are $30 for first class and $25 for tourist class. If a flight had 52 passengers who paid a total of $1360, how many first class passengers were on the trip?

*2. Wataksha's dress shop sold cheaper dresses for $20 each and more expensive ones for $45 each. They took in $1375 and sold 20 more of the cheaper dresses than the expensive dresses. How many of each kind did they sell?

3. When the mob stormed the Bastille, only 23 percent had a weapon of any kind. If 1610 had a weapon, how many were in the mob?

4. Find three consecutive odd integers such that 4 times the sum of the first two is 62 less than the product of -30 and the third.

5. Simplify: $\dfrac{(.0006 \times 10^{-23})(300 \times 10^{14})}{90,000 \times 10^{25}}$

6. Find the equations of lines (a) and (b).

Simplify:

7. $3\sqrt{5}(5\sqrt{10} - 2\sqrt{5})$

8. $2\sqrt{14}(3\sqrt{7} - 5\sqrt{2})$

Divide:

9. $x^2 - x - 6$ by $x + 2$

10. $3x^3 - 1$ by $x + 4$

Solve by factoring:

11. $x^2 - 12x + 35 = 0$

12. $-35 = x^2 + 12x$

13. $x^2 = 12x - 32$

14. $17x = -x^2 - 60$

15. $4x^2 - 9 = 0$

16. $-49 = -9p^2$

17. $x^2 + 25 = -10x$

18. $x^2 - 11x + 24 = 0$

19. $-9x^2 + 4 = 0$

Solve by graphing:

20. $\begin{cases} y = 2x - 2 \\ y = -x + 4 \end{cases}$

Use elimination:

21. $\begin{cases} 3x + 4y = 28 \\ 2x - 3y = -4 \end{cases}$

22. $2\dfrac{1}{4}$ is what part of $1\dfrac{2}{5}$?

23. Some of the sounds were susurrant, but $\dfrac{3}{17}$ of the sounds were plangent. If 2800 sounds were susurrant, how many sounds were there in all?

24. $R_M T_M + 10 = R_T T_T$, $R_M = 20$, $R_T = 55$, $T_M + T_T = 7$. Find T_M and T_T.

25. Solve: $-2(-x^0 - 3x^0) = -2(x + 5)$

26. Simplify: $-3^0 - |-3^0| - 3^2 + (-3)^2$

27. Evaluate $|-x^2| + (-x)(-y)$ if $x = -3$ and $y = 4$

28. Simplify: (a) $\dfrac{1}{-3^{-2}}$ (b) $\dfrac{1}{(-3)^{-2}}$ (c) $-(-3)^{-2}$

Simplify:

29. $\dfrac{\dfrac{xy}{a} + \dfrac{1}{y}}{\dfrac{x}{y} - \dfrac{1}{a}}$

30. $\dfrac{a^4 b^4 (2ab)^2}{(3b^{-2})^{-2}}$

LESSON 91 *Intercept-slope method of graphing*

91.A
intercept-slope method of graphing

Thus far, we have graphed lines by finding ordered pairs of x and y that lie on the line. To graph $y = -\frac{3}{5}x + 2$, we choose values for x and write them in a table.

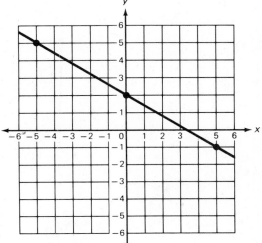

x	0	5	-5
y			

Then we use each of these numbers one at a time in the equation and find the corresponding values of y.

x	0	5	-5
y	2	-1	5

We finish by graphing the ordered pairs on the coordinate system at the right and drawing the line.

This method is dependable, but it is time-consuming. We can use the intercept and the slope of the line to get an accurate graph in less time. We will demonstrate this method by graphing the same line again.

example 91.1 Graph $y = -\frac{3}{5}x + 2$.

solution We begin by writing the slope in the form of a fraction that has a positive denominator. If we do this, the denominator will be $+5$ and the numerator will be -3.

$$y = \frac{-3}{+5}x + 2$$

Now we will graph the line in three steps as shown below. As the first step, we graph the intercept $(0, 2)$ in the left figure.

Now in the center figure, from the intercept we move to the right (the positive x direction) a distance of 5 (the denominator of the slope). Then we move up or down the distance indicated by the numerator of the slope. We move down 3 since our numerator is -3. We graph this new point, and in the right figure we draw the line through the two points.

example 91.2   Use the intercept-slope method to graph the equation $x - 2y = 4$.

solution   As the first step we write the equation in slope-intercept form by solving for y.

$$x - 2y = 4 \quad \longrightarrow \quad -2y = -x + 4 \quad \longrightarrow \quad 2y = x - 4 \quad \longrightarrow \quad y = \frac{1}{2}x - 2$$

Now we write the slope $\frac{1}{2}$ as a fraction with a positive denominator.

$$y = \frac{+1}{+2}x - 2$$

In the first figure we graph the intercept, $(0, -2)$. In the second figure we move from the intercept an x distance of $+2$ (to the right) and a y distance of $+1$ (up). In the third figure we draw the line through the two points.

When the points are close together as in this case, it is difficult to draw the line accurately. To get another point, we multiply the denominator and the numerator of the slope by a convenient integer and use the new form of the slope to get the second point. For the line under discussion, we will multiply the slope by $\frac{2}{2}$ and get

$$\frac{+1}{+2} \cdot \frac{(2)}{(2)} \quad \longrightarrow \quad \frac{+2}{+4}$$

In the figures below we use the same intercept but move an x distance of $+4$ and a y distance of $+2$ to find the new point.

**problem
set 91**

1. The total value of the pennies and nickels was $14.50. Emet and Callaway counted the coins and found that the total was 450 coins. How many coins of each type did they have?

2. Ice cream bars cost 30 cents and whifferdils cost 50 cents. Dennis and Lebeda treated all the kids, and they spent $13.50. How many kids had whifferdils if these numbered 5 less than those who had ice cream bars?

3. Only 3 golfers in 20 parred the hole. If 90 golfers parred the hole, how many played in the tournament?

4. Find four consecutive odd integers such that the opposite of the sum of the first two is 4 greater than the product of the fourth and -4.

Use the intercept-slope method to graph the following lines:

*5. $y = -\dfrac{3}{5}x + 2$ *6. $y = \dfrac{1}{2}x - 2$

7. $y = -\dfrac{3}{2}x + 3$ 8. $y = -\dfrac{1}{2}x + 2$

9. Simplify: $\dfrac{(4000 \times 10^{-23})(.00035 \times 10^{15})}{5000 \times 10^{5}}$

10. Find the equations of lines (a) and (b).

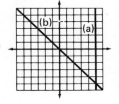

Simplify:

11. $3\sqrt{2} \cdot 4\sqrt{12} - 6\sqrt{54}$ 12. $3\sqrt{2}(5\sqrt{12} - 8\sqrt{8})$

Divide:

13. $x^3 - 3x^2 + 2x + 5$ by $x - 3$ 14. $x^3 - 1$ by $x + 3$

Solve by factoring:

15. $x^2 - 9x + 20 = 0$ 16. $42 = 13x - x^2$

17. $4x^2 - 9 = 0$ 18. $9x^2 = 4$

Solve by graphing: Use elimination:

19. $\begin{cases} y = x \\ y = -\dfrac{1}{2}x + 3 \end{cases}$ 20. $\begin{cases} 3x + 4y = 32 \\ 5x - 4y = 0 \end{cases}$

21. Solve: $4\dfrac{2}{3}x - \dfrac{1}{5} = 3\dfrac{2}{3}$

22. $\dfrac{2\sqrt{3}}{2} \in$ {What sets of numbers}?   23. Simplify: $\dfrac{4x^2 - x^2y}{x^2}$

24. $T_K R_K + 60 = T_M R_M$, $T_K = 3$, $T_M = 2$, $R_K + R_M = 125$. Find R_K and R_M.

25. Solve: $\dfrac{3x - 4}{2} + \dfrac{1}{5} = \dfrac{x}{10}$ 26. Add: $\dfrac{4x^2}{y} - \dfrac{2x}{y + 4}$

27. Evaluate $-x^0 - x(x - y^2)$ if $x = -3$ and $y = -4$

28. Simplify: $-2(-2 - 3) - (-2^0) - 3(-3^0) - 2(-2)$

29. Simplify: (a) $\dfrac{1}{-3^{-3}}$ (b) -3^{-3} (c) $-(-3)^{-3}$

30. Multiply: $x^{-2}y^{-1}\left(\dfrac{x^{-1}}{y^{-1}} - \dfrac{4x^2y^0}{(y^{-3})^2}\right)$ **ER 91-14**

LESSON 92 *Word problems with two statements of equality*

92.A
two statements of equality

The coin problems and the general value problems we have studied thus far have contained two statements about quantities that are equal. We have solved these problems by using two equations in two unknowns (variables). Many other problems contain two statements of equality and are solved the same way. We will look at some of these in this lesson. We will write the equations and give the answers. Use substitution or elimination to see if you get the same answers.

example 92.1 Together Charles and Nelle picked 92 quarts of berries. If Charles picked 6 more quarts than Nelle picked, how many quarts did each of them pick?

solution (a) The number Charles picked plus the number Nelle picked equaled 92.

$$N_C + N_N = 92$$

(b) The number Nelle picked plus 6 equaled the number Charles picked.

$$N_N + 6 = N_C$$

ANSWER $N_C = 49$ $N_N = 43$

example 92.2 The number of boys in Sarah's class exceeded the number of girls by 7. If there were a total of 29 pupils in the class, how many were boys and how many were girls?

solution (a) The number of boys equaled the number of girls plus 7.

$$N_B = N_G + 7$$

(b) The number of boys plus the number of girls equaled 29.

$$N_B + N_G = 29$$

ANSWER $N_B = 18$ $N_G = 11$

example 92.3 Phillip cut a 38-meter rope into two pieces. The long piece was 9 meters longer than the short piece. What were the two lengths?

solution (a) The length of the long piece plus the length of the short piece equaled 38 meters.

$$L + S = 38$$

(b) The long piece was 9 meters longer than the short piece.

$$S + 9 = L$$

ANSWER $L = 23.5$ $S = 14.5$

example 92.4 The sum of two numbers is 72. The difference of the numbers is 26. What are the numbers?

solution (a) The large number plus the small number equals 72.

$$L + S = 72$$

(b) The large number minus the small number equals 26.

$$L - S = 26$$

ANSWER $L = 49$ $S = 23$

example 92.5 The greater of two numbers is 16 greater than the smaller. When added together, their sum is 4 less than 3 times the smaller. What are the numbers.

solution (a)   The greater number is 16 greater than the smaller.

$$G - 16 = S$$

(b)   The sum is 4 less than 3 times the smaller.

$$G + S + 4 = 3S$$

ANSWER $G = 36$ $S = 20$

example 92.6 The ratio of two numbers is 5 to 4 and the sum of the numbers is 63. What are the numbers?

solution (a)   The ratio of the numbers is 5 to 4.

$$\frac{N_1}{N_2} = \frac{5}{4}$$

(b)   The sum of the numbers is 63.

$$N_1 + N_2 = 63$$

ANSWER $N_1 = 35$ $N_2 = 28$

problem set 92

*1.   Charles and Nelle picked 92 quarts of berries. If Charles picked 6 more quarts than Nelle picked, how many quarts did each of them pick?

*2.   The number of boys in Sarah's class exceeded the number of girls by 7. If there were 29 pupils in the class, how many were boys and how many were girls?

3.   Phillip cut a 38-meter rope in two pieces. The long piece was 8 meters longer than the short piece. How long was each piece?

*4.   The sum of two numbers is 72. The difference of the numbers is 26. What are the numbers?

5.   Shields and Jim sold tickets to the basketball game. Good seats were $5 each and poor seats cost only $2 each. If 210 people attended and paid $660, how many people bought good seats?

Find the solution to the following systems of equations by graphing. The intercept-slope method from Lesson 91 should be helpful.

6. $\begin{cases} 3x + y = 6 \\ x = -y \end{cases}$

7. $\begin{cases} y = \dfrac{2}{3}x - 3 \\ y = -x + 2 \end{cases}$

8.   Simplify: $\dfrac{(3000 \times 10^{-5})(.004 \times 10^{10})}{(200 \times 10^{14})(.000002)}$

9.   Find the equations of lines (a) and (b).

Simplify:

10. $3\sqrt{27} - 2\sqrt{3}(4\sqrt{3} - 5\sqrt{12})$ **11.** $2\sqrt{2} \cdot 3\sqrt{3} \cdot 5\sqrt{12}$

Divide:

12. $3x^3 - 2x - 4$ by $x + 1$ **13.** $2x^3 - 2x^2 - 4$ by $x + 1$

Solve by factoring:

14. $2x^2 + 20x + 50 = 0$ **15.** $3x^2 = -33x - 90$

16. $2x^2 - 18 = 0$ **17.** $27 - 3p^2 = 0$

Use substitution:

18. $\begin{cases} y = -3x + 10 \\ 2x + 2y = 8 \end{cases}$ **19.** Solve: $\dfrac{3x + 2}{5} - \dfrac{x}{2} = 5$

20. Solve: $-3x^0(-2 - 3) - (-2 - 3)4x = -2(x + 2)$

21. $-3 \in \{\text{What sets of numbers}\}$?

22. Simplify by adding like terms: $\dfrac{3x^{-2}y}{m^{-3}} - \dfrac{4y^2m}{x^2ym^{-2}} + \dfrac{6y}{x^2m^{-3}}$

23. $R_M T_M = R_B T_B$, $R_B = 5$, $R_M = 4$, $T_M + T_B = 18$. Find T_M and T_B.

24. Add: $\dfrac{x}{x + 1} + \dfrac{x^2}{x(x + 1)}$ **25.** Simplify: $\dfrac{a + \dfrac{x}{y}}{\dfrac{a}{y} - x}$

26. Graph on a number line: $-x + 4 \not> 2$, $D = \{\text{Reals}\}$

27. Evaluate $|-x| - x^0 - x^2(x - y)$ if $x = -2$ and $y = -4$

28. Simplify: $-|-2| - |-2^0| - (-2 - 4)$

29. Simplify: (a) -3^{-2} (b) $(-3)^{-2}$ (c) $-(-3)^{-2}$

30. Simplify: $\dfrac{x^{-2}(y^{-2})^2(y^0)^2}{xy^{-2}(x^{-2}y)^{-2}}$

LESSON 93   *Multiplicative property of inequality*

93.A
properties
of inequality

With one glaring exception, the rules for solving inequalities are the same as the rules for solving equations. The following two rules apply to both equalities and inequalities.

ADDITION RULE

The same quantity can be added to both sides of an equation or inequality without changing the solution set of the equation or inequality.

MULTIPLICATION RULE (POSITIVE)

Every term on both sides of an equation or inequality can be multiplied by the same positive number without changing the solution set of the equation or inequality

---

The glaring exception occurs when we multiply by a negative number! The truth of a statement of equality is not altered by multiplying by a negative number.

$$5 = 2 + 3 \qquad \text{True}$$

Now multiply every term by -2 and get

$$5(-2) = 2(-2) + 3(-2) \longrightarrow -10 = -4 - 6 \qquad \text{Still true!}$$

But the truth of a statement of inequality is altered!

$$8 > 5 \qquad \text{True}$$

Now we multiply every term by -2 and get

$$-16 > -10 \qquad \text{Now false!}$$

Thus when every term on both sides of an inequality is multiplied by a negative number, the inequality symbol must be reversed so that the solution set of the inequality will not be changed. To show this we will repeat the problem.

$$8 > 5 \qquad \text{True}$$

Now we multiply every term by -2 and reverse the inequality symbol.

$$-16 < -10 \qquad \text{Still true!}$$

example 93.1   Graph the solution of $-x \geq 2$, $D = \{\text{Reals}\}$.

solution   We solve the given inequality for $+x$ by multiplying both sides by (-1) and **reversing the inequality symbol.**

$$
\begin{array}{ll}
-x \geq 2 & \text{original inequality} \\
(-1)(-x) \leq (-1)(2) & \text{multiplied by } -1 \text{ and reversed symbol} \\
x \leq -2 & \text{simplified}
\end{array}
$$

Thus we want to graph the solution of $x \leq -2$.

The graph indicates that the number -2 and all real numbers less than -2 satisfy the stated inequality.

example 93.2   Graph the solution of $4 - x \leq 6$, $D = \{\text{Integers}\}$.

solution   We first isolate $-x$ by adding -4 to both sides. **Note that we do not reverse the inequality symbol when we add a negative quantity to both sides of an inequality**

$$
\begin{array}{r}
4 - x \leq 6 \\
-4 \qquad -4 \\
\hline
-x \leq 2
\end{array}
$$

Now we multiply both sides by -1 and **reverse the inequality symbol** to get

$$x \geq -2$$

And if we graph $x \geq -2$ over the integers, we get

example 93.3 Graph the solution of $-3x + 4 \leq 13$, $D = \{\text{Reals}\}$.

solution We add -4 to both sides to get $-3x \leq 9$ and then divide both sides by -3 (or *multiply* both sides by $-\frac{1}{3}$) **and reverse the inequality symbol** to get

$$x \geq -3$$

and we graph the solution to the inequality to indicate all real numbers that are greater than or equal to -3.

problem set 93

1. Wetumka had $6.50 in dimes and quarters. If he had 5 more quarters than dimes, how many of each type did he have?

2. Seed corn was $5 a bag, while dog food cost only $3 a bag. Wewoka bought 50 bags and spent $190. How many bags of dog food did she buy?

3. There were 90 more orchids on the first float than there were on the second float. If there were 630 orchids altogether, how many were on each float?

4. When the first frost came, the number of people wearing shoes jumped 180 percent. If 5600 people now wore shoes, how many people wore shoes before the frost?

Graph the solution to these inequalities on a number line:

*5. $-x \geq 2$; $D = \{\text{Reals}\}$ *6. $4 - x \leq 6$; $D = \{\text{Integers}\}$

*7. $-3x + 4 \leq 13$; $D = \{\text{Reals}\}$

Solve by graphing: Use elimination:

8. $\begin{cases} y = -x + 1 \\ y = 2x + 7 \end{cases}$ 9. $\begin{cases} 4x - 3y = 14 \\ 5x - 4y = 18 \end{cases}$

10. Simplify: $\dfrac{(.00004 \times 10^{15})(700 \times 10^{-5})}{14,000 \times 10^{-21}}$

11. Find the equations of lines (a) and (b).

Simplify:

12. $5\sqrt{75} - 2\sqrt{108} + 5\sqrt{12}$ 13. $2\sqrt{6}(3\sqrt{6} - 2\sqrt{12})$

14. Divide $x^4 - 2x^2 - 4$ by $x + 2$

Solve by factoring:

15. $21 = 10x - x^2$ **16.** $-49 = -4x^2$ **17.** $32 = -x^2 - 12x$

Solve:

18. $\dfrac{4x}{3} - \dfrac{x+1}{5} = 10$ **19.** $3\dfrac{1}{4}x - \dfrac{2}{3} = 7\dfrac{1}{8}$

20. Graph on a number line: $4 \leq x < 7,\, D = \{\text{Integers}\}$

21. $-3\dfrac{1}{3} \in \{\text{What sets of numbers}\}$?   **22.** Add: $\dfrac{x}{x+1} + \dfrac{4}{x}$

23. Simplify by adding like terms: $5x^2ym^2 - \dfrac{3xxy^{-2}}{y^{-3}m^{-1}} + \dfrac{2xmy}{y^3y}$

24. $R_H T_H - 125 = R_O T_O,\ T_H = 2,\ T_O = 3,\ R_H + R_O = 85.$ Find R_H and R_O.

25. Solve: $3x^0 - 2x^0 - 3(x^0 - 2x) = -2x(4-3)$

Simplify:

26. $\dfrac{\dfrac{3x}{y} - 2}{a - \dfrac{4}{y}}$ **27.** $\dfrac{3xy - 9xy^2}{3xy}$

28. (a) -2^{-3} (b) $(-2)^{-3}$ (c) $\dfrac{1}{-(-2)^{-3}}$

29. $-2^0[(-3-5)(-2-1)] - 3^0$ **30.** $\dfrac{4x^2(x^{-2})^0 xx^{-4}}{2(3x^{-2})^0(x^2)^{-2}}$

ER 93-15

LESSON 94   *Uniform motion problems about equal distances*

94.A

uniform motion word problems

Since Lesson 80 we have been using substitution to solve systems of four simultaneous equations involving four unknowns such as

$$R_F T_F = R_E T_E \qquad T_F = 16 \qquad T_E = 12 \qquad R_E = R_F + 15$$

These equations are typical of the equations that we will learn to write in this lesson to help us solve uniform motion word problems. Uniform motion problems are so named because the statement of the problem tells about objects or things that move at a uniform rate or at an average rate.

The statements of equality made in uniform motion problems are statements of equality that concern **distance**, statements of equality that concern **rate**, and statements of equality that concern **time**, and use the relationship that

$$\text{Distance} = \text{rate} \times \text{time} \qquad \text{or} \qquad D = RT$$

The statements about rate and time are not difficult to locate in the wording of the problem, but the beginner often has trouble identifying the statement that defines the relationship that concerns distances. **Since the distance equation is the troublesome one, we will consider this equation to be the key equation for this type of problem, and we will always write this equation first.** Then we will write the equations that concern time and the equations that concern rate. **When we have as many equations as we have variables, we will use the substitution method to solve for the variables.**

The statements of the distances discussed in the problems can be represented graphically by drawing diagrams in which arrows represent the distances. The problems in this book will usually describe two distances. In this lesson, we will investigate problems that describe two equal distances.

(a) $\begin{array}{c} D_1 \\ D_2 \end{array}$ (b) $\begin{array}{c} D_1 \\ D_2 \end{array}$ Distance equation: $D_1 = D_2$

One of these diagrams will result when the problem states that two distances are equal. In (a) the objects traveled in the same direction and in (b) they traveled in opposite directions. Both diagrams give us the same equation.

example 94.1 On Tuesday the express train made the trip in 12 hours. On Wednesday the freight train made the same trip in 16 hours. Find the rate of each if the rate of the freight was 15 kilometers per hour less than the rate of the express.

solution We read the problem and disregard statements about time and rate and look for the statement about distance. This information allows us to draw the diagram and from the diagram get the equation

$\begin{array}{c} D_E \\ D_F \end{array}$ Distance express = distance freight

(a) $D_E = D_F$

This is the distance equation, which is the key equation for this uniform motion problem. Now, the distance the express traveled equals the rate of the express times the time the express traveled, or

$$D_E = R_E T_E$$

and the distance the freight traveled equals the rate of the freight times the time the freight traveled, or

$$D_F = R_F T_F$$

Now, if we substitute $R_E T_E$ for D_E and $R_F T_F$ for D_F in equation (a), we get

$$R_E T_E = R_F T_F$$

which is the distance equation for this problem in final form.

The statement of the problem gives the time of the express as 12 hours and the time of the freight as 16 hours so

$$T_E = 12 \qquad T_F = 16$$

Now, we have three equations, but have four unknowns, R_E, T_E, R_F, and T_F. **Thus we need one more equation so that the number of equations will equal the number of unknowns.** We get the final equation from the statement in the problem concerning rates which says that the rate of the express is 15 kilometers per hour greater than the rate of the freight. **Writing this equation is tricky.** In an effort to avoid the common error of adding 15 to the wrong side, we first write

$$R_F = R_E \qquad \text{incorrect}$$

which we know is incorrect because the rate of the express is greater than the rate of the freight. We add 15 to R_F so that the equation will have equal quantities on both sides.

$$R_E = R_F + 15$$

We have found four equations in four unknowns,

$$R_E T_E = R_F T_F \qquad T_E = 12 \qquad T_F = 16 \qquad R_E = R_F + 15$$

and we use the substitution method to solve for R_F and R_E.

$$R_E T_E = R_F T_F \quad \longrightarrow \quad (R_F + 15)12 = R_F(16) \quad \longrightarrow \quad 12R_F + 180 = 16R_F$$

$$\longrightarrow \quad -4R_F = -180 \quad \longrightarrow \quad 4R_F = 180 \quad \longrightarrow \quad R_F = 45 \text{ kilometers per hour}$$

and since $R_E = R_F + 15$,

$$R_E = 60 \text{ kilometers per hour}$$

example 94.2   The members of the girls club hiked to Lake Tenkiller at 2 miles per hour. Mr. Ali gave them a ride back home at 12 miles per hour. Find their hiking time if it was 5 hours longer than their riding time. How far was it to Lake Tenkiller?

solution   We begin by drawing a diagram of the distances traveled and writing the distance equation.

$$D_H = D_R \qquad \text{so} \qquad R_H T_H = R_R T_R$$

Next we reread the problem and write the other three equations.

$$R_H = 2 \qquad R_R = 12 \qquad T_H = T_R + 5$$

Now we substitute these equations into the distance equation and solve.

$$2(T_R + 5) = 12T_R \qquad \text{substituted}$$
$$2T_R + 10 = 12T_R \qquad \text{multiplied}$$
$$10 = 10T_R \qquad \text{added } -2T_R \text{ to both sides}$$
$$T_R = 1 \qquad \text{and so} \qquad T_H = 6$$

Thus the distance is either 2 times 6 or 12 times 1, both of which equal **12 miles.**

example 94.3   Durant drove to the oasis in 2 hours and Madill walked to the oasis in 10 hours. How far is it to the oasis if Durant drove 16 miles per hour faster than Madill walked?

solution   We begin by drawing a distance diagram from which we can get the distance equation.

$$D_D = D_W \qquad \text{so } R_D T_D = R_W T_W$$

Now we reread the problem to get the time and rate equations.

$$T_D = 2 \qquad T_W = 10 \qquad R_D = R_W + 16$$

and now we solve:

$$(R_W + 16)2 = R_W(10) \qquad \text{substituted}$$
$$2R_W + 32 = 10R_W \qquad \text{multiplied}$$
$$32 = 8R_W \qquad \text{added } -2R_W \text{ to both sides}$$
$$4 = R_W \qquad \text{divided}$$

and thus $R_D = 20$, and since $T_D = 2$, the distance equals 2 times 20, or **40 miles.**

problem set 94

*1. On Tuesday the express train made the trip in 12 hours. On Wednesday, the freight train made the same trip in 16 hours. Find the rate of each if the rate of the freight was 15 kilometers per hour less than the rate of the express.

*2. The members of the girls club hiked to Lake Tenkiller at 2 miles per hour. Mr. Ali gave them a ride back home at 12 miles per hour. Find their hiking time if it was 5 hours longer than their riding time. How far was it to Lake Tenkiller?

*3. Durant drove to the oasis in 2 hours and Madill walked to the oasis in 10 hours. How far is it to the oasis if Durant drove 16 miles per hour faster than Madill walked?

4. Jimmy rode his bike to the conclave at 10 kilometers per hour and then walked back to school at 4 kilometers per hour. If the round trip took him 14 hours, how far was it to the site of the conclave?

5. Fustian phrases obscured .62 of the points the speaker tried to make. If he tried to make 50 points, how many was the audience able to comprehend?

6. Solve: $p - (-p) - 5(p - 3) - (2p - 5) = 3(p + 2p)$

7. Graph on a number line: $-4x + 4 \geq 8$; $D = \{\text{Reals}\}$

Solve by graphing: Use elimination:

8. $\begin{cases} y = -2x + 4 \\ y = -2 \end{cases}$ 9. $\begin{cases} 3x + y = 20 \\ 2x - 3y = -5 \end{cases}$

10. Simplify: $\dfrac{.000030 \times 10^{-18}}{(5000 \times 10^{-14})(300 \times 10^5)}$

11. Find the equations of line (a) and (b).

Simplify:

12. $3\sqrt{45} - 2\sqrt{180} + 2\sqrt{80}$ 13. $3\sqrt{2}(4\sqrt{20} - 3\sqrt{2})$

14. Divide $x^3 - 4$ by $x - 5$

Solve by factoring:

15. $x^2 = -6x - 8$ 16. $9 = 4x^2$ 17. $x^2 = -12x - 32$

Solve:

18. $\dfrac{x}{5} - \dfrac{4 + x}{7} = 5$ 19. $2\dfrac{1}{8}x - 3\dfrac{1}{4} = 2\dfrac{1}{16}$

20. $7\dfrac{2}{3}$ is what fraction of $\dfrac{5}{6}$?

21. True or false? $\{\text{Reals}\} \subset \{\text{Rationals}\}$

22. $R_W T_W + 200 = R_R T_R$, $R_W = 40$, $R_R = 60$, $T_W + T_R = 10$. Find T_W and T_R.

23. Add: $\dfrac{2x}{5} + \dfrac{3x + 5}{6x}$

Simplify:

24. $\dfrac{\dfrac{3a}{b} - 2b}{b - \dfrac{4}{b}}$

25. $\dfrac{4 + 4k}{4}$

26. Evaluate $|-x^2| - |x| + x(x - y^0)$ if $x = -3$ and $y = 4$

Simplify:

27. $-3^0 - [(-3 + 5) - (-2 - 5)]$

28. (a) -2^{-2} (b) $\dfrac{1}{-2^{-2}}$ (c) $-(-2)^{-2}$

29. Multiply: $\left(\dfrac{x^4 y}{a^2} - \dfrac{x^{-3} y^2}{y a^{-4}}\right) \dfrac{x^{-4} y}{a^{-2}}$

Enrichment Lessons 2 and 3 in the appendix are on the properties of the set of real numbers. If these topics are to be included, they should be included at this point so that the review problems in the problem sets will be effective. These lessons are prerequisites for the enrichment lesson on algebraic proofs which comes after Lesson 99.

LESSON 95 *Rational expressions*

95.A
products of rational expressions

Fractions are multiplied by multiplying the numerators to form the numerator of the product and by multiplying the denominators to form the denominator of the product. Thus to multiply

$$\frac{4x}{5} \cdot \frac{3xa}{y}$$

we multiply the numerators to get $12x^2 a$ and multiply the denominators to get $5y$.

$$\frac{4x}{5} \cdot \frac{3xa}{y} = \frac{12x^2 a}{5y}$$

Sometimes we encounter indicated products of rational expressions whose simplification is facilitated if the terms are first factored and such canceling as is possible performed before any multiplication is done.[†]

example 95.1   Simplify $\dfrac{x^2 - 25}{x^2 - 7x} \cdot \dfrac{x^2 + 3x}{x^2 - 2x - 15}$.

[†] Seldom in real life but often in algebra textbooks. These problems are really contrived problems designed to give the reader practice in factoring and canceling.

solution If we multiply the expressions in their present form, we will get a very complicated expression for a product, and it would be very difficult to simplify this product. If we factor and cancel before we multiply, however, the simplified expression can be obtained quickly and easily.

$$\frac{(x-5)(x+5)}{x(x-7)} \cdot \frac{x(x+3)}{(x+3)(x-5)} = \frac{x+5}{x-7}$$

example 95.2 Simplify $\dfrac{x^2 + x - 6}{x^2 - 4x - 21} \cdot \dfrac{x^2 - 8x + 7}{x^2 - x - 2}$.

solution We factor and cancel factors where possible.

$$\frac{(x+3)(x-2)}{(x-7)(x+3)} \cdot \frac{(x-7)(x-1)}{(x-2)(x+1)} = \frac{x-1}{x+1}$$

95.B
quotients of rational expressions

In Lesson 59 we learned to simplify expressions such as

$$\frac{\dfrac{a}{b}}{\dfrac{c}{d}}$$

by using the **denominator-numerator same-quantity theorem** to justify multiplying both the denominator and the numerator by $\dfrac{d}{c}$, which is the reciprocal of the denominator.

$$\frac{\dfrac{a}{b}}{\dfrac{c}{d}} = \frac{\dfrac{a}{b} \cdot \dfrac{d}{c}}{\dfrac{c}{d} \cdot \dfrac{d}{c}} = \frac{\dfrac{ad}{bc}}{1} = \frac{ad}{bc}$$

If the same division problem had been stated by writing

$$\frac{a}{b} \div \frac{c}{d}$$

we see that the same result can be obtained by inverting the divisor and multiplying.

$$\frac{a}{b} \div \frac{c}{d} = \frac{a}{b} \cdot \frac{d}{c} = \frac{ad}{bc}$$

We can use the same procedure to simplify quotients of more complicated rational expressions.

example 95.3 Simplify $\dfrac{x^2 - 2x}{x^2 + 2x - 8} \div \dfrac{x^2 + 5x}{x^2 + 7x + 12}$.

solution As the first step we invert the divisor and change the division symbol to a dot that indicates multiplication. Then we factor and cancel as in the two previous examples.

$$\frac{x^2 - 2x}{x^2 + 2x - 8} \cdot \frac{x^2 + 7x + 12}{x^2 + 5x} = \frac{x(x-2)}{(x+4)(x-2)} \cdot \frac{(x+4)(x+3)}{x(x+5)} = \frac{x+3}{x+5}$$

example 95.4   Simplify $\dfrac{x^2 - x - 6}{x^2 - 3x - 10} \div \dfrac{x^2 + 5x + 4}{x^2 - x - 20}$.

solution   Again as the first step we invert the divisor and indicate multiplication rather than division. Then we factor and cancel.

$$\frac{x^2 - x - 6}{x^2 - 3x - 10} \cdot \frac{x^2 - x - 20}{x^2 + 5x + 4} = \frac{(x - 3)(x + 2)}{(x - 5)(x + 2)} \cdot \frac{(x - 5)(x + 4)}{(x + 4)(x + 1)} = \frac{x - 3}{x + 1}$$

problem set 95

1. Norma and David crawled to the barn and then hopped back to the house. They crawled at 300 centimeters per minute and hopped at 400 centimeters per minute. If the round trip took 7 minutes, how long did they crawl? How far was it to the barn?

2. Annette drove to Shawnee in 4 hours and drove back in 3 hours. What were her speeds if her speed coming back was 11 miles per hour greater than her speed going?

3. When the time came to stand up and be counted, only 92 people stood up. If 460 people were present, what percent stood up and were counted?

4. Hobert and Higgs counted the boys and girls at the assembly. There were 179 students and 13 more boys than girls. How many boys and how many girls were present?

Simplify:

*5. $\dfrac{x^2 + x - 6}{x^2 - 4x - 21} \cdot \dfrac{x^2 - 8x + 7}{x^2 - x - 2}$

*6. $\dfrac{x^2 - 2x}{x^2 + 2x - 8} \div \dfrac{x^2 + 5x}{x^2 + 7x + 12}$

7. $\dfrac{x^3 - 4x}{x^2 + 7x + 10} \div \dfrac{x^2 - 2x}{x^2 - 25}$

8. Five-thirteenths of the citizens believed that the cause of their difficulty was procrastination. If 400 did not agree with this analysis, how many citizens lived in the community?

9. Solve: $-2(3x - 4^0) + 3x - 2^0 = -(x - 3^2)$

10. Graph $-x - 3 \not> 2$; $D = \{\text{Reals}\}$

Solve by graphing:

11. $\begin{cases} y = 2x - 4 \\ y = -x + 2 \end{cases}$

Use elimination:

12. $\begin{cases} 3x + 5y = -14 \\ -2x + y = 5 \end{cases}$

13. Simplify: $\dfrac{(.000004)(.003 \times 10^{21})}{(2000 \times 10^8)(.002 \times 10^{15})}$

14. Find the equations of lines (a) and (b).

Simplify:

15. $3\sqrt{6} \cdot 2\sqrt{5} - \sqrt{120}$

16. $4\sqrt{12}(3\sqrt{2} - 4\sqrt{3})$

17. Divide $3x^3 - 4$ by $x + 3$

18. Solve by factoring: $40 = -x^2 - 14x$

19. Graph $2 > x > -3$; on a number line: $D = \{\text{Reals}\}$

Solve:

20. $\dfrac{x-5}{7} + \dfrac{x}{4} = \dfrac{1}{2}$

21. $3\dfrac{1}{8}x - 2\dfrac{1}{2} = \dfrac{1}{8}$

22. $3\dfrac{1}{7}$ is what fraction of 28?

23. True or false? $\{\text{Wholes}\} \subset \{\text{Naturals}\}$

24. Add: $\dfrac{x}{ya^2} + \dfrac{xa}{a^2y^2}$

25. Simplify: $\dfrac{k + \dfrac{k}{y}}{y + \dfrac{a}{y}}$

26. Evaluate: $-y^0(-y^2 - 4y) - ay$ if $y = -2$ and $a = -5$

27. Simplify: $-2^0[(-3 - 4^0) - (-2 - 3^0) - 2^2]$

28. Simplify: (a) $\dfrac{1}{-3^{-3}}$ (b) $\dfrac{1}{-(-3)^{-3}}$ (c) $-(-3)^{-3}$

29. Simplify: $\dfrac{(-4x^{-2})^2}{(-2y^{-2})^2 x}$ **ER 95-22, 95-23, 95-24**

LESSON 96   *Uniform motion problems of the form* $D_1 + D_2 = N$

96.A
uniform motion

In the uniform motion problems we have worked up to now, two people or things have traveled equal distances. The distance diagrams have looked like one of the following.

These diagrams have indicated that our distance equation should be of the form $R_1T_1 = R_2T_2$. Now we will consider problems that state that the sum of two distances equals a given number. The diagrams and equations of these problems will have the following forms.

$$D_1 + D_2 = 352 \quad \text{so} \quad R_1T_1 + R_2T_2 = 352$$

example 96.1   A southbound bus left Fort Walton Beach at 9 a.m. Two hours later a northbound bus left the same station. If the buses traveled at the same rate and were 352 kilometers apart at 2 p.m., find the rate of the buses.

solution   The statement of the problem leads to this distance diagram.

$$\overset{\xleftarrow{\hspace{1em}} 352 \xrightarrow{\hspace{1em}}}{\underset{D_N \quad D_S}{\bullet}}$$

The distance equation is $D_N + D_S = 352$ or $R_N T_N + R_S T_S = 352$. The southbound bus traveled for 5 hours and the northbound bus traveled for 3 hours.

$$T_N = 3 \qquad T_S = 5$$

The rates were the same so $R_N = R_S$.

Now we have four equations in four unknowns.

$$\underbrace{R_N T_N + R_S T_S = 352,} \qquad \underbrace{T_N = 3} \qquad \underbrace{T_S = 5} \qquad \underbrace{R_N = R_S}$$

We use substitution to solve.

$$R_N(3) + R_N(5) = 352 \quad\longrightarrow\quad 8R_N = 352 \quad\longrightarrow\quad R_N = \textbf{44 kilometers per hour}$$

and therefore

$$R_S = \textbf{44 kilometers per hour}$$

example 96.2   A train starts from Toledo at 11 a.m. and heads for Makinaw, 332 kilometers away. At the same time a train leaves Makinaw and heads for Toledo at 65 kilometers per hour. If the trains meet at 1 p.m., what is the rate of the first train?

solution   First we draw the diagram and write the distance equation.

$$\overset{D_1 \quad D_2}{\underset{332}{\vdash\!\!\!\xrightarrow{\hspace{0.5em}}\!\!\!\dashv}} \qquad D_1 + D_2 = 332 \qquad \text{so} \qquad R_1 T_1 + R_2 T_2 = 332$$

Then we reread the problem and write the other three equations.

$$T_1 = 2 \qquad T_2 = 2 \qquad R_2 = 65$$

Now we substitute and solve:

$$R_1(2) + 65(2) = 332 \qquad\qquad \text{substituted}$$
$$2R_1 + 130 = 332 \qquad\qquad \text{multiplied}$$
$$2R_1 = 202 \qquad\qquad \text{added} -130 \text{ to both sides}$$
$$R_1 = \textbf{101 kilometers per hour} \qquad \text{divided by 2}$$

example 96.3   The ships were 400 miles apart at midnight and were headed toward each other. If they collided head-on at 8 a.m., find the speed of both ships if one was 20 miles per hour faster than the other.

solution   First we draw the diagram and write the distance equation.

$$\overset{400}{\underset{D_F \quad D_S}{\vdash\!\!\!\xrightarrow{\hspace{0.5em}}\!\!\!\dashv}} \qquad D_F + D_S = 400 \qquad \text{so} \qquad R_F T_F + R_S T_S = 400$$

The other three equations are

$$T_F = 8 \qquad T_S = 8 \qquad R_F = R_S + 20$$

Now we substitute and solve.

$$(R_S + 20)(8) + R_S(8) = 400 \qquad \text{substituted}$$
$$8R_S + 160 + 8R_S = 400 \qquad \text{multiplied}$$
$$16R_S + 160 = 400 \qquad \text{simplified}$$
$$16R_S = 240 \qquad \text{added } -160 \text{ to both sides}$$
$$R_S = \textbf{15 miles per hour} \qquad \text{divided by 16}$$

Thus $\qquad\qquad\qquad R_F = \textbf{35 miles per hour}$

problem set 96

*1. A southbound bus left Fort Walton Beach at 9 a.m. Two hours later a northbound bus left the same station. If the buses traveled at the same rate and were 352 kilometers apart at 2 p.m., find the rate of the buses.

*2. A train starts from Toledo at 11 a.m. and heads for Makinaw, 332 kilometers away. At the same time a train leaves Makinaw and heads for Toledo at 65 kilometers per hour. If the trains meet at 1 p.m., what is the rate of the first train?

*3. The ships were 400 miles apart at midnight and were headed toward each other. If they collided head-on at 8 a.m., find the speed of both ships if one was 20 miles per hour faster than the other.

4. Pitts bought pots for $5 each and Joe bought buckets for $7 each. If they spent $1140 for 192 utensils, how many of each type did they buy?

5. When the stranger came into the forest, 37 percent of the little people ran to hide. If 2520 refused to hide, how many little people lived in the forest?

Simplify:

6. $\dfrac{4x + 12}{x^2 + 11x + 30} \div \dfrac{x^3 - 4x^2 - 21x}{4x^2 + 20x}$ **7.** $\dfrac{x^2 + 11x + 24}{x^2 + 3x} \div \dfrac{x^2 + 13x + 40}{4x^2 + 20x}$

8. Nine-sixteenths of the girls believed that saltation was salubrious. If 700 girls disagreed, how many had made up their mind concerning this topic?

9. Solve: $-p^0(p - 4) - (-p^0)p + 3^0(p - 2) = -p - 6^0$

10. Graph on a number line: $-4 \le x < 1; D = \{\text{Integers}\}$

Solve by graphing: Use substitution:

11. $\begin{cases} y = x \\ x = -3 \end{cases}$ **12.** $\begin{cases} y = 2x + 4 \\ 2y - x = -1 \end{cases}$

13. Simplify: $\dfrac{(.00035 \times 10^{15})(200{,}000)}{(1000 \times 10^{-45})(.00007)}$

14. Find the equations of lines (a) and (b).

Simplify:

15. $4\sqrt{3} \cdot 5\sqrt{6} + \sqrt{5} \cdot 2$ **16.** $4\sqrt{12}(3\sqrt{2} - 3\sqrt{12})$

17. Divide $2x^3 + 5x^2 - 1$ by $2x + 1$

Solve by factoring:

18. $x^2 = 7x + 30$

19. $100 = 9p^2$

Solve:

20. $\dfrac{x-7}{4} - \dfrac{x}{2} = \dfrac{1}{8}$

21. $5\dfrac{1}{6}x + 2\dfrac{1}{4} = \dfrac{3}{8}$

22. Add: $\dfrac{a}{x^2} + \dfrac{a^2}{x^3 y} + \dfrac{a^3}{x(x+y)}$

23. Simplify: $\dfrac{1 + \dfrac{1}{y}}{y - \dfrac{1}{y}}$

24. $3\dfrac{1}{4}$ is what fraction of $\dfrac{7}{8}$?

25. $.037 \in \{$ What sets of numbers$\}$?

26. Evaluate: $ay^2 - y(-y + a^0)$ if $y = -3$ and $a = -4$

Simplify:

27. $-3[(-2^0 - 5^0) - 2 - (4 - 6)(-2)]$

28. (a) $\dfrac{4x - 4}{4}$ (b) $-(-2)^{-3}$

29. Multiply: $4x^2 y^{-1}\left(\dfrac{p^0 y}{x^2} - 3x^{-2}y^4\right)$ **ER 96-25, 96-26, 96-27**

LESSON 97   *Difference of two squares*

97.A
difference of two squares

In Lesson 64 we introduced the topic of square roots and noted that every positive real number has both a positive square root and a negative square root. Because 2 times 2 equals 4 and -2 times -2 also equals 4,

$$(2)(2) = 4 \qquad \text{and} \qquad (-2)(-2) = 4$$

we say that the two square roots of 4 are $+2$ and -2. **But when we use the radical to indicate the square root of a positive number as**

$$\sqrt{4}$$

we are designating the principal or positive square root which in this case is $+2$. If we wish to indicate the negative square root, we must use a minus sign in front of the expression as

$$-\sqrt{4} = -2$$

If we are asked to find the numbers that satisfy the equation

$$x^2 = 4$$

we know that the numbers are $+2$ and -2 because

$$(+2)^2 = 4 \qquad \text{and also} \qquad (-2)^2 = 4$$

The general form of the equation $x^2 = 4$ is

$$p^2 = q^2$$

If we add $-q^2$ to both sides, we can factor

$$p^2 = q^2 \longrightarrow p^2 - q^2 = 0 \longrightarrow (p + q)(p - q) = 0$$

and use the zero factor theorem to solve

$$
\begin{array}{ll}
\text{if} \quad p + q = 0 & \quad \text{if} \quad p - q = 0 \\
\underline{ -q \quad -q} & \quad \underline{ +q \quad +q} \\
 p = -q & \quad p = q
\end{array}
$$

and we find that if $p + q = 0$, then $p = -q$ and if $p - q = 0$, then $p = q$.

The theorem that describes the solution to an equation of the form $p^2 = q^2$ is called the **difference of two squares theorem.**

DIFFERENCE OF TWO SQUARES THEOREM

If p and q are real numbers and if $p^2 = q^2$, then

$$p = q \qquad \text{or} \qquad p = -q$$

---

example 97.1   Solve $p^2 = 16$.

solution   We know that the general equation $p^2 = q^2$ has two solutions, which are $p = q$ and $p = -q$. In the same way the given equation has two solutions, which are

$$p = +4 \quad \text{and} \quad p = -4$$

example 97.2   Solve $p^2 = 41$.

solution   To use the form $p^2 = q^2$, it is helpful to write the given equation as

$$p^2 = (\sqrt{41})^2$$

If we do this, we can write the answer as

$$p = \sqrt{41} \quad \text{or} \quad p = -\sqrt{41}$$

example 97.3   Solve $k^2 = 13$.

solution   We omit the intermediate step and write the solution by inspection.

$$k = \sqrt{13} \quad \text{or} \quad k = -\sqrt{13}$$

problem set 97

1. At 3 p.m., Brunhilde headed north at 30 kilometers per hour. Two hours later Ludwig headed south at 40 kilometers per hour. At what time will they be 340 kilometers apart?

2. Alphasia headed for the rodeo at 9 a.m. at 30 miles per hour. At 11 a.m. Bubba headed after her at 60 miles per hour. What time was it when Bubba caught Alphasia?

3. Billye H. ran to town at 8 kilometers per hour and then walked back home at 3 kilometers per hour. How far was it to town if the round trip took 11 hours?

4. Marfugge had $72.50 in quarters and half-dollars. If he had 190 coins in all, how many of each type did he have?

5. Fifteen percent of the seniors voted for Joe. If 289 seniors voted for the other candidates, how many seniors voted in the election?

6. Simplify: $\dfrac{x^3 + 2x^2 - 15x}{x^2 + 5x} \div \dfrac{x^3 - 6x^2 + 9x}{x^2 - 3x}$

Use the difference of two squares theorem to write the answers to the following equations:

*7. (a) $p^2 = 16$ (b) $p^2 = 41$ (c) $k^2 = 13$

8. Revulsion overcame $\dfrac{7}{10}$ of the audience when it was announced that the cute little animal was saprophagous. If 1200 were not affected, how many felt revulsion?

9. Solve: $m - m^0(m - 4) - (-2)m + (-2)(m - 4^0) = m - 6$

10. Graph on a number line: $-4 - x \not> -2; D = \{\text{Negative integers}\}$

Solve by graphing: Use elimination:

11. $\begin{cases} y = 2x \\ x = -1 \end{cases}$ **12.** $\begin{cases} 3x + 5y = -13 \\ 2x - 3y = 23 \end{cases}$

13. Simplify: $\dfrac{(42{,}000{,}000)(.0001 \times 10^{-5})}{(7000 \times 10^{14})(200{,}000 \times 10^{-8})}$

14. Find the equations of lines (a) and (b).

Simplify:

15. $4\sqrt{2} \cdot 3\sqrt{3} \cdot 5\sqrt{6}$ **16.** $3\sqrt{2}(2\sqrt{2} - 3\sqrt{8})$

17. Divide $x^3 - x^2 - 2$ by $x + 1$

Solve by factoring:

18. $-56 = 15x + x^2$ **19.** $-81 + 4x^2 = 0$

20. Solve: $\dfrac{k - 4}{2} - \dfrac{k + 6}{3} = 5$ **21.** $3\dfrac{1}{3}x - \dfrac{1}{6} = \dfrac{5}{12}$

22. Add: $\dfrac{4}{x} + \dfrac{x + 4}{x - 3}$

23. Graph on a number line: $-x - 4 \not\geq 2; D = \{\text{Integers}\}$

24. $5\dfrac{1}{4}$ is what fraction of $7\dfrac{1}{8}$?   **25.** $\dfrac{4\sqrt{2}}{5} \in \{\text{What sets of numbers}\}$?

26. Evaluate $-y - y^0(y - a)$ if $y = -2$ and $a = -5$

Simplify:

27. $-(-3)^0[(-3 - 2^0)(-2 - 3)]$

28. (a) $\dfrac{ax + a^2x^2}{ax}$ (b) $-(-3)^{-2}$

29. Simplify by adding like terms: $\dfrac{2x^2y^2}{xy^{-1}} + \dfrac{5xy}{y^{-2}}$ **ER 97-28, 97-29**

LESSON 98 *Pythagorean theorem*

98.A
angles
and triangles

It is interesting to note that mathematicians often disagree on definitions and on termi-
nology. Most authors use similar definitions, but not all do. The definition of an angle
is a good example. Most agree that two intersecting lines form four angles.

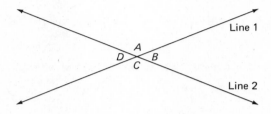

Here are shown two intersecting lines and the four angles formed. If we look only at
angle B, we see that it is formed by part of lines 1 and 2. We call these parts **half lines**
or **rays.**

European authors generally define an angle to be the opening between the rays. Thus,
to them the angle is the set of points bounded by the rays. American authors tend to
define the angle to be the rays themselves. To them the angle is the set of points that make
up the rays. Others say that the rays are the sides of the angle but don't say what the
angle is. Some don't speak of the opening at all, but define an angle to be a rotation of
a ray about its endpoint. A precise definition is not required in this book so we will just
say

   **An angle is formed by two half lines or rays that are in the same plane and that have
a common endpoint.**

If two straight lines intersect and are perpendicular to one another, we define
the measure of each of the four angles created to be 90 degrees. Instead of writing the
word degrees, it is customary to place a small elevated circle after the number that
designates the number of degrees. Thus 90 degrees can be written as 90°, and 47 degrees
and 135 degrees can be written as 47° and 135°. We see here two intersecting perpendicular lines with the resulting 90° angles

90°	90°
90°	90°

With this definition of a 90° angle and two axioms, it can be proved, by using
geometry, that **the sum of the interior angles of any triangle is 180°.** We show three

triangles and note that the sum of the three interior angles in each triangle is 180°

(a)

(b)

(c)

The triangle on the left has one angle that has a measure of 90°. It is customary to designate a 90° angle by a small square as shown in (a). Since another name for a 90° angle is a **right angle** and this triangle contains a right angle, we call this triangle a **right triangle.** Any triangle that contains a right angle is called a right triangle and the side of the triangle that is opposite the right angle is always the longest side. We call this side of a right triangle the **hypotenuse.** The other two sides are called **legs** or are just called **sides.** Right triangles have a special property that makes them very useful in mathematics, engineering, and physics.

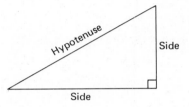

98.B
pythagorean theorem

It can be shown that **the square drawn on the hypotenuse of a right triangle has the same area as the sum of the areas of the squares drawn on the other two sides.** While this theorem was known to the Egyptians as early as the Middle Kingdom,[†] the geometric proof of the theorem is normally attributed to a Greek philosopher and mathematician named **Pythagoras.** Pythagoras was born on the Aegean island of Samos and was later associated with a school or brotherhood in the town of Crotona on the Italian peninsula in the sixth century B.C. We call the theorem for which he supposedly developed the proof the **Pythagorean theorem.**

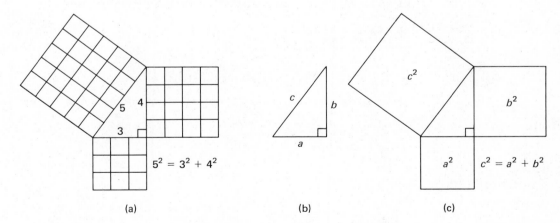

(a)   (b)   (c)

In (a) we show a right triangle whose sides have lengths of 3 and 4 units, respectively, and whose hypotenuse has a length of 5 units. A square has been drawn on each of the

---

[†] Circa 2000 B.C.

three sides, and since the area of a square whose sides have a length of L is L^2, we see that the areas of the two squares on the sides are 4^2 and 3^2 and that the sum of these two areas equals 25 square units. This is the same as the area of the square drawn on the hypotenuse, which equals 5^2 or 25 square units. Figure (b) shows another triangle whose sides have lengths of a and b and whose hypotenuse has a length of c. The area of a square drawn on the hypotenuse would be c^2, and the areas of the squares drawn on sides a and b would be a^2 and b^2, respectively, as shown in (c). We normally label the hypotenuse as c and the other two sides as a and b. Thus the general algebraic expression of the **Pythagorean theorem** is

$$a^2 + b^2 = c^2$$

where **c is the length of the hypotenuse and a and b represent the lengths of the other two sides.**

  This theorem can be used to find the length of a side of a right triangle if the lengths of the other two sides are known.

example 98.1 Given the triangle with the lengths of the sides as shown, use the Pythagorean theorem to find the length of side a.

solution The square of the hypotenuse equals the sum of the squares of the other two sides. Thus,

$$5^2 = 4^2 + a^2 \longrightarrow 25 = 16 + a^2 \longrightarrow 9 = a^2$$

We use the difference of two squares theorem to finish the solution.

$$a^2 = 9 \quad \text{which leads to} \quad a = +3 \quad \text{or} \quad a = -3$$

While -3 is a solution to the equation $a^2 = 9$, it is not a solution to the problem at hand because physical lengths are designated by positive numbers. Thus we reject this solution and say that

$$\boldsymbol{a = +3}$$

example 98.2 Find the side p in the triangle shown.

solution We apply the Pythagorean theorem to this triangle to write

$$p^2 = 5^2 + 4^2$$

  Now we simplify and use the difference of two squares theorem.

$$p^2 = 41 \longrightarrow p^2 = (\sqrt{41})^2 \longrightarrow p = \sqrt{41} \quad \text{or} \quad p = -\sqrt{41}$$

But sides of triangles do not have negative lengths so we discard the negative result and say

$$\boldsymbol{p = \sqrt{41}}$$

example 98.3 Find side k.

solution We use the Pythagorean theorem to write

$$(\sqrt{61})^2 = k^2 + 5^2$$

and now we simplify

$$61 = k^2 + 25$$

Now we finish the solution by using the difference of two squares theorem.

$$36 = k^2 \quad \longrightarrow \quad (6)^2 = k^2$$

so

$$6 = k \quad \text{or} \quad -6 = k$$

but since -6 has no meaning as the length of a side of a triangle, we say

$$k = 6$$

example 98.4   Find side m.[†]

solution   By using the Pythagorean theorem we cán write

$$m^2 = 12^2 + 8^2$$

and now we simplify and solve.

$$m^2 = 208 \quad \longrightarrow \quad m^2 = (\sqrt{208})^2$$
$$\longrightarrow \quad m = \sqrt{208} \quad \longrightarrow \quad \boldsymbol{m = 4\sqrt{13}}$$

problem set 98

1.   If the sum of -4 and the opposite of a number is multiplied by -3, the result is 6 less than the product of the number and 2. What is the number?

2.   Find four consecutive integers such that 4 times the sum of the first and fourth is 1 less than 9 times the third.

3.   Calvin could see 32. This was only 20 percent of the number that Tooley could see. How many could Tooley see?

4.   Wendy drove at 60 miles per hour. Thus, she made the trip in 1 hour less than it took Deborah because Deborah only drove at 50 miles per hour. How long did each of them drive and how long was the trip?

5.   Spann found a sack that contained $9000 in $5 bills and $10 bills. Margaret helped count the money and found that there were 1250 bills in all. How many were $5 bills and how many were $10 bills?

6.   Simplify: $\dfrac{x^2 + x - 20}{x^2 + 6x - 16} \div \dfrac{x^2 - 2x - 8}{x^2 + 10x + 16}$

Use the Pythagorean theorem to find the missing parts of these right triangles.

*7.   **8.**   **9.**

10.   Solve: $-(-3)k^0 - 3^0k + (-2)(2 - k) - (-3)(k + 2) = 0$

11.   Solve by graphing: $\begin{cases} y = x + 2 \\ y = -x + 4 \end{cases}$

[†] The Greeks must have drawn some of their right triangles as this one is drawn because the word *hypotenuse* comes from the Greek words *hupo* meaning "under" and *teinein* meaning "to stretch," so hypotenuse means "stretched under."

12. Simplify: $\dfrac{(36,000 \times 10^{-5})(400,000)}{(.0006 \times 10^{-4})(600 \times 10^5)}$

13. Find the equations of lines (a) and (b).

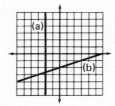

Simplify:

14. $3\sqrt{2} \cdot 5\sqrt{3} + 5\sqrt{54}$

15. $5\sqrt{2}(3\sqrt{6} - 2\sqrt{36})$

16. Solve by factoring: $100 = 25x - x^2$

17. Divide $x^3 - x$ by $x + 2$

Solve:

18. $\dfrac{x}{4} - \dfrac{x-2}{7} = 1$

19. $4\dfrac{1}{3}x + 2\dfrac{1}{4} = 7\dfrac{1}{2}$

20. Add: $\dfrac{5}{k} + \dfrac{k+3}{k+5}$

21. Simplify: $\dfrac{\dfrac{p}{k} - 4}{k - \dfrac{1}{k}}$

22. $\dfrac{1}{3}$ is what fraction of $3\dfrac{1}{8}$?

23. True or false? {Reals} ⊂ {Integers}

24. Evaluate $-xa - (-a^0) + a(x - a)$ if $x = 2$ and $a = -1$

Simplify:

25. $-(-3)^0 - 3^0 - 3^2 - (4 - 6)$

26. (a) $\dfrac{-3 - 3x}{3}$ (b) $\dfrac{-2^2}{-2^{-2}}$

27. Multiply: $\dfrac{x^{-2}}{a^2}\left(x^2a^2y^0 - \dfrac{4x^4y^2}{a^2}\right)$

ER 98-30, 98-31

LESSON 99   *Distance between two points*

99.A

distance between two points

In the last lesson, we discussed the use of the Pythagorean theorem in algebraic form to find the missing side of a triangle. To find side c of the triangle shown here, we write

$$c^2 = 4^2 + 7^2$$

and solve to find that $c = \sqrt{65}$:

$$c^2 = 16 + 49$$
$$c^2 = 65$$
$$c = \sqrt{65}$$

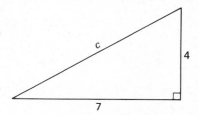

If we are given the coordinates of two points, we can find the distance between the points by graphing the points, drawing the triangle, and then solving the triangle to find the hypotenuse, which will be the missing side.

example 99.1   Find the distance between the points whose coordinates are $(4, 2)$ and $(-3, -2)$.

solutiion   The first step is to graph the points as done in the figure at the left.

 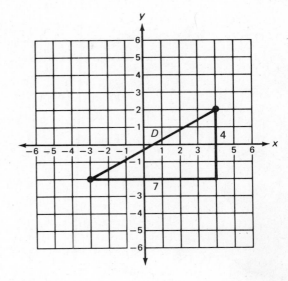

Then we connect the points with a straight line as shown in the right figure. We complete the triangle by using a vertical line for one side and a horizontal line for the other side. Next we use the Pythagorean theorem to complete the solution.

$$D^2 = 7^2 + 4^2 \longrightarrow D^2 = 65 \longrightarrow \mathbf{D = \sqrt{65}}$$

example 99.2   Find the distance between the points $(3, -4)$ and $(-5, 2)$.

solution   We graph the points and draw the required triangle as shown in the figure.

The distance between the points is found by using the Pythagorean theorem.

$$D^2 = 8^2 + 6^2$$
$$D^2 = 64 + 36$$
$$D^2 = 100$$
$$D = \sqrt{100}$$
$$\mathbf{D = 10}$$

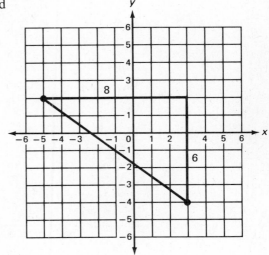

problem set 99

1.   David added 7 to twice the opposite of a number and then multiplied this sum by 3. Wade got the same result by adding 42 to 3 times the opposite of the number. What was the number?

2. Find four consecutive even integers such that if the sum of the first and the third is multiplied by 3, the result is 10 greater than 5 times the fourth.

3. The free gifts increased the crowd by 250 percent. If 180 people were there at first, how many were there after the gifts were announced?

4. At noon Wilson headed from Elk City to Idabel at 60 miles per hour. Two hours later, Johnny headed from Idabel to Elk City at 46 miles per hour. If it is 332 miles from Elk City to Idabel, what time did they meet?

5. The girls ran 20 percent longer than the boys ran. If the girls ran for 48 hours, how long did the boys run?

Use the Pythagorean theorem to:

6. Find p:

*7. Find the distance between $(4, 2)$ and $(-3, -2)$.

*8. Find the distance between $(-5, 2)$ and $(3, -4)$.

Simplify:

9. $\dfrac{x^2 + 2x}{4x + 12} \div \dfrac{x^2 - 2x - 8}{x^2 - x - 12}$

10. $\dfrac{(.00042 \times 10^{-8})(15,000)}{(5000 \times 10^7)(.0021 \times 10^{14})}$

11. Solve: $-2k^0 - 4k + 6(-k - 2^0) - (-5k) = -(2 - 5)k - 4k$

12. Solve by graphing: $\begin{cases} y = -x \\ y = -4 \end{cases}$

13. Find the equations of lines (a) and (b):

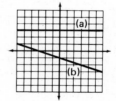

Simplify:

14. $4\sqrt{50} - 3\sqrt{8} + 2\sqrt{3}\sqrt{6}$

15. $3\sqrt{2}(6\sqrt{6} - 4\sqrt{12})$

16. Solve by factoring: $-14 = -x^2 - 5x$

17. Divide $x^3 + 6x^2 + 6x + 5$ by $x + 5$

Solve:

18. $\dfrac{y}{3} - \dfrac{y - 2}{5} = 3$

19. $4\dfrac{7}{8}p + \dfrac{2}{5} = \dfrac{3}{10}$

20. Add: $\dfrac{6}{m} + \dfrac{4m}{m + 5}$

21. What fraction of $7\dfrac{1}{4}$ is $\dfrac{5}{8}$?

22. Simplify: $\dfrac{3x - \dfrac{1}{y}}{\dfrac{2x}{y} - 4}$

23. $-.061 \in \{\text{What sets of numbers}\}$?

24. Evaluate $-x^0 a(a - x^0) - a^2$ if $x = -2$ and $a = -4$

Simplify:

25. $-2 - 2^0(-3 - 2) - (-4 + 6)(-5^0 + 2) - 2^2$

26.   (a) $\dfrac{5x^2 - 5x}{5x}$ (b) $\dfrac{-3^0}{-3^{-2}}$

27.   What is another name for the multiplicative inverse of a number?

28.   Simplify by adding like terms: $\dfrac{2xxxx}{x^{11}} - 3x^{-7} + \dfrac{4a^0}{x^7}$

29.   Graph on a number line: $-x + 2 \not< 3$; $D = \{\text{Integers}\}$

ER 99-32

Enrichment Lesson 4 is on algebraic proofs. This lesson is in the appendix and should be included at this point if this topic is to be presented.

LESSON 100   *Uniform motion—unequal distances*

100.A
unequal distances

Some uniform motion problems tell us that one person or object traveled a distance that is greater by a specified amount than the distance traveled by another person or object. The distance diagram for these problems usually takes one of the following forms:

In the left picture both started from the same place and P went 50 farther than A. In the right picture, A started out 50 in front of P, and they both ended at the same place. In either case the distance of A plus 50 equals the distance of P. The distance equation for both diagrams is the same.

$$D_A + 50 = D_P \qquad \text{so} \qquad R_A T_A + 50 = R_P T_P$$

example 100.1   At 8 p.m. Achilles left camp and headed south at 20 kilometers per hour. At 10 p.m. Patroclos headed south from the same camp. If Patroclos was 50 kilometers ahead by 3 a.m., what was his speed?

solution   Since they had the same starting point, the two arrows begin at the same point. Patroclos went farther so his arrow is longer.

Patroclos went 50 kilometers farther so we write the distance equation as

$$D_A + 50 = D_P$$

and we substitute $R_A T_A$ for D_A and $R_P T_P$ for D_P to get

$$R_A T_A + 50 = R_P T_P$$

We reread the problem to get the rate and time equations.

$$R_A = 20 \qquad T_A = 7 \qquad T_P = 5$$

Now we solve:

$$(20)(7) + 50 = R_P(5)$$
$$140 + 50 = 5R_P$$
$$190 = 5R_P$$

38 kilometers per hour $= R_P$

example 100.2   Roger has a 15-kilometer head start on Charlie. How long will it take Charlie to catch Roger if Roger travels at 70 kilometers per hour and Charlie travels at 100 kilometers per hour?

solution   Roger began 15 kilometers ahead, and they ended up in the same place so the distance diagram is

We get the distance equation from the diagram as

$$15 + D_R = D_C$$

and we replace D_R with $R_R T_R$ and D_C with $R_C T_C$ to get

$$15 + R_R T_R = R_C T_C$$

Then we reread the problem to get the other three equations:

$$R_R = 70 \qquad R_C = 100 \qquad T_R = T_C$$

Now we solve:

$$15 + 70T_R = 100\, T_C$$
$$15 = 30T_C$$
$$\frac{1}{2} = T_C$$

So Charlie will catch Roger in $\frac{1}{2}$ hour.

example 100.3   Harry and Jennet jog around a circular track that is 210 meters long. Jennet's rate is 230 meters per minute, while Harry's rate is only 200 meters per minute. In how many minutes will Jennet be a full lap ahead?

solution   This problem is simpler if we straighten it out and get the following distance diagram.

We get the distance equation from this diagram as

$$D_H + 210 = D_J \qquad \text{so} \qquad R_H T_H + 210 = R_J T_J$$

The time equation is $T_H = T_J$, and the rate equations are $R_J = 230$, $R_H = 200$. Thus our four equations are

$$\underbrace{R_H T_H + 210 = R_J T_J} \qquad \underbrace{T_H = T_J} \qquad \underbrace{R_J = 230} \qquad \underbrace{R_H = 200}$$

We use substitution to solve.

$$200 T_H + 210 = 230 T_H \quad \longrightarrow \quad 210 = 30 T_H \quad \longrightarrow \quad T_H = 7 \text{ minutes}$$

Thus $T_J = 7$ **minutes** because $T_J = T_H$.

problem set 100

*1. At 8 p.m. Achilles left camp and headed south at 20 kilometers per hour. At 10 p.m. Patroclos headed south from the same camp. If Patroclos was 50 kilometers ahead by 3 a.m., what was his speed?

*2. Roger has a 15-kilometer head start on Charlie. How long will it take Charlie to catch Roger if Roger travels at 70 kilometers per hour and Charlie travels at 100 kilometers per hour?

*3. Harry and Jennet jog around a circular track that is 210 meters long. Jennet's rate is 230 meters per minute, while Harry's rate is only 200 meters per minute. In how many minutes will Jennet be a full lap ahead?

4. When the car overturned, the jar broke and spilled 450 nickels and quarters all over the freeway. If their value was $62.50, how many coins of each type were there?

5. When the nurse gave the shots, she noticed that 34 percent of the people winced and the rest were stolid. If 3300 people were stolid, how many shots did she give?

6. Graph on a number line: $-x - 3 < 2$; $D = \{\text{Reals}\}$

Use the Pythagorean theorem to find the missing sides in the following triangles.

7.

8.

9.

10. Find the distance between $(4, 3)$ and $(7, -2)$.

11. Simplify: $\dfrac{x^2 + 11x + 28}{-x^2 + 5x} \div \dfrac{x^2 + x - 12}{x^3 - 3x^2 - 10x}$

12. Solve: $-x - (-3)(x - 5) - 2^0(2x + 3) = 5x - 7 - 7^0$

13. Solve by graphing: $\begin{cases} y = x - 4 \\ y = -x + 2 \end{cases}$

14. Simplify: $\dfrac{(22{,}000 \times 10^{-7})(500)}{(.0011)(.002 \times 10^{14})}$

15. Find the equations of lines (a) and (b).

Simplify:

16. $3\sqrt{2} \cdot 4\sqrt{3} \cdot 5\sqrt{12} + 2\sqrt{8}$ **17.** $3\sqrt{2}(5\sqrt{2} - 4\sqrt{42})$

18. Solve by factoring: $81 = 4x^2$ **19.** Divide $x^3 - 4$ by $x - 4$

Solve:

20. $\dfrac{p}{6} - \dfrac{p + 2}{4} = \dfrac{1}{3}$ **21.** $7\dfrac{1}{9}p + 3\dfrac{1}{3} = 2\dfrac{1}{6}$

22. Add: $\dfrac{x}{x + 4} + \dfrac{3}{x} - \dfrac{x + 2}{x^2}$ **23.** Simplify: $\dfrac{\dfrac{1}{a} + 4}{a^2 + \dfrac{4}{a}}$

24. $7\dfrac{3}{8}$ is what fraction of 21?   **25.** True or false? $\{\text{Integers}\} \in \{\text{Reals}\}$?

26. Evaluate $-p^2 - p^0 + p(-p + a)$ if $p = -3$ and $a = 4$

Simplify:

27. (a) $\dfrac{-2p^2a^2 - p^2a}{-p^2a}$ (b) $\dfrac{-3^2}{-(-3)^{-2}}$

28. $-2^0 - 2[(-3 - 3^0)(-2 + 6)]$

29. Simplify by adding like terms: $\dfrac{1}{(2x)^{-2}y^{-6}} - \dfrac{3x^4}{x^2y^{-6}} - 2x^2y^6$ **ER 100-33**

LESSON *101* *Square roots of large numbers*

101.A
**square roots
of large
numbers**

We have learned to simplify square roots of positive integers by first writing the integer as a product of prime numbers and then using the product of square roots theorem as we show here.

$$\sqrt{50} = \sqrt{5 \cdot 5 \cdot 2} = \sqrt{5}\sqrt{5}\sqrt{2} = \mathbf{5\sqrt{2}}$$

A similar but slightly different thought process makes the simplification of some of these expressions somewhat easier. Instead of expressing the integer as a product of prime numbers, we express it as a product of a prime number and the squares of prime numbers. For example, in the problem above, as the first step we would write 50 as the product of 25 and 2

$$\sqrt{50} = \sqrt{25 \cdot 2}$$

and then use the product of square roots theorem to complete the simplification.

$$\sqrt{25 \cdot 2} = \sqrt{25}\sqrt{2} = \mathbf{5\sqrt{2}}$$

This thought process is especially helpful when the radicand has a factor that is the square of 10 or 100 or 1000 or 10,000 or any other power of 10.

$$(10)^2 = 100 \qquad (100)^2 = 10,000 \qquad (1000)^2 = 1,000,000$$

$$\text{and} \qquad (10,000)^2 = 100,000,000$$

We note that all of these products have an even number of zeros; 100 has two; 10,000 has four; 1,000,000 has six; and 100,000,000 has eight. **Thus, when we simplify, we always write the number so that one of its factors has an even number of zeros.**

example 101.1 Simplify $\sqrt{50,000}$.

solution We see four zeros, so we write

$$\sqrt{10,000 \cdot 5} = \sqrt{10,000}\sqrt{5} = \mathbf{100\sqrt{5}}$$

example 101.2 Simplify $\sqrt{500,000}$.

solution It would not help to write $\sqrt{5 \cdot 100,000}$ because 100,000 has an odd number of zeros, so we write

$$\sqrt{50 \cdot 10,000} = \sqrt{50}\sqrt{10,000} = \mathbf{100\sqrt{50}}$$

Now we simplify $\sqrt{50}$ as $\sqrt{5 \cdot 5 \cdot 2} = \sqrt{5}\sqrt{5}\sqrt{2} = 5\sqrt{2}$. Thus

$$100\sqrt{50} = 100(5\sqrt{2}) = \mathbf{500\sqrt{2}}$$

example 101.3 Simplify $\sqrt{40,000,000}$.

solution Looking for an even number of zeros, we write

$$\sqrt{1,000,000 \cdot 40} = \sqrt{1,000,000}\sqrt{40} = 1000\sqrt{40}$$

Now we simplify $\sqrt{40}$ as $\sqrt{2 \cdot 2 \cdot 10} = \sqrt{2}\sqrt{2}\sqrt{10} = 2\sqrt{10}$. Thus $1000\sqrt{40}$ can be written as

$$1000(2\sqrt{10}) = \mathbf{2000\sqrt{10}}$$

example 101.4 Simplify $\sqrt{700,000,000}$.

solution Using eight zeros in one factor, we can write

$$\sqrt{7 \cdot 100,000,000} = \sqrt{7}\sqrt{100,000,000} = \mathbf{10,000\sqrt{7}}$$

problem set 101

1. Eleanor started out at 60 miles per hour at 9 a.m., 2 hours before Alexi started out to catch her. If she was still 60 miles ahead at 3 p.m., how fast was Alexi driving?

2. Thirty percent of the sailors wanted to turn back, but the rest of the sailors agreed with Aeneas. If 28 sailors agreed with Aeneas, how many sailors were on the ship?

3. It took Perseus 60 hours to get there with white sails and 100 hours to come back with black sails. How far was it if his speed with white sails was 2 miles per hour greater than his speed with black sails?

4. The product of 5 and the sum of a number and negative 8 is 9 greater than the product of 2 and the opposite of the number. Find the number.

5. Red shoes were $7 a pair and white shoes were $3 a pair. Lloyd G. bought 30 pairs for $130. How many pairs of each color did he buy?

6. Bobby and John found four consecutive integers such that 5 times the sum of the second and third was 6 less than 7 times the first. What were their integers?

7. Find g.

8. Find the distance between $(-4, -2)$ and $(4, -6)$.

9. Simplify: $\dfrac{4x^2 + 8x}{x^2 + 8x + 12} \div \dfrac{4x^2 - 16x}{x^2 + 3x - 18}$

10. Solve: $(-3)x^0 - (-2x) - 4(x - 4) - (2 - x) = 3^0(x - 4)$

11. Solve by graphing: $\begin{cases} y = -2x \\ y = -4 \end{cases}$

12. Simplify: $\dfrac{(1200 \times 10^{-42})(300 \times 10^{14})}{(.004 \times 10^5)(3000 \times 10^{-20})}$

13. Find the equations of lines (a) and (b).

Simplify:

14. $\sqrt{7500}$ *15. $\sqrt{50,000}$ 16. $4\sqrt{50,000} + 3\sqrt{5,000,000}$

17. Graph on a number line: $-4 < x \le 2$; $D = \{\text{Integers}\}$

18. Solve by factoring: $64 = 16x - x^2$

19. Divide: $x^4 + x^3 + 2x + 2$ by $x + 1$

Solve:

20. $\dfrac{k}{7} - \dfrac{k - 4}{3} = 2$ 21. $3\dfrac{1}{4}k - 2\dfrac{1}{2} = \dfrac{3}{4}$

22. Add: $\dfrac{4}{x^2} + \dfrac{x + 4}{x} - \dfrac{3x + 2}{x + 1}$ 23. Simplify: $\dfrac{\dfrac{1}{x} - 4}{y - \dfrac{1}{x}}$

24. $2\dfrac{1}{5}$ is what fraction of $28\dfrac{1}{10}$?   25. $.0003\sqrt{2} \in \{\text{What sets of numbers}\}$?

26. Evaluate: $p^2 - p^0(-p^2 - p - x)$ if $p = -4$ and $x = -2$

Simplify:

27. $\dfrac{pa - pa^2}{pa}$ 28. $\dfrac{-2^{-2}}{(-2)^2}$ 29. $-3^2 - 2^0 - 2[(-2 - 3) - (-4 - 6)]$

30. Multiply: $\dfrac{x^2a}{a^2}\left(\dfrac{a}{x^2} - \dfrac{x^2a}{a^2x}\right)$ **ER 101-34**

LESSON 102   *Rounding off*

102.A

place value   We use the 10 digits

$$0, 1, 2, 3, 4, 5, 6, 7, 8, 9$$

to write the decimal numerals that we use to represent numbers. The value of a digit in a numeral depends on the position of the digit with respect to the decimal point.

For instance, in the numeral

$$40{,}632{,}903.195034$$

the first 9 has a value of 900 because it is in the hundreds' place, three places to the left of the decimal point. The second 9 has a value of only 9/100 because it is written in the hundredths' place, which is two places to the right of the decimal point.

etc.	10,000,000	1,000,000	100,000	10,000	1000	100	10	1	decimal point	$\frac{1}{10}$	$\frac{1}{100}$	$\frac{1}{1000}$	$\frac{1}{10,000}$	$\frac{1}{100,000}$	$\frac{1}{1,000,000}$	etc.
	4	0	6	3	2	9	0	3	.	1	9	5	0	3	4	

It is important to note that the first place to the right of the decimal point is the tenths' place while the tens' place is not one place but is two places to the left of the decimal point. The first place to the left is the units' (ones') place.

102.B
rounding off

Often we use numbers that are approximations of other numbers. For instance, the circumference of the earth at the equator is 24,874 miles. This measurement is to the nearest mile and is a difficult number to remember. So we say that the circumference is 25,000 miles and say that we have rounded off 24,874 to the nearest thousand. This is because 24,874 is closer to 25,000 than it is to 24,000. Thus, when we round off, we change the digits on the end of a number to zeros.

Rounding off requires three steps, and mistakes can be avoided if a circle and an arrow are used as aids. To demonstrate we will round off 24,874 to the nearest thousand.

1. Circle the digit in the place to which we are rounding off and mark the digit to its right with an arrow.

$$2\ ④\overset{\downarrow}{8}\ 7\ 4$$

2. Change the arrow-marked digit and all digits to its right to zero.

$$2\ ④\overset{\downarrow}{0}\ 0\ 0$$

3. Leave the circled digit unchanged or increase it 1 unit as determined by the following rules:
   a. If the arrow-marked digit was less than 5, do not change the circled digit.
   b. If the arrow-marked digit was greater than 5 or was 5 followed somewhere by a nonzero digit, increase the circled digit 1 unit. This rule applies to the problem we are working so we finish by writing

$$2\ ⑤\overset{\downarrow}{0}\ 0\ 0$$

   c. If the arrow-marked digit was a terminal 5 or 5 followed only by zeros, the number is halfway; and the circled digit can be left unchanged or can be increased by 1 as you wish. In order to be consistent, many people do not change the circled digit if it is even and increase it 1 if it is odd. The procedure to be used in this case is really not important, and we will try to avoid this case in the problem sets. Concentrate on remembering rules a and b.

example 102.1   Round off 47,258,312.065 to the nearest ten thousand.

solution   We circle the ten-thousands' digit and mark the digit to its right with an arrow.

$$47, 2 \; ⑤ \; \overset{\downarrow}{8}, 312.065$$

Next we change the arrow-marked digit and all digits to its right to zero.

$$47, 2 \; ⑤ \; \overset{\downarrow}{0}, 000.000$$

Since the arrow-marked digit was greater than 5, we increase the circled digit 1 unit, and our answer is

$$47, 2 \; ⑥ \; \overset{\downarrow}{0}, 000.000 \qquad \text{which is } \mathbf{47,260,000}$$

example 102.2   Round off 104.06245327 to the nearest thousandth.

solution   We circle the thousandths' digit and mark the digit to its right with an arrow.

$$104.06 \; ② \; \overset{\downarrow}{4}5327$$

Then we change the arrow-marked digit and all digits to its right to zero.

$$104.06 \; ② \; \overset{\downarrow}{0}0000$$

Since the arrow-marked digit was less than 5, we do not change the circled digit. Thus our answer is as follows because the terminal zeros have no value.

104.062

example 102.3   Round off .00041378546 to the nearest one-hundred-millionth.

solution   We circle the one-hundred-millionths' place and mark the digit to its right with an arrow.

$$.0004137 \; ⑧ \; \overset{\downarrow}{5}46$$

Now we change to zero the arrow-marked digit and all digits to the right of the arrow-marked digit.

$$.0004137 \; ⑧ \; \overset{\downarrow}{0}00$$

The first digit changed to zero was 5 and it was followed by the nonzero digits 4 and 6. Thus we increase the circled digit from 8 to 9 and get

.00041379

example 102.4   Round off 2.0031664567 to five decimal places.

solution   The fifth decimal place is the hundred-thousandths' place, which we circle. Then we mark the next digit with an arrow.

$$2.0031 \; ⑥ \; \overset{\downarrow}{6}4567$$

Since the arrow-marked digit is greater than 5, when we change it to zero we increase the circled digit 1 unit from 6 to 7, and our answer is

2.00317

example 102.5   Round off 314.0$\overline{364}$ to: (a) five decimal places, (b) nine decimal places, (c) the nearest one-hundredth, (d) to the nearest ten.

solution   The line over the 364 tells us that these digits repeat, so the number is

$$314.0364364364364 \cdots$$

We round off this number as specified:

(a)	Five decimal places	314.03644
(b)	Nine decimal places	314.036436436
(c)	nearest hundredth	314.04
(d)	nearest ten	310

problem set 102

1. Boesch had a 40-meter head start. How long did it take Lebeda to catch up if Lebeda traveled at 10 meters per second while Boesch traveled at only 6 meters per second?

2. Night came and the mangroves and palmettos closed in on their victim. If 27 percent were mangroves and 511 were palmettos, how many total bushes were in on the attack?

3. Robert ran to the redoubt while Wilbur walked to the parapet. Both distances were the same and Robert's speed was 6 miles per hour while Wilbur's was 8 miles per hour. What was the time of each if Robert's time was 2 hours longer than Wilbur's?

4. Penelope and Miranda found four consecutive odd integers such that 5 times the sum of the first two is 5 less than 19 times the fourth. What were the integers?

5. The 60-foot rope was cut into two pieces. One of the pieces was 10 feet longer than 4 times the other piece. How long were the two pieces?

6. Round off:
   *(a) 47,258,312.065 to the nearest ten thousand
    (b) 104.06253527 to the nearest thousandths
   *(c) 314.0$\overline{364}$ to five decimal places
    (d) 413.0$\overline{527}$ to the nearest one hundred

7. Find s.

8. Find the distance between $(-8, -4)$ and $(-6, -2)$.

9. Simplify: $\dfrac{x^2 + 8x + 15}{x^2 + 3x} \div \dfrac{x^2 + 3x - 10}{x^3 - 6x^2 + 8x}$

10. Solve: $-2x - 3(x - 2^0) + 2x(-3 - 4^0) = x^0 - 4x - 2$

11. Solve by graphing: $\begin{cases} y = 2x \\ y = -x + 6 \end{cases}$

12. Simplify: $\dfrac{(400 \times 10^5)(.0008 \times 10^{14})}{(20,000 \times 10^{-30})(.00002)}$

13.   Find the equations of lines (a) and (b).

Simplify:

14. $3\sqrt{30{,}000} - 9\sqrt{300} + 3\sqrt{2} \cdot 5\sqrt{6}$ 15. $3\sqrt{2}(4\sqrt{8} - 3\sqrt{12})$

16. $2\sqrt{6}(4\sqrt{2} - 3\sqrt{3})$

17.   Graph on a number line: $-2 \geq -2x + 2$; $D = \{\text{Integers}\}$

18.   Solve by factoring: $35 = -12x - x^2$ 19.   Divide $x^3 + 12x + 5$ by $x + 2$

Solve:

20. $\dfrac{x}{4} - \dfrac{x+2}{6} = 4$ 21. $5\dfrac{1}{2}m + \dfrac{3}{8} = \dfrac{1}{16}$

22.   Add: $\dfrac{4}{x^2} - \dfrac{x+3}{4x} - \dfrac{2x}{x+1}$ 23.   Simplify: $\dfrac{\dfrac{ay}{x} - 4}{\dfrac{1}{x} + 5}$

24. $3\dfrac{1}{11}$ is what fraction of 22?

25. $\dfrac{3 + 4\sqrt{2}}{5} \in \{\text{What sets of numbers}\}$?

26.   Evaluate $-p^2 - p^0(-p + x)$ if $p = -2$ and $x = -4$

Simplify:

27. $\dfrac{x^2 - ax^2}{x^2}$ 28. $\dfrac{(4x)^2(y^{-2})^2xx^2y}{(2x)^2(y^2)^2x^0yx^2}$ 29. $\dfrac{-3^{-2}}{3}$

30. $-2^2 - 2[(-3 - 2)(-5 - 4)][-3^0(-2 - 5)]$ **ER 102-35, 36**

LESSON 103   *Square root tables*

103.A
square root tables

In Lesson 64 we learned how to approximate the square root of a number by a process of cut and try. Of course, the easiest way to find the square root of a particular positive real number is to insert the number into our handy pocket calculator and then hit the square root key.

    Tables of square roots can be used to find an approximation of the square root of any positive number by anyone who knows how to write a number in scientific notation and who knows how and when to use the product of square roots theorem. Practice in solving problems such as the following should increase our feel for square roots and allow us to practice scientific notation. Tables of square roots customarily give the square root of numbers between 1 and 10. Thus our procedure will be

1. Write the number in scientific notation and round off as necessary.
2. If the power of 10 is not an even power of 10, express it as 10 times an even power of 10.

example 103.1 Use the square root tables to approximate $\sqrt{416.23}$.

solution The square root table in this book has only three digits so we round off to three digits and write the number in scientific notation.

$$\sqrt{4.16 \times 10^2}$$

Now use the product of square roots theorem

$$\sqrt{4.16}\sqrt{10^2}$$

and look up $\sqrt{4.16}$ in the table and write $\sqrt{10^2}$ as 10.

n	n^2	$\sqrt{n}$
41.0	16.8100	2.02485
4.11	16.8921	2.02731
4.12	16.9744	2.02978
4.13	17.0569	2.03224
4.14	17.1396	2.03470
4.15	17.2225	2.03715
4.16	17.3056	2.03961
4.17	17.3889	2.04206
4.18	17.4724	2.04450
4.19	17.5561	2.04695

$$\sqrt{4.16}\sqrt{10^2} = \mathbf{2.03961 \times 10}$$

example 103.2 Use the square root tables to approximate $\sqrt{4156}$.

solution As the first step we round off and write the number in scientific notation.

$$\sqrt{4160} = \sqrt{4.16 \times 10^3}$$

We do not have 10 to an even power as a factor, and 10 to an odd power is never a perfect square. We write 10^3 as $10 \cdot 10^2$.

$$\sqrt{4.16 \times 10^3} = \sqrt{4.16 \cdot 10 \cdot 10^2}$$

Now we use the product of square roots theorem and the table of square roots to complete the solution.

n	n^2	$\sqrt{n}$
4.10	16.8100	2.02485
4.11	16.8921	2.02731
4.12	16.9744	2.02978
4.13	17.0569	2.03224
4.14	17.1396	2.03470
4.15	17.2225	2.03715
4.16	17.3056	2.03961
4.17	17.3889	2.04206
4.18	17.4724	2.04450
4.19	17.5561	2.04695

n	n^2	$\sqrt{n}$
9.91	98.2081	3.14802
9.92	98.4064	3.14960
9.93	98.6049	3.15119
9.94	98.8036	3.15278
9.95	99.0025	3.15436
9.96	99.2016	3.15595
9.97	99.4009	3.15753
9.98	99.6004	3.15911
9.99	99.8001	3.16070
10.00	100.0000	3.16228

$$\sqrt{4.16 \cdot 10 \cdot 10^2}$$
$$= \sqrt{4.16}\sqrt{10}\sqrt{10^2}$$
$$= \mathbf{(2.03961)(3.16228) \times 10}$$

Leave the answers in the form of an indicated multiplication. In science classes you will complete the multiplication and find a numerical answer, but at this time we are interested in the process—not the answer.

example 103.3 Use the square root tables to approximate $\sqrt{.0005273}$.

solution
$$\sqrt{.0005273} = \sqrt{5.27 \times 10^{-4}}$$

The power of 10 is even so we have no problem.

n	n^2	$\sqrt{n}$
5.20	27.0400	2.28035
5.21	27.1441	2.28254
5.22	27.2484	2.28473
5.23	27.3529	2.28692
5.24	27.4576	2.28910
5.25	27.5625	2.29129
5.26	27.6676	2.29347
5.27	27.7729	2.29565
5.28	27.8784	2.29783
5.29	27.9841	2.30000

$$\sqrt{5.27 \times 10^{-4}}$$
$$= \sqrt{5.27}\sqrt{10^{-4}}$$
$$= \mathbf{2.29565 \times 10^{-2}}$$

example 103.4 Use the square root tables to approximate $\sqrt{.00005273}$.

solution
$$\sqrt{.00005273} = \sqrt{5.27 \times 10^{-5}}$$

Here the power of 10 is again odd so we must rewrite it as 10 times an even power, **but be careful because**
$$10^{-5} \neq 10 \cdot 10^{-4} \qquad \text{but instead} \qquad 10^{-5} = 10 \cdot 10^{-6}$$

n	n^2	$\sqrt{n}$
5.20	27.0400	2.28035
5.21	27.1441	2.28254
5.22	27.2484	2.28473
5.23	27.3529	2.28692
5.24	27.4576	2.28910
5.25	27.5625	2.29129
5.26	27.6676	2.29347
5.27	27.7729	2.29565
5.28	27.8784	2.29783
5.29	27.9841	2.30000

n	n^2	$\sqrt{n}$
9.91	98.2081	3.14802
9.92	98.4064	3.14960
9.93	98.6049	3.15119
9.94	98.8036	3.15278
9.95	99.0025	3.15436
9.96	99.2016	3.15595
9.97	99.4009	3.15753
9.98	99.6004	3.15911
9.99	99.8001	3.16070
10.00	100.0000	3.16228

$$\sqrt{5.27 \times 10^{-5}}$$
$$= \sqrt{5.27 \times 10 \times 10^{-6}}$$
$$= \sqrt{5.27}\sqrt{10}\sqrt{10^{-6}}$$
$$= \mathbf{(2.29565)(3.16228) \times 10^{-3}}$$

example 103.5 Use the square root tables to approximate $\sqrt{462 \times 10^{-15}}$.

solution Again we get a negative odd power of 10. We rewrite as usual, but with care because 10^{-13} equals 10×10^{-14}.

n	n^2	$\sqrt{n}$		n	n^2	$\sqrt{n}$
4.60	21.1600	2.14476		9.91	98.2081	3.14802
				9.92	98.4064	3.14960
4.61	21.2521	2.14709		9.93	98.6049	3.15119
4.62	21.3444	2.14942		9.94	98.8036	3.15278
4.63	21.4369	2.15174				
4.64	21.5296	2.15407		9.95	99.0025	3.15436
4.65	21.6225	2.15639		9.96	99.2016	3.15595
				9.97	99.4009	3.15753
4.66	21.7156	2.15870		9.98	99.6004	3.15911
4.67	21.8089	21.6102		9.99	99.8001	3.16070
4.68	21.9024	2.16333				
4.69	21.9961	2.16564		10.00	100.000	3.16228

$$\sqrt{462 \times 10^{-15}}$$
$$= \sqrt{4.62 \times 10^{-13}}$$
$$= \sqrt{4.62 \times 10 \times 10^{-14}}$$
$$= \sqrt{4.62} \times \sqrt{10} \times \sqrt{10^{-14}}$$
$$= (2.14942)(3.16228) \times 10^{-7}$$

problem set 103

1. Louis ran to Versailles and then walked back to town. His running rate was 6 kilometers per hour and his walking rate was 3 kilometers per hour. How far was it to Versailles if the round trip took 6 hours?

2. Prince Valiant bought some replacement armor for 540 florins. He bought helmets for 4 florins each and cuirasses for 6 florins each. If he bought 100 pieces of armor, how many helmets did he buy?

3. Roger Goose had a 500-yard head start on Willy. If Willy's speed was 40 yards per second and Roger's speed was only 20 yards per second, how long did it take Willy to catch up?

4. Demosthenes had white pebbles and black pebbles. If 27 percent of his pebbles were white and he had 438 black pebbles, how many pebbles did he have in all?

5. Find four consecutive even integers such that -6 times the sum of the second and fourth is 8 less than 11 times the opposite of the third.

Use the square root tables in the appendix to approximate:

*6. $\sqrt{4156}$ *7. $\sqrt{.00005273}$ 8. $\sqrt{.003266 \times 10^{-18}}$

9. Find k.

10. Find the distance between $(4, -2)$ and $(-2, 3)$.

11. Simplify: $\dfrac{x^3 + 12x^2 + 35x}{4x^2 + 8x} \div \dfrac{x^2 + 15x + 50}{4x + 8}$

12. Solve: $3x - 2^0(x - 4) + 3(2x + 5) = 7 + x^0$

13. Solve by graphing: $\begin{cases} y = 2x + 4 \\ y = -2x \end{cases}$

14. Simplify: $\dfrac{(30,000)(.000005)}{(1500 \times 10^{-5})(10,000)}$

15. Find the equations of lines (a) and (b).

Simplify:

16. $4\sqrt{20{,}000}$

17. $4\sqrt{3}\cdot 5\sqrt{2} - 2\sqrt{2}(3\sqrt{2} - 2\sqrt{12})$

18. Graph: $2 \le x < 5$; $D = \{$Positive integers$\}$

19. Solve by factoring: $70 = x^2 + 3x$

20. Solve: $\dfrac{2x}{3} - \dfrac{x+4}{7} = 2$

21. Solve: $3\frac{1}{4}p + 3\frac{1}{4} = 7\frac{1}{8}$

22. Add: $\dfrac{4}{x^2} - \dfrac{3x+2}{x+1} + \dfrac{5}{x}$

23. Simplify: $\dfrac{\dfrac{mx}{4} + 1}{\dfrac{1}{4} + a}$

24. True or False? $\{$Integers$\} \subset \{$Naturals$\}$

25. Evaluate $-p - p(-p - ap)$ if $a = -4, p = -5$

Simplify:

26. $\dfrac{4 - 4x}{4}$

27. $\dfrac{-2^{-2}}{-2^2}$

28. $-2[|4 - 2| - (5 - 3)(-2 - 4^0)]$

29. Multiply: $\dfrac{a^{-3}y^{-2}}{x}\left(\dfrac{xa^3}{y^2} - \dfrac{2ay}{x}\right)$

30. Round off $40.\overline{37}$ to the nearest ten-millionth.   **ER 103-37**

LESSON 104 *Factorable denominators*

104.A

addition of expressions with factorable denominators

Algebra books tend to emphasize the factoring of trinomials and binomials because the ability to factor is important and also because doing these exercises provides experience in manipulating expressions that contain variables. Thus all algebra books contain problems such as this one:

$$\text{Simplify:} \quad \frac{x^2 + x - 20}{x^2 - x - 12} \div \frac{x^2 + 7x + 10}{x^2 + 9x + 14}$$

For the same reasons, algebra books present problems in addition of rational expressions whose denominators are factorable. These problems are designed so that the addition is facilitated if one or more of the denominators are factored before the addition is attempted. The key to these problems is recognizing that they are contrived problems designed to give practice in factoring.

example 104.1   Add $\dfrac{6x}{x^2 - x - 12} - \dfrac{p}{x - 4}$.

solution   We recognize this problem as a problem designed to give practice in factoring. We begin by factoring the first denominator.

$$\frac{6x}{(x - 4)(x + 3)} - \frac{p}{x - 4}$$

And now we can see that the least common multiple of the denominators is

$$(x - 4)(x + 3)$$

We use this as our new denominator and add.

$$\frac{}{(x - 4)(x + 3)} - \frac{}{(x - 4)(x + 3)} \longrightarrow \frac{6x}{(x - 4)(x + 3)} - \frac{p(x + 3)}{(x - 4)(x + 3)}$$

$$= \frac{6x - px - 3p}{x^2 - x - 12}$$

example 104.2   Add $\dfrac{7}{x^2 - 5x - 6} - \dfrac{5}{x^2 - 6x}$.

solution   As the first step we factor both denominators:

$$\frac{7}{(x - 6)(x + 1)} - \frac{5}{x(x - 6)}$$

and we see that the least common multiple of the denominators is $x(x - 6)(x + 1)$. We use this as our new denominator

$$\frac{}{x(x - 6)(x + 1)} - \frac{}{x(x - 6)(x + 1)} \longrightarrow \frac{7x}{x(x - 6)(x + 1)} - \frac{5(x + 1)}{x(x - 6)(x + 1)}$$

$$= \frac{2x - 5}{x(x - 6)(x + 1)}$$

example 104.3   Add $\dfrac{4x + 2}{x^2 + x - 6} - \dfrac{4}{x^2 + 3x}$.

solution   We begin by factoring both denominators.

$$\frac{4x + 2}{(x + 3)(x - 2)} - \frac{4}{x(x + 3)}$$

We see that the LCM of the denominators is $x(x + 3)(x - 2)$, so we get

$$\frac{x(4x + 2)}{x(x + 3)(x - 2)} - \frac{4(x - 2)}{x(x + 3)(x - 2)}$$

We now simplify the numerator, remembering that $-4(x - 2) = -4x + 8$.

$$\frac{4x^2 + 2x - 4x + 8}{x(x + 3)(x - 2)} = \frac{4x^2 - 2x + 8}{x(x + 3)(x - 2)}$$

**problem
set 104**

1. At the three-quarter pole My Bequest was only 40 feet behind Flying Lady. How long did it take My Bequest at 54 feet per second to catch Flying Lady at 46 feet per second?

2. Find three consecutive even integers such that 4 times the first equals 16 times the sum of the third and the number 2.

3. Ed and Alice walked to the dock at 5 miles per hour, jumped into the boat, and motored to Destin at 15 miles per hour. If the total distance was 20 miles and the trip took 2 hours in all, how far did they go by boat?

4. On the first day of the sale, the girls sold 20 percent of their cookies. If they still had 384 cookies left, how many cookies did they bring to the sale?

5. Charles opened the old trunk and found $6750 in $1 bills and $10 bills. If there were 150 more 1s than 10s, how many of each kind were there?

6. Use square root tables to approximate $\sqrt{108,052 \times 10^{-10}}$.

7. Find a.

8. Find the distance between $(4, -3)$ and $(-4, 2)$.

9. Simplify: $\dfrac{x^3 + 11x^2 + 24x}{x^2 + 10x + 21} \div \dfrac{4x^2 + 32x}{4x + 40}$

10. Solve: $-2x(4 - 3^0) - (2x - 5) + 3x - 2 = -2^0 x$

11. Solve by graphing: $\begin{cases} y = -2x \\ y = -2 \end{cases}$

12. Simplify: $\dfrac{(4000 \times 10^{-40})(.0003 \times 10^{-21})}{(20,000)(3000 \times 10^{-14})}$

13. Find the equations of lines (a) and (b).

Simplify:

14. $2\sqrt{60,000}$

15. $4\sqrt{5} \cdot 2\sqrt{3} + 5\sqrt{3}(\sqrt{3} + 2\sqrt{5})$

16. Graph: $-4 - x \not< 2$; $D = \{\text{Reals}\}$

17. Solve $-81 + 4x^2 = 0$ by factoring.

Solve:

18. $\dfrac{5x}{3} - \dfrac{x - 5}{2} = 14$

19. $4\dfrac{1}{8}p - \dfrac{3}{4} = 2\dfrac{1}{4}$

Add:

*20. $\dfrac{6x}{x^2 - x - 12} - \dfrac{p}{x - 4}$

*21. $\dfrac{7}{x^2 - 5x - 6} - \dfrac{5}{x^2 - 6x}$

22. $\dfrac{p}{x^2 - 9} + \dfrac{2x}{x^2 - 3x}$ **23.** $4^2 \in \{\text{What sets of numbers}\}$?

Simplify:

24. $\dfrac{-4x^2 - 8x^2a}{-4x^2}$ **25.** $\dfrac{-3^{-2}}{(-3)^2}$ **26.** $\dfrac{\dfrac{5p}{x} - 4}{\dfrac{3}{x} - x}$

27. Evaluate $-x - xk(x - k)$ if $x = -4$ and $k = 5$

Simplify:

28. $\dfrac{x^2a(x^2a)(x^{-2})^2x^0xa^2}{(a^{-3})^2ax^{-2}x^4x}$ **29.** $2^2[-(-2^0 - 3)(-2 + 5) + 3]$

30. Round off $7.\overline{185}$ to the nearest hundred-millionth.   **ER 104-38**

LESSON 105   *Absolute value inequalities*

105.A
absolute value inequalities

We review the concept of absolute value by saying that every real number except zero can be thought of as having two qualities or parts. One of the parts is designated by the plus or minus sign and the other part is the numerical part. We can think of the numerical part as designating the quality of "bigness" of the number, and we call this quality the absolute value of the number. Thus we say that the two numbers

$$3 \qquad \text{and} \qquad -3$$

both have an absolute value of 3 although one of them is a positive number and one of them is a negative number.

We designate the absolute value of a number by enclosing the number within vertical lines. Thus we designate the absolute value of 3 by writing $|3|$, and we designate the absolute value of -3 by writing $|-3|$. Of course, the absolute value of both of these numbers is 3.

$$|3| = 3 \qquad |-3| = 3$$

It is difficult to describe the absolute value of a number by using words. Most authors do not like to speak of the "bigness" of a number as we have done, for they feel that bigness can be confused with the concept of *greater than* that is used to compare numbers. For this reason, many prefer the formal definition of absolute value used in more advanced courses. This definition uses symbols and does not use words.

The definition is in three parts and we have avoided using the definition thus far because the third part is confusing to some people.

$$\begin{aligned}
&\text{(a)} \quad \text{if } x > 0, \qquad |x| = x \\
&\text{(b)} \quad \text{if } x = 0, \qquad |x| = 0 \\
&\text{(c)} \quad \text{if } x < 0, \qquad |x| = -x
\end{aligned}$$

Part (a) speaks of the absolute value of positive numbers, which are numbers that are greater than zero. Part (b) describes the absolute value of zero. Part (c) can be confusing because of the minus sign. Part (c) describes the absolute value of negative numbers, which are numbers that are less than zero. We know that the absolute value of a negative number is the opposite of the number, as

$$|-3| = -(-3) = 3$$

Thus when we write

$$\text{if } x < 0, \qquad |x| = -x$$

we are not saying that the absolute value is a negative number but that the absolute value of a negative number is the opposite of the negative number, which is a positive number.

> The absolute value of a nonzero real number is a positive number.[†] The absolute value of zero is zero.

$$|3| = +3 \qquad |-3| = +3 \qquad |0| = 0$$

In an attempt to describe the absolute value of a number by using words, some authors define the absolute value of a number to be the number that describes the distance on the number line from the origin to the graph of the number being considered. If we use this definition, we will find that $+3$ and -3 have the same absolute value, for they are both 3 units from the origin, and thus both numbers have an absolute value of 3.

Other authors note that every nonzero real number has an opposite. They say that the absolute value of either member of a pair of opposites is the positive member of the pair. Thus, the absolute value of either

$$3 \qquad \text{or} \qquad -3$$

is 3, the positive member of the pair. If we remember this we can see that there are two answers to the following question.

$$|\text{What numbers?}| = 4$$

Here we ask what numbers have an absolute value of 4. Of course, the answers are $+4$ and -4 because both of these numbers have an absolute value of 4. It is customary to use a single letter variable as the unknown so we will restate the question by writing

$$|x| = 4$$

and as we have said, the two values of x that satisfy this condition are $+4$ and -4.

Often it is desirable to display the solution of an absolute value equation or inequality in graphical form. The graph of the solution to the equation $|x| = 4$ is the graph of the numbers $+4$ and -4 as shown here.

[†] Some might like to say that the absolute value of a non-zero number is the positive value of the number, but this would be incorrect. A real number that is not zero is either a positive number or a negative number. Numbers do not have two values.

example 105.1   Graph $|x| > 2$, $D = \{$Reals$\}$.

solution   We are asked to indicate on the number line every real number whose absolute value is greater than 2.

We note that part of the solution set to this inequality consists of negative numbers. This is correct, for if we think carefully, we will realize that while these negative numbers are less than 2, their absolute values are greater than 2.

example 105.2   Graph $|x| \leq 3$, $D = \{$Integers$\}$.

solution   We are asked to indicate on the number line the location of all integers whose absolute value is equal to or less than 3.

We note that all integers that are greater than or equal to -3 and that are also less than or equal to 3 have an absolute value that is equal to or less than 3.

example 105.3   Graph $|x| < -4$, $D = \{$Reals$\}$.

solution

We have not graphed the solution for the given condition because there are no real numbers that satisfy the given condition. The statement $|x| < -4$ asks for the real number replacements for x whose absolute values are less than -4. There are no real numbers that satisfy this condition since the absolute value of any nonzero real number is greater than zero. If we use the formal language of sets, instead of saying that we have not graphed the solution, we say that the solution is the empty set $\{\ \ \}$, or the null set $\varnothing$, and we say that the bare number line shown is the graph of this set.

example 105.4   Graph $|x| > -4$, $D = \{$Reals$\}$.

solution

This one is also tricky. The absolute value of every real number is zero or a number greater than zero. Certainly, then, if the absolute value of every real number is equal to or greater than zero, the absolute value of every real number is also greater than -4. Thus the solution to the stated condition is the set of real numbers, which is graphed by indicating the entire number line.

example 105.5   Graph $-|x| \geq -2$, $D = \{$Reals$\}$.

solution   We begin by multiplying both sides of the inequality by -1 and also **reversing the sense of the inequality symbol** and we find

$$|x| \leq 2$$

which is graphed below.

example 105.6   Graph $-|x| - 2 > -5$, $D = \{\text{Integers}\}$.

solution   First we must solve the inequality for $+|x|$ by isolating $|x|$ on one side of the inequality.

$$-|x| - 2 > -5 \qquad \text{given}$$
$$-|x| > -3 \qquad \text{added } +2 \text{ to both sides}$$
$$|x| < 3 \qquad \begin{array}{l}\text{multiplied both sides by } -1 \\ \text{and } \textbf{reversed the sense of} \\ \textbf{the inequality symbol}\end{array}$$

**problem
set 105**

1. At noon the armadillo left the wild kingdom and headed north at 3 kilometers per hour. Two hours later the raccoon left the same kingdom and headed south at 5 kilometers per hour. At what time will the animals be 38 kilometers apart?

2. Stephanie threw her pennies into the sandbox and then Shannon added her nickels. If they threw in 10 more pennies than nickels and threw in $29.50 in all, how many nickels were in the box?

3. Twelve percent of the natives had white hair. If 4224 natives had colored hair, how many natives were there in all?

4. Find four consecutive integers such that 5 times the opposite of the first is 5 greater than the product of -3 and the sum of the third and fourth.

5. When the debris was cleared away, there were 52 bricks left. Some were red and the rest were white. The red bricks numbered 16 more than twice the number of white bricks. How many bricks of each color were there?

6. Use the square root table to approximate $\sqrt{8{,}372{,}150 \times 10^{-15}}$.

7. Find p.

8. Find the distance between $(5, -3)$ and $(7, -2)$.

9. Simplify: $\dfrac{x^3 + x^2 - 12x}{x^2 + 4x} \div \dfrac{x^2 - 11x + 24}{x^2 + 2x - 80}$

10. Solve: $(-x) - 3^0(x - 7) = -(-x - 5)$

11. Solve by graphing: $\begin{cases} x = 2 \\ y = -\dfrac{1}{2}x + 6 \end{cases}$

12. Simplify: $\dfrac{(.0004 \times 10^{15})(.06 \times 10^{41})}{(30{,}000{,}000)(400 \times 10^{-21})}$

13. Find the equations of lines (a) and (b).

14. Simplify: $3\sqrt{6,000,000} - 5\sqrt{60,000} + 2\sqrt{3}(3\sqrt{2} - 5\sqrt{3})$

15. Solve by factoring: $-80 = x^2 + 18x$

Graph:

*16. $-|x| - 2 > -5, D = \{\text{Integers}\}$ ***17.** $|x| > -4, D = \{\text{Reals}\}$

18. $-|x| + 3 > 0, D = \{\text{Integers}\}$

Add:

19. $\dfrac{4}{x - 4} + \dfrac{5}{x^2 - 16}$ **20.** $\dfrac{4}{x^2 + 2x + 1} + \dfrac{3x}{x + 1}$

Solve:

21. $\dfrac{5x}{2} - \dfrac{x - 3}{5} = 7$ **22.** $4\dfrac{1}{5}k + 2\dfrac{1}{4} = 7\dfrac{1}{8}$

23. Evaluate $-x^0 - x^2 - x(-x + y)$ if $x = -4$ and $y = -3$

Simplify:

24. $\dfrac{-3x - 9x^2}{-3x}$ **25.** $\dfrac{-3^0 - 4^0}{-2^{-2}}$ **26.** $\dfrac{4 + \dfrac{y}{x^2}}{y^2 + \dfrac{y}{x^2}}$

27. $.0013 \in \{\text{What sets of numbers}\}$?

28. Multiply: $\dfrac{3x^2 y^{-2}}{a^{-2}}\left(\dfrac{x^2 y}{a} - \dfrac{4x^{-5} y^{-3}}{a^5}\right)$

29. Simplify: $-3^0[(-2 - 2^0) - (-5 + 6^0)]$ ER 105-39

LESSON *106*   *Rational equations*

106.A
rational equations

In Lesson 79 we discussed the solution of rational equations in which all denominators are integers. In these problems we found that the recommended first step is to multiply the numerator of every term by the least common multiple of the denominators. Since every denominator is a factor of the least common multiple, this procedure permits us to cancel every denominator as we see in the following example.

example 106.1   Solve $\dfrac{y}{2} + \dfrac{1}{4} = \dfrac{y}{6}$.

solution   As the first step we will multiply every numerator by 12, the least common multiple of the denominators.

$$12\left(\frac{y}{2}\right) + 12\left(\frac{1}{4}\right) = 12\left(\frac{y}{6}\right) \quad \longrightarrow \quad 6y + 3 = 2y \quad \longrightarrow \quad 4y = -3 \quad \longrightarrow \quad y = -\frac{3}{4}$$

The denominators of the equation above were all real numbers. In this lesson we will discuss equations whose denominators contain variables. If there are variables in the denominator, the replacement values of the variable are restricted. For instance, if the given equation is

$$\frac{t-2}{t} = \frac{14}{3t} - \frac{1}{3}$$

The number zero would not be a permissible value for t, for if we substitute the number zero for t, we find

$$\frac{0-2}{0} = \frac{14}{3(0)} - \frac{1}{3} \quad \longrightarrow \quad \frac{-2}{0} = \frac{14}{0} - \frac{1}{3} \qquad \text{incorrect}$$

which is meaningless for division by zero is not defined.
     If our equation is

$$\frac{n}{n+2} = \frac{3}{5}$$

we cannot accept -2 as a value for n for if we try to substitute -2 for n, we obtain

$$\frac{(-2)}{(-2)+2} = \frac{3}{5} \quad \longrightarrow \quad \frac{-2}{0} = \frac{3}{5} \qquad \text{incorrect}$$

an expression in which zero is the denominator of a fraction, and division of a nonzero real number by zero is not defined.
     Thus as our first step in the solution of rational equations whose terms have variables in one or more denominators, we will list the unacceptable values of the variable which, of course, are those values of the variable that would cause any denominator to equal zero.

example 106.2   Solve $\dfrac{t-2}{t} = \dfrac{14}{3t} - \dfrac{1}{3}$.

solution   ($t \neq 0$). As the next step we multiply every numerator by $3t$, the least common multiple of the denominators. This will allow us to cancel the denominators, and then we will solve the resulting equation.

$$3t\left(\frac{t-2}{t}\right) = 3t\left(\frac{14}{3t}\right) - 3t\left(\frac{1}{3}\right) \quad \longrightarrow \quad 3t - 6 = 14 - t \quad \longrightarrow \quad 4t = 20 \quad \longrightarrow \quad t = 5$$

example 106.3   Solve $\dfrac{n}{n+2} = \dfrac{3}{5}$.

solution   ($n \neq -2$). Now we multiply every term by $(5)(n+2)$, the least common multiple of the denominators, cancel the denominators, and solve.

$$(5)(n+2)\frac{(n)}{(n+2)} = \frac{3}{5}(5)(n+2) \quad \longrightarrow \quad 5n = 3(n+2)$$

$$\longrightarrow \quad 5n = 3n + 6 \quad \longrightarrow \quad 2n = 6 \quad \longrightarrow \quad n = 3$$

example 106.4   Solve $\dfrac{2}{3n} = \dfrac{2}{n+4}$.

solution $(n \neq 0, -4)$. Now we multiply every numerator by $(3n)(n+4)$, cancel the denominators, and solve.

$$(3n)(n+4)\frac{(2)}{(3n)} = \frac{(2)}{(n+4)}(3n)(n+4) \longrightarrow (n+4)2 = 6n \longrightarrow 2n+8 = 6n$$

$$\longrightarrow 8 = 4n \longrightarrow \boldsymbol{n = 2}$$

example 106.5   Solve $\dfrac{4}{x} = \dfrac{7}{x-2}$.

solution $(x \neq 0, 2)$. Now multiply each term by $x(x-2)$, cancel the denominators, and solve.

$$x(x-2)\frac{4}{x} = \frac{x(x-2)}{1}\frac{7}{x-2} \longrightarrow 4x-8 = 7x$$

$$\longrightarrow -8 = 3x \longrightarrow \boldsymbol{x = -\frac{8}{3}}$$

example 106.6   Solve $\dfrac{4}{p} - \dfrac{3}{p-4} = 0$.

solution $(p \neq 0, 4)$. We multiply each term by $p(p-4)$, which is the least common multiple of the denominators, cancel the denominators, and solve.

$$p(p-4)\frac{4}{p} - p(p-4)\frac{3}{p-4} = 0 \longrightarrow 4p-16-3p = 0$$

$$\longrightarrow p-16 = 0 \longrightarrow \boldsymbol{p = 16}$$

problem set 106

1. When the battle began, there were 30 percent more brigantines than men-of-war. If there were 260 brigantines, how many total ships took part in the battle?

2. Frederick headed for Lutzen at 3 kilometers per hour. Later he increased his speed to 4 kilometers per hour. If it was 52 kilometers to Lutzen and the total time of travel was 15 hours, for how long did he walk at 3 kilometers per hour?

3. Josey walked to Brundig and then trotted back home. He walked at 2 miles per hour and trotted at 4 miles per hour. How far was it to Brundig if his walking time was 2 hours longer than his trotting time?

4. If the sum of a number and 10 is multiplied by 5, the result is 2 greater than 7 times the opposite of the number, what is the number?

5. Cookies sold for 10 cents and doughnuts for 20 cents. Pericles bought 25 items for $3.50. How many doughnuts did he buy and how many cookies?

Solve:

*6. $\dfrac{t-2}{t} = \dfrac{14}{3t} - \dfrac{1}{3}$

*7. $\dfrac{2}{3n} = \dfrac{2}{n+4}$

*8. $\dfrac{4}{p} - \dfrac{3}{p-4} = 0$

*9. $\dfrac{4}{x} = \dfrac{7}{x-2}$

Graph:

10. $-|x| + 4 > -2, D = \{\text{Integers}\}$ **11.** $-|x| - 4 > -2, D = \{\text{Reals}\}$

12. Find p.

13. Simplify: $\dfrac{x^2 + 10x + 25}{x^2 + 5x} \div \dfrac{x^2 + 8x + 15}{x^3 + x^2 - 6x}$

14. Solve by graphing: $\begin{cases} y = x - 2 \\ y = -\dfrac{1}{2}x + 1 \end{cases}$

15. Find the equations of lines (a) and (b).

16. Solve: $4\dfrac{1}{2}x + \dfrac{1}{2} = \dfrac{3}{4}$

Simplify:

17. (a) $\dfrac{3x - 3x^2}{3x}$ (b) $\dfrac{-4^{-2}}{-(-2)^{-2}}$

18. $\dfrac{a^2 + \dfrac{1}{a}}{ax + \dfrac{b}{a}}$ **19.** $(-2)[(-2 + 5) + (-2 - 3^0)]$

Add:

20. $\dfrac{4}{a - 2} + \dfrac{6a}{a^2 - 4}$ **21.** $\dfrac{5}{x + 4} - \dfrac{3}{x^2 + 2x - 8}$

Use square root tables to approximate:

22. $\sqrt{417{,}530 \times 10^{20}}$ **23.** $\sqrt{417{,}530 \times 10^{-60}}$

24. Simplify: $3\sqrt{2} \cdot 4\sqrt{3} - 4\sqrt{60{,}000} + 2\sqrt{3}(3\sqrt{2} - \sqrt{3})$

25. Divide $x^3 - 4$ by $x + 7$ **26.** Solve by factoring: $4x^2 - 81 = 0$

27. What fraction of $2\dfrac{1}{4}$ is $\dfrac{7}{8}$?   **28.** $\dfrac{4\sqrt{2}}{5} \in \{\text{What sets of numbers}\}$?

29. Evaluate $-x - x^2 + (-x)^3(x - y)$ if $x = -3$ and $y = -5$

30. Simplify by adding like terms: $\dfrac{x}{y} + \dfrac{3x^2y}{y^2x} - \dfrac{4x^0xx^2y^{-2}}{(x)^2yy^{-2}}$ **ER 106-40**

LESSON 107   *Abstract rational equations*

107.A
abstract rational equations

In the last lesson we discussed the fact that when we have equations with variables in the denominator

$$\frac{4}{x} + \frac{3}{x-2} = \frac{7}{2x}$$

the values we may use to replace x are restricted because a denominator can never equal zero. Thus, in the equation above, we cannot use 0 or $+2$ for x, for either of these will cause at least one denominator to equal zero. We often note the impermissible values of the variable for a problem by listing them using a notation such as $(x \neq 0)$, $(x \neq 2)$, $(x, m \neq 0)$, etc., as we do in the following examples. We will omit these notations in the problem sets.

example 107.1 $\dfrac{1}{x} + \dfrac{b}{m} = c$; find m $(x, m \neq 0)$.

solution xm is the least common multiple of the denominators. Thus we begin by multiplying† every numerator by xm, the least common multiple of the denominators, and canceling the denominators,

$$(\cancel{xm})\frac{1}{\cancel{x}} + (\cancel{xm})\frac{b}{\cancel{m}} = cxm$$

which leaves us with the following equation.

$$m + xb = cxm$$

Now we use the additive property of equality to place all terms with m's on one side and all other terms on the other side.

$$\begin{array}{r} m + xb = cxm \\ -m \qquad\qquad\quad -m \\ \hline xb = cxm - m \end{array}$$

Then we factor out the m on the right side and finish by dividing both sides by $(cx - 1)$.

$$xb = m(cx - 1) \quad \longrightarrow \quad \frac{xb}{cx - 1} = \frac{m(\cancel{cx - 1})}{(\cancel{cx - 1})} \quad \longrightarrow \quad \frac{xb}{cx - 1} = m \qquad (cx - 1 \neq 0)$$

example 107.2 $\dfrac{a}{b} + \dfrac{c}{d} = x$; find b $(b, d \neq 0)$.

solution   As the first step we multiply each numerator by bd, the least common multiple of the denominators,

$$(bd)\frac{a}{b} + (bd)\frac{c}{d} = bdx$$

and cancel the denominators

$$da + bc = bdx$$

Now we use the additive property of equality as necessary to position all terms that contain b on one side of the equation (either side) and all terms that do not contain b on the other side. We decide to position all terms that contain b on the right side of the equation by adding $-bc$ to both sides.

---

† Permitted by the multiplicative property of equality.

$$da + bc = bdx$$
$$\underline{\ -bc \qquad\qquad\qquad -bc\ }$$
$$da \qquad = bdx - bc$$

Now factor out the b on the right side

$$da = b(dx - c)$$

and as a last step divide both sides of the equation by $dx - c$.

$$\frac{da}{(dx - c)} = b\frac{(dx - c)}{(dx - c)} \longrightarrow \frac{da}{dx - c} = b \qquad (dx - c \neq 0)$$

example 107.3　　$\dfrac{a}{b} - c = \dfrac{d}{x}$, find x　　　$(b, x \neq 0)$.

solution　　First we eliminate the denominators

$$(bx)\frac{a}{b} - (bx)c = (bx)\frac{d}{x} \longrightarrow xa - bxc = bd$$

Now since all x terms are already on one side and all other terms are on the other side, we factor out the x and divide both sides by the coefficient of x.

$$x(a - bc) = bd \longrightarrow \frac{x(a - bc)}{(a - bc)} = \frac{bd}{(a - bc)} \longrightarrow x = \frac{bd}{a - bc} \qquad (a - bc \neq 0)$$

example 107.4　　$\dfrac{a}{x} - y + \dfrac{m}{n} = k$; find x　　　$(x, n \neq 0)$.

solution　　First we eliminate the denominators.

$$(xn)\frac{a}{x} - xny + (xn)\frac{m}{n} = xnk \longrightarrow na - xny + xm = xnk$$

Now we move all terms that contain an x to the right side, factor out the x and divide by the coefficient of x.

$$na - xny + xm = xnk$$
$$\underline{\ + xny - xm \qquad\qquad + xny - xm\ }$$
$$na \qquad\qquad = xnk + xny - xm \qquad \longrightarrow \quad na = x(nk + ny - m)$$

$$\longrightarrow \frac{na}{(nk + ny - m)} = \frac{x(nk + ny - m)}{(nk + ny - m)}$$

$$\longrightarrow \frac{na}{nk + ny - m} = x \qquad (nk + ny - m \neq 0)$$

problem set 107

1. Gold was worth 422 marks a gram and copper was worth 4 marks a gram. If Brother Gregory had an 8-gram mixture of gold and copper that was worth 2122 marks, how many grams of gold did he have?

2. Find three consecutive odd integers such that 4 times the first is 14 less than twice the sum of 2 and the third.

3. Thirty percent of the people did not like the king. If 81,150 people did not like the king, how many people lived in the kingdom?

4. Milton ran north at 8 kilometers per hour. Four hours later Henry set out in pursuit at 16 kilometers per hour. How long did it take Henry to catch Milton?

5. The train made the trip in 4 hours. The boys walked it in 48 hours. How far was it if the speed of the train was 55 miles per hour faster than the boys walked?

Solve:

6. $\dfrac{1 + m}{m} - \dfrac{3}{m} = 0$

7. $\dfrac{3}{4x} = \dfrac{2}{x + 5}$

8. $\dfrac{x}{5} - \dfrac{3 + x}{7} = 0$

9. $\dfrac{2}{x} - \dfrac{3}{x - 1} = 0$

*10. $\dfrac{1}{x} + \dfrac{b}{m} = c$; find m

*11. $\dfrac{a}{b} + \dfrac{c}{d} = x$; find b

*12. $\dfrac{a}{b} - c = \dfrac{d}{x}$; find x

*13. $\dfrac{a}{x} - y + \dfrac{m}{n} = k$; find x

Add:

14. $\dfrac{4}{x^2 - 4} + \dfrac{3x}{x - 2}$

15. $\dfrac{7}{x + 5} - \dfrac{2x}{x^2 - 25}$

16. Use the square root tables to approximate: $\sqrt{714{,}200 \times 10^{-15}}$

17. Find the distance between $(-4, 2)$ and $(-10, 6)$.

18. Solve: $2x^0(x - 2) - 3x - 4 - [-(-2)] - 7^0 = -2x - 4$

19. Simplify: $\dfrac{(21{,}000 \times 10^{50})(.0006 \times 10^{15})}{(.007 \times 10^{20})(9000 \times 10^{-40})}$

20. Solve by factoring : $63 = -x^2 - 16x$

21. Simplify: $4\sqrt{20{,}000} - 15\sqrt{8} + 3\sqrt{2}(4\sqrt{2} - 5)$

Add:

22. $\dfrac{3}{x - 5} + \dfrac{2}{x} + \dfrac{7}{x^2 - 25}$

23. $\dfrac{-x}{x + 5} - \dfrac{3x}{x^2 + 3x - 10}$

24. Evaluate $-x^2 - x(xy - xy^2)$ if $x = -2$ and $y = -3$

25. Simplify: $\dfrac{(x^2)^{-2}yyx^{-2}}{(x^2y^{-2})^{-3}}$

Graph:

26. $-|x| - 2 < -4, D = \{\text{Reals}\}$

27. $-x + 2 \le 7, D = \{\text{Integers}\}$

28. Simplify: (a) $\dfrac{4x^2ay - 4xay}{4xay}$ (b) $\dfrac{-2^{-2}}{-(-2^0)^{-3}}$

29. Multiply: $\dfrac{4x^{-2}}{a^2}\left(\dfrac{x^2}{4a^{-2}} - \dfrac{2x^{-2}}{a^4}\right)$ ER 107-41

LESSON 108 *Equation of a line through two points*

108.A
linear
equations

The graph of a first-degree equation in two unknowns is a straight line. This is the reason we call these equations linear equations. We also use the name *linear equation* to describe first-degree equations in more than two unknowns. The graph of a first-degree equation in three unknowns such as

$$4x + 2y - z = 5$$

is a plane, and an equation in more than three unknowns cannot be graphed because graphs are restricted to the three dimensions available for graphing.

 The standard form of the equation of a straight line is

$$ax + by + c = 0$$

where a, b, and c are constants. The following are equations of straight lines in standard form:

$$4x + y + 1 = 0 \qquad -2x - y - 11 = 0$$

 If the equation of a line is written so that y is expressed as a function of x such as

$$y = mx + b$$

we say that we have written the equation in **slope-intercept form.** In this equation m represents the slope of the line and b represents the y intercept of the line, which is the y coordinate of the point where the line in question crosses the y axis.

 Thus far, we have learned how to draw the graph of a given linear equation and have learned how to find a good approximation of the equation of a given line. Both of these exercises have helped us to understand the relationship between the equation of a line and the graph of a line.

 Algebra books usually contain three other types of straight line problems that are helpful in exploring this relationship. In the first type we are given the coordinates of two points and asked to find the equation of the line that passes through the two points.

108.B
equation of a line through two points

example 108.1   Find the equation of the line that goes through the points $(4, 2)$ and $(-5, -3)$.

solution   The slope-intercept form of the desired equation is $y = mx + b$, and **we need to find the values of m and b.** First we graph the two points and draw the line in the figure on the left. If we draw the slope triangle so that the sides of the triangle terminate on these points as we have done in the right figure, we can determine the **exact value of the slope**

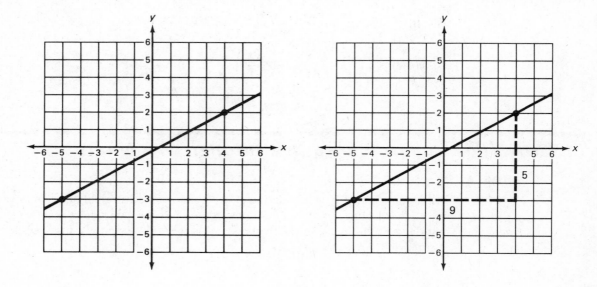

to be $+\frac{5}{9}$. We say the exact value because we were given the exact coordinates of two points on the line. We did not estimate their value by looking at a graph.

We can see from the graph that the value of the intercept is approximately $-.3$. **This estimated value of the intercept is not acceptable when the exact coordinates of two points on the graph are known, for we can find the exact value for the intercept.** We know the exact value of the slope so we can write the desired equation as

$$y = \frac{5}{9}x + b$$

We know the exact values of the coordinates of two points that lie on the line. We can use the coordinates of either of these points for x and y in the equation above and find the exact value of b algebraically.

USING $(4, 2)$

$$y = \frac{5}{9}x + b$$

$$(2) = \frac{5}{9}(4) + b$$

$$\frac{18}{9} = \frac{20}{9} + b$$

$$-\frac{2}{9} = b$$

USING $(-5, -3)$

$$y = \frac{5}{9}x + b$$

$$(-3) = \frac{5}{9}(-5) + b$$

$$-\frac{27}{9} = -\frac{25}{9} + b$$

$$-\frac{2}{9} = b$$

Now that we have the exact values of m and b, we can write the exact equation of the line that goes through the two points as

$$y = \frac{5}{9}x - \frac{2}{9}$$

example 108.2 Find the equation of the line that goes through the points $(4, -2)$, $(-3, 4)$.

solution The general form of the desired equation is $y = mx + b$. We need to find the values of m and b. We graph the points, draw the line, and draw the triangle to find that the slope m is exactly $-\frac{6}{7}$.

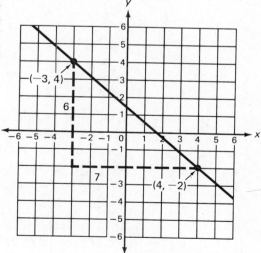

Now we have

$$y = -\frac{6}{7}x + b$$

and we can find the exact value of b by substituting either $(4, -2)$ or $(-3, 4)$ for x and y and solving algebraically for b. We will use the point $(4, -2)$.

$$-2 = -\frac{6}{7}(4) + b$$

$$-\frac{14}{7} = -\frac{24}{7} + b$$

$$\frac{10}{7} = b$$

So the desired equation is

$$y = -\frac{6}{7}x + \frac{10}{7}$$

and the values that we have found for the slope and the intercept are exact.

We see from these two examples that when we are given the coordinates of two points that lie on the line, the exact equation of the line can be determined. Estimated values of the slope and intercept are not acceptable for this type of problem.

example 108.3   Find the equation of the line through the points $(4, 3)$, $(4, -3)$.

solution   When we graph the points and draw the line, we find that the line is a vertical line. Vertical and horizontal lines can be thought of as special cases. By inspection the equation of this line is $x = 4$.

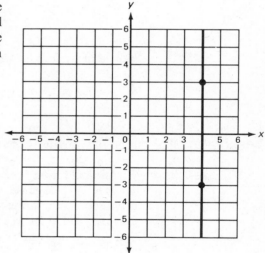

problem set 108

1.   The passenger train headed north at 70 miles per hour at 6 a.m. At 8 a.m. the freight train headed south from the same station at 30 miles per hour. At what time will the trains be 440 miles apart?

2.   The sciolist headed for town at 30 miles per hour. Two hours later, the charlatan began his pursuit at 50 miles per hour. How long did it take the charlatan to catch the sciolist?

3.   There were quarters and dimes in profusion. Their value was $9.55, and there were 64 coins in all. How many were quarters and how many were dimes?

4.   Find three consecutive even integers such that if the sum of 5 and the second is multiplied by -7, the result is 11 greater than 5 times the opposite of the third.

5.   Seventeen percent of the mob had a propensity for jogging and the rest just wanted to walk. If 3825 wanted to jog, how many were in the mob?

Find the equations of the lines that go through the following pairs of points:

*6. $(4, 2)$ and $(-5, -3)$ ***7.** $(4, -2)$ and $(-3, 4)$ ***8.** $(4, 3)$ and $(4, -3)$

Solve:

9. $\dfrac{2 + x}{4} + \dfrac{x}{2} = 5$ **10.** $\dfrac{9}{4x} = \dfrac{5}{x + 11}$ **11.** $\dfrac{12}{x} + \dfrac{1}{4x} = 7$

12. $\dfrac{x}{y} + \dfrac{1}{m} = p$; find y **13.** $\dfrac{k}{m} + \dfrac{1}{c} = x$; find c **14.** $\dfrac{1}{b} + \dfrac{k}{x} = y$; find b

15. $\dfrac{1}{m} + \dfrac{b}{c} = \dfrac{x}{y}$; find m

Add:

16. $\dfrac{4}{x^2 - 25} - \dfrac{x}{x - 5}$ **17.** $\dfrac{3x}{x^2 - x - 6} - \dfrac{3}{x - 3}$

18. Use the square root tables to approximate: $\sqrt{.000325 \times 10^{-41}}$

19. Find the distance between $(-5, 3)$ and $(4, -2)$.

20. Solve: $-3^0(x - 4) - 2x - (-2x) - [-(-3)] + 5^0 = 2(-x + 2)$

Simplify:

21. $\dfrac{(35{,}000 \times 10^{-40})(300 \times 10^{15})}{(.007 \times 10^{15})(15{,}000{,}000)}$ **22.** $\dfrac{(y^{-2})^0 y^0 y^2 yyx(xy)^2}{y^2 y^{-2}(y^2)^{-2}axy}$

23. $3\sqrt{30{,}000} - 5\sqrt{27} + 5\sqrt{3}(2\sqrt{3} - 2)$ **24.** (a) $\dfrac{6x + 6}{6}$ (b) $\dfrac{-3^{-2}}{(-2)^2}$

25. Solve by factoring: $-56 = 15x + x^2$

26. Evaluate $-x^2 - x(xy - y)$ if $x = -3$ and $y = 4$

27. Graph on a number line: $-|x| - 4 > 2$; $D = \{\text{Reals}\}$

28. Simplify by adding like terms: $\dfrac{x^2 y^3 a^2}{a^{-2}} + \dfrac{a^5 a y^4}{a^2 yx^{-2}} - \dfrac{4a^2 y^5}{y^{-2} x^{-2}}$

29. Multiply: $\dfrac{4x^2 y^{-2}}{a^2}\left(\dfrac{x^{-2} y^{-2}}{a^{-2}} + \dfrac{3xy^{-2}}{a^2}\right)$

30. Round off $45{,}732.\overline{654}$ to the nearest ten thousand.   **ER 108-42**

LESSON 109   *Functions*

109.A
dependent and independent variables (again)

If we consider the equation

$$y = x + 6$$

we see that this equation matches a value of y with any value of x. We will demonstrate this by replacing x with 4 and then with 23.

If $x = 4$ If $x = 23$

$y = (4) + 6 \longrightarrow y = 10$ $y = (23) + 6 \longrightarrow y = 29$

On the left we let $x = 4$ and find that $y = 10$, and on the right we let $x = 23$ and find that $y = 29$. **We see that the value of y depends on the value that we assign to x, so we say that y is the dependent variable and that x is the independent variable. Of course, we could assign a value to y and use the equation to find the value of x, and this would reverse the names. It is customary, however, to use the letter x to designate the independent variable and to use the letter y to designate the dependent variable. We follow this convention in this book.**

109.B
functions

We use the word **function** to describe a relationship between two sets in which

1. The first set is the domain and the domain is defined.
2. For each member of the domain there is exactly one answer in the second set.

If we consider the equation

$$y = 4x + 6 \qquad D = \{\text{Reals}\}$$

we see that this equation will give us **one answer** for y for any real number value of x. For example,

If we let $x = 2$	If we let $x = 10$
$y = 4(2) + 6$	$y = 4(10) + 6$
$\mathbf{y = 14}$	$\mathbf{y = 46}$

Thus the equation and the domain define a function because

1. **The domain is specified.**
2. **For any value of the domain the equation will give us one and only one value (answer) for y.**

Now let us consider the ordered pairs

$$(4, 7), \qquad (5, -2), \qquad (7, -4), \qquad (8, 3)$$

and ask if these ordered pairs designate a function. The answer is yes because

1. The domain is specified. The only x values we can use are the x values in the ordered pairs, which are the numbers $\{4, 5, 7, 8\}$.
2. For each value of the domain there is exactly one answer for y. If we let x be 4, then y is 7; if x is 5, then y is -2; if x is 7, then y is -4; if x is 8, then y is 3.

From this example we see that an equation is not necessary to designate[†] a function. We have fulfilled both of the requirements of a function by using the ordered pairs.

If we use proper nomenclature to describe a function, we must use the words **domain, image,** and **range.** The domain, as we know, is the set of all permissible replacement values of x. We will use the word **image** instead of using the word **answer** to describe the value of y that is matched with each value of x by the function. We will use the word **range** to describe the set of all images.

[†] The equation itself cannot be called the function because the function is the mapping or pairing between the members of the domain and the answers for each member.

DOMAIN. The set
of all permissible
values of *x*.

RANGE. The set
of all images (values
of *y* or "answers")

DEFINITION:

FUNCTION, DOMAIN, IMAGE, RANGE

A **function** is a correspondence or mapping between
two sets that associates with each element of the first
set a unique element of the second set. The first set is
called the **domain** of the function. For each element *x*
of the domain, the corresponding element *y* of the
second set is called the **image** of *x* under the function.
The set of all images of the elements of the domain is
called the **range** of the function.

**There are two accepted definitions of a function. The definition above says that a function
is the correspondence or mapping, while the other definition says that a function is the
set of ordered pairs.**

A function is a set of ordered pairs such that no two
ordered pairs have the same first element and different
second elements.

Perhaps it will help if we show a few functions and a few nonfunctions. Remember, to be a function,

1. **The domain must be specified.**
2. **A way must be provided to find the image for each member of the domain, and there
must be exactly one image for each member of the domain.**

example 109.1   Does the diagram designate a function?

solution   Yes. The elements of the domain are
specified. There is one image and only
one image for each member of the domain. We do not have an equation, but
we can find the images by looking at
the diagram.

example 109.2   Does the diagram designate a function?

solution   No. The members of the domain are
specified, but *m* has two images and it is
only allowed one.

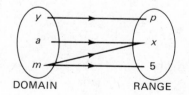

example 109.3   Does the diagram designate a function?

solution   Yes. The domain is specified and each member of the domain has exactly one image. True, 5 is the image of both *a* and 4, but this is permissible.

example 109.4   Does the diagram designate a function?

solution   Yes. The diagram says that we can use any real number for *x* so the domain is specified. The members of the range are not specified, but the equation will allow us to find the value of *y* that is paired with any real number we use for *x*.

example 109.5   Does either of the sets of ordered pairs shown designate a function?

(a) $\{(4, 6), (7, 2), (4, 5)\}$ (b) $\{(4, 8), (15, 6), (11, 7)\}$

solution   The set (a) does not designate a function, for 4 has two images. The set (b) does designate a function. The domain is specified and there is exactly one image for every member of the domain.

example 109.6   Does the diagram designate a function?

solution   Yes, it certainly does designate a function. For every value of the *x* coordinate on the line there is a corresponding *y* coordinate. For instance, when *x* equals 2, it appears that the *y* coordinate of the ordered pair that lies on the line is about 3.0. We can't tell from the graph the exact value of the *y* coordinate, but we do know that an exact value does exist.

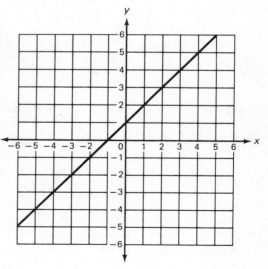

example 109.7   Does the diagram designate a function? (The domain is the set of real numbers such that $x \geq 1$.)

solution No. For every member of the domain that is greater than one, there are two images. For instance, when $x = 5$, the graph shows that y can be either $+2$ or -2.

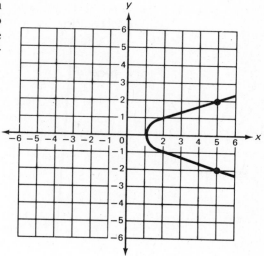

109.C

relations All functions can also be called relations because a relation is a pairing that matches each element of the domain with one or more images in the range.

example 109.8 Does the diagram designate a function?

solution No. The number 6 has two images. The diagram designates a relation.

example 109.9 Does the diagram designate a function or a relation?

solution Neither. For a correspondence to be called a relation, there must be one or more images for each element of the domain. To be called a function there must be exactly one image for each element of the domain. In the diagram the number 5 does not have even one image, and thus the diagram does not designate either a function or a relation.

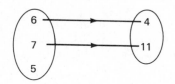

problem set 109

1. Miltiades and his army marched to Marathon, the site of the battle, at 2 miles per hour. After the battle Pheidippides ran back to Athens with the news at 13 miles per hour. If the total traveling time was 15 hours, how far was it from Marathon to Athens?

2. Leonidas walked to Thermopylae at 4 miles per hour and was carried back to Sparta at 2 miles per hour. If the total travel time was 120 hours, what was the distance from Sparta to Thermopylae?

3. The bag broke open and a veritable fortune cascaded forth. There were 5100 gold coins worth $260,000. If some were $50 gold coins and the others were $100 gold coins, how many of each kind were there?

4. The farrago contained 4000 red jelly beans. If there were 20,000 pieces of candy in the farrago, what percent were red jelly beans?

5. Once there were four consecutive integers. The product of the sum of the first and the third and -7 was 4 greater than 12 times the opposite of the fourth. What were the integers?

Find the equations of the lines through the following pairs of points:

6. $(-3, -1)$ and $(4, -6)$

7. $(2, -5)$ and $(3, -7)$

8. $(5, -7)$ and $(7, -3)$

9. Which of the following diagrams depict functions?

*(a)   (b)   (c)

(d)

10. What is the domain of 9(b)?

11. Which of these sets of ordered pairs are functions?

(a) $(4, 6), (5, 7), (3, -2), (6, -3), (2, 7)$
(b) $(4, 6), (5, 7), (3, -2), (6, -3), (5, 3)$
(c) $(-2, 6), (5, 6), (-3, -2), (-8, -6), (4, -6)$

Solve:

12. $\dfrac{4}{x} - \dfrac{2}{x-4} = 0$

13. $\dfrac{x}{4} - \dfrac{x+6}{5} = 1$

14. $\dfrac{a}{b} + \dfrac{1}{c} = d$; find b

15. $\dfrac{a}{x} - \dfrac{1}{c} = \dfrac{b}{d}$; find c

16. $\dfrac{p}{x} + \dfrac{1}{c} = k$; find c

Add:

17. $\dfrac{4}{x^2 - 9} - \dfrac{3}{x + 3}$

18. $\dfrac{5}{x + 2} - \dfrac{3x}{x^2 + 5x + 6}$

19. Use the square root tables to approximate: $\sqrt{.0052 \times 10^{-7}}$

20. Graph: $|x| - 3 > 4$; $D = \{$Negative reals$\}$

21. Solve: $-(-2x + 4) - 3^0(3 - 3x) - (-2) = 4(3 - x^0)$

Simplify:

22. $\dfrac{(42,000 \times 10^{46})(5000 \times 10^{-20})}{(.00007 \times 10^{21})(.0006 \times 10^{-14})}$

23. $\dfrac{x^2 + 6x + 9}{x^2 + 3x} \div \dfrac{x^3 + 5x^2 + 6x}{x^2 + 2x}$

24. Find k.

Simplify:

25. $\sqrt{50,000} - 25\sqrt{125} + 5\sqrt{5}(\sqrt{5} - 5)$

26. $\dfrac{xa + \dfrac{1}{a}}{\dfrac{x}{a} + a}$

27. $-2^0[(-2 + 3)(-2 - 4) - (-2 - 5)]$

28. $4\sqrt{3} \in \{$What sets of numbers$\}$?

29. Solve: $4\dfrac{1}{5}m + \dfrac{3}{4} = \dfrac{7}{8}$

30. Solve by graphing: $\begin{cases} y = x - 2 \\ y = -x + 2 \end{cases}$

31. Find the equations of lines (a) and (b).

ER 109-43

LESSON 110 *Functional notation*

110.A

functions In the last lesson we discussed functions at some length but did not discuss why functions are important.

Algebra books published in the late nineteenth century and early twentieth century did not contain the word *function*. These books spoke of equations: linear equations, quadratic equations, and other types of equations. The calculus books of the 1930s and 1940s used the word function to describe some equations. The definition used, however, permitted more than one image for each member of the domain. If two images existed for some members of the domain, the function was called a **double-valued function.** If more than two images existed for some members of the domain, the function was called a **multivalued function.** The restriction on a function that it be single-valued is a rather recent development—say, after 1950.

So we see that **a function is the word we use now to describe what used to be called a single-valued equation.** But because of the way the word is now defined, it covers any relationship that specifies for every value of the domain one and only one image in the range. For instance, a set of ordered pairs such as $(5, 4)$, $(6, 2)$, and $(3, -7)$ can be called a function even though there is no equation. The members of the domain are specified and each member of the domain is paired with exactly one image in the range. Many equations and other mathematical relationships are single-valued relationships, and it is convenient to have a word to describe them collectively.

example 110.1   Does the diagram designate a function?

solution   Yes—we assume that the figure is a portion of the entire graph and that the domain is all real numbers. If this is true, we see that the graph will match exactly one real value of y for each value of x.

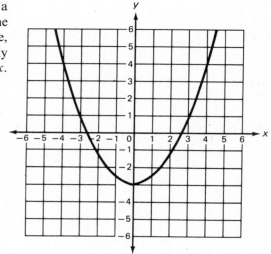

example 110.2   Does the equation $x = y^2$ describe a function?

solution   No. For example, if x is 4, then y could be either positive 2 or negative 2 because both

$$(2)^2 = 4 \qquad \text{and} \qquad (-2)^2 = 4$$

110.B

vertical line test   As we noted in the last lesson, it is customary to use the letter x to designate the independent variable in a two-variable equation, and it is customary to use the letter y to designate the dependent variable. There is no requirement that this convention be followed, but everyone does. It is also customary to graph the independent variable on the horizontal axis and the dependent variable on the vertical axis. If this convention is followed, the so-called **vertical line test** can be used on the graph of a relation to see if the relation is also a function. Remember that a relation has one or more images for each element of the domain and a function is a relation that has exactly one image for every member of the domain.

Below we see graphs of three different relations. For each value of x, the graph designates at least one value of y. **If every possible vertical line that we can draw touches the graph of the relation at only one point, then the graph is the graph of a function.**

Look at Figure (a). We have drawn several vertical lines and each of these touches the graph at only one point. If we look carefully at the figure, we can tell that no matter where we draw the vertical line, it will touch the graph of the relation at only one point. Thus by using the vertical line test, we have shown that graph (a) is the graph of a function.

In Figure (b) we drew two vertical lines at values of x arbitrarily called A and B. We see that each of these lines or any vertical line drawn anywhere between them will touch the given graph in at least two points so the graph is not the graph of a function.

In Figure (c) the graph is vertical at one value of x that we have designated by the letter A. The vertical line drawn at this value of x touches the graph for its entire vertical length. Thus this graph is not the graph of a function.

The vertical line test can be used only if the convention of graphing the independent variable on the horizontal axis is followed. If the independent variable were graphed on the vertical axis, then we would have to use a horizontal line test. We will always follow the convention and thus will not dwell on other possibilities.

example 110.3   Is the relation graphed in the accompanying figure also a function?

solution   (The domain is assumed to be the values of x greater than -2.) *No.* Any vertical line drawn at a value of x such that $x > -2$ will touch the line in at least two points. This graph is the graph of a relation, not a function.

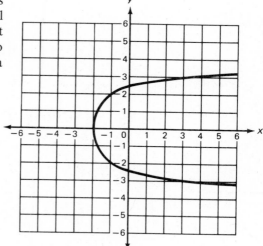

110.C

functional notation   In functional notation we use symbols such as $f(x)$, $g(x)$, $h(x)$ to designate the dependent variable instead of using the letter y. Using functional notation the equation $y = 2x + 4$

could be written as

$$f(x) = 2x + 4 \quad \text{or} \quad g(x) = 2x + 4 \quad \text{or} \quad h(x) = 2x + 4$$

We do not restrict ourselves to the letters f, g, and h in designating functions since any letter or symbol may be used.

One advantage of the functional notation is that the functional notation always indicates the element of the domain that is paired with a given image in the range. To demonstrate this on the left below we write the equation $y = 2x + 4$ and on the right we use functional notation for the same equation and write it as $f(x) = 2x + 4$. Then we replace x in each equation with the number 5 and solve to find the value of the dependent variable.

$$
\begin{array}{ll}
y = 2x + 4 & f(x) = 2x + 4 \\
y = 2(5) + 4 & f(5) = 2(5) + 4 \\
\underline{y = 10 + 4} & \underline{f(5) = 10 + 4} \\
y = 14 & f(5) = 14
\end{array}
$$

Of course, we get the same value, 14, for the dependent variable in both cases. If we cover up everything above the dotted line, however, we see in the last line on the left only the value for y is shown, but in functional notation on the right we see the value of y is shown, and the value of x that is paired with this value of y is also shown.

example 110.4   If $\phi(x) = 3x + 5$, $D = \{\text{Reals}\}$, find $\phi(-2)$.

solution   In nonfunctional notation the problem would have read "given the equation $y = 3x + 5$, find y if x equals -2," and we would proceed as follows to find y.

$$y = 3x + 5 \longrightarrow y = 3(-2) + 5 \longrightarrow y = -1$$

Now we will work the same problem again using functional notation as was used in the statement of the problem.

$$\phi(x) = 3x + 5 \longrightarrow \phi(-2) = 3(-2) + 5 \longrightarrow \phi(-2) = -1$$

example 110.5   If $g(x) = x^2 + 4$, $D = \{\text{Reals}\}$, find $g(4)$.

solution   In nonfunctional notation the problem would have read "given the equation $y = x^2 + 4$, find y if x equals 4."

$$y = x^2 + 4 \longrightarrow y = (4)^2 + 4 \longrightarrow y = 20$$

Now the same problem but this time we will use functional notation,

$$g(x) = x^2 + 4 \longrightarrow g(4) = (4)^2 + 4 \longrightarrow g(4) = 20$$

example 110.6   Given $f(x) = x + 2$, $D = \{\text{Reals}\}$, find $f(\tfrac{1}{2})$.

solution

$$f\left(\frac{1}{2}\right) = \left(\frac{1}{2}\right) + 2 \longrightarrow f\left(\frac{1}{2}\right) = \frac{5}{2}$$

example 110.7   Given $f(x) = x + 2$, $D = \{\text{Integers}\}$, find $f(\tfrac{1}{2})$.

solution   The domain of the function is the set of integers. The problem asks that we find the image of $\tfrac{1}{2}$. One-half does not have an image under this function because the replacement values of x are restricted to the set of integers. Thus we say that there is no real number that satisfies both conditions so the solution is the empty set $\{\ \}$, or the null set $\varnothing$.

problem set 110

1. The brigantine was sailing at full speed and was 30 miles at sea when Lord Nelson began to give chase at twice the speed of the brigantine. If Nelson caught up in 6 hours, how fast was the brigantine traveling?

2. The freight headed north at 9 a.m. at 40 miles per hour. Two hours later the express headed north at 60 miles per hour. What time was it when the express was 20 miles farther from town than the freight?

3. When the announcer asked the question in the shopping mall, 60 percent of the responses were fatuous. If 3000 answers were reasonable, how many answers were fatuous? How many people answered the question?

4. The sable was marked down 23 percent for the sale, yet its sale price was still $15,400. What was the original price of the sable?

5. Find three consecutive integers such that -7 times the sum of the first and the third is 12 greater than the product of 10 and the opposite of the the second.

Given that $\phi(x) = 3x + 5$, $D = \{\text{Reals}\}$; $g(x) = x^2 + 4$, $D = \{\text{Reals}\}$; $f(x) = x + 2$, $D = \{\text{Integers}\}$; find:

*6. $f(\tfrac{1}{2})$ ***7.** $g(4)$ **8.** $f(\tfrac{1}{5})$ ***9.** $\phi(-2)$

Find the equation of the lines through the following pairs of points:

10. $(5, -2)$ and $(-4, 3)$ **11.** $(-2, -2)$ and $(5, 5)$ **12.** $(3, -2)$ and $(7, -3)$

13. Which of the following diagrams depict functions?

(a) (b) (c)

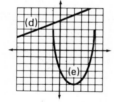

14. What are the range and domain for 13(a)?

15. Which of the following sets of ordered pairs are functions?
(a) $(-3, 2)\,(3, 2),\,(5, 2)$ (b) $(-3, -2),\,(5, -2),\,(7, -2)$
(c) $(-3, -2),\,(-3, -5),\,(-3, -6)$

Solve:

16. $\dfrac{y}{3} + \dfrac{1}{4} = 2y$ **17.** $\dfrac{4}{p} - \dfrac{3}{p - 4} = 0$ **18.** $\dfrac{a}{c} + \dfrac{1}{x} = k$; find x

19. Find f.

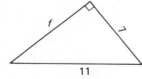

20. Simplify: $\dfrac{x^2 + 5x + 6}{x^3 + 7x^2 + 10x} \div \dfrac{x^3 + 11x^2 + 24x}{x^2 + 2x - 15}$

21. Solve by graphing: $\begin{cases} y = 2x + 2 \\ y = -x - 1 \end{cases}$

22. Find the equations of lines (a) and (b).

23. Simplify: $\dfrac{\dfrac{a}{x^2} - \dfrac{x}{a}}{\dfrac{x}{a} - \dfrac{1}{x^2}}$ **24.** Divide $x^3 + 5$ by $x - 2$

25. Simplify: $-3[(-2^0 - 3) - (-5 + 7)(-2^2 + 3)] - [(-6^0 - 2) - 2^2]$

26. What fraction of $\dfrac{3}{5}$ is $1\dfrac{1}{3}$?

27. Evaluate $-x^0 - x^2(x - xy)$ if $x = -3$ and $y = 2$

28. Add like terms: $\dfrac{x^2yya}{y^{-2}x^4} + \dfrac{3x^{-2}yy^{-5}y^9}{a^{-1}yxx^{-1}} - \dfrac{3x^2yyy^3a}{x^2yay^{-4}}$

29. $.002\sqrt{3} \in \{\text{What sets of numbers}\}$?

30. Graph: $4 - |x| < 3$; $D = \{\text{Integers}\}$

31. Add: $\dfrac{7x + 2}{x + 3} - \dfrac{x}{x^2 - 9}$ **ER 110-44**

Additional

16 Lessons

Topics

LESSON *111*   *Line parallel to a given line*

111.A

equation of a line through a point and parallel to a given line

To find the slope of a line through two specified points, we have to graph the points, draw the line, draw the slope triangle, and compute the slope. To find the slope of a line that is to be parallel to a given line, all that we need to do is to realize that two parallel lines have the same slope.

In the figure we have graphed the line whose equation is $y = 2x - 1$. We see from the equation and from the triangle drawn in the graph that the slope of this line is 2. **Any line that is parallel to this line must also have a slope of 2.** Thus if we are asked to find the equation of the line that is parallel to this line and that passes through $(-2, 3)$ we are already halfway home, for the slope of the new line has to be 2.

$$y = 2x + b$$

Now we can use the coordinates $(-2, 3)$ for x and y and find b algebraically.

$$3 = 2(-2) + b \qquad \text{replaced } x \text{ and } y \text{ with } -2 \text{ and } 3$$

$$3 = -4 + b \qquad \text{multiplied}$$

$$7 = b \qquad \text{added 4 to both sides}$$

So the equation of the line is

$$y = 2x + 7$$

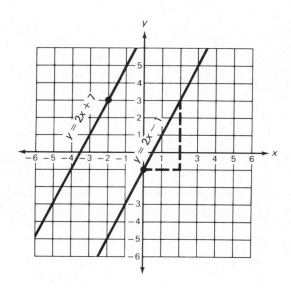

We have found the exact slope and the exact intercept. Estimated values for the slope and intercept will not be acceptable for this kind of problem.

example 111.1   Find the equation of the line through $(-1, -3)$ that is parallel to $4x + 3y = 7$.

solution   The equation of the new line is

$$y = mx + b$$

and we have to find m and b. **If the new line is to be parallel to the line $4x + 3y = 7$, the new line must have the same slope as the line $4x + 3y = 7$.** If we write $4x + 3y = 7$ in slope-intercept form, we find

$$y = -\frac{4}{3}x + \frac{7}{3}$$

and the slope of this line is $-\frac{4}{3}$. So the slope of the new line must be $-\frac{4}{3}$.

$$y = -\frac{4}{3}x + b$$

We find the value of b by using coordinates $(-1, -3)$ for x and y and solving algebraically for b.

$$-3 = -\frac{4}{3}(-1) + b \qquad \text{substituted}$$

$$-\frac{9}{3} = \frac{4}{3} + b \qquad \text{multiplied}$$

$$-\frac{13}{3} = b \qquad \text{added } -\frac{4}{3} \text{ to both sides}$$

Thus the desired equation is

$$y = -\frac{4}{3}x - \frac{13}{3}$$

and again we have found the exact value of the slope and the exact value of the intercept.

problem set 111

1. Birthe rode to town in the bus at 20 miles per hour. Then she trotted back home at 8 miles per hour. If her total traveling time was 14 hours, how far was it to town?

2. Soren had a 36-mile head start. If Erik caught him in 3 hours, how fast was Erik traveling if his speed was twice that of Soren?

3. Flexner marked the bench down $20 and sold it for 60 percent of the original price. What was the original price of the bench?

4. Weir and Max put $75 in quarters and dimes in the box. There were 400 more dimes than quarters. How many coins of each type were there?

5. Find three consecutive odd integers such that the product of -7 and the sum of the first and third is 27 greater than the product of 11 and the opposite of the second.

Find the equation of the line:

*6. That goes through the point $(-2, 3)$ and is parallel to $y = 2x - 1$.

*7. That goes through the point $(-1, -3)$ and is parallel to $4x + 3y = 7$.

8. That goes through the points $(-3, -2)$ and $(5, -3)$.

9. That goes through the points $(5, -1)$ and $(0, 0)$.

Given that $p(x) = x^2 + 5$, $D = \{$Negative integers$\}$, and $k(x) = x + 4$, $D = \{$Integers$\}$, find:

10. $k(4)$ 11. $p(4)$

12. Which of the following diagrams depict functions?

(a)   (b)

(c)   (d)

13. What are the domain and range of 12(a)?

14. Which of the following sets of ordered pairs are functions?
     (a) $(4, -7), (5, -7), (3, -2)$ (b) $(5, -3), (-5, 3), (7, -2)$
     (c) $(-7, 4), (5, -3), (-7, 6)$

Solve:

15. $\dfrac{7}{y} + \dfrac{3}{y - 2} = 0$ 16. $\dfrac{y}{7} - \dfrac{3}{4} = \dfrac{2y}{5}$ 17. $\dfrac{a}{c} - \dfrac{1}{x} = b$, find c.

18. Graph: $+2 - |x| \geq -2$; $D = \{$Negative integers$\}$

19. Find the distance between $(-3, -2)$ and $(5, -3)$.

20. Solve: $x^0 - 3x(2 - 4^0) - (-3) - 2(x - 3) = 3x - (-4)$

Simplify:

21. $\dfrac{(30{,}000 \times 10^{-42})(7000 \times 10^{15})}{(.00021 \times 10^{14})(1000 \times 10^{-23})}$

22. $\dfrac{(2x)^{-2}y^2x^2y^4y}{y^0(x^{-4})y^2y^{-2}y(x^{-4})^{-2}}$

23. $4\sqrt{50{,}000} - 3\sqrt{125}$

24. $\dfrac{-3 - 3x}{-3}$

25. $\dfrac{-2^{-2}}{(-2^0)^{-2}}$

26. Solve by factoring: $80 = -x^2 + 18x$

27. Evaluate $-x^3 - x^2 - x(x^0 - yx)$ if $x = -4$ and $y = 3$

28. Graph on a number line: $-4 - |x| \le -4$

29. Multiply: $\dfrac{x^{-3}}{y^2}\left(\dfrac{y^2}{x^3} - \dfrac{3y^{-3}}{x^{-2}}\right)$

30. Round off $.00\overline{3}18$ to the nearest thousandth.   **ER 111-45**

LESSON 112   *Equation of a line with a given slope*

112.A

equation of a line with a given slope and through a given point

This type is the easiest of all. The slope of the desired equation is *given* in the statement of the problem. All we have to do is find the value of the intercept.

example 112.1   Find the equation of the line that goes through the point $(3, 4)$ and has a slope of $-\tfrac{3}{4}$.

solution   The equation of the line in question is $y = mx + b$, and we need to determine the proper values for m and b. The statement of the problem tells us that the slope is $-\tfrac{3}{4}$ and if we use this value for m in the equation, we find

$$y = -\frac{3}{4}x + b$$

If we use $(3, 4)$ for x and y in this equation we can solve algebraically for b.

$$4 = -\frac{3}{4}(3) + b$$

$$4 = -\frac{9}{4} + b \quad \longrightarrow \quad \frac{16}{4} = -\frac{9}{4} + b \quad \longrightarrow \quad b = \frac{25}{4}$$

So the equation is

$$y = -\frac{3}{4}x + \frac{25}{4}$$

and the numbers $-\tfrac{3}{4}$ and $\tfrac{25}{4}$ are the exact values of the slope and the intercept.

example 112.2   Find the equation of the line that goes through the point $(-5, 11)$ and has a slope of $\frac{1}{7}$.

solution   The statement of the problem gives us the slope, so we can write

$$y = \frac{1}{7}x + b$$

Now we use the values -5 and 11 for x and y and solve for b.

$$11 = \frac{1}{7}(-5) + b \quad\longrightarrow\quad \frac{77}{7} = -\frac{5}{7} + b \quad\longrightarrow\quad b = \frac{82}{7}$$

So the equation is

$$y = \frac{1}{7}x + \frac{82}{7}$$

and the numbers $\frac{1}{7}$ and $\frac{82}{7}$ are the exact values of the slope and the intercept.

example 112.3   Find the equation of the line that goes through the point $(-50, 40)$ and has a slope of $-\frac{1}{2}$.

solution   Again we have been given the slope so we can write

$$y = -\frac{1}{2}x + b$$

Now to find b, we replace y with 40 and x with -50 and solve.

$$40 = -\frac{1}{2}(-50) + b$$

$$40 = 25 + b$$

$$15 = b$$

Thus the desired equation is

$$y = -\frac{1}{2}x + 15$$

problem set 112

1. The ingrates ran to the cemetery at 8 miles per hour and trotted back home at 6 miles per hour. If the total trip took 7 hours, how far was it to the cemetery?

2. At 4 a.m. the northbound train left at 40 miles per hour. At 6 a.m. the southbound train left the same station at 60 miles per hour. At what time will the trains be 880 miles apart?

3. Small pizzas were $3 and large pizzas were $5. To feed the throng, it was necessary to spend $475 for 125 pizzas. How many pizzas of each type were purchased?

4. Only 300 of the larvae metamorphosed into butterflies. If there were 2500 larvae at the outset, what percent became butterflies?

5. Mark thought of three consecutive odd integers. He added the first to twice the third and multiplied this sum by -3. The result was 3 less than the product of 8 and the opposite of the second. What were the numbers?

6. The ratio of bees to moths was 13 to 5. If there were a total of 2610 bees and moths in the bar, how many were moths and how many were bees?

Find the equation of the line that goes through:

7. The points $(-2, 3)$ and $(4, 5)$

8. The point $(-2, 3)$ and is parallel to $2x + 3y = 7$

*9. The point $(3, 4)$ and has a slope of $-\dfrac{3}{4}$

*10. The point $(-5, 11)$ and has a slope of $\dfrac{1}{7}$

11. Given $f(x) = 2x + 4, D = \{\text{Negative integers}\}$, and $p(x) = x + 5, D = \{\text{Positive integers}\}$, find: (a) $f(2)$ (b) $p(2)$

12. Which of the following diagrams or sets of ordered pairs represent functions?

(a) (b)

(c)

(d) $(4, -2), (4, -5), (3, 4)$ (e) $(1, -2), (3, -2), (6, -2)$

13. What is the range of 12(b) above?

Solve:

14. $\dfrac{5x}{3} - \dfrac{1}{3} = \dfrac{2x}{5}$

15. $\dfrac{x - 2}{3x} = \dfrac{4}{x} - \dfrac{1}{5}$

16. $\dfrac{k}{m} - \dfrac{1}{c} + \dfrac{x}{y} = p$; find c

17. Use the square root table to simplify: $\sqrt{.000178563 \times 10^{-13}}$

18. Find the distance between $(-2, 3)$ and $(4, 5)$.

19. Solve: $-x^0 - (2x - 5) + x - (-3^2) - 2 = 3x(4^0 - 2) - 2^2$

Simplify:

20. $\dfrac{(21,000 \times 10^{-42})(7,000,000)}{(.0003 \times 10^{-21})(700 \times 10^{15})}$

21. $\dfrac{x(x^{-2}y)^{-2}(x^{-2}y)x^{-2}ya^2x}{(xy^{-2})^{-2}x^{-2}y^{-4}yy^3x^2}$

22. $\sqrt{150,000} + 2\sqrt{3} \cdot 5\sqrt{5} + 2\sqrt{15}(\sqrt{15} - 3)$

23. $\dfrac{6xy + 6xy^2}{6xy}$

24. $\dfrac{-3^{-2}}{-(-3)^{-3}}$

25. Solve by factoring: $120 = -22x - x^2$

26. Evaluate $-x^0 - x^2 - xy(x - y)$ if $x = -3$ and $y = 4$

27.   Graph on a number line: $-3 - |x| \le -3$; $D = \{\text{Integers}\}$

28.   Multiply: $\dfrac{4x^{-2}}{yx}\left(\dfrac{x^3}{y} - \dfrac{3y^3}{x^3}\right)$ **ER 112-46**

LESSON *113*   *Radical equations*

113.A
square roots revisited

Both -2 and $+2$ are square roots of 4 because

$$(-2)^2 = 4 \quad \text{and} \quad (+2)^2 = 4$$

But when we write $\sqrt{4}$, we are indicating the positive or principal square root of 4, which, of course, is the number 2.

> **DEFINITION OF SQUARE ROOT**
>
> If x is greater than zero, then $\sqrt{x}$ is the unique positive real number such that
>
> $$(\sqrt{x})^2 = x$$

We can state this definition in words by saying that the principal square root of a given positive number is that positive number which, multiplied by itself, yields the given number. Thus

$$\sqrt{2}\sqrt{2} = 2 \qquad \sqrt{7}\sqrt{7} = 7 \quad \text{and} \quad \sqrt{3.14}\sqrt{3.14} = 3.14$$

and also

$$(\sqrt{2})^2 = 2 \qquad (\sqrt{7})^2 = 7 \quad \text{and} \quad (\sqrt{3.14})^2 = 3.14$$

It is necessary to remember that algebraic expressions represent particular real numbers that are determined by the values assigned to the variables. Thus, if a particular algebraic expression represents a positive real number, the definition of the square root given in the box above applies to the expression. For example,

$$(\sqrt{x^2 + 4})^2 = x^2 + 4 \qquad \left(\sqrt{\dfrac{amx^2}{p}}\right)^2 = \dfrac{amx^2}{p} \qquad (\sqrt{x + 6})^2 = x + 6$$

$$(\sqrt{x^4 + 3x^2 + 5})^2 = x^4 + 3x^2 + 5$$

113.B
radical equations

There is only one number that will satisfy the equation $x = 2$, and that number is 2. If we replace x with 2 in this equation, we find

$$2 = 2$$

which is a true statement. If we square both sides of the original equation

$$(x)^2 = (2)^2 \longrightarrow x^2 = 4$$

the result is the equation $x^2 = 4$. While the equation $x = 2$ had only one solution, the equation $x^2 = 4$ has two numbers that satisfy it, the numbers $+2$ and -2.

$$\text{REPLACING } x \text{ WITH } +2 \qquad\qquad \text{REPLACING } x \text{ WITH } -2$$
$$(+2)^2 = 4 \qquad\qquad\qquad\qquad (-2)^2 = 4$$
$$4 = 4 \quad \text{True} \qquad\qquad\qquad 4 = 4 \quad \text{True}$$

We began with the equation $x = 2$, whose only solution is 2. We squared both sides and got the equation $x^2 = 4$, which also has the number 2 as a solution but has another solution, which is the number -2. **It can be shown that if both sides of an equation are squared, all of the solutions to the original equation (if any exist) are also solutions to the resulting equation, but the reverse is not true, for all of the solutions of the resulting equation are not necessarily solutions of the original equation.**

example 113.1   Solve $\sqrt{x - 2} + 3 = 0$.

solution   We wish to **isolate the radical** on one side of the equation so that we may square both sides of the equation and use the fact that $(\sqrt{x - 2})^2 = x - 2$, so we add -3 to both sides of the equation and get

$$\sqrt{x - 2} = -3$$

Now square both sides and get

$$(\sqrt{x - 2})^2 = (-3)^2 \quad\longrightarrow\quad x - 2 = 9 \quad\longrightarrow\quad x = 11$$

Now we must check our solution in the original equation.

$$\sqrt{11 - 2} + 3 = 0 \quad\longrightarrow\quad \sqrt{9} + 3 = 0 \quad\longrightarrow\quad 3 + 3 = 0 \quad\longrightarrow\quad 6 = 0 \quad \text{False}$$

We see that while 11 is a solution of the second equation, $x - 2 = 9$, it is not a solution to the original equation $\sqrt{x - 2} + 3 = 0$. Thus we see that there is no real number replacement for x that will satisfy the first equation, and we say that the solution set of this equation is the empty set.

example 113.2   Solve $\sqrt{x - 2} - 6 = 0$.

solution   We first **isolate the radical** on one side by adding $+6$ to both sides of the equation.

$$\sqrt{x - 2} = 6$$

Now we square both sides to eliminate the radical and then solve the resulting equation.

$$(\sqrt{x - 2})^2 = (6)^2 \quad\longrightarrow\quad x - 2 = 36 \quad\longrightarrow\quad x = 38$$

Now we will check this solution in the original equation.

$$\sqrt{(38) - 2} - 6 = 0 \quad\longrightarrow\quad \sqrt{36} - 6 = 0 \quad\longrightarrow\quad 6 - 6 = 0 \quad\longrightarrow\quad 0 = 0 \quad \text{Check}$$

Thus **38** is a solution to the original equation.

example 113.3   Solve $\sqrt{x^2 + 9} - 5 = 0$.

solution   First **isolate the radical** and get

$$\sqrt{x^2 + 9} = 5$$

Now eliminate the radical by squaring both sides of the equation

$$(\sqrt{x^2 + 9})^2 = (5)^2 \quad\longrightarrow\quad x^2 + 9 = 25$$

Now we will simplify, factor, and use the zero factor theorem to solve.

$$x^2 - 16 = 0 \quad\longrightarrow\quad (x + 4)(x - 4) = 0 \quad\longrightarrow\quad x = 4, \quad \text{or} \quad x = -4$$

Since neither of these solutions to the second equation is guaranteed to be a solution of the original equation, both solutions must be checked in the original equation.

CHECK $+4$	CHECK -4
$\sqrt{(4)^2 + 9} = 5$	$\sqrt{(-4)^2 + 9} = 5$
$\sqrt{25} = 5$	$\sqrt{25} = 5$
$5 = 5$ Check	$5 = 5$ Check

Thus both $+4$ and -4 are solutions.

example 113.4 Solve $\sqrt{x - 1} - 3 + x = 0$.

solution We begin by adding $+3 - x$ to both sides of the equation to **isolate the radical** and get

$$\sqrt{x - 1} = 3 - x$$

Now we square both sides to eliminate the radical

$$(\sqrt{x - 1})^2 = (3 - x)^2 \longrightarrow x - 1 = 9 - 6x + x^2$$

Next we simplify, factor, and use the zero factor theorem to solve.

$$x^2 - 7x + 10 = 0 \longrightarrow (x - 2)(x - 5) = 0 \longrightarrow x = 2, \quad \text{or} \quad x = 5$$

Now we must check both 2 and 5 in the original equation.

CHECK $x = 2$	CHECK $x = 5$
$\sqrt{x - 1} = 3 - x$	$\sqrt{x - 1} = 3 - x$
$\sqrt{(2) - 1} = 3 - (2)$	$\sqrt{(5) - 1} = 3 - (5)$
$\sqrt{1} = 1$	$\sqrt{4} = -2$
$1 = 1$ Check	$2 \neq -2$ Does not check
Thus $x = 2$ is a solution of the original equation.	Thus $x = 5$ is not a solution of the original equation.

example 113.5 Solve $\sqrt{2x - 3} = \sqrt{x + 2}$.

solution One radical expression is isolated on each side of the equation so we begin by squaring both sides of the equation.

$$(\sqrt{2x - 3})^2 = (\sqrt{x + 2})^2 \longrightarrow 2x - 3 = x + 2 \longrightarrow \mathbf{x = 5}$$

Now we will check $x = 5$ in the original equation.

$$\sqrt{2(5) - 3} = \sqrt{5 + 2} \longrightarrow \sqrt{7} = \sqrt{7} \quad \text{Check}$$

problem set 113

1. The goliards sang songs before the banquet. If the ratio of ribald songs to scandalous songs in their repertoire of 3102 songs was 7 to 4, how many ribald songs did they know?

2. The doyenne walked to the meeting at 2 miles per hour and caught a ride home in an old truck at 10 miles per hour. How far was it to the meeting place if her total traveling time was 18 hours?

3. The gun sounded and 3 hours later Bill was 6 miles ahead of Rose. How fast did Rose run if Bill's speed was 10 miles per hour?

4. Roger and Gwenn picked 178 quarts of berries. If Roger picked 8 more quarts than Gwenn picked, how many quarts did Gwenn pick?

5. Gold bricks sold for \$400 each, while pyrite bricks were only \$3 each. To stock his booth, Grimsby bought 123 bricks for \$21,013. How many of each kind did he buy?

6. Find the equation of the line that passes through $(4, 2)$ and $(-5, -7)$.

7. Given $f(x) = x + 3$, $D = \{\text{Reals}\}$; $g(x) = x - 4$, $D = \{\text{Integers}\}$; find (a) $f(-2)$ (b) $g(-2)$

Solve:

*8. $\sqrt{x - 2} - 6 = 0$ ***9.** $\sqrt{x^2 + 9} - 5 = 0$ ***10.** $\sqrt{x - 1} - 3 + x = 0$

11. $\sqrt{x} = 5\sqrt{2}$ **12.** $\dfrac{x}{3} - \dfrac{2 + x}{5} = -3$ **13.** $\dfrac{4}{x + 3} - \dfrac{2}{2x} = 0$

14. $\dfrac{x}{y} - \dfrac{1}{c} - d = k$; find y

15. Find c.

16. Simplify: $\dfrac{x^2 - 25}{x^2 - 12x + 35} \div \dfrac{x^2 + x - 6}{x^2 - 4x - 21}$

17. Solve by graphing: $\begin{cases} y = 3x \\ y = -x + 4 \end{cases}$

18. Find the equations of lines (a) and (b).

19. Simplify: $\dfrac{\dfrac{x}{yz} - \dfrac{1}{z^2}}{\dfrac{a}{z} - \dfrac{3}{yz^2}}$ **20.** Divide: $2x^3 + 3x^2 + 5x + 4$ by $x - 1$

21. Simplify: $-[(-2^0)(-3^2) - (-7 - 2) - (-2)] - [-3(-5 + 7)]$

22. What fraction of $2\dfrac{1}{8}$ is $3\dfrac{4}{5}$?

23. Evaluate: $-x^3 + (-x)^2 - x^2 - x(x - xy^2)$ if $x = -3$ and $y = -2$

24. Add like terms: $-3x^2yy^{-2} + \dfrac{2x^2}{y} - \dfrac{3xy^{-1}}{x^{-1}} + \dfrac{4x^3x^{-1}}{y}$

25. $\dfrac{3\sqrt{2}}{5} \in \{\text{What sets of numbers}\}$?   **26.** Solve: $2\dfrac{1}{8}x + \dfrac{1}{4} = \dfrac{7}{8}$

Add:

27. $\dfrac{x}{y^2x} - \dfrac{3}{y^3x^2} - \dfrac{2}{x + y}$ **28.** $\dfrac{3x + 2}{x - 4} - \dfrac{2x}{x^2 - 16}$

29. Round off $.03\overline{74}$ to the nearest ten-millionth.   **ER 113-47**

LESSON 114   *Slope formula*

114.A
slope
formula

In Lesson 83 the slope of a straight line was defined to be the ratio of the change in the value of the y coordinate to change in the value of the x coordinate as we move from one point on the line to another point on the line. To demonstrate, we will find the slope of the line through the points $(4, 3)$ and $(-2, -2)$ by first graphing the points and drawing the line as shown on the left. The sign of the slope of this line is positive because the line segment graphed points toward the upper right. On the right we draw the triangle to determine the magnitude (absolute value) of the slope. From the triangle we see that the difference in the x coordinates of the two points is 6 and the difference in the y coordinates is 5. Thus the slope of this line is $+\frac{5}{6}$.

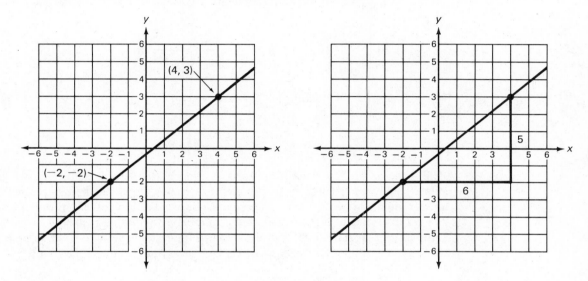

It is not necessary to graph the points to find the slope of the line. If we call the two points point 1 and point 2 and give them the coordinates (x_1, y_1) and (x_2, y_2), respectively, as shown below, we can derive a relationship from which the slope for the line through any two points can be determined algebraically.

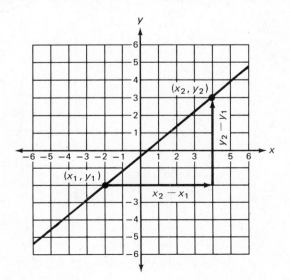

We begin at point 1 and move along the line to point 2. The change in the value of the x coordinate is $x_2 - x_1$, and the change in the value of the y coordinate is $y_2 - y_1$. The slope m is defined to be the ratio of the change in the y coordinate to the change in the x coordinate, so

$$m = \frac{y_2 - y_1}{x_2 - x_1}$$

We can use this relationship or formula to find the sign and the magnitude of the slope of the line through any two given points.

example 114.1   Find the slope of the line that passes through the points $(-3, 4)$ and $(5, -2)$.

solution   Either point can be designated as point 1. We will use $(-3, 4)$ as point 1 and $(5, -2)$ as point 2.

$$m = \frac{y_2 - y_1}{x_2 - x_1} \longrightarrow m = \frac{-2 - (4)}{5 - (-3)} \longrightarrow m = \frac{-6}{8} \longrightarrow \boldsymbol{m = -\frac{3}{4}}$$

example 114.2   Work Example 114.1 again, but this time use $(5, -2)$ as point 1 and $(-3, 4)$ as point 2.

solution $$m = \frac{y_2 - y_1}{x_2 - x_1} \longrightarrow m = \frac{4 - (-2)}{-3 - (5)} \longrightarrow m = \frac{6}{-8} \longrightarrow \boldsymbol{m = -\frac{3}{4}}$$

The slope of the line was found to be $-\frac{3}{4}$ no matter which point was designated as point 1. We see from these two examples that the slope can be determined by using this formula, and we also see that care must be exercised to prevent mistakes in handling the positive and negative signs of the numbers. In the figure below, we use the old familiar graphical method. We see from this figure that the sign of the slope is negative and the magnitude is $\frac{6}{8}$, so the slope is $-\frac{6}{8}$ or $-\frac{3}{4}$, the same slope found by using the formula.

1.   The ratio of Dobermans to terriers in the county was 2 to 7. If there were 774 Dobermans and terriers in the county, how many were Dobermans?

2. Paul and Kit had 3 hours to wait until the game began. They used the time to ride a bus into the country at 14 miles per hour and then to jog back at 7 miles per hour. How far did they go into the country?

3. Find four consecutive even integers such that -5 times the sum of the first and fourth is 10 greater than the opposite of the sum of the third and fourth multiplied by 6.

4. Smiley and Linda cut a 40-foot rope into two lengths. The ratio of the lengths was 3 to 1. How long was each length?

5. Nineteen percent of the bats had returned to the belfrey by 8 p.m. If 855 bats had returned, how many bats lived in the belfrey?

6. Find the equation of the line that passes through $(5, -2)$ and is parallel to $y = -\frac{1}{2}x - \frac{2}{3}$.

7. Find the equations of lines (a) and (b).

8. Which of the following diagrams or sets of ordered pairs depict functions?

(a) (b) (c), (d)

(e) $(4, -2), (3, 5), (7, 6)$ (f) $(4, -2), (5, -2), (3, 5)$ (g) $(4, -2), (4, 7), (3, 5)$

Solve:

9. $\sqrt{x - 1} - 4 = 0$ **10.** $\sqrt{3x} + 4 = 7$ **11.** $5\sqrt{2x} = 4$

12. Find the slope of the line that passes through $(-2, 4)$ and $(5, -3)$ by using the slope formula and letting $(-2, 4)$ be point 1 and $(5, -3)$ be point 2. Repeat and this time use $(5, -3)$ as point 1 and $(-2, 4)$ as point 2.

Solve:

13. $\dfrac{2x}{3} - \dfrac{x - 2}{5} + x = 7$ **14.** $\dfrac{x - 2}{2x} - \dfrac{3}{x} = -\dfrac{1}{5}$

15. $\dfrac{x}{y} + \dfrac{m}{n} - \dfrac{1}{c} = k$; find y

16. Use the square root tables to help simplify $\sqrt{.00052843 \times 10^{40}}$.

17. Find the distance between $(-3, 2)$ and $(5, -7)$.

18. Solve: $-3(x - 4^0) - (-2) - 3(-x - y^0) = 3(x - 2^2)$

Simplify:

19. $\dfrac{(5000 \times 10^{-15})(30{,}000 \times 10^{41})}{(6000 \times 10^{-14})(.000025 \times 10^{-50})}$ **20.** $\dfrac{x^{-2}(x^{-4}x)^3}{x^{-3}x^0xx^{-2}}$

21. $\sqrt{2} \cdot 3\sqrt{12} + 2\sqrt{3} \cdot \sqrt{2} - 2\sqrt{6}(4\sqrt{6} - \sqrt{24})$

22. $\dfrac{4x + 4}{4}$ **23.** $\left(\dfrac{-3^{-2}}{3^{-3}}\right)^{-2}$

24. Solve by factoring: $84 = -19x - x^2$

25. Evaluate: $-xy(y - x^0) - x^2 - x^0$ if $x = -3$ and $y = -5$

26. Graph on a number line: $+4 - |x| - 2 \geq 4$; $D = \{\text{Reals}\}$

27. Multiply: $xy^2\left(\dfrac{y^{-2}}{x} - \dfrac{3x^0x}{(y^{-3})^2}\right)$ **28.** True or false? $\{\text{Reals}\} \subset \{\text{Integers}\}$

29. Simplify: $\dfrac{\dfrac{a}{x^2y} - \dfrac{1}{y^2}}{\dfrac{bc}{x^2y^2} - 1}$ **30.** What fraction of $3\dfrac{1}{8}$ is $\dfrac{2}{3}$?

ER 114-48

LESSON 115   *Consistent, inconsistent, and dependent equations*

115.A
solutions of linear systems: consistent equations

We have considered three methods of finding the ordered pair that is the solution to a system of two first-degree equations in two unknowns. We call the methods the **substitution method,** the **elimination method,** and the **graphical method.** We will review these procedures by solving a system of equations by each of the three methods. The system is

$$\begin{array}{l} \text{(a)} \\ \text{(b)} \end{array} \left\{ \begin{array}{l} -x + y = 1 \\ x + y = 5 \end{array} \right.$$

SUBSTITUTION		ELIMINATION	
$x + (1 + x) = 5$	substituted $1 + x$ for y	$-x + y = 1$	equation (a)
$2x + 1 = 5$	simplified	$\underline{x + y = 5}$	equation (b)
$2x = 4$	added -1	$2y = 6$	added
$x = 2$	divided		
		$y = 3$	divided
and since $x + y = 5$		and since $x + y = 5$	
$(2) + y = 5$		$x + (3) = 5$	
$y = 3$		$x = 2$	
solution is **(2, 3)**		solution is **(2, 3)**	

On the left above we substituted for y in equation (b) the equivalent expression for y from equation (a). On the right above we added equations (a) and (b) algebraically and eliminated the variable x. The figure on the next page shows the graphs of the two linear equations. Note the coordinates of the intersection of the two lines. In each case, we find that the ordered pair $(2, 3)$ is the solution to the system of equations. If a solution exists, the solution to such a system can always be found using any one of the three methods. **Equations that have a single solution are called consistent equations.**

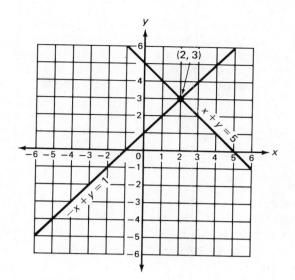

115.B

inconsistent equations

Inconsistent equations have no common solution. On the left is a system of two linear equations, and the figure shows the graph of the two equations.

$$\begin{cases} x + y = 2 \\ x + y = -3 \end{cases}$$

From the figure we see that the lines appear to be parallel and do not appear to intersect. If they do not intersect, there is no one point that lies on both of the lines and hence no ordered pair of x and y will satisfy both equations. If we try to use either the substitution method or the elimination method, we will end up with a false numerical statement. To demonstrate this result, we will attempt to solve the given system by both methods.

SUBSTITUTION	ELIMINATION
$(-3 - y) + y = 2$	$-x - y = -2$
$-3 - y + y = 2$	$\dfrac{x + y = -3}{0 = -5}$ False
$-3 = 2$ False	

In both cases the final result is a false numerical statement.

In Lessons 58 and 70 we discussed the substitution and elimination methods in some detail. We saw that before either method was used, an assumption was necessary! We assumed that a value of x and y existed that would simultaneously satisfy both equations and that the symbols x and y in the equations represented these numbers. From the graph of the parallel lines on the last page it is obvious that in this problem our assumption was invalid, for in the graph we see that there is no point that lies on both lines and whose coordinates therefore satisfy both equations. Since our assumption was invalid, the substitution and elimination methods do not produce a solution.

115.C
dependent equations

We have defined equivalent equations to be equations that have the same solution set. If we multiply every term on both sides of an equation by the same nonzero quantity, the resulting equation is an equivalent equation to the original equation. In the left figure, we have graphed the equation

$$y - x = 2$$

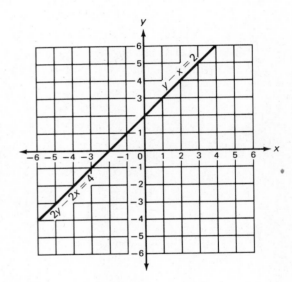

There is an infinite number of ordered pairs of x and y whose coordinates satisfy the equation, and the graph of these points is the line in the figure. If we multiply every term in the given equation by some quantity, say the number 2,

$$y - x = 2 \qquad \text{multiplied by (2) yields} \qquad 2y - 2x = 4$$

we get the equation $2y - 2x = 4$, which is an equivalent equation to the original equation. Thus all ordered pairs of x and y that satisfy the original equation also satisfy the new equation, and the graph of the new equation is the same as the graph of the original equation. If we are asked to find a graphical solution to a system that consists of a pair of equivalent equations such as the pair being discussed,

(a) $\begin{cases} y - x = 2 \\ (b) \quad 2y - 2x = 4 \end{cases}$

we find that the graph of both equations is the single line of either figure above. Thus any ordered pair that is a solution to one of the equations is also a solution to the other equation. **Equivalent linear equations are called dependent equations.** If we try to find the solution to our pair of dependent equations, we find that the result is a true numerical statement. We will attempt to solve the system above by using both substitution and elimination.

	SUBSTITUTION		ELIMINATION
(a)	$y - x = 2$	(a)	$y - x = 2$
(b)	$2y - 2x = 4$	(b)	$2y - 2x = 4$
	$2(x + 2) - 2x = 4$		$-2y + 2x = -4$
	$2x + 4 - 2x = 4$		$\underline{2y - 2x = 4}$
	$4 = 4 \qquad$ True		$0 = 0 \qquad$ True

On the left we solved equation (a) for y and substituted this expression for y in equation (b). The result reduced to the true statement that 4 equals 4. On the right we multiplied each term in equation (a) by -2 and then added the resulting equation to equation (b). The result was the true statement that 0 equals 0. This indicates that any ordered pair of x and y that satisfies one of the equations will satisfy the other equation. Thus there is an infinite number of ordered pairs that will satisfy two dependent equations.

115.D
summary

Two linear equations in two unknowns fall into one of three categories.

1. **Consistent equations,** which are equations that have a single ordered pair as a common solution. The graphs of consistent equations intersect, as shown in Figure (a).
2. **Inconsistent equations,** which are equations that have no common solution. The graphs of inconsistent equations are parallel lines, as shown in Figure (b).
3. **Dependent equations,** which are equivalent equations, or if you will, the same equation in two different forms. The graph of two dependent equations is a single line, as shown in Figure (c) on the next page.

(a)   (b)

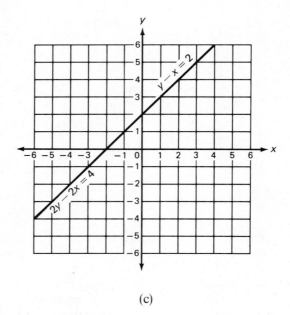

(c)

1. Poltavia walked for a while at 3 miles per hour and then rode a bus at 15 miles per hour. Her time on the bus was twice her time walking. How long did she ride if the total distance covered was 66 miles?

2. Marlow headed south at 20 miles per hour at noon. At 4 p.m. Mangum headed north from the same station at 60 miles per hour. At what time will they be 880 miles apart?

3. Rabbits were $2 each and cats were $7 each. Marcia bought 52 animals for $259. How many were rabbits and how many were cats?

4. The ratio of coons to mongooses was 5 to 3. If there were 968 total in the pen, how many were coons and how many were mongooses?

5. Thirty seven percent of the women had flowers in their hair. If 4788 women in the parade did not wear flowers, how many women were in the parade?

6. Find the equation of the line that passes through the points $(4, 2)$ and $(7, -3)$.

7. Given that $f(x) = x^2 + 4$, $D = \{\text{Reals}\}$, find $f(\frac{1}{2})$.

Solve:

8. $7\sqrt{x} = 14\sqrt{2}$ 9. $\sqrt{x - 3} = 5$ 10. $\sqrt{x} + 2 = 7$

11. Are the equations of lines that are parallel called inconsistent or consistent?

Solve:

12. $\dfrac{4x}{3} - \dfrac{x + 2}{5} = 7$

13. $\dfrac{p - 3}{p} = \dfrac{5}{3p} - \dfrac{1}{4}$

14. $\dfrac{k}{x} - \dfrac{1}{c} + p = m$; find x

15. Find b.

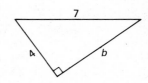

16. Simplify: $\dfrac{x^4 + 7x^3 + 12x^2}{x^2 - 16} \div \dfrac{x^4 - 2x^3 - 15x^2}{x^2 - 9x + 20}$

17. Solve by graphing: $\begin{cases} y = -x - 2 \\ y = \dfrac{1}{2}x + 1 \end{cases}$

18. Find the equations of lines (a) and (b).

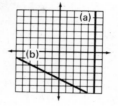

19. Simplify: $\dfrac{\dfrac{x}{y^2 a} - \dfrac{a}{y^2}}{\dfrac{p}{y} - \dfrac{3}{ay^2}}$ 20. Divide $x^3 - 2x^2 - 5$ by $x - 1$.

21. What fraction of $\dfrac{5}{16}$ is $14\dfrac{3}{8}$?

22. Simplify: $-2^0 - (-2^0) - [(-3 - 2) - (-5 + 7)]$

23. Evaluate $-x^0 - x^2 - x^3 + xy(x - xy)$ if $x = -3$ and $y = 2$

24. $-3 \in \{$What sets of numbers$\}$?

25. Solve by factoring: $x^2 + 40 = -22x$

26. Simplify by adding like terms: $x^2 y - \dfrac{3x^3 y}{x} + \dfrac{2x^4 y^2}{yx^{-2}} + \dfrac{5x^2}{y^{-1}}$

Add:

27. $\dfrac{x}{y^2} + \dfrac{2x^2}{xy^2} - \dfrac{x^2}{y - 1}$ 28. $\dfrac{4x - 2}{x - 3} - \dfrac{x + 3}{x^2 - 9}$

29. Round off $32.07\overline{581}$ to the nearest hundred-millionth.   **ER 115-49**

LESSON 116 *Conjunctions and Disjunctions*

116.A
absolute value
inequalities
revisited

We have discussed three ways to think of the absolute value of a nonzero real number. One way is to think of the absolute value as the "bigness" of the number. Another way is to think of the absolute value as representing the distance of the graph of the number from the origin. The third way is to think of the absolute value as being the positive member of a pair of opposites. In any case we say that the absolute value of a nonzero real number is a positive number and that the absolute value of zero is zero.

In Lesson 105 we studied the graphing of the solutions of absolute value inequalities. We can classify absolute value inequalities into four categories as determined by the solution sets of the inequalities:

1. The first type is an absolute value inequality whose solution set is the empty set.

$$|x| < -6 \qquad |x| < -2 \qquad |x| < -4$$

No real number replacement for the unknown will satisfy any of the inequalities above because the absolute value of any real number is equal to or greater than zero. Thus there is no real number whose absolute value is less than zero.

2. The second type is an absolute value inequality whose solution set is the set of real numbers.

$$|x| > -6 \qquad |x| > -2 \qquad |x| > -4$$

Any real number replacement for the unknown will satisfy each of the inequalities above because the absolute value of any real number is equal to or greater than zero. If the absolute value of every real number is equal to or greater than zero, certainly the absolute value is greater than any negative number.

3. The third type is an absolute value inequality whose solution set consists of those numbers whose absolute value is less than a given positive number.

$$|x| < 3 \qquad D = \{\text{Reals}\}$$

The solution of this inequality is graphed below and consists of all real numbers whose absolute value is less than 3.

4. The fourth type is an absolute value inequality whose solution set consists of those numbers whose absolute value is greater than a given positive number.

$$|x| > 3 \qquad D = \{\text{Reals}\}$$

The solution to this inequality is graphed below and consists of all real numbers whose absolute value is greater than 3.

If we dismiss cases 1 and 2, in which no real number satisfies or all real numbers satisfy, we see that absolute value inequalities fall into *two categories*.

3. The solution is all numbers that are greater than one specified number *and* also are less than another specified number.
4. The solution is all numbers that are greater than one specified number *or* are less than another specified number.

116.B
conjunctions and disjunctions

In both cases 3 and 4 we note that there are two restrictions on the numbers that can be used. In 3 both of the restrictions must be satisfied, and we use the word *and* to designate that compliance with both restrictions is required. We call a dual restriction of this type a *conjunction*. In 4 a number satisfies the stated condition if it satisfies either one of the restrictions. We use the word *or* to designate that a solution of either restriction is all that is necessary and we call this type of dual restriction a **disjunction**.

example 116.1   Graph $x > -1$ and $x \leq 4$, $D = \{\text{Reals}\}$.

solution   **The use of the word** *and* **designates this dual restriction on the variable as a conjunction and indicates that both restrictions must be satisfied.**

The solution designates all numbers that are greater than -1 *and* are less than or equal to 4.

example 116.2   Graph $-3 \leq x < 2$, $D = \{\text{Integers}\}$.

solution   This notation is an alternate notation for the designation of a conjunction in that both of the restrictions must be satisfied. We can break the notation into two notations by beginning in the middle and reading to the left as "x is greater than or equal to -3." Now we go back to the middle and read to the right as "x is less than 2."

$$x \geq -3 \qquad and \qquad x < 2$$

In the figure we indicate all integers that are equal to or greater than -3 and are also less than 2.

example 116.3   Graph $0 \leq x + 2 < 5$, $D = \{\text{Reals}\}$.

solution   This notation designates a conjunction but places the restrictions on $x + 2$ instead of x. To find the restrictions on x, we must simplify both of the stated conditions.

$$
\begin{array}{ccc}
0 \leq x + 2 & \quad and \quad & x + 2 < 5 \\
\underline{-2 \qquad -2} & & \underline{-2 \quad -2} \\
-2 \leq x & & x \quad < \quad 3
\end{array}
$$

In the graph we designate all real numbers that are equal to or greater than -2 and are also less than 3.

example 116.4   Graph $x < -1$ or $x \geq 2$, $D = \{\text{Integers}\}$.

solution   **The use of the word** *or* **between the two restrictions tells us that this dual restriction is a disjunction and is satisfied by any number that satisfies either one of the conditions.**

In the graph we designate all integers that are less than -1 *or* are greater than or equal to 2.

example 116.5   Graph $x > 15$ or $x \leq 10$, $D = \{\text{Reals}\}$.

solution   This dual restriction is a disjunction as indicated by the word *or*.

We indicate all real numbers that are less than or equal to 10 *or* are greater than 15.

problem
set 116

1. Homer drove as fast as he could to get 60 miles ahead of Mae. If Homer drove at 17 miles per hour and the race lasted 20 hours, how fast did Mae drive?

2. Wanatobe walked to the moot at 4 miles per hour and then rode back home in a bus at 24 miles per hour. If her total traveling time was 14 hours, how far was it to the moot?

3. The sum of 13 and the opposite of a number was multiplied by 3. This result was 11 less than twice the number. What was the number?

4. The dress was marked down 20 percent for the sale, and it still sold for $120. What would it sell for if the markdown was only 10 percent? (*Hint:* Find the original price and take off 10 percent of this price.)

5. Peaches were $7 a bushel and apples were $6 a bushel. Harry sold $346 worth and sold 29 more bushels of peaches than apples. How many bushels of each did he sell?

6. Find the equation of the line that passes through the point $(-3, 2)$ and has a slope of $-\dfrac{1}{7}$.

7. Which of the following diagrams or sets of ordered pairs depict functions?

   (a)   (b)   (c)

   (d)   (e) $(4, 3), (3, 2), (3, -2)$ (f) $(4, 2), (4, 3), (4, 3)$

   (g) $(4, 2), (3, -2), (7, -8)$

Solve:

8. $4\sqrt{y} = 20$ 　　　　　　　　　　9. $\sqrt{x - 4} - 5 = 0$

Graph the following conjunctions and disjunctions on a number line:

*10. $0 \le x + 2 < 5, D = \{\text{Reals}\}$ 　　　*11. $x < -1 \text{ or } x \ge 2, D = \{\text{Integers}\}$

*12. $x > 15 \text{ or } x \le 10, D = \{\text{Reals}\}$

Solve:

13. $\dfrac{x}{4} - \dfrac{x - 2}{3} = 7$ 　　14. $\dfrac{p - 5}{p} = \dfrac{5}{3p} - \dfrac{1}{5}$ 　　15. $\dfrac{a}{x} - \dfrac{1}{c} = \dfrac{1}{d}$; find x

16. Use the square root tables to simplify: $\sqrt{.0004168521 \times 10^{-30}}$

17. Find the distance between $(-5, 2)$ and $(3, -7)$.

18. Solve: $p - 3p^0 - 2(p - 4^0) - (-3) - 2 = -3^0(2 - p)$

Simplify:

19. $\dfrac{(2000 \times 10^{15})(.0004 \times 10^{21})}{(4000 \times 10^{-23})(1000 \times 10^{14})}$

20. $\dfrac{3(x^{-2}y)^{-3}x^{-3}y^0y^2}{(x^{-3}y^{-2})^{-2}xx^0}$

21. $2\sqrt{800} - 3\sqrt{18}$

22. $4\sqrt{2}(5\sqrt{2} - 2\sqrt{12})$

23. $\dfrac{4x^2 - 4x}{4x}$

24. $\dfrac{-3^{-3}(-3)^{-2}}{3^{-2}}$

25. Solve by factoring: $45 = x^2 + 4x$

26. Evaluate: $-x - ab(x - b^0)$ if $x = -3, a = 4, b = -5$

27. Graph on a number line: $-4 - |x| \le -4; D = \{\text{Reals}\}$

28. Multiply: $\dfrac{x^{-2}a^2}{y}\left(\dfrac{ya^{-2}}{x^{-2}} - \dfrac{3x^{-2}a^2}{y}\right)$

29. Round off $478,325.06\overline{3}$ to the nearest thousand.   **ER 116-50**

LESSON 117   *More on multiplication of radical expressions*

117.A **multiplication of radical expressions**

Thus far, our most advanced radical multiplication problem has been of the form

$$\sqrt{3}(5 + \sqrt{12})$$

This notation indicates that $\sqrt{3}$ is to be multiplied by both of the terms inside the parentheses and that the products are to be added.

$$\sqrt{3}(5 + \sqrt{12}) = 5\sqrt{3} + \sqrt{36} = \mathbf{5\sqrt{3} + 6}$$

When we have a problem of the form

$$(4 + \sqrt{3})(5 + \sqrt{12})$$

we have four multiplications indicated. Both 4 and $\sqrt{3}$ must be multiplied by each term in the second parentheses and the four products simplified as shown here.

$$(4 + \sqrt{3})(5 + \sqrt{12}) = 4 \cdot 5 + 4\sqrt{12} + 5\sqrt{3} + \sqrt{3} \cdot \sqrt{12}$$
$$= 20 + 8\sqrt{3} + 5\sqrt{3} + 6$$
$$= \mathbf{26 + 13\sqrt{3}}$$

example 117.1   Multiply $(2 + \sqrt{2})(3 + \sqrt{8})$.

solution   We will multiply 2 and $\sqrt{2}$ by both numbers in the second parentheses and simplify the result.

$$2 \cdot 3 + 2\sqrt{8} + 3\sqrt{2} + \sqrt{2} \cdot \sqrt{8} = 6 + 4\sqrt{2} + 3\sqrt{2} + 4$$
$$= \mathbf{10 + 7\sqrt{2}}$$

example 117.2   Multiply $(4 + \sqrt{5})(2 - 2\sqrt{5})$.

solution   We have four multiplications to perform, which are

$$4(2) + 4(-2\sqrt{5}) + \sqrt{5}(2) + \sqrt{5}(-2\sqrt{5})$$

Now we multiply and simplify the results.

$$8 - 8\sqrt{5} + 2\sqrt{5} - 10 = -2 - 6\sqrt{5}$$

example 121.3   Multiply $(2 + \sqrt{2})(3 + 2\sqrt{2})$.

solution   We perform the multiplications and then simplify by adding like terms.

$$6 + 4\sqrt{2} + 3\sqrt{2} + 4 = 10 + 7\sqrt{2}$$

problem set 117

1. Of the 1200 displays at the flower show, 300 had at least 1 rose. What percent of the displays did not contain any roses?

2. The northbound bus had been on the road at 50 miles per hour for 4 hours before the southbound bus left the same station at 45 miles per hour. How long was the southbound bus on the road before the buses were 580 miles apart?

3. Judy walked into the country at 5 kilometers per hour and then rode a bus back home at 30 kilometers per hour. It was a long trip as she was gone 21 hours. How far did she walk?

4. The ratio of haves to have-nots was 11 to 2. If the king heard requests for boons from 195 subjects, how many of these subjects were haves?

5. Maxine bought 176 stamps for $10.75. She bought some 5-cent stamps and some 20-cent stamps. How many of each kind did she buy?

6. Is the following set of ordered pairs a function: $(4, -2), (3, -5), (8, -11), (-2, 4),$ $(-5, 3), (-11, -5)$?

7. Find the equation of the line that passes through the point $(-2, -3)$ that is parallel to the line $y = -\dfrac{2}{3}x + 5$.

8. Find the equations of lines (a) and (b).

9. Given that $g(x) = x + 4, D = \{\text{Integers}\}$, find $g\left(\dfrac{1}{2}\right)$.

Solve:

10. $2\sqrt{x} - 4 = 3$ 11. $\sqrt{x + 5} - 3 = 2$

Graph:

12. $4 \le x - 3 < 6, D = \{\text{Integers}\}$

13. $x + 2 < 5$ or $x + 2 \ge 6, D = \{\text{Reals}\}$

Multiply:

*14. $(4 + \sqrt{3})(5 + \sqrt{12})$ ***15.** $(4 + \sqrt{5})(2 - 2\sqrt{5})$

*16. $(2 + \sqrt{2})(3 + 2\sqrt{2})$ **17.** $\sqrt{3}(2 - 3\sqrt{12})$

18. Solve: $\dfrac{k - 5}{k} = \dfrac{1}{5k} - \dfrac{1}{5}$ **19.** $\dfrac{a}{b} + \dfrac{x}{m} - \dfrac{1}{c} = p$; find m

20. Solve by graphing: $\begin{cases} y = -x + 3 \\ y = -2 \end{cases}$

21. Simplify: $\dfrac{\dfrac{xy}{a^2} - \dfrac{1}{a}}{\dfrac{1}{a} + \dfrac{x^2 y}{a^2}}$

22. Divide: $x^3 - 2x - 4$ by $x + 2$

23. What fraction of $3\dfrac{1}{7}$ is $\dfrac{3}{8}$?

24. $-3\sqrt{2} \in \{$What sets of numbers$\}$?

25. Simplify: $-[(-3 - 2)(-2) - 3^0(-3)] - [(-2^0 - 3^2) - (-2 - 4)]$

26. Evaluate: $-y^2 - y^0 - y(xy - y)$ if $x = 4$ and $y = -3$

27. Solve: $3\dfrac{1}{8}m + 2\dfrac{1}{4} = \dfrac{3}{8}$ **28.** Add: $\dfrac{5x + 2}{x - 3} - \dfrac{2x + 2}{x^2 - 9}$

29. Round off $.00\overline{371}$ to the nearest billionth.

Graph on a number line:

30. $-2 < x \le 2$; $D = \{$Reals$\}$

31. $-|x| + 2 \ge -1$; $D = \{$Positive integers$\}$ **ER 117-51**

LESSON 118 *Direct variation*

118.A
direct variation

In every classroom of a particular school, there are twice as many boys as there are girls. We can express this relationship mathematically by writing the equation

$$B = 2G$$

Then we can use the equation to find the number of boys in any classroom if we are told the number of girls in that room, or to find the number of girls in a room if we are told the number of boys in that room.

In another school there are three times as many boys in each room as there are girls. For this school, the relationship can be expressed mathematically by writing

$$B = 3G$$

In the equation for the first school we had the variables B and G and the constant was the number 2. This last equation is the same equation except that the constant is the number 3. The general form of the relationship is

$$B = kG$$

where k represents the constant for any particular school.

We call a relationship such as this, where one variable is expressed as a constant times another variable, a direct variation or a direct proportion and the constant in the equation is called the constant of proportionality.

If we consider a third school and are told that

1. The number of boys in any room **varies directly** as the number of girls in the room
   or
2. That the number of boys in any room **is directly proportional to** the number of girls in the room,

we have been told that the relationship between the number of boys and girls in any room in the school may be stated mathematically as

$$B = kG$$

Before we can solve any problem, we need to know the constant of proportionality for this school. If we are told the number of boys and girls in any room in the school, we can use these values in the equation to solve for k. If there are 30 boys and 5 girls in one room and we use these values for B and G in the equation, we find

$$30 = k(5) \qquad \text{and thus} \qquad k = 6$$

Since we have found the constant for this school, we may write the relationship for this school as

$$B = 6G$$

Now, if we are given the number of girls in any room in this school, we can solve for the number of boys and conversely can find the number of girls if we are given the number of boys.

The key to this type of problem is recognizing the verbal statement that denotes a direct variation or a direct proportion and realizing the equation that is implied by this statement. We will give some examples here. On the left is the key verbal statement and on the right is the equation that is implied by the statement.

STATEMENT	EQUATION
The weight of a substance **varies directly** as the volume of the substance.	$W = kV$
Force **varies directly** as the current.	$F = kC$
The distance traveled **varies directly** as the time.	$D = kT$

The words **directly proportional** imply the same equation as do the words **varies directly**.

STATEMENT	EQUATION
The circumference of a circle is **directly proportional** to the length of the radius.	$C = kR$
The volume of a right circular cylinder of fixed radius is **directly proportional** to its height.	$V = kH$

In each of the relationships just discussed, the equation contained an unknown **constant of proportionality *k*. We begin the solution of any direct variation problem by finding the constant of proportionality for that problem.** Then we can solve for the value of one unknown if we are given the value of the other unknown.

example 118.1 The mass of a substance varies directly as the volume of the substance. If the mass of 2 liters of the substance is 10 kilograms, what will be the volume of 35 kilograms of the substance?

solution We will solve the problem in *four steps*.

Step 1: Recognize that the words **varies directly** imply the relationship

$$M = kV$$

Step 2: Reread the problem to find the values of M and V that can be used to find the value of k. Use these values to solve for k.

$$(10) = k(2) \longrightarrow k = 5$$

Step 3: Replace k in the equation with the value we have found

$$M = 5V$$

Step 4: Reread the problem to find that we are asked to find the value of V if M is 35. We replace M in the equation with 35 and solve for V by dividing by 5.

$$35 = 5V \longrightarrow \frac{35}{5} = \frac{5V}{5} \longrightarrow \textbf{7 liters} = V$$

example 118.2 The distance traversed by a car traveling at a constant speed is directly proportional to the time spent traveling. If the car goes 75 kilometers in 5 hours, how far will it go in 7 hours?

solution We will use the same four steps.

Step 1:	$D = kT$	write the equation
Step 2:	$75 = k(5) \longrightarrow k = 15$	solve for k
Step 3:	$D = 15T$	put k in the equation
Step 4:	$D = (15)(7) \longrightarrow \textbf{D = 105 kilometers}$	solve for D

example 118.3 Under certain conditions the pressure of a gas varies directly as the temperature. When the pressure is 800 torr, the temperature is 400°K. What is the temperature when the pressure is 400 torr?

solution Again we will use four steps in our solution.

Step 1:	$P = kT$	write the equation
Step 2:	$800 = k(400) \longrightarrow k = 2$	solve for k
Step 3:	$P = 2T$	put k in the equation
Step 4:	$400 = 2T \longrightarrow \textbf{T = 200°K}$	solve for T

problem set 118

1. Wormley ran to the store at 8 kilometers per hour and rode a bus home at 20 kilometers per hour. If his round trip traveling time was 7 hours, how far was it to the store?

*2. The mass of a substance varies directly as the volume of the substance. If the mass of 2 liters of the substance is 10 kilograms, what will be the volume of 35 kilograms of the substance?

*3. Under certain conditions the pressure of a gas varies directly as the temperature. When the pressure is 800 torr, the temperature is 400°K. What is the temperature when the pressure is 400 torr?

4. The red carnations sold for 50 cents a bunch and the white ones for 40 cents a bunch. Mike bought 27 bunches for $12.30. How many of each kind did he get?

*5. The distance traversed by a car traveling at a constant speed is directly proportional to the time spent traveling. If the car goes 75 kilometers in 5 hours, how far will it go in 7 hours?

6. Find the equation of the line that passes through $(-2, -5)$ that is parallel to $y = 2x + 4$.

7. Which of the following sets of ordered pairs are functions?

    (a) $(4, -3), (5, -3), (6, -3)$ (b) $(4, -3), (-3, 4), (-2, 4)$
    (c) $(4, -3), (4, 3), (-4, 6)$ (d) $(4, -3), (-2, -4), (8, 3)$

Solve:

8. $2\sqrt{x} + 2 = 5$ 9. $\sqrt{x - 4} - 2 = 6$

10. Graph: $-2 \le x + 5 < 3$; $D = \{\text{Reals}\}$

Multiply:

11. $(2 + \sqrt{3})(4 - 5\sqrt{12})$ 12. $(2 + \sqrt{2})(4 - 3\sqrt{8})$ 13. $(5 + \sqrt{6})(2 - 3\sqrt{24})$

Solve:

14. $\dfrac{x}{4} - \dfrac{x - 3}{2} = \dfrac{1}{5}$ 15. $\dfrac{m - 2}{m} = \dfrac{1}{2m} - \dfrac{1}{3}$ 16. $\dfrac{x}{a} - \dfrac{1}{k} = \dfrac{m}{c}$; find a

17. Use the square root tables to simplify: $\sqrt{.000416852 \times 10^{-13}}$

18. Find the distance between $(-5, -2)$ and $(3, 7)$.

19. Solve: $3x^0 - 2(-x - 5) - (-3) + 2(x - 5) = -2(x + 3^0)$

Simplify:

20. $\dfrac{(35{,}000 \times 10^{-41})(700 \times 10^{14})}{(7000 \times 10^{21})(.00005 \times 10^{15})}$ 21. $\dfrac{(x^{-2})^{-3}(x^{-2}y^2)}{x^2yy^0(x^0y)^{-2}}$

22. $(3 + 3\sqrt{2})(4 - \sqrt{2})$ 23. $2\sqrt{5}(\sqrt{5} - 2\sqrt{75})$

24. $\dfrac{2x^2yz - 2x^2yz^2}{2x^2yz}$ 25. $\dfrac{3^{-2}}{-2^{-3}}$

26. Solve by factoring: $-8 = -x^2 + 7x$

27. Evaluate: $-x^2 - x^0 - x^3 + xy(y - xy)$ if $x = -3$ and $y = 2$

28. Graph on a number line: $-3 - |-x| \ge -2$; $D = \{\text{Integers}\}$

29. Multiply: $\dfrac{a^{-2}x}{y}\left(\dfrac{ya^2}{x} - \dfrac{4x^2y}{a^2}\right)$

30. Round off $4.0\overline{60}$ to the nearest ten-thousandth.

LESSON *119* *Inverse variation*

When the problem states that one variable **varies inversely** as the other variable or that the value of one variable is **inversely proportional** to the value of the other variable, an equation of the form

$$V = \frac{k}{W}$$

is implied where k is the constant of proportionality and V and W are the two variables. If we look at the equations for direct variation and inverse variation

DIRECT VARIATION EQUATION   INVERSE VARIATION EQUATION

$$V = kW \qquad\qquad\qquad V = \frac{k}{W}$$

we see that each equation contains two variables and one constant of proportionality k. In both equations, the constant k is in the numerator! In a direct variation equation, both variables are in the numerator; and in an inverse variation equation, one variable is in the numerator and the other variable is in the denominator!

STATEMENT	EQUATION
The pressure of a perfect gas **varies inversely** as the volume.	$P = \dfrac{k}{V}$
The current **varies inversely** as the resistance.	$C = \dfrac{k}{R}$
The velocity is **inversely proportional** to the time.	$V = \dfrac{k}{T}$

Inverse variation problems are solved in the same way as the direct variation problems. First we recognize the equation implied by the statement of inverse variation. Then we find the constant of proportionality for this problem and use this constant in the equation to find the final solution.

example 119.1 Under certain conditions, the pressure of a perfect gas varies inversely as the volume. When the pressure of a quantity of gas is 7 torr, the volume is 75 liters. What would be the volume if the pressure is increased to 15 torr?

solution We will use the same four steps that we used for direct variation problems.

Step 1: $P = \dfrac{k}{V}$ write the equation

Step 2: $7 = \dfrac{k}{75} \longrightarrow k = 525$ solve for k

Step 3: $P = \dfrac{525}{V}$ substitute 525 for k

Step 4: $15 = \dfrac{525}{V} \longrightarrow V = \dfrac{525}{15}$ substitute 15 for P and solve

$\longrightarrow V = \textbf{35 liters}$

example 119.2 To travel a fixed distance, the rate is inversely proportional to the time required. When the rate is 60 kilometers per hour, the time required is 4 hours. What would be the time required for the same distance if the rate were increased to 80 kilometers per hour?

solution We will use the same four steps.

Step 1: $R = \dfrac{k}{T}$ write the equation

Step 2: $60 = \dfrac{k}{4} \longrightarrow k = 240$ solve for k

Step 3: $R = \dfrac{240}{T}$ substitute 240 for k

Step 4: $80 = \dfrac{240}{T} \longrightarrow T = \dfrac{240}{80} \longrightarrow T = \textbf{3 hours}$

problem set 119

*1. Under certain conditions, the pressure of a perfect gas varies inversely as the volume. When the pressure of a quantity of gas is 7 torr, the volume is 75 liters. What would be the volume if the pressure is increased to 15 torr?

*2. To travel a fixed distance, the rate is inversely proportional to the time required. When the rate is 60 kilometers per hour, the time required is 4 hours. What would be the time required for the same distance if the rate were increased to 80 kilometers per hour?

3. The number of girls in a class varied directly as the number of boys. One class had 3 boys and 21 girls. If another class had 5 boys, how many girls were in this class?

4. Peaches varied directly as apples. When there were 40 peaches, there were 120 apples. How many apples went with 500 peaches?

5. Silent Steve planted his farm acreage in cotton and peanuts in the ratio of 4 to 5. If his farm had 1278 acres, how many acres were in cotton?

6. Find the equation of the line that passes through $(-2, -3)$ and has a slope of $-\dfrac{1}{5}$.

7. Which of the following diagrams depict functions?

(a)   (b)   (c)   (d)

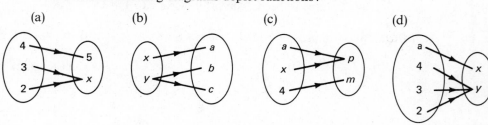

Solve:

8. $\sqrt{x} - 4 = 2$ **9.** $\sqrt{x - 5} - 3 = 2$

10. Graph: $x + 2 \geq 5$ or $x + 3 \leq 0$, $D = \{$Integers$\}$

Multiply:

11. $(4 - 3\sqrt{2})(2 + 6\sqrt{2})$ **12.** $(2 - \sqrt{5})(3 - 2\sqrt{5})$

13. $(3 - \sqrt{3})(2 - 2\sqrt{3})$

Solve:

14. $\dfrac{x}{4} - \dfrac{x - 5}{7} = 2$ **15.** $\dfrac{m + 5}{m} = \dfrac{3}{2m} - \dfrac{2}{5}$

16. $\dfrac{x}{y} - m + \dfrac{1}{c} = k$; find y

17. Find k.

18. Simplify: $\dfrac{x^5 - 5x^4}{x^2 - 25} \div \dfrac{x^4 + 4x^3 - 32x^2}{x^2 + x - 20}$

19. Solve by graphing: $\begin{cases} y = \dfrac{1}{2}x - 2 \\ x = -4 \end{cases}$

20. Find the equations of lines (a) and (b).

21. Simplify: $\dfrac{a + \dfrac{a}{x}}{\dfrac{1}{x} + a^2}$ **22.** Divide: $x^3 - 2$ by $x^2 - 1$.

23. Simplify: $-2^0(-3^0 - 5)[(7 - 2^2)(-3 - 4^0) - (-2)][-(-2)]$

Graph on a number line:

24. $+1 - |x| \leq -1$; $D = \{$Positive integers$\}$

25. $-1 < x + 2 \leq 1$; $D = \{$Reals$\}$

26. Evaluate: $-p^0 - (p^0)^2 - p^2 - p^3 - p(p - y)$ if $p = -3$ and $y = 2$

27. Simplify by adding like terms: $x^2yz^{-1} + \dfrac{3y}{x^{-2}z} - \dfrac{4yx^2}{z} + 2y^2x^2z^{-1}$

28. Solve: $3\dfrac{1}{6}p + \dfrac{1}{4} = \dfrac{7}{8}$

29. Add: $\dfrac{3x - 2}{x - 3} - \dfrac{2x + 5}{x^2 - 9}$

LESSON *120* *Linear inequalities*

120.A

linear inequalities

Below we have graphed the line whose equation is $y = \frac{1}{2}x + 1$. We see that this line divides the set of all the points of the plane into three mutually exclusive subsets:

1. The set of points that lie on the line.
2. The set of points that lie above the line (region A).
3. The set of points that lie below the line (region B).

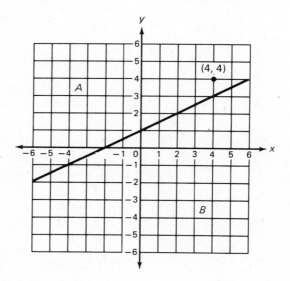

It can be shown that the coordinates of any point in the plane either will satisfy the equation of a given line or will satisfy one of the two linear inequalities that define the regions on either side of the line. For this particular line, we can say that the coordinates of any point in the plane will satisfy one and only one of the following:

$$y > \frac{1}{2}x + 1 \qquad y = \frac{1}{2}x + 1 \qquad y < \frac{1}{2}x + 1$$

The points in the regions denoted by A and B above do not lie on the line, and the coordinates of any point in region A or region B will satisfy one and only one of the inequalities. To see which of these inequalities defines the region above the line, we will choose a test point that clearly lies on one side of the line and test the coordinates of this point in both inequalities. We choose the point $(4, 4)$.

$$y > \frac{1}{2}x + 1 \qquad\qquad y < \frac{1}{2}x + 1$$

$$4 > \frac{1}{2}(4) + 1 \qquad\qquad 4 < \frac{1}{2}(4) + 1$$

$$4 > 2 + 1 \qquad\qquad 4 < 2 + 1 \qquad \text{False}$$

$$4 > 3 \qquad \text{True}$$

Thus the coordinates of all points above the line satisfy the inequality $y > \frac{1}{2}x + 1$. If we choose a point below the line in region B, we can show that all points in this region satisfy the other inequality, $y < \frac{1}{2}x + 1$. We choose the point $(0, 0)$.

$$y < \frac{1}{2}x + 1 \quad \longrightarrow \quad 0 < \frac{1}{2}(0) + 1 \quad \longrightarrow \quad 0 < 1 \quad \text{True}$$

We indicate in the next figure that the coordinates of all points above the line satisfy the inequality $y > \frac{1}{2}x + 1$ and that the coordinates of all points below the line satisfy the inequality $y < \frac{1}{2}x + 1$.

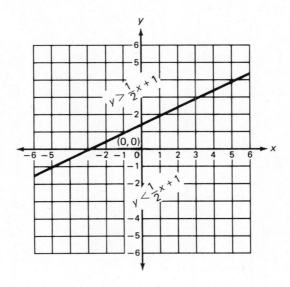

example 120.1   Graph the inequality $y > -x + 2$.

solution   We first graph the line $y = -x + 2$ in the figure on the left below. We show the line as a dotted line to indicate that the points on the line do not satisfy the stated inequality, which uses a *greater than* symbol. Had the inequality used an *equal to or greater than* symbol, the line would have been drawn as a solid line. We choose the point $(0, 0)$ as a test point because it clearly lies on one side of the line and also because it is the easiest test point to use.

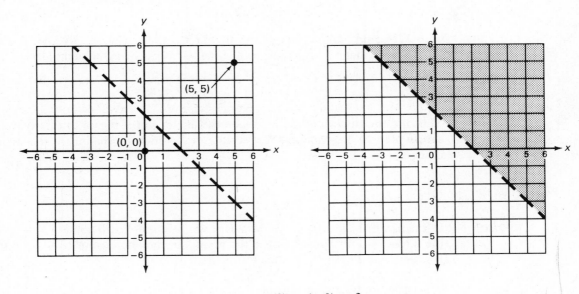

$$(0) > (-0) + 2$$

$$0 > 2 \quad \text{False}$$

The coordinates of this point do not satisfy the inequality so the coordinates of any point above the line must satisfy the inequality. We will use the point $(5, 5)$.

$$(5) > (-5) + 2$$

$$5 > -3 \qquad \text{True}$$

We indicate that the coordinates of all points above the line satisfy the given condition by shading the region above the line in the figure at the right.

example 120.2 Graph $y \le 2x - 1$.

solution In the figure at the left we have graphed the line $y = 2x - 1$. We show the line as a solid line because we wish to indicate that the points on the line satisfy the stated inequality because the inequality symbol here is read from left to right as "equal to or less than."

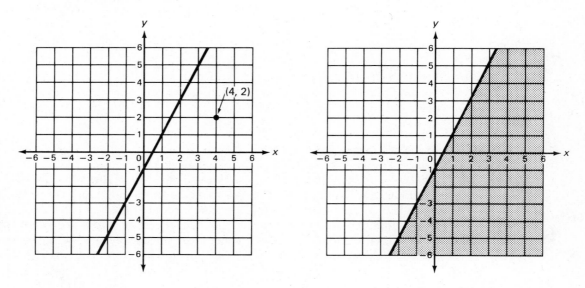

We could use the test point $(1, 0)$, but it is rather close to the line. To be sure we have a point that is well on one side of the line, we choose the point $(4, 2)$.

$$2 < 2(4) - 1$$

$$2 < 7 \qquad \text{True}$$

Thus the coordinates of the points on the line or in the region to the right of the line satisfy the stated inequality. We indicate the solution by shading in this region in the other figure.

example 120.3 Graph $y > -2$.

solution In the figure at the left on the next page we have graphed the line $y = -2$. We show the line as a dotted line because the points on the line do not satisfy the inequality $y < -2$. We choose $(0, 0)$ as the test point.

$$0 > -2 \qquad \text{True}$$

We shade the region above the line in the next figure to indicate that the coordinates of any point above the line will satisfy the stated inequality.

example 120.4   Graph $x < 3$.

solution   In the figure at the left we graph the equation $x = 3$. We draw the line as a dotted line because points on the line do not satisfy $x < 3$, the given inequality. We will use the point $(0, 0)$ as our test point.

$$0 < 3 \qquad \text{True}$$

We shade in the region to the left of the dotted line in the figure at the right to indicate that all points in this region will satisfy the inequality $x < 3$.

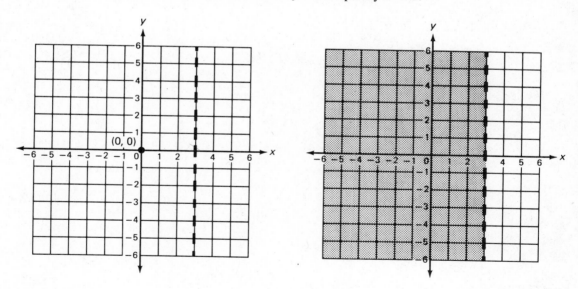

problem set 120

1.   Pressure varies inversely as the volume. When the pressure is 10 torr, the volume is 150 liters. What would the volume be if the pressure is reduced to 3 torr?

2.   Attractiveness varies inversely with the wiggles. At 300 wiggles the attractiveness is 10. What would the attractiveness be at 150 wiggles?

3.   Mickey and Yarberry saved nickels and dimes. They had a total of 34 coins whose value was $2.70. How many of each kind of coin did they have?

4. Rosie ran to the park at 7 miles per hour and walked back home at 3 miles per hour. How far was it to the park if the round trip took 20 hours?

*5. Graph $y > -x + 2$ on a rectangular coordinate system.

*6. Graph $x < 3$ on a rectangular coordinate system.

7. Given $p(x) = x^2 + 2x + 5$, $D = \{\text{Reals}\}$; find $p(-2)$.

Solve:

8. $-2\sqrt{x} + 4 = -1$

9. $2\sqrt{p + 2} - 4 = 3$

10. Graph on a number line: $4 \leq x + 3 < 7$; $D = \{\text{Integers}\}$

Multiply:

11. $(3 + 2\sqrt{2})(5 - 3\sqrt{2})$

12. $(4 + \sqrt{3})(2 - 4\sqrt{3})$

13. $(2 + \sqrt{8})(3 - 2\sqrt{2})$

Solve:

14. $\dfrac{2x}{5} - \dfrac{x + 4}{2} = 3$

15. $\dfrac{k - 3}{2k} = \dfrac{3}{6k} - \dfrac{1}{4}$

16. $\dfrac{x}{m} - \dfrac{c}{d} = d$; find m

17. Use the square root tables to simplify: $\sqrt{.0001234567 \times 10^{-15}}$

18. Find the distance between $(-4, -4)$ and $(3, 0)$.

19. Solve: $-(-3)x^0 - (-2)(x - 4) = -3(-x^0)$

Simplify:

20. $\dfrac{(21{,}000 \times 10^{-40})(5000 \times 10^{-20})}{(.00003 \times 10^{15})(.0007 \times 10^{28})}$

21. $\dfrac{4x^2y^{-2}(x^2)^{-2}y^2xy}{(2x^0)^2x^2y^{-2}(xy)}$

22. $3\sqrt{2} \cdot \sqrt{3} - 5\sqrt{24} + 3\sqrt{54}$

23. $3\sqrt{2}(\sqrt{2} - 4\sqrt{8})$

24. $\dfrac{xy - 4xy^2}{xy}$

25. $\dfrac{-2^{-4}}{(-2)^{-3}}$

26. Solve by factoring: $45 = x^2 - 4x$

27. Evaluate: $xy - a - ya(y - a)$ if $x = -2$, $y = -5$, $a = -1$

28. Graph on a number line: $4 \geq |x|$; $D = \{\text{Positive integers}\}$

29. Multiply: $\dfrac{3ax}{y}\left(\dfrac{y}{3ax} - \dfrac{a^{-1}x}{3y}\right)$

30. True or false? $4 \in \{\text{Naturals}\}$

LESSON 121 *Quotient theorem for square roots*

121.A
quotient
theorem for
square roots

The product theorem for square roots tells us that the square root of a product equals the product of the square roots.

$$\sqrt{3 \cdot 2} = \sqrt{3}\sqrt{2}$$

In a similar fashion, the quotient theorem for square roots tells us that the square root of a quotient (fraction) equals the quotient of the square roots.

$$\sqrt{\frac{3}{2}} = \frac{\sqrt{3}}{\sqrt{2}}$$

The expression on the right has the irrational number $\sqrt{2}$ in the denominator. We can change the denominator to the rational number 2 by multiplying both the numerator and the denominator by $\sqrt{2}$.

$$\frac{\sqrt{3}}{\sqrt{2}} = \frac{\sqrt{3}}{\sqrt{2}} \cdot \frac{\sqrt{2}}{\sqrt{2}} = \frac{\sqrt{6}}{2}$$

This process of changing a denominator to a rational number is called rationalizing the denominator. Many people prefer fractions with rational denominators. If we go along with their preference we can say that: **An expression containing square roots is in the simplified form when no square roots are in the denominator and no radicand has a factor that is a perfect square.**

example 121.1 Write $\sqrt{\dfrac{5}{3}}$ in simplified form.

solution We begin by writing the radical as a fraction of radicals,

$$\sqrt{\frac{5}{3}} = \frac{\sqrt{5}}{\sqrt{3}}$$

and we finish by multiplying both top and bottom by $\sqrt{3}$.

$$\frac{\sqrt{5}}{\sqrt{3}} = \frac{\sqrt{5}}{\sqrt{3}} \cdot \frac{\sqrt{3}}{\sqrt{3}} = \frac{\sqrt{15}}{3}$$

example 121.2 Simplify $\dfrac{4 + \sqrt{3}}{\sqrt{2}}$.

solution To simplify we must change the denominator to a rational number. Thus, we multiply by $\sqrt{2}$ over $\sqrt{2}$.

$$\frac{4 + \sqrt{3}}{\sqrt{2}} \cdot \frac{\sqrt{2}}{\sqrt{2}} = \frac{4\sqrt{2} + \sqrt{6}}{2}$$

This may appear to some as more complicated than the original expression, but this form is preferred by many people because no radical appears in the denominator.

example 121.3 Simplify $\dfrac{2 + \sqrt{15}}{\sqrt{5}}$.

solution We will multiply top and bottom by $\sqrt{5}$.

$$\frac{2 + \sqrt{15}}{\sqrt{5}} \cdot \frac{\sqrt{5}}{\sqrt{5}} = \frac{2\sqrt{5} + \sqrt{75}}{5} = \frac{2\sqrt{5} + 5\sqrt{3}}{5}$$

problem set 121

1. RPM (revolutions per minute) varies inversely as the number of gear teeth. If 100 teeth result in 10 RPM, what RPM would result if the number of teeth were reduced to 25?

2. Hannibal rode the elephant to the outskirts of Rome at 2 kilometers per hour and then took a chariot back to camp at 10 kilometers per hour. If the total trip took 18 hours, how far was it from camp to the outskirts of Rome?

3. Bustles just weren't moving so Julie marked them down 40 percent so that the sale price would be $3.60 each. What was the price before the sale?

4. The cheap dresses were $15 and the more expensive ones were $50. On Saturday the store took in $550 and sold 15 more cheap dresses than expensive dresses. How much dresses of each type did the store sell?

5. The ratio of acres planted in peanuts to the acres planted in cotton was 6 to 5. If 2640 acres were planted, how many were planted in peanuts?

6. Find the equations of lines (a) and (b).

7. Which of the following sets of ordered pairs are functions?
    (a) $(-5, -3), (5, 3), (-3, 5)$ (b) $(-5, -3), (-5, 3), (-3, 5)$
    (c) $(-5, -3), (5, -3), (6, -3)$ (d) $(4, -2), (-4, 2), (-2, 4)$

8. Solve: $\sqrt{x - 3} - 2 = 5$

9. Graph on a number line: $x + 2 > 6$ or $x - 3 \le -6$; $D = \{\text{Reals}\}$

Multiply:

10. $(3 + 2\sqrt{2})(2 - 4\sqrt{2})$ 11. $(2 + 3\sqrt{3})(2 - \sqrt{3})$ 12. $(3 + \sqrt{5})(2 - 4\sqrt{5})$

Simplify:

*13. $\sqrt{\dfrac{5}{3}}$ 14. $\sqrt{\dfrac{2}{5}}$ 15. $\sqrt{\dfrac{3}{7}}$

*16. $\dfrac{4 + \sqrt{3}}{\sqrt{2}}$ *17. $\dfrac{2 + \sqrt{15}}{\sqrt{5}}$ 18. $\dfrac{2 + 3\sqrt{6}}{\sqrt{2}}$

Solve:

19. $\dfrac{m + 2}{2m} = \dfrac{7}{3m} + \dfrac{2}{5}$ 20. $\dfrac{a}{x} - \dfrac{m}{c} + b = k$; find c

21. Simplify: $\dfrac{\dfrac{ax^2}{y} - \dfrac{x}{y^2}}{\dfrac{p}{y^2} - \dfrac{3k}{y}}$ 22. Divide: $4x^3 - 2x^2 + 4$ by $x + 2$

23. Simplify: $-2[(-2^0 - 2^2)(-2^3 - 2) + (-2)][-(-3)(-2^2)]$

24. Evaluate: $-xy - y^2 - y^0(x - y)$ if $x = 3$ and $y = -2$

25. Solve: $3\dfrac{1}{5}k + \dfrac{2}{3} = \dfrac{1}{9}$

26. Find the equation of the line through $(5, -2)$ that is parallel to $y = -3x + 2$.

27. Find the equation of the line that goes through $(4, -3)$ and $(-5, 2)$.

28.   Find the equation of the line whose slope is -5 that passes through the point $(-4, -3)$.

29.   Graph $y \leq -x - 3$ on a rectangular coordinate system.

LESSON 122   *Advanced trinomial factoring*

122.A

**advanced
trinomial
factoring**

Thus far, we have restricted our trinomial factoring to trinomials such as

$$x^2 - x - 6$$

in which the coefficient of the x^2 term is 1 and to trinomials that can be reduced to this form by factoring a common factor. Trinomials whose leading coefficient is not 1 can be formed by multiplying binomials as we see here.

$$
\begin{array}{rr}
2x & + 3 \\
x & - 5 \\
\hline
2x^2 + & 3x \\
 & - 10x - 15 \\
\hline
2x^2 - & 7x - 15
\end{array}
\qquad
\begin{array}{rr}
3x & + 2 \\
2x & - 3 \\
\hline
6x^2 + & 4x \\
 & - 9x - 6 \\
\hline
6x^2 - & 5x - 6
\end{array}
$$

We note that the first term of the trinomial is the product of the first two terms of the binomials and that the last term of the trinomial is the product of the last terms of the binomials, **but alas the coefficient of the middle term of the trinomial is not the sum of the last two terms of the binomials.** The middle term is the sum of the product of the first term of the first binomial and the last term of the second binomial and the product of the last term of the first binomial and the first term of the second binomial. It is easier to see this if we write the original indicated multiplication in horizontal form and note that the middle term is the sum of the products of the means and the extremes.[†]

$$
\overset{\displaystyle\text{extremes}}{(2x + 3)(\underset{\text{means}}{x - 5})} = 2x^2 - 7x - 15
\qquad
\overset{\displaystyle\text{extremes}}{(3x + 2)(\underset{\text{means}}{2x - 3})} = 6x^2 - 5x - 6
$$

$$
\begin{array}{ll}
\text{product of means} & = 3x \\
\text{product of extremes} & = -10x \\
\text{sum} & = {-7x}
\end{array}
\qquad
\begin{array}{ll}
\text{product of means} & = 4x \\
\text{product of extremes} & = -9x \\
\text{sum} & = {-5x}
\end{array}
$$

example 122.1   Factor $-7x - 15 + 2x^2$.

solution   We begin by writing the trinomial in descending powers of the variable

$$2x^2 - 7x - 15$$

Now, to factor $2x^2 - 7x - 15$, we note that the product of the first terms of the binomials is $2x^2$; the product of the last terms of the binomials is -15; and the middle term is the

---

[†] Mathematicians sometimes use the word *mean* to mean middle and the word *extreme* to mean end. Thus the mean terms in the multiplication shown are the middle terms, and the extreme terms are the end terms.

sum of the products of the means and extremes. Since the term $2x^2$ is the product of the first terms of the binomial we write

$$(2x \quad)(x \quad)$$

Now, the four pairs of integral factors of -15 are (a) $+15$ and -1, (b) -15 and $+1$, (c) $+5$ and -3, and (d) -5 and $+3$. Now we must try each pair **twice** and see what middle term will result in each case.

for $(15, -1)$	$(2x + 15)(x - 1)$	middle term is $13x$
	$(2x - 1)(x + 15)$	middle term is $29x$
for $(5, -3)$	$(2x + 5)(x - 3)$	middle term is $-x$
	$(2x - 3)(x + 5)$	middle term is $7x$
for $(-15, 1)$	$(2x - 15)(x + 1)$	middle term is $-13x$
	$(2x + 1)(x - 15)$	middle term is $-29x$
for $(-5, 3)$	$(2x - 5)(x + 3)$	middle term is x
	$(2x + 3)(x - 5)$	middle term is $-7x$

We will use the last entry because the sum of the products of the means and the extremes is $-7x$. Thus we see that $2x^2 - 7x - 15$ can be factored over the integers as $(2x + 3)(x - 5)$.

example 122.2   Factor $3x^2 - x - 2$.

solution   We begin by writing as follows.

$$(3x \quad)(x \quad)$$

The second terms of the binomials must have a product of -2. The two pairs of integral factors of -2 are (a) -2 and $+1$, and (b) $+2$ and -1. We will try each pair *twice* and check to see what middle term results.

$$(3x + 1)(x - 2) \qquad \text{middle term is } -5x$$
$$(3x - 2)(x + 1) \qquad \text{middle term is } +x$$
$$(3x - 1)(x + 2) \qquad \text{middle term is } +5x$$
$$(3x + 2)(x - 1) \qquad \text{middle term is } -x$$

The sum of the products of the means and extremes of the last multiplication gives us a middle term of $-x$. Thus, these are the desired factors.

$$3x^2 - x - 2 = (3x + 2)(x - 1)$$

example 122.3   Factor $5x^2 - 13x - 6$.

solution   To begin we write as follows:

$$(5x \quad)(x \quad)$$

The pairs of integral factors of -6 are (a) $+1$ and -6, (b) -1 and $+6$, (c) $+3$ and -2, and (d) $+2$ and -3. We try each pair twice if necessary to find that the pair we need are $+2$ and -3 because

$$(5x + 2)(x - 3) = 5x^2 - 13x - 6$$

problem set 122

1. Up to a point, the yield varied directly as the amount of fertilizer used. If 500 pounds of fertilizer resulted in 2000 tons of produce, how much produce would be harvested if only 400 pounds of fertilizer were used?

2. The ratio of boys to girls in every class in the school was 7 to 5. If there were 2160 students in the school, how many were boys and how many were girls?

Graph the solution to the following linear inequalities:

3. $y \geq x + 2$ 4. $y > x$ 5. $y < -1$

Factor the following trinomials. Always begin by writing the trinomial in descending powers of the variable.

*6. $-7x - 15 + 2x^2$ *7. $3x^2 - x - 2$ *8. $5x^2 - 13x - 6$

9. $3x^2 - 14x - 5$ 10. $2x^2 + 8 + 10x$ 11. $18 - 15x + 2x^2$

12. $-15 + 7x + 2x^2$ 13. $8x - 24 + 2x^2$ 14. $2x^2 - 24 - 8x$

15. $2x^2 - 6x + 4$ 16. $2x^2 - 18 + 9x$ 17. $2x^2 + 4 + 6x$

18. $3x^2 - 7 - 20x$ 19. $3x^2 - 8 - 23x$ 20. $4 - 7x + 3x^2$

Multiply:

21. $(4\sqrt{2} + 2)(3\sqrt{2} - 4)$ 22. $(2\sqrt{3} + 4)(5\sqrt{6} - 2)$

Simplify:

23. $\sqrt{\dfrac{3}{7}}$ 24. $\sqrt{\dfrac{5}{8}}$ 25. $\dfrac{2 + \sqrt{3}}{\sqrt{5}}$ 26. $\dfrac{4 + \sqrt{2}}{\sqrt{3}}$

27. Graph on a number line: $+3 - |x| \geq 1$; $D = \{\text{Reals}\}$

LESSON 123 *Factoring by grouping*

123.A

factoring by grouping

Sometimes it is necessary to factor polynomials that are the product of two binomials such as

$$(x - 4)(a + b)$$

When we multiply these binomials and look at the product

$$(x - 4)(a + b) = xa + xb - 4a - 4b$$

we see that it consists of four terms. Thus again, we encounter a factorable polynomial that can be factored only if we recognize the form.

example 123.1   Factor $xya - 4a + xyb - 4b$.

solution   We recognize the four-term format and note that the first two terms have a as a factor and that the last two terms have b as a factor. We begin by using parentheses to **group** these terms.

$$(xya - 4a) + (xyb - 4b)$$

Now we factor a from the first group and b from the second.

$$a(xy - 4) + b(xy - 4)$$

Lastly, we recognize that both of these terms have $xy - 4$ as a factor, so we factor this expression.

$$(xy - 4)(a + b)$$

example 123.2   Factor $ac + 2ad + 2bc + 4bd$.

solution   We recognize the form and note that the first and third terms have a common factor of c, and that the second and fourth terms have a common factor of $2d$; so we rearrange the terms and use parentheses as follows:

$$(ac + 2bc) + (4bd + 2ad)$$

Now we factor these terms as

$$c(a + 2b) + 2d(2b + a)$$

and complete the problem by factoring $2b + a$. The final result is

$$(a + 2b)(c + 2d)$$

There is no set procedure for rearranging the terms so that the expression can be factored. Adeptness comes after much practice and frustration.

problem set 123

1. In a particular experiment, the pressure varied inversely as the volume. When the pressure was 15 pounds per square inch, the volume was 20 liters. What was the pressure when the volume was reduced to 10 liters?

2. In another experiment, the pressure varied directly as the temperature. When the pressure was 1000 pounds per square inch, the temperature was 250 degrees. What was the pressure when the temperature was 1000 degrees?

3. The ratio of the pigs to chickens in the barnyard was 2 to 11. If there were 169 chickens and pigs in the barnyard, how many were chickens and how many were pigs?

4. David and Wade had a race. David started out at 9 a.m. at 40 miles per hour. Wade waited until 11 a.m. to start and he drove at 70 miles per hour. What time was it when Wade got 10 miles in front?

5. Calvin and Tooley hoarded nickels and pennies. They had 280 coins whose value was $6.80. How many nickels did they have?

6. Graph $y < 2x + 2$ on a rectangular coordinate system.

Factor the following trinomials. Begin by writing the terms in descending order of the variable.

7. $-14 + 3x^2 - 19x$ 8. $-14 + 19x + 3x^2$ 9. $2x^2 - 15 + 7x$

10. $2x^2 - 18 + 9x$ 11. $3x^2 + 14 + 23x$ 12. $2x^2 - 17x + 21$

13. $3x^2 + 16 - 26x$ 14. $18 + 15x + 2x^2$ 15. $3x^2 + 13x + 14$

Factor the following expressions by grouping:

*16. $xya - 4a + xyb - 4b$ *17. $ac + 2ad + 2bc + 4bd$

18. $ac - ad + bc - bd$ 19. $ab + 4a + 2b + 8$

20. $ab + ac + xb + xc$ 21. $2mx - 3m + 2pcx - 3pc$

22. $4k - kxy + 4pc - pcxy$ 23. $ac - axy + dc - dxy$

Simplify:

24. $\sqrt{\dfrac{3}{11}}$ 25. $\dfrac{2\sqrt{3} + 2}{\sqrt{5}}$

26.   Graph on a number line: $-2 < x + 2 \le 3$; $D = \{$Positive integers$\}$

LESSON 124   *Direct and inverse variation squared*

124.A
variation

Direct and inverse variation statements are not always simple statements of direct or inverse variation. Often one variable will vary as the other variable squared or the other variable cubed.

STATEMENT	IMPLIED EQUATION
The weight of a body varies inversely with the square of the distance to the center of the earth.	$W = \dfrac{k}{D^2}$
The distance required to stop is directly proportional to the square of the velocity.	$D = kV^2$
The price of a diamond varies directly as the square of its weight.	$P = kW^2$
The strength of the field is inversely proportional to the cube of the radius.	$S = \dfrac{k}{R^3}$

These problems are solved in the same way that simple variation problems are solved. First we recognize the statement of the implied variation and write down the indicated equation. Then we find k, insert its value in the equation, and solve for the required unknown.

example 124.1   The distance required for an automobile to stop is directly proportional to the square of its velocity. If a car can stop in 200 meters at 20 kilometers per hour, what will be the required distance at 28 kilometers per hour?

solution
Step 1 $D = kV^2$ write equation

Step 2 $200 = k(20)^2 \longrightarrow k = .5$ find k

Step 3 $D = .5V^2$ put k in equation

Step 4 $D = .5(28)^2 \longrightarrow$ **$D = 392$ meters**   solve for D

example 124.2   The distance a body falls varies directly as the square of the time that it falls. If it falls 144 feet in 3 seconds, how far will it fall in 10 seconds?

solution Step 1 $D = kt^2$ write equation

Step 2 $144 = k(3)^2 \longrightarrow k = 16$ find k

Step 3 $D = 16t^2$ put k in the equation

Step 4 $D = 16(10)^2 \longrightarrow \mathbf{D = 1600\ feet}$ solve for D

example 124.3 The weight of a body on or above the surface of the earth varies inversely with the square of the distance from the body to the center of the earth. If a body weighs 10,000 pounds at a distance of 5000 miles from the center of the earth, how much would it weigh 50,000 miles from the center of the earth?

solution Step 1 $W = \dfrac{k}{D^2}$ write equation

Step 2 $10,000 = \dfrac{k}{(5000)^2} \longrightarrow 10,000(5000)^2 = k$ find k

$\longrightarrow 25 \times 10^{10} = k$

Step 3 $W = \dfrac{25 \times 10^{10}}{D^2}$ put k in the equation

Step 4 $W = \dfrac{25 \times 10^{10}}{(50,000)^2} \longrightarrow W = \dfrac{25 \times 10^{10}}{25 \times 10^8}$ solve for W

$\longrightarrow \mathbf{W = 100\ pounds}$

problem set 124

*1. The distance required for an automobile to stop is directly proportional to the square of its velocity. If a car can stop in 200 meters from a velocity of 20 kilometers per hour, what will be the required distance at 28 kilometers per hour?

*2. The distance a body falls varies directly as the square of the time that it falls. If it falls 144 feet in 3 seconds, how far will it fall in 10 seconds?

3. Greens vary directly as purples squared. When there were 4 greens, there were 2 purples. How many greens would be present if there were 6 purples?

4. The reds varied inversely as the square of the blues. When there were 4 reds, there were 20 blues. How many reds would there be if there were only 4 blues?

5. The ratio of rabbits to squirrels in the forest was 7 to 5. If there were 16,800 animals total, how many were rabbits and how many were squirrels?

6. Graph $y \geq \dfrac{1}{2}x - 2$ on a rectangular coordinate system.

7. $\dfrac{a}{b} + \dfrac{c}{x} = m$; find x

Factor the following trinomials. Begin by writing the terms in descending order of the variable.

8. $3x^2 + 25x - 18$ 9. $3x^2 - 4 - x$ 10. $2x^2 - 6 - 4x$

11. $3x^2 + 28x - 20$ 12. $2x^2 + 15x + 25$ 13. $2x^2 - 5x - 25$

Factor by grouping:

14. $ab + 15 + 5a + 3b$ 15. $ay + xy + ac + xc$

16. $3mx - 2p + 3px - 2m$ 17. $kx - 15 - 5k + 3x$

18. $xpc + pc^2 + 4x + 4c$ 19. $acb - ack + 2b - 2k$

Graph on a number line:

20. $-2 - |x| > -4; D = \{\text{Reals}\}$

21. $4 \le x + 2 < 7; D = \{\text{Integers}\}$

22. $x \le 2$ or $x > 5; D = \{\text{Reals}\}$

Solve:

23. $\sqrt{x + 2} - 4 = 1$

24. $\sqrt{x - 3} - 5 = 3$

25. Find the equation of the line through $(-2, 5)$ and $(3, -2)$.

26. Find the equation of the line through $(-2, 5)$ that has a slope of $-\dfrac{1}{4}$.

27. Find the equation of the line through $(-2, 5)$ that is parallel to $y = -\dfrac{1}{3}x + 2$.

Simplify:

28. $\sqrt{\dfrac{3}{5}}$

29. $\sqrt{\dfrac{7}{3}}$

30. $\dfrac{2\sqrt{2} + \sqrt{2}}{\sqrt{2}}$

LESSON 125   *Completing the square*

125.A
difference of two squares

To review the difference of two squares theorem, we note that if $p^2 = q^2$, we can use the additive property of equality to justify adding $-q^2$ to both sides of the equation.

$$
\begin{aligned}
p^2 &= q^2 \\
-q^2 & \quad -q^2 \\
\hline
p^2 - q^2 &= 0
\end{aligned}
$$

Now we factor the left side of the equation and use the zero factor theorem to find that $p = q$ or $p = -q$.

$$p^2 - q^2 = 0 \quad \longrightarrow \quad (p + q)(p - q) = 0 \quad \longrightarrow \quad p = q, -q$$

DIFFERENCE OF TWO SQUARES THEOREM

If p and q are real numbers and if $p^2 = q^2$, then

$$p = q \qquad \text{or} \qquad p = -q$$

We review the use of this theorem by finding the values of the unknowns that satisfy the following equations.

	EQUATION	SOLUTION
(a)	$p^2 = q^2$	$p = \pm q$
(b)	$x^2 = 4$	$x = \pm 2$
(c)	$m^2 = 3$	$m = \pm\sqrt{3}$
(d)	$(x + 2)^2 = 3$	$x + 2 = \pm\sqrt{3}$ so $x = -2 \pm \sqrt{3}$
(e)	$(x - 3)^2 = 5$	$x - 3 = \pm\sqrt{5}$ so $x = 3 \pm \sqrt{5}$

125.B
completing
the square

The values of x that satisfy $x^2 + 5x + 6 = 0$ can be found by factoring this quadratic equation and using the zero factor theorem.

$$x^2 + 5x + 6 = 0 \longrightarrow (x + 3)(x + 2) = 0 \longrightarrow x = -3, -2$$

If we wish to find the replacement values of x that will satisfy the quadratic equation

$$2x^2 - x - 5 = 0$$

we are stumped because this equation cannot be factored over the integers. In this lesson we will learn that this equation can be rewritten as

$$\left(x - \frac{1}{4}\right)^2 = \frac{41}{16}$$

and that the solution can be completed by using the difference of two squares theorem as we show here.

$$x - \frac{1}{4} = \pm \sqrt{\frac{41}{16}} \longrightarrow x = \frac{1}{4} \pm \frac{\sqrt{41}}{4}$$

To understand the development of this method of solution, it is helpful to observe the relationship between the terms of a particular binomial and the terms of the square of the same binomial.

BINOMIAL	BINOMIAL SQUARED
$x + 3$	$x^2 + 6x + 9$
$x - 3$	$x^2 - 6x + 9$
$x + 5$	$x^2 + 10x + 25$
$x - 5$	$x^2 - 10x + 25$
$x + 7$	$x^2 + 14x + 49$
$x - 7$	$x^2 - 14x + 49$

In each of the expressions above there is a definite relationship between the constant term of the binomial and the two constants of the trinomial. In every example the coefficient of the second term of the trinomial is twice the constant term of the binomial and the third term of the trinomial is the square of the constant term of the binomial. But more important is the relationship between the constants in the trinomial. **In each case the last term is the square of one-half of the coefficient of x in the middle term.** In the first trinomial

$$x^2 + 6x + 9$$

we note that 9 is $\left(\frac{6}{2}\right)^2$. In the third trinomial

$$x^2 + 10x + 25$$

we note that 25 is $\left(\frac{10}{2}\right)^2$. In the last trinomial

$$x^2 - 14x + 49$$

we note that 49 is $\left(\frac{-14}{2}\right)^2$.

We can use these patterns to help us write a given quadratic equation in the form $(x + a)^2 = k$, which can be solved for x by using the difference of two squares theorem. This procedure is called **completing the square.**

example 125.1   Solve $-2x^2 = -x - 5$ by completing the square.

solution   We begin by writing the equation in standard form as

$$2x^2 - x - 5 = 0$$

Now we will use five steps to solve.

   Step 1:   Divide every term in the equation by the coefficient of the x^2 term so that the resulting coefficient of the x^2 term will be 1.

$$\frac{2x^2}{2} - \frac{x}{2} - \frac{5}{2} = \frac{0}{2} \longrightarrow x^2 - \frac{1}{2}x - \frac{5}{2} = 0$$

   Step 2:   Move the constant term to the right side of the equation.

$$
\begin{array}{rl}
x^2 - \dfrac{1}{2}x - \dfrac{5}{2} = & 0 \\[2mm]
+\dfrac{5}{2} \quad +\dfrac{5}{2} & \quad \text{added } \dfrac{5}{2} \text{ to both sides} \\[2mm]
\hline
x^2 - \dfrac{1}{2}x \quad = & \dfrac{5}{2}
\end{array}
$$

   Step 3:   Square the product of the coefficient of x and $\frac{1}{2}$

$$\left(-\frac{1}{2} \cdot \frac{1}{2}\right)^2 = \frac{1}{16}$$

and add the result of $\frac{1}{16}$ to both sides of the equation.

$$x^2 - \frac{1}{2}x + \frac{1}{16} = \frac{5}{2} + \frac{1}{16}$$

   Step 4:   Write the left side as the square of a binomial and simplify the right side.

$$\left(x - \frac{1}{4}\right)^2 = \frac{41}{16}$$

   Step 5:   Use the difference of two squares theorem to complete the solution.

$$x - \frac{1}{4} = \pm\sqrt{\frac{41}{16}} \longrightarrow x = \frac{1}{4} \pm \sqrt{\frac{41}{16}} \longrightarrow x = \frac{1}{4} \pm \frac{\sqrt{41}}{4}$$

example 125.2   Solve $x^2 + 3x - 5 = 0$ by completing the square.

solution   (1)   Since the coefficient of the x^2 term is already 1, the first step is omitted.

   (2)
$$
\begin{array}{rl}
x^2 + 3x - 5 = & 0 \\
+5 \quad +5 & \\
\hline
x^2 + 3x \quad = & 5
\end{array}
$$

   (3)
$$\left(3 \cdot \frac{1}{2}\right)^2 = \frac{9}{4}, \text{ so add } \frac{9}{4} \text{ to both sides.}$$

$$x^2 + 3x + \frac{9}{4} = 5 + \frac{9}{4}$$

Now write the left side as a binomial squared and simplify the right side.

   (4)
$$\left(x + \frac{3}{2}\right)^2 = \frac{29}{4}$$

Use the difference of two squares theorem to complete the solution.

(5) $x + \dfrac{3}{2} = \pm\sqrt{\dfrac{29}{4}} \longrightarrow x = -\dfrac{3}{2} \pm \sqrt{\dfrac{29}{4}} \longrightarrow x = -\dfrac{3}{2} \pm \dfrac{\sqrt{29}}{2}$

example 125.3   Solve $x^2 + 5x - 8 = 0$ by completing the square.

solution   (1)   Step 1 is not required since the coefficient of the x^2 term is already 1.

(2)
$$
\begin{array}{rcc}
x^2 + 5x - 8 = & 0 \\
+8 & +8 \\
\hline
x^2 + 5x \quad = & 8
\end{array}
$$

(3) $\left(5 \cdot \dfrac{1}{2}\right)^2 = \dfrac{25}{4}$, so add $\dfrac{25}{4}$ to both sides.

$$x^2 + 5x + \dfrac{25}{4} = 8 + \dfrac{25}{4}$$

Now write the left side as a binomial squared and simplify the right side.

(4) $\left(x + \dfrac{5}{2}\right)^2 = \dfrac{57}{4}$

Use the difference of two squares theorem to complete the solution.

(5) $x + \dfrac{5}{2} = \pm\sqrt{\dfrac{57}{4}} \longrightarrow x = -\dfrac{5}{2} \pm \sqrt{\dfrac{57}{4}} \longrightarrow x = -\dfrac{5}{2} \pm \dfrac{\sqrt{57}}{2}$

example 125.4   Solve $2x^2 + \dfrac{1}{3}x - 6 = 0$ by completing the square.

solution   (1) $2x^2 + \dfrac{1}{3}x - 6 = 0 \longrightarrow x^2 + \dfrac{1}{6}x - 3 = 0$

(2)
$$
\begin{array}{rcc}
x^2 + \dfrac{1}{6}x - 3 = & 0 \\
\\
+3 & +3 \\
\hline
x^2 + \dfrac{1}{6}x \quad = & 3
\end{array}
$$

(3) $\left(\dfrac{1}{6} \cdot \dfrac{1}{2}\right)^2 = \dfrac{1}{144}$, so add $\dfrac{1}{144}$ to both sides.

$$x^2 + \dfrac{1}{6}x + \dfrac{1}{144} = 3 + \dfrac{1}{144}$$

Now write the left side as a binomial squared and simplify the right side.

(4) $\left(x + \dfrac{1}{12}\right)^2 = \dfrac{433}{144}$

Now use the difference of two squares theorem to complete the solution.

(5) $x + \dfrac{1}{12} = \pm\sqrt{\dfrac{433}{144}} \longrightarrow x = -\dfrac{1}{12} \pm \dfrac{\sqrt{433}}{12}$

problem set 125

1.   The freight headed south at 9 a.m. and the express headed north from the same station at noon. At 3 p.m., the trains were 420 miles apart. What was the speed of each if the speed of the express was 20 miles per hour greater than the speed of the freight?

2. The ratio of greens to whites was 2 to 11. If the total was 2340, how many were greens and how many were whites?

3. Ninety-eight percent of the citizens favored the resolution. If 480 were against it or did not care, how many citizens lived in the town?

4. There was $2900 in the pot. If there were 293 more $1 bills than $10 bills, how many bills of each kind were there?

5. Reds varied inversely as yellows squared. When there were 10 reds, there were 100 yellows. How many reds were present when the yellows were reduced to 5?

6. Find four consecutive even integers such that the product of -12 and the sum of the first and fourth is 6 less than the product of 19 and the opposite of the third.

Solve the following quadratic equations by completing the square.

*7. $x^2 + 3x - 5 = 0$ *8. $x^2 + 5x - 8 = 0$ 9. $x^2 + 2x - 4 = 0$

10. $x^2 + 3x - 8 = 0$ 11. $3x^2 + 2x - 5 = 0$ 12. $2x^2 + 4x - 7 = 0$

Factor these trinomials. Begin by writing the terms in descending order of the variables.

13. $3x^2 - 35 - 16x$ 14. $-2x + 3x^2 - 5$ 15. $2x^2 - 5x - 12$

Factor by grouping:

16. $p^2c - ab + p^2b - ac$ 17. $ax^2 - ca + cx^2 - c^2$

18. $2y + mx^3 + my + 2x^3$ 19. $4ab + 4x + abc + cx$

Graph on a number line:

20. $4 \leq x - 2 \leq 8; D = \{\text{Reals}\}$ 21. $3 - |x| > 2; D = \{\text{Integers}\}$

22. $x < -2$ or $x \geq 4; D = \{\text{Integers}\}$

23. Find the equation of the line through $(-3, 2)$ and $(5, -3)$.

Solve:

24. $\sqrt{4x + 1} - 1 = 2$ 25. $\sqrt{5m - 5} + 6 = 7$

26. Find the equation of the line through $(2, 4)$ that is parallel to $y = \frac{1}{5}x - 6$.

27. Graph: $\begin{cases} y \leq -x + 2 \\ y \geq x \end{cases}$

Simplify:

28. $\sqrt{\dfrac{2}{7}}$ 29. $\sqrt{\dfrac{5}{12}}$ 30. $\dfrac{4 + \sqrt{3}}{\sqrt{6}}$

LESSON 126   *The quadratic formula*

126.A
the
quadratic
formula

In Lesson 89 we learned how to find solutions to quadratic equations by factoring over the integers and using the zero factor theorem. In Lesson 125 we learned to use the method of completing the square to find solutions of quadratic equations that we cannot factor over the integers. **If we write a quadratic equation in standard form using a, b, and c to represent the constants and complete the square, we can find a formula that can be used to find the values of x that will satisfy any quadratic equation.**

$$ax^2 + bx + c = 0 \qquad \text{general form of a quadratic equation}$$

$$x^2 + \frac{b}{a}x + \frac{c}{a} = 0 \qquad \text{divide by } a$$

$$x^2 + \frac{b}{a}x = \quad -\frac{c}{a} \qquad \text{add } -\frac{c}{a} \text{ to both sides}$$
$$\text{(additive property of equality)}$$

$$x^2 + \frac{b}{a}x + \frac{b^2}{4a^2} = \frac{b^2}{4a^2} - \frac{c}{a} \qquad \text{add } \left(\frac{b}{2a}\right)^2 = \frac{b^2}{4a^2} \text{ to both sides}$$

$$\left(x + \frac{b}{2a}\right)^2 = \frac{b^2 - 4ac}{4a^2} \qquad \text{simplification of both sides}$$

$$x + \frac{b}{2a} = \pm \sqrt{\frac{b^2 - 4ac}{4a^2}} \qquad \text{square root of both sides}$$

$$x + \frac{b}{2a} = \pm \frac{\sqrt{b^2 - 4ac}}{\sqrt{4a^2}} \qquad \text{quotient of square roots theorem}$$

$$x = -\frac{b}{2a} \pm \frac{\sqrt{b^2 - 4ac}}{2a} \qquad \text{added } -\frac{b}{2a} \text{ to both sides}$$

$$x = \frac{-b \pm \sqrt{b^2 - 4ac}}{2a} \qquad \text{addition of rational expressions}$$

126.B
use of the quadratic formula

By completing the square on $ax^2 + bx + c = 0$, we have derived the **quadratic formula**

$$x = \frac{-b \pm \sqrt{b^2 - 4ac}}{2a} \qquad (a \neq 0)$$

This formula expresses x in terms of the constants a, b, and c of the equation. If we wish to use this formula to find the values of x that satisfy a particular quadratic equation, it is necessary to compare the particular quadratic equation with the equation $ax^2 + bx + c$ to determine the values of a, b, and c.

example 126.1 Use the quadratic formula to determine the roots of the equation $3x^2 + 2x - 7 = 0$.

solution The general form of the quadratic equation is $ax^2 + bx + c = 0$. If we write our equation and the general equation one over the other

$$ax^2 + bx + c = 0 \qquad \text{general equation}$$
$$3x^2 + 2x - 7 = 0 \qquad \text{our equation}$$

we see that 3 corresponds to a, 2 corresponds to b, and -7 corresponds to c. If we use these numbers as replacements for a, b, and c in the quadratic formula, we find

$$x = \frac{-b \pm \sqrt{b^2 - 4ac}}{2a} \longrightarrow x = \frac{-(2) \pm \sqrt{(2)^2 - 4(3)(-7)}}{2(3)}$$

$$\longrightarrow \frac{-2 \pm \sqrt{4 + 84}}{6} \longrightarrow \frac{-2 \pm \sqrt{88}}{6} \longrightarrow \frac{-2 \pm 2\sqrt{22}}{6} \longrightarrow \frac{-1 \pm \sqrt{22}}{3}$$

Thus the two real numbers that will satisfy the given equation are

$$x = \frac{-1 + \sqrt{22}}{3} \quad \text{and} \quad x = \frac{-1 - \sqrt{22}}{3}$$

We can verify that these numbers are solutions by using them as replacements for x in the original equation.

USING $\dfrac{-1+\sqrt{22}}{3}$ $\qquad\qquad\qquad$ USING $\dfrac{-1-\sqrt{22}}{3}$

$$3\left(\frac{-1+\sqrt{22}}{3}\right)^2 + 2\left(\frac{-1+\sqrt{22}}{3}\right) - 7 = 0 \qquad 3\left(\frac{-1-\sqrt{22}}{3}\right)^2 + 2\left(\frac{-1-\sqrt{22}}{3}\right) - 7 = 0$$

$$\frac{3(23-2\sqrt{22})}{9} + \frac{-2+2\sqrt{22}}{3} - 7 = 0 \qquad \frac{3(23+2\sqrt{22})}{9} + \frac{-2-2\sqrt{22}}{3} - 7 = 0$$

$$\frac{23}{3} - \frac{2\sqrt{22}}{3} - \frac{2}{3} + \frac{2\sqrt{22}}{3} - 7 = 0 \qquad \frac{23}{3} + \frac{2\sqrt{22}}{3} - \frac{2}{3} - \frac{2\sqrt{22}}{3} - 7 = 0$$

$$\frac{21}{3} - 7 = 0 \qquad\qquad\qquad\qquad \frac{21}{3} - 7 = 0$$

$$7 - 7 = 0 \quad \text{True} \qquad\qquad\qquad 7 - 7 = 0 \quad \text{True}$$

example 126.2   Use the quadratic formula to determine the roots of $-6 = -x^2 + x$.

solution   We begin by writing the given equation in standard form as

$$x^2 - x - 6 = 0$$

Now we compare this equation to the equation $ax^2 + bx + c = 0$ to determine the numbers that correspond to a, b, and c.

$$ax^2 + bx + c = 0 \qquad \text{general equation}$$
$$x^2 - x - 6 = 0 \qquad \text{our equation}$$

We see that $a = 1$, $b = -1$, and $c = -6$.

$$x = \frac{-b \pm \sqrt{b^2 - 4ac}}{2a} \longrightarrow x = \frac{-(-1) \pm \sqrt{(-1)^2 - 4(1)(-6)}}{2(1)}$$

$$\longrightarrow x = \frac{1 \pm \sqrt{1 + 24}}{2} \longrightarrow x = \frac{1 \pm \sqrt{25}}{2} \longrightarrow x = \frac{1 \pm 5}{2} \longrightarrow x = 3, -2$$

example 126.3   Use the quadratic formula to find the values of x that satisfy the equation $-x = 7 - x^2$.

solution   We begin by writing the given equation in standard form and comparing it to the equation $ax^2 + bx + c = 0$

$$ax^2 + bx + c = 0 \qquad \text{general equation}$$
$$x^2 - x - 7 = 0 \qquad \text{our equation}$$

We see that $a = 1$, $b = -1$, and $c = -7$.

$$x = \frac{-b \pm \sqrt{b^2 - 4ac}}{2a} \longrightarrow x = \frac{-(-1) \pm \sqrt{(-1)^2 - 4(1)(-7)}}{2(1)}$$

$$\longrightarrow x = \frac{1 \pm \sqrt{1 + 28}}{2} \longrightarrow x = \frac{1 \pm \sqrt{29}}{2}$$

example 126.4 Use the quadratic formula to find the roots of $5x^2 = 2x + 1$.

solution Again we begin by writing the given equation in standard form and comparing it to the equation $ax^2 + bx + c = 0$.

$$ax^2 + bx + c = 0 \qquad \text{general equation}$$
$$5x^2 - 2x - 1 = 0 \qquad \text{our equation}$$

We see that $a = 5$, $b = -2$, and $c = -1$.

$$x = \frac{-b \pm \sqrt{b^2 - 4ac}}{2a} \longrightarrow x = \frac{-(-2) \pm \sqrt{(-2)^2 - 4(5)(-1)}}{2(5)}$$

$$\longrightarrow x = \frac{2 \pm \sqrt{4 + 20}}{10} \longrightarrow x = \frac{2 \pm \sqrt{24}}{10}$$

$$\longrightarrow x = \frac{2 \pm 2\sqrt{6}}{10} \longrightarrow x = \frac{1 \pm \sqrt{6}}{5}$$

problem set 126

1. The express made the trip in 20 hours. The freight took 25 hours because it was 10 miles per hour slower than the express. What was the speed of each train?

2. Inflation took its toll and the merchant had to mark the suit up 30 percent so it would sell for $156. What was the original price of the suit?

3. Greens varied inversely as blues squared. When there were 5 greens, there were 50 blues. How many greens were there, when there were only 10 blues?

4. The ratio of friendlies to enemies was 11 to 5. If there were 800 people hiding in the ruins, how many were friendlies?

5. Good ones were $7 each and sorry ones were only $3 each. If Fred spent $414 and bought 2 more good ones than sorry ones, how many of each kind did he buy?

6. Find three consecutive odd integers such that -3 times the sum of the first and third is 50 greater than 8 times the opposite of the second.

The quadratic formula is:

$$x = \frac{-b \pm \sqrt{b^2 - 4ac}}{2a}$$

Use the formula to solve the following quadratic equations. Begin by writing each equation in standard form so that it can be compared to $ax^2 + bx + c = 0$ in order to determine the values of a, b, and c.

*7. $-6 = -x^2 + x$ *8. $3x^2 + 2x - 7 = 0$ *9. $-x = 7 - x^2$

*10. $5x^2 = 2x + 1$ 11. $2x^2 + 2x - 11 = 0$ 12. $5x^2 - 6x - 4 = 0$

Now solve these by completing the square. They are the same as Problems 7 through 9, so the answers should match.

13. $-6 = -x^2 + x$ 14. $3x^2 + 2x - 7 = 0$ 15. $-x = 7 - x^2$

Factor. Begin by writing terms in descending order of the variable.

16. $3x^2 - 5 + 14x$ 17. $-27 + 24x + 3x^2$ 18. $9x - 5 + 2x^2$

Factor by grouping:

19. $km^2 + 2c - 2m^2 - kc$ 20. $6a - xya - xyb + 6b$

21. $abx - 2yc + xc - 2yab$ 22. $4xn + abn - abm - 4xm$

23. Find the equation of the line through $(2, 5)$ and $(-3, -4)$.

24. Find the equation of the line through $(-2, 5)$ that is parallel to $y = \dfrac{2}{5}x - 3$.

25. Graph: $\begin{cases} y \geq x \\ y \leq -x + 2 \end{cases}$

Simplify:

26. $(4 + 2\sqrt{2})(\sqrt{2} + 2)$

27. $\sqrt{\dfrac{3}{8}}$

28. $\dfrac{\sqrt{2} + 1}{\sqrt{2}}$

Appendix

Appendix

ENRICHMENT
LESSON **1**

Closure

EL1.A
closure

The concept of **closure** of a given set under a given operation is interesting, important, and simple. In fact, it is so simple that the simplicity itself often causes students difficulty, for they believe that surely there must be more to it than there is. The word **closure** will appear in almost all higher-level algebra books and will not cause apprehension if we devote sufficient time to understanding what it means now. We will try to be as straightforward as possible with the explanation.

Suppose we want to use a particular operation and only work with a designated set of numbers. If we restrict ourselves to this set of numbers, it seems reasonable to say that we also restrict our answers to members of this set of numbers. If we don't do this, then we are not restricting ourselves to working with the designated set of numbers because we are accepting numbers for answers that are outside of that set.

If we use two of the numbers from this set in an operation, and the answer is always a member of the set, we say that the set is **closed** under the operation. A set cannot be partially closed under an operation. It is either closed, or it is not closed. **If the use of any two of the members of the set in an operation results in an answer that is not a member of the set, we say that the set is not closed under the operation.** Let's investigate the set of positive integers. We will draw a circle and consider it to be an enclosure.

We mentally place all of the positive integers inside the enclosure and all the other numbers outside the enclosure.

Now let's see if the set of positive integers is closed under the operation of addition.

$$2 + 5 = 7, \qquad 51 + 93 = 144, \qquad 7 + 13 = 20$$

In fact, the sum of any two positive integers is also a positive integer and lies inside our enclosure. Thus we say that the set of positive integers is closed under the operation of addition.

Let's try subtraction.

$$20 - 6 = 14, \qquad 14 - 2 = 12, \qquad 8 - 12 = -4$$

The first two answers are positive integers and thus lie inside our enclosure, but the third answer does not lie inside the enclosure because -4 is not a member of the set of positive integers. **This one example is all that is needed to state that the set of positive integers is not closed under the operation of subtraction.**

Let's try multiplication.

$$4 \cdot 2 = 8, \qquad 71 \cdot 3 = 213, \qquad 48 \cdot 2 = 96$$

All these products are positive integers. It can be proved (a topic for a more advanced course) that all products of positive integers are also positive integers. Thus we say that the set of positive integers is closed under the operation of multiplication.

What about division? One example of a quotient of two integers that is not an integer is $\frac{1}{2}$. Thus we say that the set of positive integers is not closed under the operation of division.

example EL1.1   For what operations is the set $\{0, 1\}$ closed?

solution   We will begin by writing down A, S, M, D to represent the operations of addition, subtraction, multiplication, and division. When we investigate a particular operation and find that the set is not closed, we will draw a slash line through the letter that represents the particular operation. We use a check mark to designate closure.

ADDITION:	Not closed. $1 + 1 = 2$	A̸SMD†
SUBTRACTION:	Not closed. $0 - (1) = -1$	A̸S̸MD
MULTIPLICATION:	Closed. The product of any two numbers of this set is also a member of the set.	A̸S̸M̌D
DIVISION:	Not closed. $\frac{1}{0}$ is not defined.	A̸S̸M̌D̸

Thus the set $\{0, 1\}$ is closed for the operation of multiplication and is not closed for the operations of addition, subtraction, and division.

example EL1.2   For what operations is the set of negative integers closed?

solution

ADDITION:	Closed. The sum of any two negative integers is a negative integer.	ǍSMD
SUBTRACTION:	Not closed. $-4 - (-8) = +4$	ǍS̸MD
MULTIPLICATION:	Not closed. $(-4)(-2) = +8$	ǍS̸M̸D
DIVISION:	Not closed. $\dfrac{-4}{-2} = +2$	ǍS̸M̸D̸

Thus the set of negative integers is closed for the operation of addition and is not closed for subtraction, multiplication, or division.

example EL1.3   For what operations is the set $\{-1, 0, 1\}$ closed?

solution

ADDITION:	Not closed. $1 + 1 = 2$	A̸SMD
SUBTRACTION:	Not closed. $1 - (-1) = 2$	A̸S̸MD

---

† Note that if a given number is a member of a set, we may use the number more than once, in fact, as often as we desire.

| MULTIPLICATION: | Closed. The product of any combination of the numbers. $-1, 0, 1$ is one of these numbers. | A$̶MD |
| DIVISION: | Not closed, $\dfrac{-1}{0}$ is not defined. | A$̶MD̶ |

Thus the set $\{-1, 0, 1\}$ is closed for the operation of multiplication but is not closed for addition, subtraction, or division.

example EL1.4 For what operations is the set of integers closed?

solution

ADDITION:	Closed. The sum of any two integers is an integer.	ASMD
SUBTRACTION:	Closed. The difference of any two integers is an integer.	ASMD
MULTIPLICATION:	Closed. The product of any two integers is an integer.	ASMD
DIVISION:	Not closed. $\frac{7}{13}$ is not an integer.	ASMD̶

Thus the set of integers is closed for the operations of addition, subtraction, and multiplication but is not closed for division.

example EL1.5 For what operations is the set of real numbers closed?

ADDITION:	Closed. The sum of any two real numbers is a real number.	ASMD
SUBTRACTION:	Closed. The difference of any two real numbers is a real number.	ASMD
MULTIPLICATION:	Closed. The product of any two real numbers is a real number.	ASMD
DIVISION:	Not closed. $\frac{23}{0}$ is not defined.	ASMD̶

problem set enrichment 1

1. Three percent of the caterpillars metamorphosed into butterflies. If Ramona could count 120 butterflies, how many caterpillars had there been?

2. Sakahara socked it to them. If 4800 were present and Sakahara socked 34 percent of them, how many did he sock?

3. Muhammad counted the tents and found that 784 were patched. If there were 1400 tents in all, what percent were patched? What percent were not patched?

4. What fraction of 210 is $5\frac{1}{4}$?

5. $-3 \in \{\text{What sets of numbers}\}$?

6. Graph $x = -2\frac{1}{2}$ on a rectangular coordinate system.

7. Add: $\dfrac{4}{x^2 c} + \dfrac{5}{xc} + \dfrac{6}{x(c + x)}$

8. Simplify: $\dfrac{x^{-2} y^0 (x^{-2})^{-2} y^2}{(y^2 x^{-4})^2 (y^3 x^2)}$

Simplify:

9. $\dfrac{1}{-3^{-3}}$

10. $\dfrac{5}{-3^{-2}}$

11. Simplify by adding like terms: $\dfrac{4ax^2}{x} - \dfrac{3a^{-4}x}{a^{-5}} - \dfrac{2x}{a^{-1}} + \dfrac{6a^2a^{-1}}{x}$

Write these numbers in scientific notation:

12. $.00123 \times 10^{-5}$ **13.** $.00123 \times 10^8$

For what operations are the following sets closed?

*14.   {Integers}   *15. $\{-1, 0, 1\}$

16. Graph: $y = -\dfrac{1}{3}x - 3$ on a rectangular coordinate system.

Use substitution to solve: Use elimination to solve:

17. $\begin{cases} y + 3x = -2 \\ 2x - 4y = 22 \end{cases}$ **18.** $\begin{cases} 4x - 5y = -1 \\ 2x + 3y = 5 \end{cases}$

19. Expand: $\dfrac{4x^2}{y^2}\left(\dfrac{x^{-2}}{4y^2} - \dfrac{a^{-2}x^{-1}}{y^{-1}}\right)$ **20.** Simplify: $15\sqrt{8} - 30\sqrt{18} + 4\sqrt{50}$

Factor these trinomials. If there is a common factor, factor it as the first step.

21. $x^2 + 3x - 10$ **22.** $4x + x^2 - 21$ **23.** $18 + 9x + x^2$

24. $5x^2 - 15x - 50$ **25.** $x^3 - 3x^2 + 2x$ **26.** $18x - x^3 + 3x^2$

Factor these binomials. If there is a common factor, factor it as the first step.

27. $b^3x^2 - 4b^3$ **28.** $16x^2 - a^2$ **29.** $-m^2 + 9p^2$ **30.** $4b^3x^3 - b^3x$

ENRICHMENT LESSON *2* — *Properties of the real numbers*

EL2.A
the properties of real numbers

In this book we restrict our investigation of numbers to the set of **real numbers.** We multiply, divide, add, and subtract real numbers. We take the roots of real numbers and raise real numbers to powers. We use real numbers as coefficients of variables in expressions, equations, and inequalities. We find the real numbers that make particular equations and inequalities true statements and say that these real numbers satisfy the equations and inequalities and thus are said to be solutions of or roots of the equations or inequalities.

In advanced mathematics courses, we will be introduced to and work with numbers that are not called real numbers. We will work with numbers that are called imaginary numbers and we will work with numbers that are called complex numbers. We will also learn to work with arrays of numbers that are called matrices. We will find that while matrices consist of more than one number, they can often be treated as if they were a single number in that they can be added, subtracted, multiplied and divided.

When we add, subtract, multiply, and divide matrices, real numbers, imaginary numbers, and complex numbers, we find that the same rules do not necessarily apply to all. We find that each set has its own peculiarities or properties when its members are used as elements in a particular operation. One of the tasks of the mathematician is to discover the peculiarities or properties of the various sets of numbers and to name, define, and classify these peculiarities or properties. In this lesson we will review the properties of the set of real numbers that we have already discussed and also introduce several new properties. We will begin by discussing the word *operation*, which of course is not a property.

EL2.B
binary operations

The four basic operations of algebra are addition, subtraction, multiplication, and division. It is interesting to note that each of these operations is a **binary operation**[†] in that **only two numbers can be used in the operation at one time.**

$$\text{(a)} \quad 10 + 3 + 2 \qquad \text{(b)} \quad 10 \cdot 3 \cdot 2$$

Neither of the expressions above can be simplified in **one step.** In (a) we can compute the sum of the three numbers by adding 10 and 3 to get 13 and then adding 2 to get the final result of 15. This calculation could not be performed in one step because addition is a binary operation. The notation in (b) can be used to demonstrate that multiplication is also a binary operation. We can compute the product by multiplying 10 by 3 to get 30

[†] *Bi* from the Latin word *bis* meaning "two," as in bicycle.

and then multiplying 30 by 2 to get 60. This calculation could not be performed in one step because multiplication is a binary operation. Subtraction and division are also binary operations.

EL2.C
closure

If we exclude division by zero, the set of real numbers is closed for the operations of addition, subtraction, multiplication, and division. The emphasis that is given to closure may be bewildering to some, for it seems that we are making a mountain out of a mole hill. Maybe we are, but it is necessary to realize that the result of using two real numbers in any of the four basic operations (division by zero excluded) will be a real number.

EL2.D
properties of zero and 1

Now we begin our discussion of properties. We note that the numbers 1 and 0 have some special properties that no other numbers have. Because no other numbers have these properties, we can use the word **unique** to describe them and speak of the **unique properties of the number 1 and the unique properties of the number 0.**

EL2.E
multiplication and division properties of 1

The product of a particular real number and the number 1 is the particular real number.

$$4 \cdot 1 = 4$$

The quotient of any particular real number divided by the number 1 is the particular real number.

$$\frac{4}{1} = 4$$

These properties are not properties of any other real number and because they are important, we give these properties special names. We show two forms of the multiplication property to emphasize that the order of the factors does not affect the product

---

MULTIPLICATIVE PROPERTY OF 1

The product of any real number a and the number 1 is the real number itself.

$$a \cdot 1 = a \qquad \text{also} \qquad 1 \cdot a = a$$

DIVISION PROPERTY OF 1

The quotient of any real number a divided by the number 1 is the real number itself.

$$\frac{a}{1} = a$$

EL2.F
multiplicative and additive properties of zero

The product of any particular real number and zero is zero.

$$4 \cdot 0 = 0$$

The sum of any particular real number and zero is the particular real number.

$$4 + 0 = 4$$

We also give special names to these properties of the number zero.

> MULTIPLICATIVE PROPERTY OF ZERO
>
> The product of any real number a and the number zero is zero.
>
> $$a \cdot 0 = 0 \quad \text{also} \quad 0 \cdot a = 0$$

> ADDITIVE PROPERTY OF ZERO
>
> The sum of any real number a and the number zero is the real number itself.
>
> $$a + 0 = a \quad \text{also} \quad 0 + a = a$$

problem set enrichment 2

1. Paula ran to Hugo in 6 hours while Busking walked the same distance in 72 hours. How fast did Paula run if her speed was 11 kilometers per hour faster than Busking's? How far was it to Hugo?

2. Stephens ran to town at 8 miles per hour and Beck drove him back home at 32 miles per hour. How long did it take Stephens to run to town if the total time he was gone was 5 hours? How far was it to town?

3. Horses sold for $400 each and ponies for only $100. Weir spent $4500 and bought 5 more horses than ponies. How many horses did she buy?

4. Beaty and George judged the pigs at the fair. Beaty gave the black pigs only 35 percent of the points that George gave the pink pigs. If Beaty gave the black pigs 1351 points, how many points did the pink pigs get?

5. Factor: $ax^2 + 20a + 9xa$

6. Solve: $-3(-x - 2) - (-x - 5) + 2^0 - (-x) = 5(3 - 2x)$

7. Graph on a number line: $4 \le x < 10$; $D = \{\text{Integers}\}$

Solve by graphing:

8. $\begin{cases} y = -3x + 2 \\ x = 3 \end{cases}$

Use elimination:

9. $\begin{cases} 4x - 5y = -3 \\ 2x + y = 9 \end{cases}$

10. Simplify: $\dfrac{(.0005)(.08 \times 10^{14})}{(40,000)(200 \times 10^{-5})}$

11. Find the equations of lines (a) and (b).

Simplify:

12. $3\sqrt{2} \cdot 4\sqrt{3} - 7\sqrt{24} + \sqrt{54}$

13. $-3\sqrt{12}(2\sqrt{6} - 5\sqrt{8})$

14. Divide: $3x^3 - 5$ by $x + 3$

Solve by factoring:

15. $-x^2 = 12 + 7x$ **16.** $-4 + 9x^2 = 0$ **17.** $5 = -x^2 - 6x$

18. Solve: $\dfrac{4x}{3} - \dfrac{2x + 4}{2} = 5$ **19.** $4 \in \{$What sets of numbers$\}$?

20. Solve: $\dfrac{1}{2}x - 2\dfrac{3}{5} = \dfrac{1}{10}$ **21.** $R_H T_H = R_R T_R,\ R_H = 2,\ R_R = 12,$ $T_H = T_R + 5$; find T_H and T_R

22. Add: $\dfrac{x}{x + 4} + \dfrac{3}{x} - \dfrac{2}{x^2}$ **23.** Simplify: $\dfrac{a + \dfrac{4}{a^2}}{4 - \dfrac{1}{a^2}}$

24. Graph on a number line: $-x - 3 \nleq 2;\ D = \{$Integers$\}$

25. Simplify $-2 - 3|-4| + 2^0[(5 - 2)(-3 - 5) - (-2)] - [(-3 - 2) - 2]$

26. Demonstrate the additive property of zero.

Simplify:

27. (a) $\dfrac{-3^{-2}}{3^2}$ (b) $\dfrac{(-3)^{-2}}{3^2}$ **28.** $\dfrac{(x^0)^{-2}(xx^2)^3(y^{-2})^4 y}{(xyy)^{-2}(yx^{-2})^4}$

29. Simplify by adding like terms: $x^2 y^{-2} - \dfrac{3x^2}{y^2} + \dfrac{12x^4 xy^{-2}}{x^3} - \dfrac{3x^2 y^2}{x^{-4}}$

ER-16, 17, 18

Properties of a field

The set of real numbers has nine special properties that apply when real numbers are used in the operations of multiplication and addition, and we say that any set that has these properties under the closed operations of multiplication and addition constitutes a **field.** There is no special significance attached to the word field. We have to use some name and field is possibly as good a name as any other. It is the word that we use to describe a set that, when used in the operations of addition and multiplication, has these nine properties. The chart below gives the names of the properties and both a numerical and an abstract example of each property.

PROPERTIES OF A FIELD		
Addition	*Name*	*Multiplication*
$4 + 0 = 4$ $a + 0 = a$	IDENTITY	$4 \cdot 1 = 4$ $a \cdot 1 = a$
$4 + (-4) = 0$ $a + (-a) = 0$	INVERSE	$4 \cdot \dfrac{1}{4} = 1$ $a \cdot \dfrac{1}{a} = 1$
$4 + (3 + 2) = (4 + 3) + 2$ $a + (b + c) = (a + b) + c$	ASSOCIATIVE PROPERTY	$(4 \cdot 3) \cdot 2 = 4 \cdot (3 \cdot 2)$ $(a \cdot b) \cdot c = a \cdot (b \cdot c)$
$4 + 2 = 2 + 4$ $a + b = b + a$	COMMUTATIVE PROPERTY	$4 \cdot 2 = 2 \cdot 4$ $ab = ba$

DISTRIBUTIVE PROPERTY

$$4(3 + 2) = 4 \cdot 3 + 4 \cdot 2$$
$$a(b + c) = ab + ac$$

We note that there are four properties listed under addition and that the same four properties are listed under multiplication. These are the **identity,** the **inverse,** the **associative property,** and the **commutative property.** At the bottom of the chart we see the **distributive property** in the middle of the page since the **distributive property includes both addition and multiplication.**

EL3.B

identity

In the chart the first entry in the middle column is the word *identity*. To the left and right of this word are examples under the column labeled addition and under the column labeled multiplication.

To have a mathematical system all that is necessary is a set of numbers to work with and an operation to use under which the particular set is closed. But to have a useful system, it is necessary to be able to perform an operation on a particular number and have the result equal the particular number. We begin with a number, then perform the operation with a special number and have the result **identical** to the number with which we began. We can add the number zero to any particular real number and the sum is **identical** to the particular real number. We can multiply any particular real number by the number 1 and the product is **identical** to the particular real number. **Thus we call the numbers 0 and 1 the identities for the operations of addition and multiplication, respectively.**

IDENTITY

The number zero is the additive identity because:

The number 1 is the multiplicative identity because:

For any real number a,

$$a + 0 = a \qquad\qquad a \cdot 1 = a$$

We note that the numbers 0 and 1 are the identities for all real numbers under the operations of addition and multiplication, respectively.

EL3.C

inverse

In the chart on the last page, the second entry in the middle column is the word *inverse*. To the left and right of this word are examples under the column labeled addition and under the column labeled multiplication.

Another feature of a useful system is the ability to perform an operation and have the result equal the identity for that operation. We give the name **inverse** to a number that when used with a particular number in an operation yields a result that is the **identity** for that operation. We note that if we add to a particular number the *opposite of* the number, the sum is zero, the identity for addition.

$$4 + (-4) = 0 \qquad -3 + (3) = 0 \qquad -\frac{1}{5} + \left(\frac{1}{5}\right) = 0$$

Thus every number has its own unique **additive inverse** under the operation of addition. Further, we see that another name for the **opposite of a number** is the **additive inverse** of the number.

We note that the product of a particular number and the **reciprocal** of the number is the number 1, the **identity** for multiplication.

$$-5 \cdot \left(\frac{1}{-5}\right) = 1 \qquad 13 \cdot \left(\frac{1}{13}\right) = 1 \qquad -\frac{1}{14}(-14) = 1$$

Thus every number except zero has its own unique **multiplicative inverse** under the operation of multiplication, and we see that another name for the **reciprocal** of a number is the **multiplicative inverse** of the number.

INVERSE

For any real number a the
additive inverse is $-a$

For any nonzero real
number a, the multipli-
cative inverse is $\dfrac{1}{a}$

because

$$a + (-a) = 0$$
$$a \cdot \left(\dfrac{1}{a}\right) = 1$$

It is important to remember that the numbers 0 and 1 are the additive and multiplicative identities, respectively, for all real numbers. Further we must remember that every real number has its own unique additive inverse and that every real number except zero has its own unique multiplicative inverse.

EL3.D
associative
property

The third entry in the chart on page 397 deals with the **associative property.** If we are asked to compute the sum of

$$4 + 3 + 2$$

we get the same result if we **associate** the first two numbers together for addition

$$(4 + 3) + 2$$
$$7 + 2$$
$$9$$

or if we **associate** the second two numbers together.

$$4 + (3 + 2)$$
$$4 + (5)$$
$$9$$

This property of real numbers is called the **associative property of real numbers under addition.**

Real numbers can be **associated** in the same manner when computing a product.

$4 \cdot 3 \cdot 2$	$4 \cdot 3 \cdot 2$
$(4 \cdot 3) \cdot 2$	$4 \cdot (3 \cdot 2)$
$12 \cdot 2$	$4 \cdot 6$
24	24

We obtain the same result whether we **associate** the first two numbers as shown on the left or whether we **associate** the second two numbers as shown on the right. This property of real numbers is called the **associative property of real numbers under multiplication.**

ASSOCIATIVE PROPERTY

For any three real numbers a, b, and c,

$$a + (b + c) = (a + b) + c \quad \text{and} \quad a(bc) = (ab)c$$

EL3.E
commutative property

An operation performed with any two members of a particular set is said to be **commutative** if the order in which the operation is performed does not affect the result. Since the order of addition or multiplication of two real numbers does not affect the result, we can say that addition and multiplication of real numbers are **commutative** operations.

$$4 + 2 = 6 \qquad 2 + 4 = 6 \qquad\qquad 4 \cdot 2 = 8 \qquad 2 \cdot 4 = 8$$

COMMUTATIVE PROPERTY

For any two real numbers a and b,

$$a + b = b + a \qquad \text{and} \qquad ab = ba$$

EL3.F
distributive property

Lastly we define the **distributive property** of real numbers. This property of real numbers permits multiplication to be distributed over addition as shown here.

$$4(3 + 2) = 4 \cdot 3 + 4 \cdot 2$$

DISTRIBUTIVE PROPERTY

For any three real numbers a, b, and c,

$$a(b + c) = ab + ac$$

We note that while there are statements of the identity, inverse, associative property, and commutative property for both multiplication and addition, there is only one statement of the distributive property and that this single statement includes both multiplication and addition.

problem set enrichment 3

1. Gaskin and Sloan raced to the cotton patch. Gaskin's speed was 12 kilometers per hour while Sloan's was only 8 kilometers per hour. What was the time of each if Gaskin's time was 5 hours less than Sloan's time?

2. At noon Joyce F. headed for the lake at 30 miles per hour. Her car broke down and she made the long walk back home at 4 miles per hour. How long did she walk if she was gone for 17 hours? How far did she walk?

3. Tickets to the carnival were $3 for adults and $2 for kids. Tommy was a big spender as he took 77 people to the carnival and spent $209 for their tickets. How many kids did he pay for?

4. Eighty percent of the boys preferred the wild goose ride to the ferris wheel. If 300 boys preferred the ferris wheel, how many preferred the wild goose ride?

Use the letters a, b, and c as required to state:

5. The distributive property

6. The associative property of addition

7. What is another name for the multiplicative inverse of a number?

8. Factor: $-m^3 - 11m^2 - 24m$

9. Solve: $-4(x - 2)(-2) + 3(x - 4) = 2x(4 - 2^0)$

10. Graph: $-x + 4 \leq 2$; $D = \{$Integers$\}$

Solve by graphing:

11. $\begin{cases} y = x + 2 \\ y = -x \end{cases}$

Use elimination:

12. $\begin{cases} 5x - 2y = 18 \\ 3x + y = 24 \end{cases}$

13. Simplify: $\dfrac{(.00042 \times 10^{-15})(300,000)}{(180,000 \times 10^{14})(7000 \times 10^{-23})}$

14. Find the equations of line (a) and line (b).

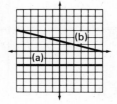

Simplify:

15. $3\sqrt{2} \cdot 4\sqrt{3} \cdot 4\sqrt{6} - 3\sqrt{2}$

16. $3\sqrt{2}(5\sqrt{12} - 6\sqrt{36})$

17. Divide $x^4 - x - 4$ by $x - 1$

Solve by factoring:

18. $24 = -x^2 - 10x$

19. $-4 + 9x^2 = 0$

Solve:

20. $\dfrac{3x}{2} - \dfrac{x - 5}{6} = 3$

21. $3\dfrac{2}{5}x - \dfrac{1}{10} = \dfrac{1}{20}$

22. Add: $\dfrac{p}{p + 4} - \dfrac{p - 2}{p}$

23. Simplify: $\dfrac{xy - \dfrac{1}{y}}{\dfrac{x}{y} - 4}$

24. $15\dfrac{2}{5}$ is what fraction of $20\dfrac{1}{3}$?

25. $\dfrac{5\sqrt{2}}{7} \in \{$What sets of numbers$\}$?

26. Evaluate: $-|x^0| - (x - y)(y - x)$ if $x = -4$ and $y = -5$

27. Simplify: $-2^0 - 2[(-3 - 2) - (-2)] - [(-3 + 2) - (-2)]$

28. Simplify: (a) $\dfrac{1}{-3^{-2}}$ (b) $\dfrac{1}{-(-3)^{-2}}$ (c) $-(-2)^{-2}$

29. Multiply: $x^2y^2\left(\dfrac{x^{-2}}{y^{-2}} + 4x^4y^{-2}\right)$ **ER-19, 20, 21**

Algebraic proofs

EL4.A
**definitions,
axioms,
and proofs**

The development of algebra rests on a foundation of definitions and axioms. We must first define what we mean when we write certain symbols. In Lesson 17 we defined the notation

$$x^4$$

to mean $x \cdot x \cdot x \cdot x$. It was not necessary to choose the notation x^4. We could just as easily have chosen the notation

$$^4x$$

—but we didn't. Later on we defined the notation

$$xc$$

to indicate the product of x and c. We could have defined this notation to indicate the quotient of x divided by c—but we didn't. Thus we began our development by carefully defining all notations and symbols.

Unfortunately, we cannot build a mathematical system using definitions alone. To the definitions we must add statements of observations that everyone accepts as true without proof. These statements are called **axioms** or **postulates.** The statements that concern real numbers make up part of what we call the properties of real numbers as we discussed in Enrichment Lesson 2. For instance, we accept without proof that the order of addition of two real numbers does not affect the resultant sum. We state this abstractly by writing

$$a + b = b + a$$

In Enrichment Lesson 3 we said that would call this property the commutative property of addition for real numbers. Also in the same lesson we stated the associative property of real numbers under addition by writing

$$(a + b) + c = a + (b + c)$$

In mathematics we take pride in holding our list of definitions and axioms to an absolute minimum. Then we use the definitions and axioms to prove all other statements and assertions.

You will note that when we stated the commutative property we restricted our statement of this property to a statement about only two numbers and we restricted our statement of the associative property to a statement about only three numbers.

In Lesson 3 we said that the order of addition of any number of signed numbers did not affect the resultant sum. This statement cannot be justified by the commutative property alone because the commutative property is applicable only to the order of addition of two numbers. In the following examples we will take these two restricted statements and use them to prove relationships that involve three or more terms or three or more factors.

example EL4.1   If a, b, and c represent unspecified real numbers, use the associative and commutative properties to show that

$$a + b + c = c + b + a$$

solution   We will begin with the expression $a + b + c$ as it appears on the left and use the associative and commutative properties to rearrange the terms so that the resulting expression is $c + b + a$, the same as the right side. We begin by associating b and c by using a set of parentheses.

$$a + (b + c)$$

Next we change the order inside the parentheses as permitted by the commutative property.

$$a + (c + b)$$

Now c is the middle term instead of the last term. Next we enclose a and c in parentheses and then change the order of a and c.

$$(a + c) + b = (c + a) + b$$

Now we have c as the first term but since our goal is $c + b + a$, we must change the order of a and b so we need one more step. As before, we use parentheses, change the order inside the parentheses, and then remove the parentheses.

$$c + (a + b) = c + (b + a) = \boldsymbol{c + b + a}$$

example EL4.2   If x, y, and m represent unspecified real numbers, use the commutative and associative properties to show that

$$x + y + m = m + x + y$$

solution   We will work with the left side and transform it into the same form as the right side.

$x + y + m$	given
$x + (y + m)$	associative property
$x + (m + y)$	commutative property
$(x + m) + y$	associative property
$(m + x) + y$	commutative property
$m + x + y$	removed parentheses

example EL4.3   If a, b, c, and d represent unspecified real numbers, show that

$$a + b + c + d = b + a + d + c$$

solution   The expression on the right is the same as that on the left except that the order of the first two symbols and the last two symbols is reversed. We need only four steps.

$a + b + c + d$	given
$(a + b) + (c + d)$	associative property
$(b + a) + (d + c)$	commutative property (twice)
$b + a + d + c$	removed parentheses

In Lesson 8 we stated that the order in which real numbers are multiplied does not affect the value of the product. In the following examples we will use the associative and commutative properties of multiplication to prove this assertion for some expressions that have a finite number of factors.

example EL4.4 If a, b, c, and d represent unspecified real numbers, show that

$$abcd = bdac$$

solution This one looks more difficult. Let's go about it step by step and feel our way along. We will transform the expression on the left into the same form as the expression that we have on the right. On the right the first factor is b, so let's move b to the front.

$abcd$	given
$(ab)cd$	associative property
$(ba)cd$	commutative property
$bacd$	removed parentheses

Now we have moved b to the front; on the right above we note that c is the last factor, so

$bacd$	from above
$ba(cd)$	associative property
$ba(dc)$	commutative property
$badc$	removed parentheses

We wanted to end up with $bdac$. Thus far, we have moved b to the front and c to the end, and all that is amiss is the order of the two middle factors.

$badc$	from above
$b(ad)c$	associative property
$b(da)c$	commutative property
$bdac$	removed parentheses

As you see, these examples are almost the same as those involving addition. In these we use the properties of associativity and commutativity for multiplication instead of the same properties for addition.

example EL4.5 If P, Q, and L are unspecified real numbers, show that $PQL = LPQ$.

solution We want to prove that $PQL = LPQ$. We will begin with PQL and change this to LPQ

PQL	given
$P(QL)$	associative property
$P(LQ)$	commutative property
$(PL)Q$	associative property
$(LP)Q$	commutative property
LPQ	removed parentheses

problem set enrichment 4

1. Steely carried the rock part of the way and Paul carried it the rest of the way. Steely traveled at 20 miles per hour and Paul traveled at 35 miles per hour. If they carried the rock 490 miles in 17 hours, how long did Paul carry the rock?

2. Mike made the trip in 8 hours while it took Gene 10 hours for the same trip. Find the rate of each if Mike was 10 kilometers per hour faster than Gene.

3. Find four consecutive even integers such that 6 times the sum of the first and the fourth is 108 greater than the product of 8 and the opposite of the third.

4. The 3000 hawks and eagles filled the sky. The number of hawks was 1800 greater than 3 times the number of eagles. How many of each kind were there?

5. Twenty-three percent of the newcomers thought there was no difference between a thaumaturge and a prestidigitator. If 3465 people believed that there was a difference, how many newcomers were there in all?

If the letters represent unspecified real numbers, use the associative and commutative properties to show that:

*6. $x + y + m = m + x + y$

*7. $PQL = LPQ$ *8. $abcd = bdac$

9. Find d.

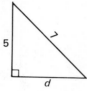

10. Find the distance between $(-3, -4)$ and $(-1, 2)$.

11. Simplify: $\dfrac{x^2 + 5x + 6}{-x^2 - 3x} \div \dfrac{x^2 + 7x + 10}{x^3 + 8x^2 + 15x}$

12. Solve: $-3(-2 - p) - p^0(-4) - 2^0(-5p - 6) = -3 - (-2p)$

13. Solve by graphing: $\begin{cases} y = -2x - 3 \\ x = 2 \end{cases}$

14. Simplify: $\dfrac{(.000075)(200 \times 10^{-15})}{(.025 \times 10^{45})(300 \times 10^{-23})}$

15. Find the equations of lines (a) and (b).

Simplify:

16. $5\sqrt{75} \cdot 2\sqrt{3} + 7\sqrt{3} \cdot 2\sqrt{6}$ 17. $4\sqrt{6}(3\sqrt{6} - 2\sqrt{2})$

18. Solve by factoring: $50 = x^2 + 5x$

19. Divide: $2x^3 + x^2 - 3x$ by $2x + 3$

Solve:

20. $\dfrac{x}{4} - \dfrac{x - 5}{8} = 2$ 21. $7\dfrac{1}{2}x - 4\dfrac{1}{3} = 14\dfrac{1}{8}$

22. Add: $\dfrac{4}{x} + \dfrac{6x + 2}{x^2} + \dfrac{3}{x(x + 1)}$ 23. Simplify: $\dfrac{x + \dfrac{x}{y}}{y - \dfrac{1}{y}}$

24. $7\dfrac{3}{5}$ is what fraction of $14\dfrac{1}{8}$?

25. $(3\sqrt{2} - 5) \in \{$What sets of numbers$\}$?

26. Evaluate: $-p^0 - p^2(p - a^0) - ap + |-ap|$ if $a = -3$ and $p = -4$

27. Simplify: (a) $\dfrac{4xym + 4x^2ym^2}{4xym}$ (b) $\dfrac{-5^0}{-2^{-4}}$

28. Multiply: $4x^2y\left(\dfrac{x^{-2}}{4y} - \dfrac{5x^3y^{-3}}{a^{-4}}\right)$

29. Simplify: $-3^0[(-3^2 + 4)(-2^2 - 2) - (-2) + 4]$

Enrichment problems

The following problems are provided to permit a continuing review of the enrichment topics. The first part of the number for each problem designates the problem set to which this problem should be appended.

77-1. For what operations is the set of negative integers closed?

77-2. For what operations is the set of whole numbers closed?

78-3. For what operations is the set {4, 3, 2} closed?

78-4. For what operations is the set of positive real numbers closed?

79-5. If we exclude division by zero, for what operations is the set of real numbers closed?

80-6. For what operations is the set of real numbers closed?

81-7. For what operations is the set of negative real numbers closed?

82-8. For what operations is the set of negative even integers closed?

83-9. For what operations is the set of positive odd integers closed?

84-10. For what operations is the set of counting numbers closed?

86-11. For what operations is the set of rational numbers closed?

87-12. For what operations is the set of irrational numbers closed?

89-13. For what operations is the set of positive rational numbers closed?

91-14. For what operations is the set of negative irrational numbers closed?

93-15. For what operations is the set of positive irrational numbers closed?

EL2-16. For what operations is the set of whole numbers closed?

EL2-17. Which of the four basic operations are binary operations?

EL2-18. List and demonstrate the special properties of the numbers 1 and 0.

EL3-19. Use the letters a, b, and c to state the distributive property.

EL3-20. What is (a) the additive inverse of -2? (b) the multiplicative inverse of $-b$?

EL3-21. A field is closed. What are the other nine properties of a field?

95-22. Use the letters a, b, and c as necessary to state (a) the associative property of multiplication, and (b) the commutative property of addition.

95-23. If the reciprocal of a number is $-\dfrac{1}{9}$, what is the additive inverse of the number?

95-24. $a + (b + c) = (a + b) + c$ is a statement of which property of the set of real numbers?

96-25. What is the additive identity?

96-26. What is the multiplicative identity?

96-27. If the reciprocal of a number is -9, what is the multiplicative inverse of the same number?

97-28. The additive inverse of a number is 7. What is the multiplicative inverse of the number?

97-29. $a(b + c) = ab + ac$ is a statement of which property of the set of real numbers?

98-30. For what operations is the set of positive real numbers closed?

98-31. $a + b = b + a$ is a statement of which property of the set of real numbers?

99-32. List the properties of a field.

100-33. If the letters represent unspecified real numbers, use the associative and commutative properties to prove that $a + c + x = x + c + a$.

101-34. If the numbers represent unspecified real numbers, use the associative and commutative properties to prove that $mpa = map$.

102-35. If the reciprocal of a number is $-\dfrac{1}{101}$, what is the additive inverse of the number?

102-36. $(a + x) + y = a + (x + y)$ is a statement of what property of the set of real numbers?

103-37. If the letters represent unspecified real numbers, use the associative and commutative properties to prove that $abxy = yabx$.

104-38. The set of negative irrational numbers is closed for what operations?

105-39. What is the product of -42 and the additive identity?

106-40. What is the sum of -56 and the multiplicative identity?

107-41. If the letters represent unspecified real numbers, use the associative and commutative properties to prove that $abxy = xbya$.

108-42. If the additive inverse of a number is $\dfrac{2}{3}$, what is the reciprocal of the same number?

109-43. The set of positive odd integers is closed for what operations?

110-44. What is the product of -105 and the multiplicative identity?

111-45. Why do we say that addition is a binary operation?

112-46. Use the letters a, b, and c as necessary to state (a) the associative property of addition, and (b) the commutative property of multiplication.

113-47.   Use the letters a, b, and c as necessary to state the distributive property.

114-48.   If the letters represent unspecified real numbers, use the associative and commutative properties to prove that $abcd = bdac$.

115-49.   If the reciprocal of a number is $-\dfrac{1}{5}$, what is the product of the number and the multiplicative identity?

116-50.   The set of negative odd integers is closed for what operations?

117-51.   If the letters represent unspecified real numbers, use the associative and commutative properties to prove that $a + x + y + m = y + a + m + x$.

Table of squares and square roots

TABLE OF SQUARES AND SQUARE ROOTS

n	n^2	$\sqrt{n}$	n	n^2	$\sqrt{n}$	n	n^2	$\sqrt{n}$	n	n^2	$\sqrt{n}$	n	n^2	$\sqrt{n}$	n	n^2	$\sqrt{n}$
1.00	1.0000	1.00000	1.50	2.2500	1.22474	2.00	4.0000	1.41421	2.50	6.2500	1.58114	3.00	9.0000	1.73205	3.50	12.2500	1.87083
1.01	1.0201	1.00499	1.51	2.2801	1.22882	2.01	4.0401	1.41774	2.51	6.3001	1.58430	3.01	9.0601	1.73494	3.51	12.3201	1.87350
1.02	1.0404	1.00995	1.52	2.3104	1.23288	2.02	4.0804	1.42127	2.52	6.3504	1.58745	3.02	9.1204	1.73781	3.52	12.3904	1.87617
1.03	1.0609	1.01489	1.53	2.3409	1.23693	2.03	4.1209	1.42478	2.53	6.4009	1.59060	3.03	9.1809	1.74069	3.53	12.4609	1.87883
1.04	1.0816	1.01980	1.54	2.3716	1.24097	2.04	4.1616	1.42829	2.54	6.4516	1.59374	3.04	9.2416	1.74356	3.54	12.5316	1.88149
1.05	1.1025	1.02470	1.55	2.4025	1.24499	2.05	4.2025	1.43178	2.55	6.5025	1.59687	3.05	9.3025	1.74642	3.55	12.6025	1.88414
1.06	1.1236	1.02956	1.56	2.4336	1.24900	2.06	4.2436	1.43527	2.56	6.5536	1.60000	3.06	9.3636	1.74929	3.56	12.6736	1.88680
1.07	1.1449	1.03441	1.57	2.4649	1.25300	2.07	4.2849	1.43875	2.57	6.6049	1.60312	3.07	9.4249	1.75214	3.57	12.7449	1.88944
1.08	1.1664	1.03923	1.58	2.4964	1.25698	2.08	4.3264	1.44222	2.58	6.6564	1.60624	3.08	9.4864	1.75499	3.58	12.8164	1.89209
1.09	1.1881	1.04403	1.59	2.5281	1.26095	2.09	4.3681	1.44568	2.59	6.7081	1.60935	3.09	9.5481	1.75784	3.59	12.8881	1.89473
1.10	1.2100	1.04881	1.60	2.5600	1.26491	2.10	4.4100	1.44914	2.60	6.7600	1.61245	3.10	9.6100	1.76068	3.60	12.9600	1.89737
1.11	1.2321	1.05357	1.61	2.5921	1.26886	2.11	4.4521	1.45258	2.61	6.8121	1.61555	3.11	9.6721	1.76352	3.61	13.0321	1.90000
1.12	1.2544	1.05830	1.62	2.6244	1.27279	2.12	4.4944	1.45602	2.62	6.8644	1.61864	3.12	9.7344	1.76635	3.62	13.1044	1.90263
1.13	1.2769	1.06301	1.63	2.6569	1.27671	2.13	4.5369	1.45945	2.63	6.9169	1.62173	3.13	9.7969	1.76918	3.63	13.1769	1.90526
1.14	1.2996	1.06771	1.64	2.6896	1.28062	2.14	4.5796	1.46287	2.64	6.9696	1.62481	3.14	9.8596	1.77200	3.64	13.2496	1.90788
1.15	1.3225	1.07238	1.65	2.7225	1.28452	2.15	4.6225	1.46629	2.65	7.0225	1.62788	3.15	9.9225	1.77482	3.65	13.3225	1.91050
1.16	1.3456	1.07703	1.66	2.7556	1.28841	2.16	4.6656	1.46969	2.66	7.0756	1.63095	3.16	9.9856	1.77764	3.66	13.3956	1.91311
1.17	1.3689	1.08167	1.67	2.7889	1.29228	2.17	4.7089	1.47309	2.67	7.1289	1.63401	3.17	10.0489	1.78045	3.67	13.4689	1.91572
1.18	1.3924	1.08628	1.68	2.8224	1.29615	2.18	4.7524	1.47648	2.68	7.1824	1.63707	3.18	10.1124	1.78326	3.68	13.5424	1.91833
1.19	1.4161	1.09087	1.69	2.8561	1.30000	2.19	4.7961	1.47986	2.69	7.2361	1.64012	3.19	10.1761	1.78606	3.69	13.6161	1.92094
1.20	1.4400	1.09545	1.70	2.8900	1.30384	2.20	4.8400	1.48324	2.70	7.2900	1.64317	3.20	10.2400	1.78885	3.70	13.6900	1.92354
1.21	1.4641	1.10000	1.71	2.9241	1.30767	2.21	4.8841	1.48661	2.71	7.3441	1.64621	3.21	10.3041	1.79165	3.71	13.7641	1.92614
1.22	1.4884	1.10454	1.72	2.9584	1.31149	2.22	4.9284	1.48997	2.72	7.3984	1.64924	3.22	10.3684	1.79444	3.72	13.8384	1.92873
1.23	1.5129	1.10905	1.73	2.9929	1.31529	2.23	4.9729	1.49332	2.73	7.4529	1.65227	3.23	10.4329	1.79722	3.73	13.9129	1.93132
1.24	1.5376	1.11355	1.74	3.0276	1.31909	2.24	5.0176	1.49666	2.74	7.5076	1.65529	3.24	10.4976	1.80000	3.74	13.9876	1.93391
1.25	1.5625	1.11803	1.75	3.0625	1.32288	2.25	5.0625	1.50000	2.75	7.5625	1.65831	3.25	10.5625	1.80278	3.75	14.0625	1.93649
1.26	1.5876	1.12250	1.76	3.0976	1.32665	2.26	5.1076	1.50333	2.76	7.6176	1.66132	3.26	10.6276	1.80555	3.76	14.1376	1.93907
1.27	1.6129	1.12694	1.77	3.1329	1.33041	2.27	5.1529	1.50665	2.77	7.6729	1.66433	3.27	10.6929	1.80831	3.77	14.2129	1.94165
1.28	1.6384	1.13137	1.78	3.1684	1.33417	2.28	5.1984	1.50997	2.78	7.7284	1.66733	3.28	10.7584	1.81108	3.78	14.2884	1.94422
1.29	1.6641	1.13578	1.79	3.2041	1.33791	2.29	5.2441	1.51327	2.79	7.7841	1.67033	3.29	10.8241	1.81384	3.79	14.3641	1.94679
1.30	1.6900	1.14018	1.80	3.2400	1.34164	2.30	5.2900	1.51658	2.80	7.8400	1.67332	3.30	10.8900	1.81659	3.80	14.4400	1.94936
1.31	1.7161	1.14455	1.81	3.2761	1.34536	2.31	5.3361	1.51987	2.81	7.8961	1.67631	3.31	10.9561	1.81934	3.81	14.5161	1.95192
1.32	1.7424	1.14891	1.82	3.3124	1.34907	2.32	5.3824	1.52315	2.82	7.9524	1.67929	3.32	11.0224	1.82209	3.82	14.5924	1.95448
1.33	1.7689	1.15326	1.83	3.3489	1.35277	2.33	5.4289	1.52643	2.83	8.0089	1.68226	3.33	11.0889	1.82483	3.83	14.6689	1.95704
1.34	1.7956	1.15758	1.84	3.3856	1.35647	2.34	5.4756	1.52971	2.84	8.0656	1.68523	3.34	11.1556	1.82757	3.84	14.7456	1.95959
1.35	1.8225	1.16190	1.85	3.4225	1.36015	2.35	5.5225	1.53297	2.85	8.1225	1.68819	3.35	11.2225	1.83030	3.85	14.8225	1.96214
1.36	1.8496	1.16619	1.86	3.4596	1.36382	2.36	5.5696	1.53623	2.86	8.1796	1.69115	3.36	11.2896	1.83303	3.86	14.8996	1.96469
1.37	1.8769	1.17047	1.87	3.4969	1.36748	2.37	5.6169	1.53948	2.87	8.2369	1.69411	3.37	11.3569	1.83576	3.87	14.9769	1.96723
1.38	1.9044	1.17473	1.88	3.5344	1.37113	2.38	5.6644	1.54272	2.88	8.2944	1.69706	3.38	11.4244	1.83848	3.88	15.0544	1.96977
1.39	1.9321	1.17898	1.89	3.5721	1.37477	2.39	5.7121	1.54596	2.89	8.3521	1.70000	3.39	11.4921	1.84120	3.89	15.1321	1.97231
1.40	1.9600	1.18322	1.90	3.6100	1.37840	2.40	5.7600	1.54919	2.90	8.4100	1.70294	3.40	11.5600	1.84391	3.90	15.2100	1.97484
1.41	1.9881	1.18743	1.91	3.6481	1.38203	2.41	5.8081	1.55242	2.91	8.4681	1.70587	3.41	11.6281	1.84662	3.91	15.2881	1.97737
1.42	2.0164	1.19164	1.92	3.6864	1.38564	2.42	5.8564	1.55563	2.92	8.5264	1.70880	3.42	11.6964	1.84932	3.92	15.3664	1.97990
1.43	2.0449	1.19583	1.93	3.7249	1.38924	2.43	5.9049	1.55885	2.93	8.5849	1.71172	3.43	11.7649	1.85203	3.93	15.4449	1.98242
1.44	2.0736	1.20000	1.94	3.7636	1.39284	2.44	5.9536	1.56205	2.94	8.6436	1.71464	3.44	11.8336	1.85472	3.94	15.5236	1.98494
1.45	2.1025	1.20416	1.95	3.8025	1.39642	2.45	6.0025	1.56525	2.95	8.7025	1.71756	3.45	11.9025	1.85742	3.95	15.6025	1.98746
1.46	2.1316	1.20830	1.96	3.8416	1.40000	2.46	6.0516	1.56844	2.96	8.7616	1.72047	3.46	11.9716	1.86011	3.96	15.6816	1.98997
1.47	2.1609	1.21244	1.97	3.8809	1.40357	2.47	6.1009	1.57162	2.97	8.8209	1.72337	3.47	12.0409	1.86279	3.97	15.7609	1.99249
1.48	2.1904	1.21655	1.98	3.9204	1.40712	2.48	6.1504	1.57480	2.98	8.8804	1.72627	3.48	12.1104	1.86548	3.98	15.8404	1.99499
1.49	2.2201	1.22066	1.99	3.9601	1.41067	2.49	6.2001	1.57797	2.99	8.9401	1.72916	3.49	12.1801	1.86815	3.99	15.9201	1.99750
1.50	2.2500	1.22474	2.00	4.0000	1.41421	2.50	6.2500	1.58114	3.00	9.0000	1.73205	3.50	12.2500	1.87083	4.00	16.0000	2.00000
n	n^2	$\sqrt{n}$	n	n^2	$\sqrt{n}$	n	n^2	$\sqrt{n}$	n	n^2	$\sqrt{n}$	n	n^2	$\sqrt{n}$	n	n^2	$\sqrt{n}$

TABLE OF SQUARES AND SQUARE ROOTS (Continued)

n	n^2	$\sqrt{n}$	n	n^2	$\sqrt{n}$	n	n^2	$\sqrt{n}$	n	n^2	$\sqrt{n}$	n	n^2	$\sqrt{n}$	n	n^2	$\sqrt{n}$
4.00	16.0000	2.00000	4.50	20.2500	2.12132	5.00	25.0000	2.23607	5.50	30.2500	2.34521	6.00	36.0000	2.44949	6.50	42.2500	2.54951
4.01	16.0801	2.00250	4.51	20.3401	2.12368	5.01	25.1001	2.23830	5.51	30.3601	2.34734	6.01	36.1201	2.45153	6.51	42.3801	2.55147
4.02	16.1604	2.00499	4.52	20.4304	2.12603	5.02	25.2004	2.24054	5.52	30.4704	2.34947	6.02	36.2404	2.45357	6.52	42.5104	2.55343
4.03	16.2409	2.00749	4.53	20.5209	2.12838	5.03	25.3009	2.24277	5.53	30.5809	2.35160	6.03	36.3609	2.45561	6.53	42.6409	2.55539
4.04	16.3216	2.00998	4.54	20.6116	2.13073	5.04	25.4016	2.24499	5.54	30.6916	2.35372	6.04	36.4816	2.45764	6.54	42.7716	2.55734
4.05	16.4025	2.01246	4.55	20.7025	2.13307	5.05	25.5025	2.24722	5.55	30.8025	2.35584	6.05	36.6025	2.45967	6.55	42.9025	2.55930
4.06	16.4836	2.01494	4.56	20.7936	2.13542	5.06	25.6036	2.24944	5.56	30.9136	2.35797	6.06	36.7236	2.46171	6.56	43.0336	2.56125
4.07	16.5649	2.01742	4.57	20.8849	2.13776	5.07	25.7049	2.25167	5.57	31.0249	2.36008	6.07	36.8449	2.46374	6.57	43.1649	2.56320
4.08	16.6464	2.01990	4.58	20.9764	2.14009	5.08	25.8064	2.25389	5.58	31.1364	2.36220	6.08	36.9664	2.46577	6.58	43.2964	2.56515
4.09	16.7281	2.02237	4.59	21.0681	2.14243	5.09	25.9081	2.25610	5.59	31.2481	2.36432	6.09	37.0881	2.46779	6.59	43.4281	2.56710
4.10	16.8100	2.02485	4.60	21.1600	2.14476	5.10	26.0100	2.25832	5.60	31.3600	2.36643	6.10	37.2100	2.46982	6.60	43.5600	2.56905
4.11	16.8921	2.02731	4.61	21.2521	2.14709	5.11	26.1121	2.26053	5.61	31.4721	2.36854	6.11	37.3321	2.47184	6.61	43.6921	2.57099
4.12	16.9744	2.02978	4.62	21.3444	2.14942	5.12	26.2144	2.26274	5.62	31.5844	2.37065	6.12	37.4544	2.47386	6.62	43.8244	2.57294
4.13	17.0569	2.03224	4.63	21.4369	2.15174	5.13	26.3169	2.26495	5.63	31.6969	2.37276	6.13	37.5769	2.47588	6.63	43.9569	2.57488
4.14	17.1396	2.03470	4.64	21.5296	2.15407	5.14	26.4196	2.26716	5.64	31.8096	2.37487	6.14	37.6996	2.47790	6.64	44.0896	2.57682
4.15	17.2225	2.03715	4.65	21.6225	2.15639	5.15	26.5225	2.26936	5.65	31.9225	2.37697	6.15	37.8225	2.47992	6.65	44.2225	2.57876
4.16	17.3056	2.03961	4.66	21.7156	2.15870	5.16	26.6256	2.27156	5.66	32.0356	2.37908	6.16	37.9456	2.48193	6.66	44.3556	2.58070
4.17	17.3889	2.04206	4.67	21.8089	2.16102	5.17	26.7289	2.27376	5.67	32.1489	2.38118	6.17	38.0689	2.48395	6.67	44.4889	2.58263
4.18	17.4724	2.04450	4.68	21.9024	2.16333	5.18	26.8324	2.27596	5.68	32.2624	2.38328	6.18	38.1924	2.48596	6.68	44.6224	2.58457
4.19	17.5561	2.04695	4.69	21.9961	2.16564	5.19	26.9361	2.27816	5.69	32.3761	2.38537	6.19	38.3161	2.48797	6.69	44.7561	2.58650
4.20	17.6400	2.04939	4.70	22.0900	2.16795	5.20	27.0400	2.28035	5.70	32.4900	2.38747	6.20	38.4400	2.48998	6.70	44.8900	2.58844
4.21	17.7241	2.05183	4.71	22.1841	2.17025	5.21	27.1441	2.28254	5.71	32.6041	2.38956	6.21	38.5641	2.49199	6.71	45.0241	2.59037
4.22	17.8084	2.05426	4.72	22.2784	2.17256	5.22	27.2484	2.28473	5.72	32.7184	2.39165	6.22	38.6884	2.49399	6.72	45.1584	2.59230
4.23	17.8929	2.05670	4.73	22.3729	2.17486	5.23	27.3529	2.28692	5.73	32.8329	2.39374	6.23	38.8129	2.49600	6.73	45.2929	2.59422
4.24	17.9776	2.05913	4.74	22.4676	2.17715	5.24	27.4576	2.28910	5.74	32.9476	2.39583	6.24	38.9376	2.49800	6.74	45.4276	2.59615
4.25	18.0625	2.06155	4.75	22.5625	2.17945	5.25	27.5625	2.29129	5.75	33.0625	2.39792	6.25	39.0625	2.50000	6.75	45.5625	2.59808
4.26	18.1476	2.06398	4.76	22.6576	2.18174	5.26	27.6676	2.29347	5.76	33.1776	2.40000	6.26	39.1876	2.50200	6.76	45.6976	2.60000
4.27	18.2329	2.06640	4.77	22.7529	2.18403	5.27	27.7729	2.29565	5.77	33.2929	2.40208	6.27	39.3129	2.50400	6.77	45.8329	2.60192
4.28	18.3184	2.06882	4.78	22.8484	2.18632	5.28	27.8784	2.29783	5.78	33.4084	2.40416	6.28	39.4384	2.50599	6.78	45.9684	2.60384
4.29	18.4041	2.07123	4.79	22.9441	2.18861	5.29	27.9841	2.30000	5.79	33.5241	2.40624	6.29	39.5641	2.50799	6.79	46.1041	2.60576
4.30	18.4900	2.07364	4.80	23.0400	2.19089	5.30	28.0900	2.30217	5.80	33.6400	2.40832	6.30	39.6900	2.50998	6.80	46.2400	2.60768
4.31	18.5761	2.07605	4.81	23.1361	2.19317	5.31	28.1961	2.30434	5.81	33.7561	2.41039	6.31	39.8161	2.51197	6.81	46.3761	2.60960
4.32	18.6624	2.07846	4.82	23.2324	2.19545	5.32	28.3024	2.30651	5.82	33.8724	2.41247	6.32	39.9424	2.51396	6.82	46.5124	2.61151
4.33	18.7489	2.08087	4.83	23.3289	2.19773	5.33	28.4089	2.30868	5.83	33.9889	2.41454	6.33	40.0689	2.51595	6.83	46.6489	2.61343
4.34	18.8356	2.08327	4.84	23.4256	2.20000	5.34	28.5156	2.31084	5.84	34.1056	2.41661	6.34	40.1956	2.51794	6.84	46.7856	2.61534
4.35	18.9225	2.08567	4.85	23.5225	2.20227	5.35	28.6225	2.31301	5.85	34.2225	2.41868	6.35	40.3225	2.51992	6.85	46.9225	2.61725
4.36	19.0096	2.08806	4.86	23.6196	2.20454	5.36	28.7296	2.31517	5.86	34.3396	2.42074	6.36	40.4496	2.52190	6.86	47.0596	2.61916
4.37	19.0969	2.09045	4.87	23.7169	2.20681	5.37	28.8369	2.31733	5.87	34.4569	2.42281	6.37	40.5769	2.52389	6.87	47.1969	2.62107
4.38	19.1844	2.09284	4.88	23.8144	2.20907	5.38	28.9444	2.31948	5.88	34.5744	2.42487	6.38	40.7044	2.52587	6.88	47.3344	2.62298
4.39	19.2721	2.09523	4.89	23.9121	2.21133	5.39	29.0521	2.32164	5.89	34.6921	2.42693	6.39	40.8321	2.52784	6.89	47.4721	2.62488
4.40	19.3600	2.09762	4.90	24.0100	2.21359	5.40	29.1600	2.32379	5.90	34.8100	2.42899	6.40	40.9600	2.52982	6.90	47.6100	2.62679
4.41	19.4481	2.10000	4.91	24.1081	2.21585	5.41	29.2681	2.32594	5.91	34.9281	2.43105	6.41	41.0881	2.53180	6.91	47.7481	2.62869
4.42	19.5364	2.10238	4.92	24.2064	2.21811	5.42	29.3764	2.32809	5.92	35.0464	2.43311	6.42	41.2164	2.53377	6.92	47.8864	2.63059
4.43	19.6249	2.10476	4.93	24.3049	2.22036	5.43	29.4849	2.33024	5.93	35.1649	2.43516	6.43	41.3449	2.53574	6.93	48.0249	2.63249
4.44	19.7136	2.10713	4.94	24.4036	2.22261	5.44	29.5936	2.33238	5.94	35.2836	2.43721	6.44	41.4736	2.53772	6.94	48.1636	2.63439
4.45	19.8025	2.10950	4.95	24.5025	2.22486	5.45	29.7025	2.33452	5.95	35.4025	2.43926	6.45	41.6025	2.53969	6.95	48.3025	2.63629
4.46	19.8916	2.11187	4.96	24.6016	2.22711	5.46	29.8116	2.33666	5.96	35.5216	2.44131	6.46	41.7316	2.54165	6.96	48.4416	2.63818
4.47	19.9809	2.11424	4.97	24.7009	2.22935	5.47	29.9209	2.33880	5.97	35.6409	2.44336	6.47	41.8609	2.54362	6.97	48.5809	2.64008
4.48	20.0704	2.11660	4.98	24.8004	2.23159	5.48	30.0304	2.34094	5.98	35.7604	2.44540	6.48	41.9904	2.54558	6.98	48.7204	2.64197
4.49	20.1601	2.11896	4.99	24.9001	2.23383	5.49	30.1401	2.34307	5.99	35.8801	2.44745	6.49	42.1201	2.54755	6.99	48.8601	2.64386
4.50	20.2500	2.12132	5.00	25.0000	2.23607	5.50	30.2500	2.34521	6.00	36.0000	2.44949	6.50	42.2500	2.54951	7.00	49.0000	2.64575
n	n^2	$\sqrt{n}$	n	n^2	$\sqrt{n}$	n	n^2	$\sqrt{n}$	n	n^2	$\sqrt{n}$	n	n^2	$\sqrt{n}$	n	n^2	$\sqrt{n}$

TABLE OF SQUARES AND SQUARE ROOTS (Continued)

n	n^2	$\sqrt{n}$	n	n^2	$\sqrt{n}$	n	n^2	$\sqrt{n}$	n	n^2	$\sqrt{n}$	n	n^2	$\sqrt{n}$	n	n^2	$\sqrt{n}$
7.00	49.0000	2.64575	7.50	56.2500	2.73861	8.00	64.0000	2.82843	8.50	72.2500	2.91548	9.00	81.0000	3.00000	9.50	90.2500	3.08221
7.01	49.1401	2.64764	7.51	56.4001	2.74044	8.01	64.1601	2.83019	8.51	72.4201	2.91719	9.01	81.1801	3.00167	9.51	90.4401	3.08383
7.02	49.2804	2.64953	7.52	56.5504	2.74226	8.02	64.3204	2.83196	8.52	72.5904	2.91890	9.02	81.3604	3.00333	9.52	90.6304	3.08545
7.03	49.4209	2.65141	7.53	56.7009	2.74408	8.03	64.4809	2.83373	8.53	72.7609	2.92062	9.03	81.5409	3.00500	9.53	90.8209	3.08707
7.04	49.5616	2.65330	7.54	56.8516	2.74591	8.04	64.6416	2.83549	8.54	72.9316	2.92233	9.04	81.7216	3.00666	9.54	91.0116	3.08869
7.05	49.7025	2.65518	7.55	57.0025	2.74773	8.05	64.8025	2.83725	8.55	73.1025	2.92404	9.05	81.9025	3.00832	9.55	91.2025	3.09031
7.06	49.8436	2.65707	7.56	57.1536	2.74955	8.06	64.9636	2.83901	8.56	73.2736	2.92575	9.06	82.0836	3.00998	9.56	91.3936	3.09192
7.07	49.9849	2.65895	7.57	57.3049	2.75136	8.07	65.1249	2.84077	8.57	73.4449	2.92746	9.07	82.2649	3.01164	9.57	91.5849	3.09354
7.08	50.1264	2.66083	7.58	57.4564	2.75318	8.08	65.2864	2.84253	8.58	73.6164	2.92916	9.08	82.4464	3.01330	9.58	91.7764	3.09516
7.09	50.2681	2.66271	7.59	57.6081	2.75500	8.09	65.4481	2.84429	8.59	73.7881	2.93087	9.09	82.6281	3.01496	9.59	91.9681	3.09677
7.10	50.4100	2.66458	7.60	57.7600	2.75681	8.10	65.6100	2.84605	8.60	73.9600	2.93258	9.10	82.8100	3.01662	9.60	92.1600	3.09839
7.11	50.5521	2.66646	7.61	57.9121	2.75862	8.11	65.7721	2.84781	8.61	74.1321	2.93428	9.11	82.9921	3.01828	9.61	92.3521	3.10000
7.12	50.6944	2.66833	7.62	58.0644	2.76043	8.12	65.9344	2.84956	8.62	74.3044	2.93598	9.12	83.1744	3.01993	9.62	92.5444	3.10161
7.13	50.8369	2.67021	7.63	58.2169	2.76225	8.13	66.0969	2.85132	8.63	74.4769	2.93769	9.13	83.3569	3.02159	9.63	92.7369	3.10322
7.14	50.9796	2.67208	7.64	58.3696	2.76405	8.14	66.2596	2.85307	8.64	74.6496	2.93939	9.14	83.5396	3.02324	9.64	92.9296	3.10483
7.15	51.1225	2.67395	7.65	58.5225	2.76586	8.15	66.4225	2.85482	8.65	74.8225	2.94109	9.15	83.7225	3.02490	9.65	93.1225	3.10644
7.16	51.2656	2.67582	7.66	58.6756	2.76767	8.16	66.5856	2.85657	8.66	74.9956	2.94279	9.16	83.9056	3.02655	9.66	93.3156	3.10805
7.17	51.4089	2.67769	7.67	58.8289	2.76948	8.17	66.7489	2.85832	8.67	75.1689	2.94449	9.17	84.0889	3.02820	9.67	93.5089	3.10966
7.18	51.5524	2.67955	7.68	58.9824	2.77128	8.18	66.9124	2.86007	8.68	75.3424	2.94618	9.18	84.2724	3.02985	9.68	93.7024	3.11127
7.19	51.6961	2.68142	7.69	59.1361	2.77308	8.19	67.0761	2.86182	8.69	75.5161	2.94788	9.19	84.4561	3.03150	9.69	93.8961	3.11288
7.20	51.8400	2.68328	7.70	59.2900	2.77489	8.20	67.2400	2.86356	8.70	75.6900	2.94958	9.20	84.6400	3.03315	9.70	94.0900	3.11448
7.21	51.9841	2.68514	7.71	59.4441	2.77669	8.21	67.4041	2.86531	8.71	75.8641	2.95127	9.21	84.8241	3.03480	9.71	94.2841	3.11609
7.22	52.1284	2.68701	7.72	59.5984	2.77849	8.22	67.5684	2.86705	8.72	76.0384	2.95296	9.22	85.0084	3.03645	9.72	94.4784	3.11769
7.23	52.2729	2.68887	7.73	59.7529	2.78029	8.23	67.7329	2.86880	8.73	76.2129	2.95466	9.23	85.1929	3.03809	9.73	94.6729	3.11929
7.24	52.4176	2.69072	7.74	59.9076	2.78209	8.24	67.8976	2.87054	8.74	76.3876	2.95635	9.24	85.3776	3.03974	9.74	94.8676	3.12090
7.25	52.5625	2.69258	7.75	60.0625	2.78388	8.25	68.0625	2.87228	8.75	76.5625	2.95804	9.25	85.5625	3.04138	9.75	95.0625	3.12250
7.26	52.7076	2.69444	7.76	60.2176	2.78568	8.26	68.2276	2.87402	8.76	76.7376	2.95973	9.26	85.7476	3.04302	9.76	95.2576	3.12410
7.27	52.8529	2.69629	7.77	60.3729	2.78747	8.27	68.3929	2.87576	8.77	76.9129	2.96142	9.27	85.9329	3.04467	9.77	95.4529	3.12570
7.28	52.9984	2.69815	7.78	60.5284	2.78927	8.28	68.5584	2.87750	8.78	77.0884	2.96311	9.28	86.1184	3.04631	9.78	95.6484	3.12730
7.29	53.1441	2.70000	7.79	60.6841	2.79106	8.29	68.7241	2.87924	8.79	77.2641	2.96479	9.29	86.3041	3.04795	9.79	95.8441	3.12890
7.30	53.2900	2.70185	7.80	60.8400	2.79285	8.30	68.8900	2.88097	8.80	77.4400	2.96648	9.30	86.4900	3.04959	9.80	96.0400	3.13050
7.31	53.4361	2.70370	7.81	60.9961	2.79464	8.31	69.0561	2.88271	8.81	77.6161	2.96816	9.31	86.6761	3.05123	9.81	96.2361	3.13209
7.32	53.5824	2.70555	7.82	61.1524	2.79643	8.32	69.2224	2.88444	8.82	77.7924	2.96985	9.32	86.8624	3.05287	9.82	96.4324	3.13369
7.33	53.7289	2.70740	7.83	61.3089	2.79821	8.33	69.3889	2.88617	8.83	77.9689	2.97153	9.33	87.0489	3.05450	9.83	96.6289	3.13528
7.34	53.8756	2.70924	7.84	61.4656	2.80000	8.34	69.5556	2.88791	8.84	78.1456	2.97321	9.34	87.2356	3.05614	9.84	96.8256	3.13688
7.35	54.0225	2.71109	7.85	61.6225	2.80179	8.35	69.7225	2.88964	8.85	78.3225	2.97489	9.35	87.4225	3.05778	9.85	97.0225	3.13847
7.36	54.1696	2.71293	7.86	61.7796	2.80357	8.36	69.8896	2.89137	8.86	78.4996	2.97658	9.36	87.6096	3.05941	9.86	97.2196	3.14006
7.37	54.3169	2.71477	7.87	61.9369	2.80535	8.37	70.0569	2.89310	8.87	78.6769	2.97825	9.37	87.7969	3.06105	9.87	97.4169	3.14166
7.38	54.4644	2.71662	7.88	62.0944	2.80713	8.38	70.2244	2.89482	8.88	78.8544	2.97993	9.38	87.9844	3.06268	9.88	97.6144	3.14325
7.39	54.6121	2.71846	7.89	62.2521	2.80891	8.39	70.3921	2.89655	8.89	79.0321	2.98161	9.39	88.1721	3.06431	9.89	97.8121	3.14484
7.40	54.7600	2.72029	7.90	62.4100	2.81069	8.40	70.5600	2.89828	8.90	79.2100	2.98329	9.40	88.3600	3.06594	9.90	98.0100	3.14643
7.41	54.9081	2.72213	7.91	62.5681	2.81247	8.41	70.7281	2.90000	8.91	79.3881	2.98496	9.41	88.5481	3.06757	9.91	98.2081	3.14802
7.42	55.0564	2.72397	7.92	62.7264	2.81425	8.42	70.8964	2.90172	8.92	79.5664	2.98664	9.42	88.7364	3.06920	9.92	98.4064	3.14960
7.43	55.2049	2.72580	7.93	62.8849	2.81603	8.43	71.0649	2.90345	8.93	79.7449	2.98831	9.43	88.9249	3.07083	9.93	98.6049	3.15119
7.44	55.3536	2.72764	7.94	63.0436	2.81780	8.44	71.2336	2.90517	8.94	79.9236	2.98998	9.44	89.1136	3.07246	9.94	98.8036	3.15278
7.45	55.5025	2.72947	7.95	63.2025	2.81957	8.45	71.4025	2.90689	8.95	80.1025	2.99166	9.45	89.3025	3.07409	9.95	99.0025	3.15436
7.46	55.6516	2.73130	7.96	63.3616	2.82135	8.46	71.5716	2.90861	8.96	80.2816	2.99333	9.46	89.4916	3.07571	9.96	99.2016	3.15595
7.47	55.8009	2.73313	7.97	63.5209	2.82312	8.47	71.7409	2.91033	8.97	80.4609	2.99500	9.47	89.6809	3.07734	9.97	99.4009	3.15753
7.48	55.9504	2.73496	7.98	63.6804	2.82489	8.48	71.9104	2.91204	8.98	80.6404	2.99666	9.48	89.8704	3.07896	9.98	99.6004	3.15911
7.49	56.1001	2.73679	7.99	63.8401	2.82666	8.49	72.0801	2.91376	8.99	80.8201	2.99833	9.49	90.0601	3.08058	9.99	99.8001	3.16070
7.50	56.2500	2.73861	8.00	64.0000	2.82843	8.50	72.2500	2.91548	9.00	81.0000	3.00000	9.50	90.2500	3.08221	10.00	100.0000	3.16228
n	n^2	$\sqrt{n}$	n	n^2	$\sqrt{n}$	n	n^2	$\sqrt{n}$	n	n^2	$\sqrt{n}$	n	n^2	$\sqrt{n}$	n	n^2	$\sqrt{n}$

Index

Answers to problem sets

problem set A

1. $\dfrac{3}{5}$ 2. $\dfrac{1}{8}$ 3. $3\dfrac{2}{3}$ 4. $1\dfrac{3}{7}$ 5. $1\dfrac{4}{11}$ 6. $\dfrac{8}{15}$ 7. $\dfrac{7}{40}$ 8. $\dfrac{13}{24}$ 9. $\dfrac{18}{65}$

10. $\dfrac{7}{15}$ 11. $\dfrac{43}{45}$ 12. $\dfrac{3}{40}$ 13. $\dfrac{9}{52}$ 14. $\dfrac{1}{20}$ 15. $\dfrac{11}{17}$ 16. $\dfrac{11}{26}$ 17. $\dfrac{6}{35}$

18. $1\dfrac{11}{56}$ 19. $\dfrac{17}{20}$ 20. $\dfrac{21}{22}$ 21. $\dfrac{19}{40}$ 22. $7\dfrac{3}{10}$ 23. $5\dfrac{7}{10}$ 24. $13\dfrac{17}{24}$

25. $8\dfrac{21}{40}$ 26. $7\dfrac{11}{15}$ 27. $9\dfrac{14}{15}$ 28. $387\dfrac{1}{5}$ 29. $7\dfrac{8}{15}$ 30. $20\dfrac{5}{8}$ 31. $8\dfrac{14}{15}$

32. $402\dfrac{25}{33}$ 33. $63\dfrac{7}{10}$ 34. $36\dfrac{47}{104}$ 35. $5\dfrac{43}{65}$ 36. $13\dfrac{77}{190}$ 37. $21\dfrac{47}{170}$

problem set 1

1. A number is an idea, and a numeral is the symbol that we use to represent the idea.

2. The decimal system. 3. The Hindus of India 4. 0, 1, 2, 3, 4, 5, 6, 7, 8, 9

5. 1, 2, 3, 4, . . . 6. Counting numbers are also called natural numbers.

7. A positive real number is a number that can be used to describe a physical distance greater than zero.

8. Yes 9. The mark we place on the number line to indicate the location of the number

10. The origin 11. The number associated with the point

12. No, because zero has no sign 13. A positive number

14. All real numbers can be graphed on the number line. 15. Yes 16. $+2$

17.

18. That we are discussing both positive and negative numbers.

19. If its graph is farther to the right on the number line

20. No. Its graph is to the left of the graph of –9.

21. $7\dfrac{8}{15}$ 22. $14\dfrac{13}{40}$ 23. $1\dfrac{13}{15}$ 24. $\dfrac{32}{45}$ 25. $7\dfrac{7}{9}$ 26. $2\dfrac{1}{4}$ 27. $13\dfrac{1}{3}$

28. $5\dfrac{9}{13}$ 29. $10\dfrac{7}{8}$ 30. $15\dfrac{19}{20}$ 31. $1\dfrac{2}{5}$ 32. $\dfrac{24}{35}$ 33. $1\dfrac{11}{28}$ 34. $1\dfrac{17}{33}$

35. $1\dfrac{3}{16}$ 36. $3\dfrac{9}{17}$ 37. $4\dfrac{23}{24}$ 38. $397\dfrac{28}{39}$ 39. $\dfrac{1}{3}$ 40. $3\dfrac{1}{3}$

problem set 2

1. $\{\ldots, -3, -2, -1, 0, 1, 2, 3, \ldots\}$ 2. $\{0, 1, 2, 3, \ldots\}$ 3. $\{1, 2, 3, 4, \ldots\}$

4. $\{1, 2, 3, 4, \ldots\}$ 5. Not equal to   6. Members of the set

7. The numbers are addends, and the answer is a sum.

8. A number that can be used to describe a physical distance greater than zero

9. Yes   10. If its graph is farther to the right on the number line

11. No, because its graph is to the left of the graph of -1.   12. No, because zero has no sign.

13. A positive number (unless the number is 0).   14. The Hindus of India

15. Yes because they can both be graphed on a number line.

16. The numbers are called factors, and the answer is called a product.   17. A quotient.

18. The product of any number and zero is zero. The sum of zero and any number is the number itself.

19. The top is called the numerator and the bottom is called the denominator.

20. 4.00723   21. .06496   22. $\dfrac{43}{58}$ 23. 201.3   24. $1\dfrac{7}{25}$ 25. $15\dfrac{11}{24}$

26. $18\dfrac{5}{12}$ 27. $2\dfrac{5}{12}$ 28. 12.37   29. $1\dfrac{29}{39}$

problem set 3

1. 7   2. 7   3. 5   4. -18 5. -10

6.

7.

8.

9.

10.

11.

12. $\{1, 2, 3, 4, \ldots\}$ 13. $\{\ldots, -3, -2, -1, 0, 1, 2, 3, \ldots\}$ 14. A quotient

15. The number that is associated with the point

16. The mark that we make on the number line to indicate the location of the numbers

17. A product **18.** One or two or more quantities that are multiplied

19. 10 is the dividend and 5 is the divisor.

20. All real numbers can be graphed on the number line.

21. $39\frac{9}{16}$ **22.** $3\frac{5}{16}$ **23.** $\frac{1}{6}$ **24.** $11\frac{11}{24}$ **25.** 4.003 **26.** 46.8636

27. .12704808 **28.** $\frac{18}{25}$ **29.** $27\frac{17}{35}$ **30.** $11\frac{27}{40}$ **31.** 4.002 **32.** 443.61413

33. .1465712 **34.** $1\frac{1}{4}$ **35.** $\frac{51}{104}$

problem set 4

Answers 1 to 3 may be looked up in the lessons.
4. 3 **5.** -4 **6.** -10 **7.** -15 **8.** -13 **9.** -11 **10.** -17
11. -35 **12.** -28 **13.** -31 **14.** -2 **15.** 1 **16.** -9 **17.** -13
18. 1 **19.** -38 **20.** -14 **21.** 6 **22.** 0 **23.** -5 **24.** -11
25. -7 **26.** -15 **27.** 0 **28.** -12 **29.** -3 **30.** 0 **31.** -11
32. -10 **33.** -19 **34.** 2 **35.** 0 **36.** 3 **37.** -9 **38.** -7
39. -2 **40.** 3 **41.** $7\frac{23}{40}$ **42.** $50\frac{10}{11}$ **43.** .00148785 **44.** .022

problem set 5

1. $+45,654$ **2.** 0 **3.** -4 **4.** 4 **5.** -4 **6.** 4 **7.** 8 **8.** 3
9. -4 **10.** 8 **11.** -2 **12.** -2 **13.** 1 **14.** 8 **15.** -4 **16.** 8
17. 3 **18.** 5 **19.** 9 **20.** 0 **21.** -3 **22.** -19 **23.** 10 **24.** 7
25. 20 **26.** -11 **27.** -14 **28.** 0 **29.** -9 **30.** 4 **31.** $26\frac{109}{120}$
32. $1\frac{3}{65}$ **33.** $1\frac{32}{45}$ **34.** 5.423 **35.** .402 **36.** 6.314408 **37.** .06

problem set 6

1. The number that is associated with the point **2.** $+2$
3. If its graph is farther right on the number line **4.** A product **5.** A quotient
6. $\{\ldots, -2, -1, 0, 1, 2, \ldots\}$ **7.** 12 **8.** 12 **9.** -12 **10.** -6 **11.** 15
12. -10 **13.** -10 **14.** -4 **15.** 0 **16.** -2 **17.** 5 **18.** -3
19. -13 **20.** 2 **21.** -17 **22.** -14 **23.** -7 **24.** 1 **25.** -5
26. 15 **27.** 6 **28.** -2 **29.** 7 **30.** -7 **31.** 1 **32.** -10 **33.** 41.265
34. $13\frac{21}{40}$ **35.** $\frac{176}{245}$ **36.** $\frac{33}{40}$ **37.** $8\frac{5}{6}$ **38.** $\frac{18}{49}$ **39.** $-1\frac{3}{40}$ **40.** $1\frac{43}{85}$
41. 8100

problem set 7

1. Division **2.** Multiplication **3.** Operations that "undo" each other **4.** -3
5. -48 **6.** 5 **7.** 20 **8.** 10 **9.** 12 **10.** -20 **11.** -24 **12.** -2
13. -4 **14.** -11 **15.** -17 **16.** -3 **17.** -13 **18.** -5 **19.** 1
20. -8 **21.** -8 **22.** -9 **23.** 3 **24.** 2 **25.** -7 **26.** 4 **27.** -4
28. 4 **29.** -7 **30.** -4 **31.** -15 **32.** -4 **33.** $\frac{17}{37}$ **34.** .02
35. $1\frac{1}{2}$

problem set 8

1. Yes   2. No. For example: $(5) + (-2) = 3$
3. Because we are trying to undo a multiplication that was never performed   4. 24
5. 48   6. -24 7. 5   8. -2 9. -2 10. 18   11. -24 12. -24
13. 48   14. -5 15. 2   16. 13   17. -6 18. -13 19. -4
20. -4 21. -4 22. -10 23. 2   24. 1   25. 1   26. -13 27. 5
28. 8   29. -5 30. 4   31. 6   32. $\dfrac{51}{110}$ 33. $-3\dfrac{14}{15}$ 34. $2\dfrac{8}{51}$
35. 3.00093   36. 3.03   37. $-\dfrac{14}{15}$ 38. $-4\dfrac{43}{88}$

problem set 9

1. Zero does not because $\dfrac{1}{0}$ is not defined.   2. (a) 1; (b) no
3. (a) Answer to a division; (b) Answer to a multiplication   4. 8   5. -1 6. 5
7. -2 8. 51   9. 4   10. 1   11. -18 12. 32   13. 0   14. -20
15. 1   16. 18   17. 22   18. -20 19. 2   20. 2   21. -9 22. 6
23. -16 24. -12 25. -2 26. 0   27. 0   28. 8   29. 18   30. 4
31. $-4\dfrac{4}{35}$ 32. $\dfrac{16}{37}$ 33. 8   34. $13\dfrac{3}{10}$

problem set 10

1. $\{0, 1, 2, 3, \ldots\}$ 2. $\{\ldots, -2, -1, 0, 1, 2, \ldots\}$
3. (a) One of the things that is multiplied to form a product; (b) Answer to a division; (c) Answer to an addition
4. 5   5. 33   6. 28   7. -2 8. -1 9. -40 10. 2   11. $-\dfrac{1}{13}$
12. 79   13. 80   14. 6   15. -25 16. 42   17. 0   18. 8   19. 4
20. -31 21. -27 22. -12 23. -74 24. 62   25. -4 26. 118
27. 35   28. -17 29. -19 30. 145   31. -6 32. -22 33. -4.02
34. $-\dfrac{22}{31}$ 35. $-1\dfrac{13}{15}$

problem set 11

1. $\{0, 1, 2, 3, \ldots\}$ 2. Any number that can be graphed on the number line
3. (a) One of the things that is multiplied to form a product; (b) Answer to a multiplication (c) Answer to a division.
4. 18   5. -17 6. 10   7. 6   8. -12 9. -2 10. -20 11. -6
12. -4 13. 15   14. 36   15. 13   16. -17 17. 25   18. 78   19. 5
20. $-\dfrac{1}{13}$ 21. 1   22. $\dfrac{5}{7}$ 23. -2 24. $-\dfrac{3}{4}$ 25. -20 26. -9 27. 3
28. -16 29. $-\dfrac{9}{10}$ 30. 1.05   31. $-1\dfrac{1}{50}$ 32. $-\dfrac{27}{35}$

problem set 12

1. Negative   2. Negative   3. $-\dfrac{1}{5}$ 4. Multiplicative inverse   5. 3   6. -2
7. 2   8. 16   9. 10   10. 30   11. 84   12. 30   13. 3   14. 2
15. $-\dfrac{5}{4}$ 16. $-\dfrac{18}{11}$ 17. -13 18. 3   19. $-\dfrac{10}{7}$ 20. 16   21. 13
22. 13   23. -1 24. 2   25. -6 26. -16 27. -23 28. -12
29. 61   30. -15 31. $-1\dfrac{1}{30}$ 32. -2.2 33. $-1\dfrac{31}{65}$

problem set 13

1. A numerical expression contains only numbers, while an algebraic expression may contain only numbers, only letters, or both numbers and letters.
2. The number the expression represents
3. (a) A letter that represents an unspecified number; (b) A variable   4. 40   5. 75
6. 4   7. -9 8. 12   9. 14   10. -30 11. -4 12. 0   13. -5
14. -8 15. -2 16. 46   17. -2 18. 1   19. -11 20. 5
21. -16 22. -126 23. -4 24. 6   25. 1   26. -32 27. -7
28. -40 29. -20 30. 12   31. -3 32. $-\dfrac{43}{44}$ 33. $-\dfrac{4}{5}$

problem set 14

1. (a) One of the things that is multiplied to form a product; (b) Answer to a division; (c) Answer to an addition
2. -8 3. -2 4. -25 5. 0   6. -1 7. 48   8. 18   9. -30
10. 70   11. 30   12. 9   13. 1   14. 32   15. 96   16. 40   17. -30
18. -54 19. -6 20. -14 21. 5   22. 15   23. $\dfrac{29}{3}$ 24. $-\dfrac{9}{14}$
25. $-\dfrac{11}{6}$ 26. 14   27. -13 28. 30   29. -34 30. 7   31. 0
32. 15

problem set 15

1. (a) A coefficient is a factor. Usually this word is reserved for the numerical factor of a term.
2. A factor made up of letters   3. $a(b + c) = ab + ac$ 4. 8   5. 8   6. 9
7. $mnx + mny + 2mnp$ 8. $4xmn - 12x$ 9. $-6x + 12$ 10. $2xa + 2xbc$
11. $3ax + 6ay$ 12. 10   13. 9   14. 35   15. 4   16. 22   17. -60
18. 0   19. -2 20. -20 21. -40 22. -15 23. 4   24. -48
25. -2 26. 36   27. $\dfrac{11}{8}$ 28. -7 29. -9 30. 20   31. $-.0162$
32. $-1\dfrac{17}{33}$ 33. $-1\dfrac{33}{40}$

problem set 16

1. A single symbol, a product, or a quotient   2. Like terms
3. (a) One of the things that is multiplied to form a product; (b) Answer to a multiplication; (c) Answer to a division
4. $-2x + xy + 5$ 5. $-7yx - 3xyz$ 6. $-6 + 10mxy$ 7. $9mxy - 4y - 22$
8. $-2xyz + 2xy$ 9. $2yx - x - 4$ 10. $4x + 2xy$ 11. $3xy - 6xm$
12. $2pxy - 6pk$ 13. 27   14. 26   15. -10 16. 0   17. -2 18. 0
19. -180 20. -162 21. -21 22. -7 23. -7 24. -7 25. 8
26. -49 27. 10   28. $-\dfrac{4}{3}$ 29. $-\dfrac{7}{6}$ 30. $-\dfrac{32}{57}$ 31. $2\dfrac{7}{15}$ 32. -2.03

problem set 17

1. One solution is: $2(3 + 4) = (2)(3) + (2)(4)$ 2. All integers are real numbers.   3. 9
4. -81 5. -32 6. 94   7. -41 8. $2xmy - my$ 9. $-5pxk - kp - 3kx$
10. $-3mc + 2m - 2$ 11. $-2ax - 4a - 8$ 12. $2ax - 6p - 4$ 13. $4x - apx$
14. $20xyp - 8xyc$ 15. $8kc - 4ka + 12km$ 16. -5 17. 4   18. 4   19. 64
20. 31   21. 36   22. 9   23. -32 24. -9 25. -10 26. -17
27. 140   28. -11 29. 1   30. -4 31. 3   32. 6

problem set 18

1. (a) $\{0, 1, 2, 3, \ldots\}$ (b) $\{\ldots, -3, -2, -1, 0, 1, 2, 3, \ldots\}$ 2. Reciprocal
3. Positive   4. 18   5. -16 6. 12   7. -1 8. -38 9. 16
10. -6 11. -23 12. 24   13. -1 14. 22   15. 5   16. 8   17. -20
18. -10 19. 3   20. 36   21. $-2xy + x + 3$ 22. $-3bpx + 2pb - 3$
23. $15 - 5k + kx$ 24. $-4 + py - y$ 25. $16x - 8xp$ 26. $3x + 12$
27. $-2xa + 6px$ 28. 40   29. -27 30. 39   31. $-\dfrac{8}{15}$ 32. $1\dfrac{5}{16}$
33. $-.004$

problem set 19

1. A letter that stands for an unspecified number
2. Terms whose letters represent the same number no matter what numbers are used to replace the letters
3. $x^5 y^5 m^5$ 4. $x^4 y^3$ 5. $m^6 x^4 p^6$ 6. $x^5 m^6$ 7. $k^6 y^7$ 8. $a^5 b^8$
9. $4x^2 yp^5 - 5xy^2 p^5$ 10. $4x^2 y - 3x^2 y^2$ 11. $8b^2 a - 3ab$ 12. $2x^2 y - 4xy + x^2$
13. $5m^2 pxy - 3y^2 pxm$ 14. $3px - 2xy$ 15. $3a^2 - 2a^2 b$ 16. $10 - 20p$ 17. 5
18. 15   19. -35 20. 124   21. -49 22. -72 23. -1 24. 2
25. -1 26. -3 27. 5   28. 14   29. -35 30. 26   31. $-\dfrac{3}{4}$
32. $-2\dfrac{7}{30}$ 33. $.000048$

problem set 20

1. (a) $6 = 4 + 1 + 1$; (b) $x + 2 = x$; (c) $x + 2 = 4$
2. -2 is a solution and 5 is not a solution.   3. Both -2 and 5 are solutions.
4. 5 is a solution and -2 is not a solution.   5. $x^5 y^5$ 6. $p^6 y^3 m^6$ 7. $8k^{12} n^6$
8. $a^{11} b^3$ 9. $m^3 p^3 a^7$ 10. $4p^3 x^5 k^4$ 11. $x^2 + 3x - 2$ 12. $-3xy + 2xy^2$
13. $-8ymx^2 + 23x$ 14. 0   15. $-4x^2 + 6x - 13$ 16. $-8m^2 y + 6y$
17. $3x^2 - 5x + 8$ 18. $3ax - 2a$ 19. $20xy - 8axy$ 20. $18ax - 24x$ 21. 1
22. -13 23. -37 24. 45   25. -10 26. -35 27. -5 28. -18
29. -60 30. -17 31. $-1\dfrac{15}{76}$ 32. $-\dfrac{15}{28}$ 33. $.12$

problem set 21

1. Find the value(s) of the variable that will make the equation a true equation.
2. Equations that have the same answer(s)
3. 2   4. 45   5. 36   6. -22 7. 86   8. 19   9. -11 10. $-\dfrac{5}{8}$
11. $-\dfrac{19}{9}$ 12. $x^4 y^5 m^5$ 13. $x^4 y^3 m^2$ 14. $x^5 y^6$ 15. $-3x^2 y + 3yx - 2y^2 x$
16. $-2x^3 + 5x^2 + x + 4$ 17. -5 18. $-4x^2 + 6x - 9$ 19. $4ax + 8bx$
20. $6x + 12$ 21. $4pxmy - 12pxab^2$ 22. -35 23. -30 24. -26 25. 7
26. -16 27. -18 28. 22   29. -5 30. $-\dfrac{57}{10}$

problem set 22

1. Reciprocal   2. 1   3. 4   4. 27   5. 7   6. 35   7. 10   8. 15
9. $\dfrac{35}{2}$ 10. $\dfrac{4}{3}$ 11. $\dfrac{35}{2}$ 12. 2   13. $\dfrac{17}{10}$ 14. 5   15. -2 16. 57
17. 2   18. $\dfrac{3}{2}$ 19. $\dfrac{9}{8}$ 20. 4   21. $\dfrac{5}{34}$ 22. $m^2 x^4 y^6 p^2$ 23. $3x^3 p^6 y^3$

24. $-5a^2 - 2a + 8$ **25.** $2x^2ym - 2my^2x$ **26.** $3axy - 5pxy$ **27.** -11 **28.** 4
29. -17 **30.** 24

problem set 23

1. $\dfrac{3}{2}$ **2.** $\dfrac{1}{6}$ **3.** 4 **4.** $-\dfrac{26}{5}$ **5.** 16 **6.** $\dfrac{33}{4}$ **7.** $-\dfrac{7}{2}$ **8.** $\dfrac{35}{2}$ **9.** $\dfrac{25}{4}$

10. $\dfrac{5}{3}$ **11.** $\dfrac{27}{28}$ **12.** $-\dfrac{1}{14}$ **13.** .1 **14.** 65.15 **15.** x^7k^7y **16.** $a^8b^5x^5$

17. $6a^2xy$ **18.** $c - 4$ **19.** $-x^2a^2$ **20.** $8xy - 12x + 8xa$ **21.** $3x - 12$

22. 19 **23.** -14 **24.** -3 **25.** -10 **26.** -24 **27.** -40 **28.** $\dfrac{39}{8}$

29. -44

problem set 24

1. Equations that have the same solution sets **2.** Yes. The number 4 satisfies both equations.

3. $\dfrac{6}{35}$ **4.** -3 **5.** $\dfrac{11}{3}$ **6.** 120.3 **7.** -5 **8.** 3 **9.** $\dfrac{4}{5}$

10. -6 **11.** -3 **12.** 3 **13.** 1 **14** $\dfrac{9}{5}$ **15.** -1 **16.** m^6y^9

17. $k^{10}m^6a^3$ **18.** $-5m^2xyz$ **19.** $5a^2bc - bc$ **20.** $xa - 2a - 3$ **21.** $28 - 12x$
22. $3amxy + 6apxy$ **23.** 0 **24.** 24 **25.** 136 **26.** -9 **27.** 16 **28.** 4
29. 27 **30.** 0

problem set 25

1. $a(b + c) = ab + ac$ **2.** -4 **3.** $\dfrac{5}{3}$ **4.** 210 **5.** 2000 **6.** $-\dfrac{33}{14}$

7. $\dfrac{33}{20}$ **8.** $\dfrac{9}{35}$ **9.** $-\dfrac{1}{5}$ **10.** $\dfrac{1}{2}$ **11.** $-\dfrac{6}{5}$ **12.** 1 **13.** $-\dfrac{3}{4}$ **14.** 1

15. $p^2x^6y^4$ **16.** $3p^7x^6y^4$ **17.** $-4x^2y + xy - 8$ **18.** $8x^2 - 7x - 5$
19. $-5p^2xy$ **20.** $xy^3 - x^3yz$ **21.** $4ax^3 - 8x^2$ **22.** $4x^5y^4 - 20x^2y^3$
23. $2ax^2y^2 - 8x^2y^6$ **24.** $40m^5x^2 - 24m^2x^6 + 16m^2x^2$ **25.** -8 **26.** 32

27. -11 **28.** 0 **29.** -31 **30.** $-\dfrac{1}{7}$

problem set 26

1. 92 **2.** $\dfrac{3}{5}$ **3.** 600 **4.** 10 **5.** $\dfrac{5}{6}$ **6.** 98 **7.** $\dfrac{17}{5}$ **8.** $\dfrac{13}{4}$ **9.** -279

10. $-\dfrac{5}{4}$ **11.** $\dfrac{17}{6}$ **12.** $-\dfrac{5}{2}$ **13.** $-\dfrac{16}{3}$ **14.** $\dfrac{10}{3}$ **15.** 4 **16.** x^3y^{11}

17. m^7y^4 **18.** $-3m^2xy + 8mxy^2$ **19.** $4pc - p - 6c$ **20.** $4a^2xy + 3ax^2y$
21. $3x^3y^4 - 5x^2y^4$ **22.** $3x^7a - 6x^8a^4$ **23.** $p^3x^2y^2 - 3p^3x^2y$ **24.** 108 **25.** 59

26. -43 **27.** -30 **28.** -17 **29.** -23 **30.** $\dfrac{20}{7}$

problem set 27

1. (a) $\{\ldots, -3, -2, -1, 0, 1, 2, 3, \ldots\}$; (b) $\{0, 1, 2, 3, \ldots\}$ **2.** $\dfrac{1000}{3}$ **3.** $\dfrac{7}{8}$

4. 225 **5.** $\dfrac{44}{5}$ **6.** 2 **7.** $\dfrac{3}{2}$ **8.** 0 **9.** 4 **10.** $\dfrac{9}{2}$ **11.** 1577.2

12. -5 **13.** $-\dfrac{3}{13}$ **14.** $\dfrac{1}{9}$ **15.** 27 **16.** $\dfrac{1}{9}$ **17.** 4 **18.** x^2m^{-3}

19. $x^3y^{-1}z^6$ **20.** $y^2a^{-3}x^{-4}$ **21.** p^2m^{-1} **22.** $6px^2y - 4$ **23.** $3m^2x^2y + 8m^2xy^2$
24. $6x^4y^4p^2 - 4x^2y^3p^2 + 2x^2y^3p$ **25.** $4cm^4z^3 - 20m^2z^6$ **26.** 18   **27.** 85
28. 36   **29.** -5 **30.** 6

problem set 28

1. 76   **2.** .62   **3.** 36.12   **4.** $\dfrac{49}{3}$ **5.** $\dfrac{44}{5}$ **6.** $-\dfrac{11}{2}$ **7.** 5.05   **8.** -2

9. $-\dfrac{7}{5}$ **10.** $-\dfrac{5}{6}$ **11.** $\dfrac{3}{2}$ **12.** $-\dfrac{5}{6}$ **13.** $\dfrac{1}{9}$ **14.** 16   **15.** -64 **16.** 8

17. $-\dfrac{1}{125}$ **18.** $x^5zp^{-3} - 4$ **19.** $1 + 4x^2$ **20.** $3 - 12x^3y^{-5}$ **21.** $2 - 8x^{-8}y^4$

22. $1 - 3y^2x$ **23.** $1 - y^2x$ **24.** $7m^3x^2y - 2x^2y$ **25.** $7abc^2 - 6ab^2c$ **26.** -24

27. -12 **28.** 132   **29.** -43 **30.** $\dfrac{16}{21}$

problem set 29

1. 2000   **2.** 105   **3.** 22,214   **4.** $5(2N - 5)$ **5.** $7(N + 10)$ **6.** $5N - 8$

7. $3(-N - 7)$ **8.** $-\dfrac{11}{2}$ **9.** 4   **10.** -45 **11.** 15   **12.** -3 **13.** $\dfrac{6}{5}$

14. $\dfrac{1}{36}$ **15.** $\dfrac{1}{27}$ **16.** 16   **17.** 125   **18.** $1 - \dfrac{2}{x}$ **19.** $12x^3p^5 - 8$

20. $1 + 3xp^2$ **21.** $8x^{-10} - 12x^{-5}$ **22.** $2x^4 - 6p^{-2}x^3y^{-2}$ **23.** $4y^5p^3 - 20x^6y^2p^{-6}$

24. $3xmp^{-2} - 5mx$ **25.** $-7k^2yp^{-4}$ **26.** -31 **27.** -21 **28.** -16 **29.** 27

30. -35

problem set 30

1. $7(x - 5)$ **2.** $-2x - 7$ **3.** $7x - 51$ **4.** $4x - 15$ **5.** 180   **6.** $\dfrac{2}{55}$

7. $\dfrac{441}{8}$ **8.** 66   **9.** $-\dfrac{1}{36}$ **10.** $\dfrac{11}{3}$ **11.** $-\dfrac{1}{5}$ **12.** $\dfrac{2}{3}$ **13.** -3 **14.** $\dfrac{22}{3}$

15. $\dfrac{69}{20}$ **16.** -64 **17.** $\dfrac{1}{64}$ **18.** $1 - 5y^4$ **19.** $1 + 3x^5y^7$ **20.** $-6x^6 + 12p$

21. $5x^6y^6 - 25y^{-2}$ **22.** $3m^6n^3 - 3$ **23.** $2p^5x^5 - 6x^5p^{-5}$ **24.** $3xym^2z^3 + x^3ym^2z^2$

25. $11xy$ **26.** -10 **27.** 52   **28.** -56 **29.** 3   **30.** -57

problem set 31

1. 31   **2.** 75   **3.** 2   **4.** 3   **5.** -1 **6.** $\dfrac{9}{5}$ **7.** 25   **8.** 49   **9.** $-\dfrac{1}{18}$

10. 2   **11.** $\dfrac{13}{2}$ **12.** -12 **13.** $\dfrac{4}{3}$ **14.** $\dfrac{28}{3}$ **15.** 4   **16.** $\dfrac{1}{36}$ **17.** -27

18. $-\dfrac{15}{4}$ **19.** -64 **20.** $15xp + 6$ **21.** $2x^{-3}p^{-3} - 6x^3p^{-6}$

22. $2p^9xym^4 - 10p^4x^{-2}y^{-3}$ **23.** $3x^4p^{-4} - 6p^{-2}$ **24.** $5x^{-9}y^{-2} - x^3y^2$

25. $2xym^2 - x^2ym$ **26.** 4   **27.** -327 **28.** -14 **29.** -129 **30.** 5

problem set 32

1. 12   **2.** 8   **3.** -3 **4.** $\dfrac{77}{15}$ **5.** $\dfrac{9}{4}$ **6.** 51.25   **7.** $\dfrac{28}{5}$ **8.** $\dfrac{119}{4}$

9. 92   **10.** 4   **11.** $-\dfrac{4}{7}$ **12.** $\dfrac{9}{2}$ **13.** $-\dfrac{7}{2}$ **14.** $2 \cdot 2 \cdot 2 \cdot 2 \cdot 5$ **15.** $3 \cdot 7 \cdot 7$

16. $2 \cdot 5 \cdot 5 \cdot 5$ **17.** $2 \cdot 3 \cdot 3 \cdot 5 \cdot 5$ **18.** $-\dfrac{1}{27}$ **19.** $\dfrac{1}{8}$ **20.** $2x^{-4} + 2y^5$

21. $x^3 p^5 - 3$ **22.** $4x^2 y^2 - 12x^4 y^{-2}$ **23.** $4 - 3p^7 x^{-3}$ **24.** $5x^2 y^3$ **25.** xyz

26. 2 **27.** -72 **28.** 62 **29.** -101 **30.** $\dfrac{5}{4}$

problem set 33

1. $-\dfrac{110}{3}$ **2.** 5 **3.** 42 **4.** $\dfrac{1}{16}$ **5.** 4 **6.** -1 **7.** 2 **8.** -460

9. $-\dfrac{9}{5}$ **10.** 2 **11.** $3xy^2$ **12.** $4z^3 m^2 p$ **13.** $2x^2 yp$ **14.** $2ab^2 c$

15. $5xy^2 m^2$ **16.** $2 \cdot 3 \cdot 3 \cdot 5 \cdot 7$ **17.** $2 \cdot 2 \cdot 2 \cdot 3 \cdot 5 \cdot 5$ **18.** $\dfrac{1}{16}$ **19.** 64

20. $3 - 9x^6 y^2$ **21.** $2p^{-3} x^5 - 6$ **22.** $4x^{-6} - 8xy^6$ **23.** $1 - 2y^{12} x^5$

24. $2x^2 yz - xyz^2$ **25.** $7x^2 + x^3 y^3$ **26.** 32 **27.** -37 **28.** -4 **29.** -49

30. $\dfrac{40}{13}$

problem set 34

1. $-\dfrac{49}{5}$ **2.** $-\dfrac{5}{8}$ **3.** $\dfrac{87}{104}$ **4.** $.012$ **5.** $\dfrac{85}{8}$ **6.** $\dfrac{11}{21}$ **7.** $\dfrac{13}{7}$ **8.** 2.55

9. $-\dfrac{13}{4}$ **10.** $\dfrac{13}{7}$ **11.** $(2a)(x + c)$ **12.** $a^2 xm^2(ax + m - a^2 x^2)$

13. $2a^2 bz^3(2ab^3 + z)$ **14.** $3x^2 y^2 p(x^2 - 2y^3 p^3)$ **15.** $2a^2 x^2 m(3am^4 + a^2 x^3 m^4 + 2)$

16. $2 \cdot 5 \cdot 5 \cdot 5$ **17.** $2 \cdot 2 \cdot 2 \cdot 3 \cdot 3 \cdot 5$ **18.** $\dfrac{1}{81}$ **19.** 64 **20.** $1 - 4x^2 y^6$

21. $p^5 y^{10} - 1$ **22.** $x^5 - 3x^7 y^{-1}$ **23.** $2 - 4x^2 p^5 y^5$ **24.** $x^2 y^{-2} p^4 + 7x^2 y^2 p^4$

25. $6y^2 - 10x^2$ **26.** 48 **27.** -16 **28.** -12 **29.** -18 **30.** 3

problem set 35

1. 11 **2.** -1 **3.** $\dfrac{5}{3}$ **4.** $\dfrac{19}{6}$ **5.** $.04515$ **6.** $\dfrac{235}{48}$ **7.** 8 **8.** $\dfrac{3}{4}$

9. $\dfrac{10}{3}$ **10.** $2a^2 x(2y^4 p - 3x^3)$ **11.** $3ax^2 y^4(ax^2 y^2 + 3 - 2ax^2 yz)$ **12.** $x - 1$

13. $x + 1$ **14.** $1 - 3x$ **15.** $x + 3$ **16.** $1 + y$ **17.** $\dfrac{1 - k}{y}$ **18.** $2 \cdot 3 \cdot 3 \cdot 3 \cdot 5$

19. 256 **20.** $1 - x^{-1} y^4$ **21.** $12z - 21x^2 y^{-3}$ **22.** $3x^5 y^{-2} - 9y^7$

23. $8xy - 6x^5 y^{-1}$ **24.** $7x^2 ym^5 - 8xym^5$ **25.** $6x^4 y^{-4} + x^2 y^{-3} - x^4 y^{-3}$ **26.** -3

27. 125 **28.** 24 **29.** -13 **30.** $-\dfrac{34}{7}$

problem set 36

1. 5 **2.** 20 **3.** $\dfrac{16}{3}$ **4.** 6 **5.** 3.05 **6.** $\dfrac{68}{65}$ **7.** $\dfrac{11}{7}$ **8.** $\dfrac{1}{2}$ **9.** 3.15

10. $3ax^2 y^2(4x^3 ay^5 - 1)$ **11.** $3a^2 x^3 y(5a^3 xy^5 + a^2 y^6 - 3x^3)$ **12.** $x - 1$ **13.** $1 - 3x$

14. $x + 1$ **15.** $\dfrac{x - 1}{m}$ **16.** $2 \cdot 3 \cdot 5 \cdot 5 \cdot 5$ **17.** $-\dfrac{1}{9}$ **18.** $\dfrac{1}{9}$ **19.** -9

20. $\dfrac{1}{9}$ **21.** $\dfrac{x^4}{y^2} - \dfrac{3x^2 y^2}{m}$ **22.** $\dfrac{axp}{zk} - \dfrac{2m^5 p^4}{z}$ **23.** $\dfrac{ab^4}{c^2 k} - \dfrac{2axb^2}{c^2}$ **24.** $1 - 4p^{-6} x^{-1} y^5$

25. $5m^2 xy - 12x^2 ym^2$ **26.** $-3x$ **27.** -34 **28.** -15 **29.** -2 **30.** -1

J. C. Warren

problem set 37

1. 15 2. 4 3. $\frac{3}{2}$ 4. .06 5. $\frac{43}{18}$ 6. $\frac{22}{25}$ 7. 7 8. -1 9. -2

10. $4a^2x^2y^4(xy - 2a^2)$ 11. $2axm^5(3a + x^3m - 9a^4x^2)$ 12. $x - 3$ 13. $\frac{4x - 8}{x}$

14. 15. 16.

17. 18. $-\frac{1}{27}$ 19. $-\frac{1}{27}$ 20. -25 21. $\frac{p^3k^2}{x^3} - \frac{p^4}{x^2}$

22. $4ax - \frac{8x^3}{a^2}$ 23. $\frac{a^2x^3}{c^3} - \frac{3a^2}{x^2}$ 24. $1 - 5x^2p^7$ 25. $-2x^3y^3p$

26. $13a^2x - 2ax^2$ 27. 0 28. -23 29. 9 30. $\frac{5}{9}$

problem set 38

1. 65 2. 48 3. 1560 4. 5 5. .06 6. $\frac{216}{77}$ 7. -1.7 8. $-\frac{13}{28}$

9. 0 10. -5 11. -5 12. $3xy^3z^5(x - 3y^3z)$ 13. $4xy(x - 3y + 6x^2y^2)$

14. $x + 1$ 15. $1 - 2kp$ 16. 17. 18. $\frac{1}{4}$

19. $-\frac{1}{4}$ 20. $-\frac{1}{8}$ 21. $4 - \frac{p^4}{x^2}$ 22. $\frac{4a^3}{x} - 12a$ 23. $\frac{m^2p^2}{k^2} - 1$

24. $x^{-3}y^{-2} - 3x^{-2}k^5$ 25. $-\frac{2p^6}{x^3} - \frac{p^7}{x^3}$ 26. $3xy$ 27. 11 28. 301 29. 322

30. -6

problem set 39

1. -20 2. $-\frac{12}{5}$ 3. 65 4. 91 5. $\frac{28}{3}$ 6. $\frac{33}{7}$ 7. $\frac{5}{4}$ 8. $\frac{3}{2}$

9. -30 10. $4ax(3a + x^3)$ 11. $ay(2x^2a^2 - x + 4y)$ 12. $x - 3$ 13. $1 + 4y$

14. 15. 16.

17. 18. $-\frac{1}{4}$ 19. $\frac{1}{4}$ 20. $-\frac{1}{4}$ 21. $\frac{p^3}{a^2} - \frac{p^3}{xa^2}$

22. $\frac{4x^5y}{m^3} - \frac{4x^2y^2}{m^2}$ 23. $\frac{mpa}{x^2} - \frac{mp}{x^3}$ 24. $1 - 3p^3x^5$ 25. $3xyp^2$ 26. $-8ay - 4a^4y$

27. 1 28. 28 29. -3 30. $-\frac{7}{4}$

problem set 40

1. -3 2. $-\frac{5}{2}$ 3. 980 4. $\frac{48}{17}$ 5. 3.856 6. $\frac{21}{2}$ 7. 0 8. $\frac{8}{5}$

9. -2.1 10. $3x^2yp^3(y^4p^3 - 3y^3 + 4p)$ 11. $2xy^2(x - 3x^3 - 6y^3)$ 12. $x - 5$

13. $1 + 4xy$ 14. 15. $2 \cdot 2 \cdot 2 \cdot 3 \cdot 3 \cdot 5$ 16. $-\frac{1}{16}$ 17. 27

18. $x^{-6}y^4z^4$ 19. $p^{-6}y^2z^6$ 20. $m^5z^{-1}x^5y^{-9}$ 21. $\frac{p^2a}{xbc} - \frac{p^2}{xc}$ 22. $\frac{3x^3y}{m^3} - \frac{6x^3y}{m^2}$

23. $1 - 4p^2k^5$ 24. $3x^6y^{-6} - 6y^{-2}$ 25. 30 26. 3 27. -12 28. -19

29. $\frac{10}{3}$

problem set 41

1. -6 2. $-\dfrac{1}{3}$ 3. 918 4. $\dfrac{2}{9}$ 5. $2 \cdot 5 \cdot 13$ 6. 25 7. $\dfrac{1}{2}$ 8. $\dfrac{1}{3}$

9. 667 10. $2x^2m^3y(2m^2 - x^2y^2)$ 11. $2m^2x^2(2x^3 - 1 + 3m^3)$ 12. $1 - x$

13. $3 - x$ 14. $-6\ -5\ -4$ 15. $-3\ -2\ -1$ 16. $-\dfrac{1}{9}$ 17. $\dfrac{1}{9}$

18. $\dfrac{x^2}{y^2}$ 19. $\dfrac{x}{y^4z^4}$ 20. $\dfrac{x}{y^4p^2}$ 21. $\dfrac{y^5}{x^4} - \dfrac{x^2}{y}$ 22. $ap^2 - 4k^2p^4$ 23. $\dfrac{1}{a^6x^5} - \dfrac{1}{x^2}$

24. $\dfrac{b}{m^5} - \dfrac{4a}{b^5}$ 25. $-6x^2yp$ 26. $2m^2x^2y - 2x^2my$ 27. 78 28. 48 29. -8

30. -6

problem set 42

1. -4 2. 8 3. 1848 4. 2.02 5. $\dfrac{261}{32}$ 6. 120 7. 2 8. $-\dfrac{8}{3}$

9. 3.5 10. $mk(6k^4m - 2k^2 - 1)$ 11. $x^3y^2m(x - ym + 5x^3m)$ 12. $1 - 3x$

13. $\dfrac{y-1}{x}$ 14. $0\quad 5$ 15. $2\quad 4$ 16. $2 \cdot 2 \cdot 2 \cdot 5 \cdot 7$

17. $x^{-2}y^8m^{-2}$ 18. xy^8m^{-5} 19. $x^3y^5p^2$ 20. $\dfrac{z}{xy^2} - \dfrac{y^3}{x^2}$ 21. $\dfrac{b}{a} - \dfrac{2b^3}{a^3}$

22. $2bx^{-2}y^2$ 23. $3a^{-3}b^4$ 24. $2a^{-3}b^2c + 3a^{-3}b^2c^{-1}$ 25. $-2m^2y^{-2}$ 26. -36

27. -1 28. -2 29. $\dfrac{1}{27}$ 30. -2

problem set 43

1. -3 2. 8000 farthings   3. $\dfrac{3}{5}$ 4. 6 5. $\dfrac{3}{8}$ 6. $\dfrac{9}{4}$ 7. 1.45

8. $y = \dfrac{4}{3}x + 2$ 9. $p = \dfrac{4}{3}a + \dfrac{5}{3}$ 10. $x = 1 - \dfrac{3}{4}y$ 11. $y = \dfrac{3}{2} - \dfrac{3}{2}x$ 12. $y = \dfrac{5}{3} - x$

13. $y = \dfrac{1}{4}x + 1$ 14. $2x^2y^2(3x^2 - 2z)$ 15. $xyz(8x^4y - 16xyz - 1)$ 16. $1 - 2y$

17. $5 - 25xy$ 18. $0\ 2\ 4$ 19. $5 \cdot 5 \cdot 5 \cdot 3 \cdot 3$ 20. $\dfrac{1}{x^9y}$ 21. $\dfrac{4m^5}{x^7y^{11}}$

22. $x^5z^{-6} - 3$ 23. xy^5 24. 0 25. $-2x^2y^{-2}$ 26. 18 27. 120 28. 10

29. -12 30. -2

problem set 44

1. 7 2. 600 3. $\dfrac{8}{85}$ 4. $\dfrac{54}{25}$ 5. 39 6. $\dfrac{24}{11}$ 7. $-\dfrac{4}{3}$ 8. $y = \dfrac{4}{3} - \dfrac{1}{3}x$

9. $y = \dfrac{1}{4}x + \dfrac{7}{4}$ 10. $y = 2 - k - 2x$ 11. $y = \dfrac{2}{3}x + \dfrac{7}{3}$ 12. $3a^2b^4c^5(1 - 2b^2c)$

13. $2xa(4x - 2xa + a)$ 14. $z - 1$ 15. $2x$ 16. $0\qquad 5$ 17. 600

18. 1200 19. 1200 20. $p^{-2}z^6$ 21. $a^{-1}pk^{-3}$ 22. p^{-1} 23. $1 - 3m^4z$

24. $3a^2x^2y^{-3} - 4ax^2y^{-3}$ 25. $2m^2xy^{-2} - 3m^2x^{-1}y^2$ 26. 1 27. 3 28. 3

29. $-\dfrac{127}{8}$ 30. $\dfrac{2}{3}$

problem set 45

1. $\dfrac{38}{7}$ 2. 450 3. $.622$ 4. $\dfrac{15}{2}$ 5. 8 6. $-\dfrac{5}{2}$ 7. $-.5$

8. $y = \dfrac{2}{5}x + \dfrac{4}{5}$ 9. $2 - x = y$ 10. $5x^2y^2m^2(y^3 - 2x^2m)$ 11. $xyz(-x + 2z)$

12. $x + 1$ **13.** $y + 1$ **14.**   **15.** 600   **16.** 270

17. $30a^2b^3$ **18.** $12mx^3$ **19.** $6x^2y^5z^2$ **20.** $\dfrac{p^4}{x^4y^2}$ **21.** $\dfrac{m^7}{k^2}$ **22.** $\dfrac{b^4}{a^2}$

23. $y^2 + 4m^3y^7$ **24.** $5ay^2x^{-1}$ **25.** $5m^2k^5 - 3mk^5$ **26.** 12   **27.** -18 **28.** 6

29. $-\dfrac{1}{27}$ **30.** $-\dfrac{2}{3}$

---

problem set 46

1. $\dfrac{40}{3}$ **2.** 21,530   **3.** $\dfrac{1}{3}$ **4.** 14   **5.** $-\dfrac{13}{11}$ **6.** $y = \dfrac{5}{2} - \dfrac{3}{2}x$

7. $y = \dfrac{1}{3}x + \dfrac{7}{3}$ **8.** $\dfrac{3}{11}$ **9.** $\dfrac{7 + a + b}{a + 6}$ **10.** $\dfrac{4 - 6ax}{2x^2 + y}$ **11.** $\dfrac{2 + z}{a^2 + 7y}$

12. $\dfrac{c^2 + 4}{b}$ **13.**   **14.** 1125   **15.** $2c^3$ **16.** $12c^4$ **17.** b^3c^2

18. $x^2y^4p^2(4y - 3x^3)$ **19.** $x - 1$ **20.** $\dfrac{1}{x^{-2}y^3}$ **21.** $\dfrac{1}{m^2p^{-8}}$ **22.** $\dfrac{1}{x^{-8}y^{-8}}$

23. $y^{-4}p^{-4} - x^2p^6y^{-4}$ **24.** $-6x^2y^{-2}$ **25.** $3a^2xy - 6xy^{-1}$ **26.** 1   **27.** 5

28. -6 **29.** -16 **30.** $-\dfrac{15}{2}$

---

problem set 47

1. -2 **2.** 1278   **3.** .023   **4.** $-\dfrac{100}{11}$ **5.** $\dfrac{7}{8}$ **6.** $\dfrac{5}{4}$ **7.** .9

8. $y = \dfrac{5}{3}x + \dfrac{4}{3}$ **9.** $y = \dfrac{1}{2}x + \dfrac{5}{2}$ **10.** $\dfrac{3}{4}$ **11.** $\dfrac{29}{35}$ **12.** $\dfrac{59}{28}$ **13.** $\dfrac{8c + 2b + bc}{2bc}$

14. $\dfrac{2m + 8c - c^3}{2c^3}$ **15.** $\dfrac{bp - 2ab + 4c}{4b}$ **16.** $\dfrac{12ac^2 + 9c + 4m}{12c^4}$ **17.** $\dfrac{4x + b + 4cy}{4y}$

18. $\dfrac{16 + c + 20a}{4a}$ **19.**   **20.** 2520   **21.** $x^2ym^5(xy - 3m)$

22. $1 - 4a^4x^3$ **23.** $4 - y$ **24.** x^6y^5 **25.** $x^{-1}y^2$ **26.** $3a^2x^3y^{-1} - 3a^2x^3$

27. 5   **28.** 4   **29.** $-\dfrac{1}{27}$ **30.** -2

---

problem set 48

1. -3 **2.** 252   **3.** $\dfrac{20}{3}$ **4.** $-\dfrac{27}{10}$ **5.** $-\dfrac{10}{3}$ **6.** -7 **7.** $y = \dfrac{3}{2} - \dfrac{1}{2}x$

8. $y = \dfrac{2}{3}x + \dfrac{4}{3}$ **9.** 1800   **10.** x^2c^2d **11.** $\dfrac{31}{30}$ **12.** $\dfrac{62}{63}$ **13.** $\dfrac{c^2xa + b + dc^2x^2}{c^2x^2}$

14. $\dfrac{16x^2 + 12x - 3}{4x^4}$ **15.** $\dfrac{16x + c + 4mx^3}{4x^3}$ **16.** $\dfrac{8ab^2 + 2bc + a^2}{8b^3}$

17. $\dfrac{4m + 2ak - 3a^2}{4a^5}$ **18.** $\dfrac{4b^2 + 6a^2 + ca}{8a^3b^2}$ **19.**

20.   **21.** $4m^3xp(2xy^4 - 1)$ **22.** $1 - axy^7$ **23.** $x^4 - 1$

24. $m^8p^6y^{-7}$ **25.** $x^{-5}y^8$ **26.** $-7my^{-1} + 2m^{-1}$ **27.** -22 **28.** 50   **29.** $\dfrac{1}{9}$

30. 16

problem set 49

1. -5 2. 560 3. $\dfrac{2}{17}$ 4. 75

of 75
Before, 100%

15 is 20%
60 is 80%
After

5. 45

of 300
Before, 100%

45 is 15%
255 is 85%
After

6. 1600

of 1600
Before, 100%

640 is 40%
960 is 60%
After

7. 12 percent

of 8300
Before, 100%

996 is 12%
7304 is 88%
After

8. 1400

of 1400
Before, 100%

280 is 20%
1120 is 80%
After

9. 14 10. -9.2

11. $y = \dfrac{3}{4}x - \dfrac{7}{4}$

12. $y = \dfrac{3}{2}x + \dfrac{5}{2}$ 13. 840 14. $8x^2ym^3$ 15. 1 16. $\dfrac{ac + xb + dxc}{xc}$

17. $\dfrac{k^2c^2m + kd^2 - 3p}{xk^2c^3}$ 18. $\dfrac{4ad - a^2b^2c - bdm}{a^3db^2}$ 19. ├──────⊕───► 0 2 4

20. ●━━━━━⊕ 2 5   21. $9x^3ym(2x^2y - m^4)$ 22. $x^{-2} - 3y^{-6}p^{-6}$ 23. $x - 1$

24. k^3p^{-8} 25. $m^8x^{-6}y^4$ 26. $m^2x^2 - 3m^{-6}x^2 - 3mx^{-1}$ 27. 11 28. -13

29. -9 30. 40

problem set 50

1. -8 2. 294 3. $\dfrac{2}{9}$ 4. 900

of 900
Before, 100%

72 is 8%
828 is 92%
After

5. 5 percent

of 860
Before, 100%

43 is 5%
817 is 95%
After

6. 672

of 4200
Before, 100%

672 is 16%
3528 is 84%
After

7. 5000

Before, 100%   After

8. 2 percent

Before, 100%   After

9. $-\dfrac{19}{7}$ **10.** -9 **11.** $y = \dfrac{3}{4}x + \dfrac{7}{4}$ **12.** 1500   **13.** $\dfrac{103}{105}$

14. $\dfrac{xym^2 + xmb + ac}{x^2ym}$ **15.** $\dfrac{24cd + 10d + 7c}{8c^2d^2}$ **16.** $\dfrac{pm + 5 + x^2am}{xam}$

17. $3x^3 + x^2 + 6x + 9$ **18.** $-3x^4 + 3x^3 + x^2 - 6x + 3$

19. $-2x^5 - x^4 + 3x^2 + 3x - 15$ **20.** $-3x^3 - 7x^2 + 11$ **21.**

22. $x^2ym(1 - 4m^2 + 2x^2y^2m^5)$ **23.** $1 - x^2$ **24.** x^3ym^{-5} **25.** $y^{-5}p^3 - 3y^{-3}p^3$

26. $4x^2y^{-2}p + 2xy^2$ **27.** -120 **28.** -36 **29.** $-\dfrac{1}{27}$ **30.** 18

problem set 51

1. -8 **2.** 1377   **3.** $\dfrac{184}{399}$ **4.** 5600

Before, 100%   After

5. 20 percent

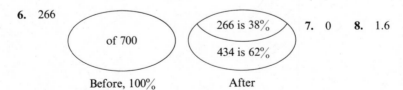

Before, 100%   After

6. 266

of 700   266 is 38%   **7.** 0   **8.** 1.6

434 is 62%

Before, 100%   After

9. $7x^5 + 6x^4 - 4x + 8$ **10.** $12x^2 + 7x - 10$ **11.** $4x^2 - 18x - 10$

12. $12x^2 - 14x - 10$ **13.** $9x^2 + 12x + 4$ **14.** $4x^3 + 2x^2 + 14x - 8$

15. $4x^2 - 4x + 1$ **16.** 1575   **17.** $\dfrac{4m^2a + 2km^2 + xa}{8m^2a^2}$ **18.** $\dfrac{x^2m + 4x^2 - 3y^2}{x^4y}$

19. $\dfrac{mc + x + dmc^2}{mc^2}$ **20.** $\dfrac{x^2a + by - kxya^2}{xya^2}$ **21.**

22. $xy(1 - 4xym - 3x)$ **23.** $x^2 - 1$ **24.** x^5y^{-7} **25.** $-2x^2$

26. $-8x^2yp^3 - 2xyp^{-1}$ **27.** 8   **28.** -8 **29.** -16 **30.** 6

problem set 52

1. -6 **2.** 2940 **3.** 2.03 **4.** 96

of 60 | 96 is 160%
Before, 100% | After

5. 187.5

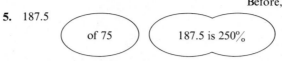

of 75 | 187.5 is 250%
Before, 100% | After

6. 1400 percent

of 50 | 700 is 1400%
Before, 100% | After

7. $-\dfrac{11}{14}$ **8.** $\dfrac{13}{4}$ **9.** $-\dfrac{3}{4}$

10. $-2x^4 + x^3 - x^2 - 12x + 25$ **11.** $10x^2 + 14x - 12$ **12.** $x^2 + 6x + 9$

13. $3x^2 + 5x - 12$ **14.** $4x^2 - 49$ **15.** 1800 **16.** $\dfrac{cxa + b + dcx^2}{cx^2}$

17. $\dfrac{15km - 20apk + 18p}{15p^2k^2}$ **18.** $\dfrac{amx^2 + bcx^2 - 4m^2c}{mx^2}$ **19.** $\dfrac{2kxc + 2bkmc - 4m}{2kmc^2}$

20. ●———⊕ (4, 9) **21.** $2xym(2x - 3 + xym)$ **22.** $x - 1$ **23.** $x^5p^{-2}y^{-2}$

24. $1 + 4x^8a^{-2}y^{-4}$ **25.** $-xym^2$ **26.** -66 **27.** 126 **28.** $-\dfrac{1}{81}$ **29.** -16

30. -21

problem set 53

1. -6 **2.** 9380 **3.** $\dfrac{2}{17}$ **4.** 60

of 60 | 78 is 130%
Before, 100% | After

5. 70

of 70 | 10.5 is 15% / 59.5 is 85%
Before, 100% | After

6. 64 percent

of 450 | 162 is 36% / 288 is 64%
Before, 100% | After

7. $\dfrac{285}{136}$ **8.** -4.25

9. $3x^2 + 10x - 8$ **10.** $25x^2 - 30x + 9$ **11.** $6x^2 - 19x + 10$

12. $30x^2 - 47x + 7$ **13., 14., 15., 16.**

17. $\dfrac{4a^2x^2 + 2a^2 - 5x}{a^2x^2}$ **18.** $\dfrac{p - 4am - ka^2m}{a^2m}$ **19.** $\dfrac{3a^2mx^2 + 4x^2 + 2am}{am^2x}$

20. $\dfrac{2ap^2 - 5 - 3mp^2x}{p^2x}$ **21.**   **22.** $\dfrac{x-1}{a}$ **23.** $y^{-4}m^4$

24. $6x^{-8}y^{-8} - 9x^{-2}y^{-6}a^{-1}$ **25.** $3x^2y^2m^{-5} - x^2y^{-2}m^{-5}$ **26.** -210 **27.** -378

28. $-\dfrac{1}{4}$ **29.** -8 **30.** 4

problem set 54

1. $\dfrac{1}{2}$ **2.** 70 **3.** $\dfrac{55}{64}$ **4.** 4875

of 1300 — Before, 100%   4875 is 375% — After

5. 400

of 400 — Before, 100%   140 is 35% / 260 is 65% — After

6. 150 percent

of 18 — Before, 100%   27 is 150% — After

7. $\dfrac{40}{99}$ **8.** $\dfrac{1}{7}$

9. $10x^2 + 7x - 12$ **10.** $5x^2 - 18x - 8$ **11.** $49x^2 - 70x + 25$

12., 13., 14., 15., 16.

17. $\dfrac{4xy^3 + x^2y - 2a}{xy^3}$

18. $\dfrac{xca - 3a - 4x^2c + 3ac + 2c}{x^3c^2}$ **19.** $\dfrac{mc^2b - p + 4m^3pc^2}{m^3pc^2}$ **20.** $\dfrac{4x^3p^5 + 8 + x^2yp^4}{4x^2p^5}$

21.   **22.** $3y$ **23.** $p^{14}x^{-2}$ **24.** 1 **25.** $1 - 12y^{-9}p^{-1}$

26. $-2xy^{-1} - 2xy^{-2}$ **27.** -17 **28.** -4 **29.** -39 **30.** -24

problem set 55

1. 4 **2.** 826 **3.** $\dfrac{115}{8}$ **4.** 3220

of 700 — Before, 100%   3220 is 460% — After

5. $700{,}000$

of 700,000 — Before, 100%   49,000 is 7% / 651,000 is 93% — After

6. .5 percent

of 2000 — Before, 100%   10 is .5% / 1990 is 99.5% — After

7. $\dfrac{87}{190}$ **8.** 1

9. $6x^2 + 31x + 35$ **10.** $16x^2 - 24x + 9$ **11.** 3150

12., 13., 14., 15., 16.

17. $\dfrac{x^4m^2 + 2x^4 - 3}{m^2x^3}$ **18.** $\dfrac{-2x + 7k^2p}{k^2p}$ **19.** $\dfrac{-3x^4p^2 + 2x^2 - 5xp}{x^3p^2}$

20. $\dfrac{2k + k^2m^2 - 24m^2}{8m^2}$ **21.**
$\quad -5 \;\; -3 \;\; -1$ **22.** $2kp(2kz - 3k^2pz^5 - kpz^2 - 2)$

23. $1 - 4y$ **24.** $k^8m^{-6}p$ **25.** $1 - 15x^3m^2$ **26.** $-a^2k^2y^{-1} - 6k^{-6}y^{-1}$ **27.** 59

28. 52   **29.** -26 **30.** -47

problem set 56

1. -5 **2.** 280   **3.** .003   **4.** 700

of 700   2429 is 347%

Before, 100%   After

5. 261

of 300   261 is 87% / 39 is 13%

Before, 100%   After

6. 250 percent

of 460   1150 is 250%

Before, 100%   After

7. $-\dfrac{2}{3}$ **8.** $-\dfrac{8}{9}$ **9.** $12x^2 + 14x - 10$ **10.** $25x^2 - 10x + 1$ **11.** $35x^2 - 74x + 35$

12., 13., 14., 15.

16. $\dfrac{6x - ab - 4cx^3y}{4x^3y}$ **17.** $\dfrac{ax + ay + bx}{x^2 + xy}$ **18.** $\dfrac{4x^2 + ax + ac + bc}{xa + xb}$

19. $\dfrac{ma + mb + cx + dxm}{xm}$ **20.** $\dfrac{ab^2 - xb^3 - xb^2c + 4b + 4c}{b^3 + b^2c}$ **21.**
$\quad\quad 4 \quad\quad 7$

22. $3kbm^2(3km^2 - b^3 + 4b^2m)$ **23.** $1 - 2m$ **24.** $m^{-3}x^{-5}$

25. $x^2p^{-2}m^4k^{-1} - p^2m^4$ **26.** $3mx + 4m^{-1}x^{-1}$ **27.** 9   **28.** 3   **29.** -6

30. 8

problem set 57

1. $-\dfrac{13}{6}$ **2.** 546 **3.** $\dfrac{2}{31}$ **4.** 60

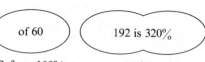

of 60 | 192 is 320%

Before, 100% | After

5. 4 percent

of 98 | 3.92 is 4%
| 94.08 is 96%

Before, 100% | After

6. 165.6

of 72 | 165.6 is 230%

Before, 100% | After

7. $-\dfrac{625}{396}$ **8.** $\dfrac{1}{6}$

9. $2x^2 - 10x + 12$ **10.** $9x^2 + 30x + 25$ **11., 12., 13.**

14. $\dfrac{x^3 + x^2y + y^2}{x^4y}$ **15.** $\dfrac{7y - 3x}{y(x - y)}$ **16.** $\dfrac{9xy + 2bxy - 3x^2y + 6}{x^2y}$

17. [number line: 0 ... 4, arrow] **18.** $4x^2py(3x^2p^2 - xyz - 2py)$ **19.** -4

20. $\dfrac{3^{-2}x^4y^{-10}}{z^{-8}}$ **21.** $2^{-5}a^{10}b^{-10}z^{50}$ **22.** $2^{-2}y^8z^{-2}$ **23.** $25x^{-5}y^{10}$ **24.** $x^{-3}y^3$

25. $a - 9x^3y^6$ **26.** $3xy - y^2x - x^2y$ **27.** 39 **28.** -16 **29.** -10 **30.** 32

problem set 58

1. 400 **2.** 140 percent **3.** 248 **4.** $-\dfrac{20}{7}$ **5.** $(2, 3)$ **6.** $(1, -5)$

7. $(-1, 3)$ **8.** $(4, -4)$ **9.** $(1, 1)$ **10.** 11,025 **11.** $x^2 - 6x + 9$

12., 13., 14.

15. $\dfrac{-x + a^3 - a^2b}{a^2b}$ **16.** $\dfrac{ax + bx - 3}{ax}$

17. $\dfrac{4a^2 - 6a + b}{a^2}$ **18.** $5x^2y^4p^2(5y - 3x^3p^2 + 2x^2p^2z)$ **19.** $1 - pq$ **20.** $x^{-8}y^2$

21. x^6y^3 **22.** x^{-15} **23.** $\dfrac{4y^{-12}p^{-6}}{9}$ **24.** $x^{-5}y^7$ **25.** $1 - 8x^{-4}y^{-2}$

26. $4xp^3y - 2xp^{10}y - 2x^2p^{10}y$ **27.** 15 **28.** -45 **29.** 0 **30.** 12

problem set 59

1. -3 2. 70 3. $\dfrac{31}{2}$ 4. 300 5. 300 6. 168 7. $-\dfrac{95}{128}$ 8. $-\dfrac{4}{7}$

9. $(1, 2)$ 10. $(-2, -2)$ 11. $(3, 3)$ 12. $-3x^2 + 10x - 8$ 13. $x^3 + 3x^2 + 4x + 2$

14., 15., 16.

17. $\dfrac{ad}{bc}$ 18. $\dfrac{a}{bc}$ 19. $\dfrac{b}{c}$ 20. $a^2 + ax$

21. $\dfrac{4y^2 - 3x - 3y}{xy^2 + y^3}$ 22. $\dfrac{ax + ay + 4ax^3y + 4ax^2y^2 - mx^2y}{x^3y + x^2y^2}$ 23. $\dfrac{x + y + ay^2}{a^2}$

24. $x^{-9}y^{10}$ 25. $x^{-3}y^{-9}$ 26. $9p^9x^5y^7$ 27. $x^2 - 1$ 28. $x^{-8}p^{-2} + 2x^{-8}p^{-4}y^2$

29. $5xy^3z^{-1}$ 30. 22

problem set 60

1. 40 2. 20 percent 3. $.36$ 4. 272 5. $K = \{0, 1, 2, 4, 5, 7\}$

6. (a) True; (b) False; (c) True; (d) False 7. $(8, -1)$ 8. $(2, 2)$ 9. $(-15, -27)$

10. $x^2 + 24x + 144$ 11. $4x^3 - 12x^2 + 17x - 12$ 12., 13.

14. $\dfrac{1}{ax}$ 15. $\dfrac{ab + b^2}{c}$ 16. $\dfrac{x}{xc + yc}$ 17. $\dfrac{m^2c^2}{a}$ 18. $\dfrac{16xpy - 6xp + 2ay}{8x^3py}$

19. $\dfrac{m + kb + kc}{b^2 + bc}$ 20. $\dfrac{6apm^2 - a^2xm + 8kx}{2xm^4a}$ 21. $4kax(k - 2ax + 3k^2a^3x)$

22. $\dfrac{x - xz}{z}$ 23. $x^4y^{-20}m^{-2}$ 24. $k^{-3}p^{10}$ 25. $x^{-5}y^2p^8$ 26. $1 - 2m^{-4}k$

27. $6x^2y^2m - 3x^3y^3$ 28. 4 29. 8 30. -2

problem set 61

1. $-\dfrac{1}{3}$ 2. 25 percent 3. 15 4. 375 5. $\dfrac{3}{82}$ 6. $.02$ 7. $-\dfrac{155}{136}$

8. $-\dfrac{1}{7}$ 9. ⊶———●—— 10. $K = \{0, 1, 2, 5, 7\}$
                        2   4

11. (a) False; (b) False; (c) True 12. $(-1, -3)$ 13. $(-2, -3)$ 14. $(-3, -3)$

15. $-6x^2 + 14x + 40$ 16. $16x^2 + 16x + 4$ 17., 18.

19. $xa + xb$ 20. $\dfrac{1}{xa + xb}$ 21. $\dfrac{ax}{b}$ 22. $a + b$ 23. $\dfrac{xy + x + y}{y^2 + y}$ 24. $\dfrac{y + x}{y}$

25. $\dfrac{y^2 - 1}{y}$ **26.** $5x^2y^2z(2y^3 - x^3z^4 - 2x^2y^2z^3)$ **27.** $x^2m^{-4}y^3$ **28.** x^4y^3

29. $8x^3y^6m^6$ **30.** 12

problem set 62

1. 14,000

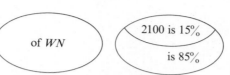

Before, 100%   After

2. 80 pounds

Before, 100%   After

3. $380

Before, 100%   After

4. 2000

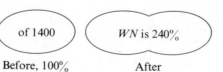

Before, 100%   After

5. 854 **6.** (a) False;
(b) True;
(c) True

7. $(11, 6)$ **8.** $(7, 4)$ **9.** $(11, 3)$ **10.** $(2, -6)$ **11.** $2x^2 - 12x + 16$

12. $-2x^3 - 2x^2 + 8x + 8$ **13., 14., 15.** **16.** $\dfrac{a^2 + ab}{b}$

17. $\dfrac{a}{cd + c^2}$ **18.** 5040 **19.** $\dfrac{x + xy + y^2}{x + y}$ **20.** $\dfrac{a + c^4x^2}{x^2c}$

21. $\dfrac{mc + mb - c^2 - cb + bc^2}{c^3 + c^2b}$ **22.** $3x^2yp^2(xp^4 - y + 3p^3)$ **23.** $1 + 3z$ **24.** y^8

25. $8x^{-4}y^{-8}p^{-9}$ **26.** $x^5y^3m^7$ **27.** $1 - 3xy^{-4}$ **28.** $-xym$ **29.** -32 **30.** $\dfrac{8}{7}$

problem set 63

1. 273 **2.** 3360

of 1400   WN is 240%

Before, 100%   After

3. 800

of WN   120 is 15%
is 85%

Before, 100%   After

4. $\dfrac{17}{74}$ **5.** $\dfrac{1}{38}$ **6.** $-\dfrac{79}{111}$

7. -1 **8.** **9.** Reals, rationals **10.** Reals, irrationals

11. (a) False; (b) True; (c) False; (d) True **12.** $(-1, -1)$ **13.** $(-1, 2)$

14. $(-1, 3)$ **15.** $4x^2 + 5x - 6$ **16.** $16x^2 + 24x + 9$

17., 18., 19. **20.** $\dfrac{a}{cb + bx}$ **21.** $\dfrac{a(c + x)}{b}$ **22.** $a + b$

23. $\dfrac{4y(c + 1) - 5myc - 3ck(c + 1)}{xy^2c(c + 1)}$ **24.** $\dfrac{k^2 + 1}{k}$ **25.** $\dfrac{my^2 + p}{y}$

26. $10xm^4k^4(2xmk^2 - 1 + 3x^4k^2)$ **27.** $16x^4k^{22}y^{-1}$ **28.** x^2 **29.** $p^{-2}x^{-4}y$

30. -12

problem set 64

1. 3000 **2.** \$450 **3.** 4000 **4.** Reals, rationals **5.** Reals, rationals, integers

6. 2.2 **7.** 2.8 **8.** $(-3, -4)$ **9.** $(-1, -2)$ **10.** $(-3, -3)$ **11.** $4x^2 - 25$

12. $9x^2 - 12x + 4$ **13., 14., 15.** **16.** $\dfrac{m}{a}$ **17.** $\dfrac{x}{x + y}$

18. m^2 **19.** $\dfrac{a(x + y) + x^2y(m + c)}{x^2y(x + y)}$ **20.** $\dfrac{x + y}{x}$ **21.** $\dfrac{xy + y}{x}$

22. $3ym^2(3x^3m^3 + 2y^3p^4 - y^2m)$ **23.** $1 + x$ **24.** $m^3x^3y^3$ **25.** a^5p^{-2}

26. $9x^3my^{-2}$ **27.** $-1 + 3x^{-6}y^{-3}$ **28.** $4y^2 - y^2x^{-2}$ **29.** 8 **30.** -5

problem set 65

1. 4 **2.** 6000

3. 36 **4.** 4.02 **5.** $\dfrac{14}{25}$

6. $\dfrac{35}{6}$ **7.** 2.5 **8.** **9.** Reals, irrationals

10. Reals, rationals, integers, wholes, naturals **11.** $(2, 4)$ **12.** $(-29, -18)$

13. $(5, -6)$ **14.** $5\sqrt{2}$ **15.** $10\sqrt{2}$ **16.** $7\sqrt{3}$ **17.** $6\sqrt{3}$

18. $-9x^2 + 21x - 10$ **19., 20., 21.**   **22.** $a(a + b)$ **23.** $\dfrac{x^2}{x + y}$

24. $\dfrac{4x + y}{x^2 + xy}$ **25.** $\dfrac{xy + 1}{y}$ **26.** $\dfrac{y + 1}{y}$ **27.** $5m^2xk^4(3x^4 - m^4x^5k^2 + 4m^2k)$

28. $2x^4m^6$ **29.** $16p^{-1}x^5$ **30.** $\dfrac{x^{12}}{9y^{18}}$ **31.** -6

problem set 66

1. $114{,}000$ **2.** 6000 **3.** 460 **4.** $6\sqrt{2}$ **5.** $15\sqrt{3}$ **6.** 72

7. $48x^2 - 28x - 6$ **8., 9., 10.**

11.   **12.**

13. $(3, -3)$ **14.** $(4, 4)$ **15.** 5.2 **16.** x **17.** y **18.** y **19.** $\dfrac{4x + 4a + 7a^2}{a^2x(x + a)}$

20. $\dfrac{2y + 3}{y}$ **21.** $\dfrac{y + x}{y}$ **22.** $20xym(2x^3m^6z - x^4y^4mz + y)$ **23.** $1 + x$ **24.** p

25. $9m^5y^7$ **26.** $2p^6x^{-2}y^4$ **27.** 29 **28.** $-3x^4$ **29.** $-1 + 3x^{-4}y^{-8}$ **30.** -4

problem set 67

1. 2040 **2.** $2{,}875{,}000$ **3.** 5000 **4.** $\dfrac{1}{34}$ **5.** $\dfrac{1}{10}$ **6.** $\dfrac{145}{96}$ **7.** 3

8. 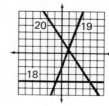   **9.**   **10.** $(1, 2)$ **11.** $(-3, -3)$

12. $(4, 2)$ **13.** $20\sqrt{2}$ **14.** $18\sqrt{5}$ **15.** $4\sqrt{3}$ **16.** $20\sqrt{2}$ **17.** $4x^2 - 20x + 25$

18., 19., 20.   **21.** $x(xy + b)$ **22.** $\dfrac{xb}{a}$ **23.** -10

24. $\dfrac{xy + m + cx^3}{x^3y}$ **25.** $\dfrac{xy + 1}{x}$ **26.** $\dfrac{xy + 1}{xy}$ **27.** $15ab^3c^4(2a - bc + 3b)$

28. x^2y^2 **29.** $2xy^8p^{-2}$ **30.** $4x^{12}y^5$

problem set 68

1. 250 2. 500 ounces 3. 700 4. $20\sqrt{5}$ 5. $6\sqrt{30}$ 6. $11\sqrt{2}$

7. $6\sqrt{5} + 2\sqrt{3}$ 8. $6p^2 + 7p - 20$ 9., 10., 11.

12.

13. $(5,5)$ 14. $(2,1)$ 15. $(4,4)$ 16. $\dfrac{a^3}{x}$

17. $\dfrac{1}{a+b}$ 18. x 19. $\dfrac{ax^2 + 5x + 5y}{x^2(x+y)}$ 20. $\dfrac{b+a}{b}$ 21. $\dfrac{x^2+1}{x}$

22. $4x^2y^2z(1 - 2z^4)$ 23. $-3 - 12y^3$ 24. $1 - x$ 25. $p^{-4}x^4y^7$

26. $16x^{-6}p^{-6}y^2$ 27. x^8 28. $4a^2xm^{-1}$ 29. 19 30. -55

problem set 69

1. -2 2. 75 3. 111 4. $\dfrac{287}{60}$ 5. .05 6. $\dfrac{133}{85}$ 7. 1

8.

9. $(4,2)$ 10. $(-4,2)$ 11. $(-4,-2)$ 12. $5\sqrt{2}$

13. $9\sqrt{3}$ 14. $10\sqrt{5} - 24\sqrt{2}$ 15. 0 16. $16x^2 + 40x + 25$

17., 18.

19. Real, irrational 20. ax 21. $\dfrac{ac}{b}$ 22. $\dfrac{a}{bc}$

23. $\dfrac{a(x+y) + bx^2y}{x^2y(x+y)}$ 24. $\dfrac{ky^2 + 1}{y^2}$ 25. $\dfrac{m^3 + 1}{m^2}$ 26. $4x^2y^2p^4(3y - xp^2 + 4x^2y^2)$

27. $9y^8m^4$ 28. $8x^{-1}y^5$ 29. $16x^{-12}y^{-3}$ 30. -4

problem set 70

1. 220 2. 2610 3. 5000 tons 4. $24\sqrt{5}$ 5. $-8\sqrt{2}$ 6. $4\sqrt{3} - 9\sqrt{2}$

7., 8. 9. $(5,5)$ 10. 11. $(5,-3)$

12. $(-2,-3)$ 13. $(-1,-1)$ 14. $(4,3)$ 15. $(3,2)$ 16. x^2a^2 17. $\dfrac{a+b}{a}$

18. $\dfrac{ma + mx + 3x^2}{x^2a(a+x)}$ 19. $\dfrac{4xy + 1}{y}$ 20. $\dfrac{y+x}{y}$ 21. $1 + y$ 22. $x^{13}y^{-1}$

23. $64x^6y^9p^{12}$ 24. $x^{-4}y^{-10} - 3x^{-4}p^{-1}$ 25. 11 26. 12

problem set 71

1. 5   2. 150   3. 3000   4. $\dfrac{240}{49}$ 5. $\dfrac{16}{315}$ 6. $\dfrac{25}{324}$ 7. $\dfrac{5}{2}$

8. ●───●───●───● →
    1 2 3 4   9. True   10. $(5, 5)$ 11. $(2, -2)$

12., 13.

14. $(4, 5)$ 15. $(2, -3)$ 16. $(1, 1)$ 17. $\dfrac{ay}{bx}$

18. $\dfrac{x+1}{x-1}$ 19. $\dfrac{x+1}{7x}$ 20. $\dfrac{1}{x-1}$ 21. $\dfrac{a+b}{x+4b}$ 22. $\dfrac{3b-a}{1+b^2}$ 23. $8\sqrt{2} - 6\sqrt{3}$

24. $-26\sqrt{3}$ 25. $-9x^2y$ 26. $\dfrac{x^4}{16m^2y^3}$ 27. $x^{-14}y^{-2}p^{-6}$ 28. $x^{10}y^{-10}$

29. $x^2y^{-6}p^6$ 30. 19

problem set 72

1. $(x-6)(x+3)$ 2. $(x-15)(x+1)$ 3. $(x-8)(x+2)$ 4. $(x+8)(x-2)$
5. $(x-3)(x-3)$ 6. $(x-9)(x+3)$ 7. $(p-5)(p+4)$ 8. $(x-5)(x+3)$
9. $(p-7)(p+3)$ 10. $(p+5)(p-4)$ 11. $(k-8)(k+5)$ 12. $(m+4)(m+5)$
13. $(x+3)(x+11)$ 14. $(p-4)(p-9)$ 15. $(m-6)(m+5)$ 16. $(n+9)(n+2)$
17. $(x+3)(x+9)$ 18. $(x-9)(x-10)$ 19. $(x+12)(x-11)$
20. $(a-45)(a-2)$ 21. $(m+8)(m+2)$ 22. $(2, 3)$ 23. $(1, 1)$ 24. $(3, -4)$
25. $(4, 5)$ 26. $14\sqrt{5} - 20\sqrt{2}$ 27. $16\sqrt{2}$ 28. $y+1$ 29. $\dfrac{a-4b}{x-b^2}$

30. $\dfrac{a-ax}{x^2+y}$

problem set 73

1. $(x-5)(x+2)$ 2. $(x+4)(x+3)$ 3. $(x-6)(x+5)$ 4. $(x+2)(x+5)$
5. $(x+2)(x+6)$ 6. $(x-2)(x-2)$ 7. $(x+2)(x+7)$ 8. $(x-7)(x+2)$
9. $(x-6)(x+3)$ 10. $(x+2)(x+4)$ 11. $(x+4)(x-2)$ 12. $(x-4)(x+2)$
13. $2(x+2)(x+3)$ 14. $5(x+2)(x+4)$ 15. $4bx(x-5)(x+4)$
16. $a(x+3)(x+3)$ 17. $ab(x+3)(x-2)$ 18. $x(x+4)(x+5)$
19. $-(x-5)(x+4)$ 20. $-x(x+5)(x-4)$ 21. $-3x(x+6)(x-4)$
22. $-2(p-11)(p+5)$ 23. $-3(m+2)(m+8)$ 24. $-b(b-8)(b+3)$
25. $N_N = 8, N_Q = 10$ 26. $N_N = 7, N_D = 9$ 27. $-60\sqrt{2}$ 28. $\dfrac{a+x^2}{1-x}$

problem set 74

1. $(m+8)(m+2)$ 2. $(n-12)(n+4)$ 3. $(y-8)(y-7)$ 4. $(p-11)(p+5)$
5. $(t+7)(t+5)$ 6. $(y+50)(y+1)$ 7. $(r-7)(r-11)$ 8. $(m+15)(m+6)$
9. $(v+5)(v+11)$ 10. $(h-9)(h+7)$ 11. $(x-15)(x+2)$ 12. $(w-2)(w-11)$
13. $-(x-3)(x-4)$ 14. $-(s+3)(s+5)$ 15. $-(a+8)(a-5)$
16. $2x(x+3)(x+5)$ 17. $4(a+8)(a-5)$ 18. $ab(x-8)(x+3)$
19. $(x-1)(x+2)(x+5)$ 20. $m(x-1)(x+2)(x+5)$ 21. $(a+b)(x+3)(x+5)$
22. $(5, 5)$ 23. $N_D = 7, N_Q = 17$ 24. $(3, 3)$ 25. $(2, 1)$ 26. $\sqrt{2}$ 27. $-6\sqrt{5}$
28. $\dfrac{a^2+b}{1-4a}$ 29. $\dfrac{m+p^2}{1-xp}$ 30. -2000

problem set 75

1. $(x + y)(x - y)$ 2. $(2p - 5)(2p + 5)$ 3. $(m - 4)(m + 4)$ 4. $(x + 2)(x - 2)$

5. $(7m - a)(7m + a)$ 6. $(5y - 6a)(5y + 6a)$ 7. $(2px - k)(2px + k)$

8. $(5px + 2m)(5px - 2m)$ 9. $(2y - 3x)(2y + 3x)$ 10. $(3ka - 7)(3ka + 7)$

11. $(p - 2k)(p + 2k)$ 12. $(6ax - k)(6ax + k)$ 13. $(x - 5)(x + 4)$

14. $4(x - 5)(x + 4)$ 15. $2(b - 8)(b + 3)$ 16. $3(x - 15)(x + 2)$

17. $(a + b)(x + 2)(x + 5)$ 18. $p(m + 4)(m + 5)$ 19. $5(k + 2)(k + 3)$

20. $-(x + 7)(x + 1)$ 21. $5(m - 1)(m - 1)$ 22. $(4, 4)$ 23. $(3, -3)$ 24. $(1, 1)$

25. $N_P = 100, N_N = 75$ 26. $21\sqrt{5}$ 27. $4\sqrt{3}$ 28. $\dfrac{1 + x}{y + x^2}$ 29. $\dfrac{2x + y}{x(x + y)}$

30. $\dfrac{6}{5}$

problem set 76

1. 1200 2. $\$88,800$ 3. $\$7000$ 4. 117 5.
                                                                                    5   7   9

6. $x^3 + 6x^2 + 11x + 12$ 7. $\dfrac{xc + x^2 + bxc^2 + 5c + 5x}{x^2c^2(c + x)}$ 8. $\dfrac{xb + a}{b - 1}$ 9. 18

10. 4.78×10^4 11. 4.78×10^{-3} 12. 4.78×10^{11} 13. 4.78×10^{-4}

14.   15. $(-2, -4)$ 16. $N_D = 35, N_Q = 5$

17. $1 + 32x^4y^{-2}a^6$ 18. $x^{-2}y^{-4}$ 19. $-13\sqrt{15}$ 20. $40\sqrt{5}$

21. $(x + 4)(x + 5)$ 22. $(x + 3)(x + 5)$ 23. $(x + 7)(x + 4)$ 24. $x(x + 6)(x + 4)$

25. $a(x - 5)(x + 3)$ 26. $5(x + 7)(x - 4)$ 27. $5(x + y)(x - y)$

28. $5(3x - 2m)(3x + 2m)$ 29. $(2a - 3b)(2a + 3b)$ 30. $(7ap - k)(7ap + k)$

problem set enrichment 1

1. 4000 2. 1632 3. 56 percent patched, 44 percent not patched   4. $\dfrac{1}{40}$

5. Reals, rationals, integers   6.   7. $\dfrac{4c + 4x + 11xc + 5x^2}{x^2c(c + x)}$

8. x^8y^{-5} 9. -27 10. -45 11. $6ax^{-1} - ax$ 12. 1.23×10^{-8}

13. 1.23×10^5 14. A, S, M 15. M 16.   17. $(1, -5)$

18. $(1, 1)$ 19. $y^{-4} - 4a^{-2}xy^{-1}$ 20. $-40\sqrt{2}$ 21. $(x + 5)(x - 2)$

22. $(x + 7)(x - 3)$ **23.** $(x + 3)(x + 6)$ **24.** $5(x - 5)(x + 2)$ **25.** $x(x - 2)(x - 1)$

26. $-x(x - 6)(x + 3)$ **27.** $b^3(x + 2)(x - 2)$ **28.** $(4x - a)(4x + a)$

29. $(3p - m)(3p + m)$ **30.** $b^3x(2x - 1)(2x + 1)$

problem set 77

1. 72, 73, 74 **2.** $-2, -1, 0$ **3.** 10, 11, 12, 13 **4.** $-12, -11, -10, -9$ **5.** $\dfrac{83}{156}$

6. $-\dfrac{2}{5}$ **7.** False **8.**   **9.** 11 **10.** 4.3×10^{-15}

11. 4.3×10^{10} **12.** 14,352 **13.** $\dfrac{a^2cx + 2cx + ab}{acx^3}$ **14.** $\dfrac{ax + 6a + bx}{x(x + 6)}$

15. x^4a^{-4} **16.** $2a^2x^5y^{-1} - 3a^2x^3y^{-3}$ **17.** $N_N = 30, N_Q = 15$ **18.** $(2, 5)$

19. $3x^6a^{-1}y^{-4} - 6xy^{-4}$ **20.** $8\sqrt{5}$ **21.** $(x - 7)(x + 2)$ **22.** $-x(x - 6)(x + 2)$

23. $a(x + 2)(x + 5)$ **24.** $2(x + 7)(x + 5)$ **25.** $3(x + 8)(x + 1)$

26. $p(x - 2)(x + 1)$ **27.** $(m - 3x)(m + 3x)$ **28.** $(2x - 3m)(2x + 3m)$

29. $5(5m - x)(5m + x)$ **30.** $2(x - 6k)(x + 6k)$

problem set 78

1. $-8, -6, -4$ **2.** $-13, -11, -9$ **3.** 5, 7, 9, 11

4. 140 **5.** 4.2

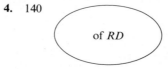

of *RD*

Before, 100%

112 is 4/5

is 1/5

After

6.   **7.** $\dfrac{x + c + ac(x + c) + mxc}{xc^2(x + c)}$ **8.** $\dfrac{xy + 1}{x^2 - 5y}$

4 5 6 7 8 9

9. $\dfrac{12}{5}$ **10.** 7×10^{-12} **11.** 7×10^{-21} **12.** 1680 **13.** A, M, D

14.   **15.** $(4, 4)$ **16.** $N_N = 400, N_D = 100$

13

14

17. $3x^2y^{-4} - 12a^3x^2y^{-3}$ **18.** $\dfrac{x^4y^7}{4}$ **19.** $-2\sqrt{15}$ **20.** $-8\sqrt{3}$

21. $(x - 3)(x - 3)$ **22.** $2(x - 2)(x - 2)$ **23.** $2(x + 2)(x + 2)$

24. $2(x + 5)(x + 5)$ **25.** $3(x - 5)(x - 5)$ **26.** $a(x - 6)(x - 6)$

27. $(2x - 7)(2x + 7)$ **28.** $(k - 3xy)(k + 3xy)$ **29.** $3(p - 2k)(p + 2k)$

30. $(k - 2m)(k + 2m)$

**problem
set 79**

1. $-10, -9, -8$ **2.** $3, 5, 7, 9$ **3.** 4000

of noses   9120 is 228%

4. 6000

of FN

4200 is $\frac{7}{10}$

is $\frac{3}{10}$

Before, 100%   After

5. $\frac{1}{4}$ **6.** Real, irrational

7. $\dfrac{ay^2 + by + c}{x^2y^3}$ **8.** x^8y^4 **9.** $-\dfrac{1}{27}$ **10.** -7 **11.** $-5m^2y$ **12.** 3×10^{-19}

13. 4×10^{-37} **14.**

15. $N_D = 20, N_Q = 320$ **16.** $(-3, -3)$

17. $2ax^{-1}y^{-4} - 6x^{-3}y^{-4}$ **18.** $12\sqrt{7} - 10\sqrt{14}$ **19.** $-\dfrac{3}{4}$ **20.** 2 **21.** 7

22. $\dfrac{161}{11}$ **23.** $(x - 4)(x - 5)$ **24.** $2a(x - 7)(x - 3)$ **25.** $m(x + 6)(x + 7)$

26. $(4x - 3a)(4x + 3a)$ **27.** $(5m - 2)(5m + 2)$ **28.** $9(my - 2k)(my + 2k)$

**problem
set 80**

1. $-8, -6, -4, -2$ **2.** 2 **3.** 145 pounds

of 125 WN is 116%

4. 400

of WN

268 is 67%

is 33%

Before, 100%   After

5. 3200

6. Reals, rationals, integers, naturals, wholes   **7.** $\dfrac{476}{125}$ **8.** $-\dfrac{3}{4}$ **9.** 1.35×10^{-21}

10. 1.35×10^{-12} **11.** $\dfrac{ax + ay + 4y}{xy(x + y)}$ **12.** $\dfrac{x^{-2}a^{-2}}{64}$ **13.** $4x^2y - 3x^2y^4$

14. $(4, 4)$ **15.** $N_N = 17, N_D = 5$ **16.** $1 - 3x^2y^{-6}$ **17.** $\sqrt{5}$ **18.** -152

19. $\dfrac{21}{2}$ **20.** $\dfrac{2}{3}$ **21.** 40 **22.** $T_L = 10, T_S = 5$ **23.** $R_F = 45, R_E = 60$

24. $T_R = 15, T_T = 15$ **25.** $(p - 11)(p + 5)$ **26.** $(x - 15)(x + 2)$

27. $2(m - 7)(m - 5)$ **28.** $-x(x - 10)(x - 4)$ **29.** $(2m - 7xp)(2m + 7xp)$

30. $a(5a - 4m)(5a + 4m)$

**problem
set 81**

1. $34, 36, 38$ **2.** $-3, -1, 1, 3$ **3.** 200

of WN 460 is 230%

4. 50

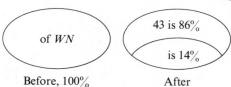

of WN

43 is 86%

is 14%

Before, 100%   After

5. $10,725$ pounds

6.

7. $\dfrac{3a^2 + 3ax + 4a + 4x + 7}{a^2(a + x)}$ **8.** -2 **9.** $\dfrac{4y^2 + 1}{xy + m}$

10. 2×10^{10} **11.** 1×10^{-19} **12.** 3400   **13.**

14. $N_N = 42, N_D = 30$ **15.** $(3, -2)$ **16.** $a^{-4} + 3x^{-6}a^3$ **17.** $x^{-6}y^{30}$ **18.** $12\sqrt{2}$

19. $\dfrac{52}{11}$ **20.** $\dfrac{11}{2}$ **21.** $R_F = 96, R_S = 80$ **22.** $T_M = 1, T_R = 4$

23. $T_G = 5, T_B = 8$ **24.** $(x - 6)(x - 8)$ **25.** $(x - 10)(x - 8)$ **26.** $x(x + 8)(x + 1)$

27. $-a(x + 16)(x - 3)$ **28.** $bc(x + a)(x - a)$ **29.** $m(k - m)(k + m)$

problem set 82

1. 14, 16, 18   **2.** $-3, -2, -1, 0$ **3.** 420 pounds   **4.** 195   **5.** $\dfrac{20}{3}$

6. Reals, irrationals   **7.** $\dfrac{x - y}{x + ym}$ **8.** $2x^6y^6$ **9.** -1 **10.** a

11. 2×10^{15} **12.** 1×10^{-5} **13.** 700   **14.** $(1, 2)$ **15.** $(2, 2)$ **16.** $(2, 3)$

17. $N_N = 10, N_D = 20$ **18.** $\dfrac{21}{17}$ **19.** $\dfrac{8}{5}$ **20.** $x^2a^3y^{-2} - 3x^{-6}a^2$ **21.** $4\sqrt{7}$

22. $T_G = 10, T_B = 8$ **23.** $T_N = 16, T_K = 8$ **24.** $x(x - 5)(x - 1)$

25. $2(x - 4)(x - 1)$ **26.** $a(x - 6)(x - 1)$ **27.** $-m(x + 2)(x + 4)$

28. $m(x - 3a)(x + 3a)$ **29.** $(2mx - k)(2mx + k)$

problem set 83

1. 0, 1, 2   **2.** 32, 34, 36, 38   **3.** \$4000, \$560   **4.** 5180   **5.** 4600

6. Reals, irrationals   **7.** 14   **8.** $\dfrac{59}{18}$ **9.** -1.1 **10.** $x = 4$ **11.** $x = -3$

12. $y = -x + 2$ **13.** $y = \dfrac{3}{4}x - 3$ **14.** 7×10^{-36} **15.** $\dfrac{x^2 + y^2}{ay - x}$

16. $\dfrac{10ax + 6ay - 4x - 4y}{ax(x + y)}$ **17.** $x^{-3}a^5y^{-4} - 3a^2y^{-4}$

18. $4xya^{-2} + 5x^{-1}ya^{-2} - 2xya^2$ **19.** $(-1, 3)$ **20.** $N_N = 10, N_D = 40$ **21.** $(1, -1)$

22. -44 **23.** $-\dfrac{5}{2}$ **24.** $-12\sqrt{2}$ **25.** $T_B = 24, T_T = 6$ **26.** $T_L = 14, T_S = 6$

27. $4(x + 5)(x + 5)$ **28.** $-x(x + 6)(x + 5)$ **29.** $a(x + 7)(x - 5)$

30. $(2k + 4xy)(2k - 4xy)$

problem set 84

1. $N_N + N_D = 28; 5N_N + 10N_D = 195; N_N = 17, N_D = 11$

2. $10N_D + 25N_Q = 445; N_Q = N_D + 8; N_D = 7, N_Q = 15$

3. $N_N + N_D = 22; 5N_N + 10N_D = 135; N_N = 17, N_D = 5$ **4.** 6800   **5.** 4×10^{-8}

6. 1×10^2 **7.** $x = 4$ **8.** $y = 2x$ **9.** $(-1, 1)$ **10.** $(5, -2)$ **11.** $6\sqrt{2}$

12. $\dfrac{62}{13}$ **13.**   **14.** $T_S = 3, T_M = 4$ **15.** $T_P = -4, T_M = -12$

16. $R_G = 40, R_P = 85$ **17.** $\dfrac{21}{4}$ **18.** 2   **19.** $-\dfrac{122}{13}$ **20.** $-\dfrac{38}{3}$ **21.** $\dfrac{4a + 5b}{4a^3}$

22. $\dfrac{4a^2 + 6a + 6b}{a^2(a + b)}$ **23.** 0 **24.** $\dfrac{xy + x}{ax + y}$ **25.** $\dfrac{abc - 1}{4c^2 - a}$ **26.** $-\dfrac{1}{16}$

27. $x(x + 7)(x + 4)$ **28.** $x(2a - y)(2a + y)$ **29.** $1 - 3y^{-4}$ **30.** $x^{-10}y^4$

31. x^2y^5

problem set 85

1. $N_P + N_N = 175;\ N_P + 5N_N = 475;\ N_P = 100,\ N_N = 75$

2. $N_N = N_D + 12;\ 5N_N + 10N_D = 510;\ N_N = 42,\ N_D = 30$ **3.** 160 **4.** $-6, -4, -2$

5. 4×10^1 **6.** $y = 4$ **7.** $y = -2x - 2$ **8.** $12\sqrt{6}$ **9.** $20\sqrt{6} + 72$

10. $10\sqrt{6} + 10\sqrt{3}$ **11.** $-6\sqrt{2}$ **12.** $(1, -4)$ **13.** $(-2, -2)$ **14.** $\dfrac{9}{25}$

15. Reals, integers, rationals   **16.** -51 **17.** $-\dfrac{65}{4}$ **18.** $T_K = 12,\ T_M = 4$

19. $\dfrac{x - a - 3ax^2 + 2x^2 - 2ax}{x^2(x - a)}$ **20.** $\dfrac{bcx - x^2c^2a - d}{x^3c^2}$ **21.** $\dfrac{a^3 + a}{3 - ba^2}$ **22.** $\dfrac{abx - 4c}{1 + ac}$

23. -47 **24.** -5 **25.** 9 **26.** 1.6 **27.** $3y^{-10} - 3x^{-5}y^{-4}a^{-3}$ **28.** x^{-13}

29. xy^{-3}

problem set 86

1. $N_D + N_Q = 40;\ 10N_D + 25N_Q = 475;\ N_D = 35,\ N_Q = 5$

2. $N_N = N_Q + 15;\ 5N_N + 25N_Q = 525;\ N_N = 30,\ N_Q = 15$ **3.** $-13, -11, -9, -7$

4. 350 **5.** 6×10^{-5} **6.** $y = -2x + 4$ **7.** $x = -3$ **8.** $72\sqrt{2}$

9. $42 - 6\sqrt{3}$ **10.** $9\sqrt{5}$ **11.** $3x^2 + 7x - 1$ **12.** $3x^2 + 7x + 26 + \dfrac{128}{x - 3}$

13. $2x^2 + 3x + 5 + \dfrac{14}{x - 2}$ **14.** $-\dfrac{8}{3}$ **15.** $\dfrac{8x^2 + 3y + 2x^3 + 2xy}{x^2(x^2 + y)}$ **16.** $\dfrac{x^2 + 4y^2}{x + 2y}$

17. $1 - 3a^{-4}x^{-8}$ **18.** $3x - 4xy^{-1}$ **19.** $(1, 1)$ **20.** $(-4, 2)$ **21.** $(5, -5)$

22. $\dfrac{11}{3}$ **23.** 6 **24.** $T_D = 2,\ T_M = 6$ **25.** $8x^3y^2$ **26.** $a(x + 2)(x + 2)$

27. $(x - 5)(x + 2)$ **28.** $a(3 - 2x)(3 + 2x)$ **29.** $x(x + 2)(x + 10)$

problem set 87

1. $N_N + N_D = 500;\ 5N_N + 10N_D = 3000;\ N_N = 400,\ N_D = 100$

2. $N_Q = N_D + 300;\ 10N_D + 25N_Q = 8200;\ N_D = 20,\ N_Q = 320$ **3.** $11, 12, 13, 14$

4. 1737 **5.** 1.2×10^{19} **6.** $y = 3$ **7.** $y = -\dfrac{1}{3}x - 4$ **8.** $24 + 36\sqrt{3}$

9. -150 **10.** $x^2 + 4 + \dfrac{8}{x - 2}$ **11.** $2x^2 - 9x + 29 - \dfrac{91}{x + 3}$ **12.** $.018$

13. $\dfrac{35ax + 12y}{840a^2}$ **14.** $\dfrac{k - 7x}{42}$ **15.** $\dfrac{x - y}{a + by}$ **16.** $2x^2y^2 - 5xy^{-1}$ **17.** $-\dfrac{17}{4}$

18.

(0, −2)

    19. $(1, 1)$ **20.** $(4, 4)$ **21.** $1 - 3x^{-6}ay^{-2}$ **22.** -3

23. $T_W = 4,\ T_E = 5$ **24.** $T_1 = 4,\ T_2 = 6$ **25.** y^6x^{-5} **26.** $ma(x + 7)(x + 2)$

27. $-x(x + 5)(x + 7)$ **28.** $(a^3m^3 + y)(a^3m^3 - y)$ **29.** $a^3(2x - y)(2x + y)$

problem set 88

1. 151 2. 20 3. 228 4. $12, 13, 14$ 5. 5×10^{-21} 6. $y = -x$

7. $y = x$ 8. $30 - 20\sqrt{6}$ 9. $30 - 90\sqrt{3}$ 10. $3x^2 + 9x + 25 + \dfrac{80}{x-3}$

11. $3x^2 + 15x + 74 + \dfrac{363}{x-5}$ 12. $(2, 0)$ 13. $(1, -1)$ 14. $(1, -2)$

15. $24 - 6\sqrt{6}$ 16. $20 - 12\sqrt{10}$ 17. $6 - 12\sqrt{5}$ 18. $\dfrac{133}{90}$

19.   20. -88 21. 66 22. $\dfrac{6a + 6b - 4ab - b^2}{b^2(a+b)}$

23. $\dfrac{3x + 3a + 2bax}{a^2 x(x+a)}$ 24. $\dfrac{xy - 4}{a + 3y}$ 25. 50 26. $\dfrac{1}{2}$ 27. $a(x+8)(x+7)$

28. $T_P = 3, T_D = 6$ 29. $a^{-2}y^{-4}x^{-5} - 3y^{-3}x^{-3}$ 30. $\dfrac{yx^8}{12}$

problem set 89

1. $N_P = 120, N_N = 30$ 2. $N_P = 90, N_N = 36$ 3. $31,200$ 4. $7, 9, 11, 13$

5. 6×10^{-2} 6. $y = -2$ 7. $y = -\dfrac{1}{2}x + 3$ 8. $50\sqrt{2} - 75$ 9. $56 - 84\sqrt{2}$

10. $2x^2 - 4x + 3 - \dfrac{2}{x+2}$ 11. $3x^2 + 15x + 75 + \dfrac{371}{x-5}$ 12. $-3, 6$ 13. $7, -6$

14. $\pm\dfrac{5}{2}$ 15. $7, -8$ 16. $\pm\dfrac{4}{3}$ 17. $7, -4$ 18. ± 5 19. $3, -2$ 20. -4

21. $(-4, -11)$ 22. $(2, 2)$ 23. ———●——●——●———   24. $\dfrac{7}{92}$ 25. -1

26. $R_K = 30, R_P = 40$ 27. $\dfrac{93}{4}$ 28. $\dfrac{4a^2 + ax + 2a + 2x}{a^2(a+x)}$ 29. -4

30. $a^{-2}x^{-13}y^{-2}$

problem set 90

1. $N_T = 40, N_F = 12$ 2. $N_C = 35, N_E = 15$ 3. 7000 4. $-5, -3, -1$

5. 2×10^{-40} 6. (a) $y = 4$; (b) $y = -\dfrac{1}{2}x - 2$ 7. $75\sqrt{2} - 30$

8. $42\sqrt{2} - 20\sqrt{7}$ 9. $x - 3$ 10. $3x^2 - 12x + 48 - \dfrac{193}{x+4}$ 11. $5, 7$

12. $-5, -7$ 13. $8, 4$ 14. $-5, -12$ 15. $\pm\dfrac{3}{2}$ 16. $\pm\dfrac{7}{3}$ 17. -5

18. $8, 3$ 19. $\pm\dfrac{2}{3}$ 20. $(2, 2)$ 21. $(4, 4)$ 22. $\dfrac{45}{28}$ 23. 3400

24. $T_T = 2, T_M = 5$ 25. -9 26. -2 27. -3 28. (a) -9; (b) 9; (c) $-\dfrac{1}{9}$

29. $\dfrac{xy^2 + a}{ax - y}$ 30. $36a^6b^2$

problem set 91

1. $N_P = 200, N_N = 250$ 2. $N_I = 20, N_W = 15$ 3. 600 4. $-9, -7, -5, -3$

5., 6.   7., 8.   9. 2.8×10^{-17}

10. (a) $x = 5$; (b) $y = -x$ **11.** $6\sqrt{6}$ **12.** $30\sqrt{6} - 96$ **13.** $x^2 + 2 + \dfrac{11}{x - 3}$

14. $x^2 - 3x + 9 - \dfrac{28}{x + 3}$ **15.** $4, 5$ **16.** $6, 7$ **17.** $\pm\dfrac{3}{2}$ **18.** $\pm\dfrac{2}{3}$ **19.** $(2, 2)$

20. $(4, 5)$ **21.** $\dfrac{29}{35}$ **22.** Reals, irrationals **23.** $4 - y$ **24.** $R_K = 38, R_M = 87$

25. $\dfrac{9}{7}$ **26.** $\dfrac{4x^2y + 16x^2 - 2xy}{y(y + 4)}$ **27.** -58 **28.** 18

29. (a) -27; (b) $-\dfrac{1}{27}$; (c) $\dfrac{1}{27}$ **30.** $x^{-3} - 4y^5$

problem set 92

1. $N_N = 43, N_C = 49$ **2.** $N_B = 18, N_G = 11$ **3.** $L_S = 15, L_L = 23$ **4.** $49, 23$
5. 80 **6.** $(3, -3)$ **7.** $(3, -1)$ **8.** 3×10^{-5} **9.** (a) $y = -2x$; (b) $y = -3$

10. $9\sqrt{3} + 36$ **11.** $180\sqrt{2}$ **12.** $3x^2 - 3x + 1 - \dfrac{5}{x + 1}$ **13.** $2x^2 - 4x + 4 - \dfrac{8}{x + 1}$

14. -5 **15.** $-6, -5$ **16.** ± 3 **17.** ± 3 **18.** $(3, 1)$ **19.** 46 **20.** $-\dfrac{19}{22}$

21. Reals, rationals, integers **22.** $\dfrac{5m^3y}{x^2}$ **23.** $T_B = 8, T_M = 10$ **24.** $\dfrac{2x^2}{x(x + 1)}$

25. $\dfrac{ay + x}{a - xy}$ **26.** **27.** -7 **28.** 3 **29.** (a) $-\dfrac{1}{9}$; (b) $\dfrac{1}{9}$;

(c) $-\dfrac{1}{9}$ **30.** x^{-7}

problem set 93

1. $N_D = 15, N_Q = 20$ **2.** 30 **3.** $N_F = 360, N_S = 270$ **4.** 2000
5. **6.** **7.**

8. $(-2, 3)$ **9.** $(2, -2)$ **10.** 2×10^{25} **11.** (a) $y = -3$; (b) $y = \dfrac{1}{2}x + 3$

12. $23\sqrt{3}$ **13.** $36 - 24\sqrt{2}$ **14.** $x^3 - 2x^2 + 2x - 4 + \dfrac{4}{x + 2}$ **15.** $3, 7$ **16.** $\pm\dfrac{7}{2}$

17. $-4, -8$ **18.** 9 **19.** $\dfrac{187}{78}$ **20.** **21.** Reals, rationals

22. $\dfrac{x^2 + 4x + 4}{x(x + 1)}$ **23.** $5x^2ym^2 - 3x^2ym + 2xmy^{-3}$ **24.** $R_H = 76, R_O = 9$ **25.** $\dfrac{1}{4}$

26. $\dfrac{3x - 2y}{ay - 4}$ **27.** $1 - 3y$ **28.** (a) $-\dfrac{1}{8}$; (b) $-\dfrac{1}{8}$; (c) 8 **29.** -25 **30.** $2x^3$

problem set 94

1. $T_ER_E = T_FR_F, T_E = 12, T_F = 16, R_F = R_E - 15; R_E = 60, R_F = 45$

2. $R_HT_H = R_RT_R, R_H = 2, R_R = 12, T_H = T_R + 5, T_H = 6; 12$ miles

3. $R_DT_D = R_WT_W, T_D = 2, T_W = 10, R_D = R_W + 16; 40$ miles

4. $R_CT_C = R_ST_S, T_C + T_S = 14, R_C = 10, R_S = 4; 40$ kilometers **5.** 19

6. $\dfrac{10}{7}$ **7.** **8.** $(3, -2)$ **9.** $(5, 5)$ **10.** 2×10^{-20}

11. (a) $x = -3$; (b) $y = \frac{1}{3}x + 2$ **12.** $5\sqrt{5}$ **13.** $24\sqrt{10} - 18$

14. $x^2 + 5x + 25 + \frac{121}{x - 5}$ **15.** $-2, -4$ **16.** $\pm\frac{3}{2}$ **17.** $-4, -8$ **18.** $\frac{195}{2}$

19. $\frac{5}{2}$ **20.** $\frac{46}{5}$ **21.** False   **22.** $T_W = 4, T_R = 6$ **23.** $\frac{12x^2 + 15x + 25}{30x}$

24. $\frac{3a - 2b^2}{b^2 - 4}$ **25.** $1 + k$ **26.** 18   **27.** -10 **28.** (a) $-\frac{1}{4}$; (b) -4; (c) $-\frac{1}{4}$

29. $y^2 - x^{-7}y^2a^6$

problem set enrichment 2

1. $R_P T_P = R_B T_B, T_P = 6, T_B = 72, R_P = R_B + 11; R_P = 12; 72$ kilometers

2. $R_S T_S = R_B T_B, R_S = 8, R_B = 32, T_S + T_B = 5; 4$ hours, 32 miles   **3.** 10

4. 3860   **5.** $a(x + 4)(x + 5)$ **6.** $\frac{1}{5}$ **7.**

8. $(3, -7)$ **9.** $(3, 3)$ **10.** 5×10^7 **11.** (a) $y = 4$; (b) $y = -\frac{2}{3}x - 2$ **12.** $\sqrt{6}$

13. $60\sqrt{6} - 36\sqrt{2}$ **14.** $3x^2 - 9x + 27 - \frac{86}{x + 3}$ **15.** $-3, -4$ **16.** $\pm\frac{2}{3}$

17. $-5, -1$ **18.** 21   **19.** Reals, rationals, naturals, wholes, integers   **20.** $\frac{27}{5}$

21. $T_R = 1, T_H = 6$ **22.** $\frac{x^3 + 3x^2 + 10x - 8}{x^2(x + 4)}$ **23.** $\frac{a^3 + 4}{4a^2 - 1}$ **24.**

25. -29 **26.** $a + 0 = a$ **27.** (a) $-\frac{1}{81}$; (b) $\frac{1}{81}$ **28.** $x^{19}y^{-7}$

29. $10x^2y^{-2} - 3x^6y^2$

problem set enrichment 3

1. $R_G T_G = R_S T_S, R_G = 12, R_S = 8, T_G = T_S - 5; T_S = 15$ hours, $T_G = 10$ hours

     $R_L T_L = R_H T_H, R_L = 30, R_H = 4, T_L + T_H = 17; 15$ hours, 60 miles   **3.** 22

4. 1200   **5.** $a(b + c) = ab + ac$ **6.** $(a + b) + c = a + (b + c)$ **7.** Reciprocal

8. $-m(m + 8)(m + 3)$ **9.** $\frac{28}{5}$ **10.**   **11.** $(-1, 1)$ **12.** $(6, 6)$

13. 1×10^{-13} **14.** (a) $y = -2$; (b) $y = -\frac{1}{5}x + 1$ **15.** $288 - 3\sqrt{2}$

16. $30\sqrt{6} - 108\sqrt{2}$ **17.** $x^3 + x^2 + x - \frac{4}{x - 1}$ **18.** $-4, -6$ **19.** $\pm\frac{2}{3}$

20. $x = \frac{13}{8}$ **21.** $\frac{3}{68}$ **22.** $\frac{8 - 2p}{p(p + 4)}$ **23.** $\frac{xy^2 - 1}{x - 4y}$ **24.** $\frac{231}{305}$

25. Reals, irrationals   **26.** 0   **27.** 4   **28.** (a) -9; (b) -9; (c) $-\frac{1}{4}$

29. $y^4 + 4x^6$

problem set 95

1. $R_C T_C = R_H T_H, R_C = 300, R_H = 400, T_C + T_H = 7; T_C = 4; 1200$ centimeters

2. $R_T T_T = R_F T_F, T_T = 4, T_F = 3, R_F = R_T + 11; R_T = 33, R_F = 44$ **3.** 20 percent

4. 83 girls, 96 boys   **5.** $\dfrac{x-1}{x+1}$ **6.** $\dfrac{x+3}{x+5}$ **7.** $x-5$ **8.** (a) 650; (b) 1

9. -4 **10.** [number line: point at -5, between -6 and -4]   **11.** $(2,0)$ **12.** $(-3,-1)$ **13.** 3×10^{-11}

14. (a) $x=-5$; (b) $y=-2x$ **15.** $4\sqrt{30}$ **16.** $24\sqrt{6}-96$

17. $3x^2 - 9x + 27 - \dfrac{85}{x+3}$ **18.** $-4, -10$ **19.** [number line: segment from -3 to 2, open circles]   **20.** $\dfrac{34}{11}$

21. $\dfrac{21}{25}$ **22.** $\dfrac{11}{98}$ **23.** False   **24.** $\dfrac{xy+xa}{a^2y^2}$ **25.** $\dfrac{ky+k}{y^2+a}$ **26.** -14 **27.** 5

28. (a) -27; (b) 27; (c) $\dfrac{1}{27}$ **29.** $4y^4x^{-5}$

problem set 96

1. [diagram: 352] $R_S T_S + R_N T_N = 352, R_S = R_N, T_S = 5, T_N = 3$; 44 kilometers per hour

2. [diagram: 332] $R_1 T_1 + R_2 T_2 = 332, R_2 = 65, T_1 = T_2 = 2$; $R_1 = 101$ kilometers per hour

3. [diagram: 400] $R_F T_F + R_S T_S = 400, T_F = 8, T_S = 8, R_F = R_S + 20$; $R_S = 15$ miles per hour,

$R_F = 35$ miles per hour   **4.** 90 buckets, 102 pots   **5.** 4000   **6.** $\dfrac{16}{(x+6)(x-7)}$

7. 4   **8.** 1600   **9.** $-\dfrac{3}{2}$ **10.** [number line: points at $-4, -3, -2, -1, 0$]   **11.** $(-3,-3)$

12. $(-3,-2)$ **13.** 1×10^{63} **14.** (a) $y=-2$; (b) $y=\dfrac{3}{2}x+3$

15. $60\sqrt{2}+2\sqrt{5}$ **16.** $24\sqrt{6}-144$ **17.** x^2+2x-1 **18.** $10, -3$

19. $\pm\dfrac{10}{3}$ **20.** $-\dfrac{15}{2}$ **21.** $-\dfrac{45}{124}$ **22.** $\dfrac{axy(x+y)+a^2(x+y)+a^3x^2y}{x^3y(x+y)}$

23. $\dfrac{1}{y-1}$ **24.** $\dfrac{26}{7}$ **25.** Reals, rationals   **26.** -24 **27.** 24

28. (a) $x-1$; (b) $\dfrac{1}{8}$ **29.** $4-12y^3$

problem set 97

1. [diagram: 340] $R_B T_B + R_L T_L = 340, T_B = T_L + 2, R_B = 30, R_L = 40$; 9 p.m.

2. [diagram] $R_A T_A = R_B T_B, R_A = 30, R_B = 60, T_A = T_B + 2$; 1 p.m.

3. [diagram] $R_R T_R = R_W T_W, T_R + T_W = 11, R_R = 8, R_W = 3$; 24 kilometers

4. $N_Q = 90, N_H = 100$ **5.** 340   **6.** 1   **7.** (a) ± 4; (b) $\pm\sqrt{41}$; (c) $\pm\sqrt{13}$

8. 2800   **9.** 12   **10.** [number line: points at $-2, -1$]   **11.** $(-1,-2)$ **12.** $(4,-5)$

13. 3×10^{-17} **14.** (a) $x=-5$; (b) $y=-\dfrac{3}{2}x+3$ **15.** 360   **16.** -24

17. $x^2 - 2x + 2 - \dfrac{4}{x+1}$ **18.** $-7, -8$ **19.** $\pm\dfrac{9}{2}$ **20.** 54   **21.** $\dfrac{7}{40}$

22. $\dfrac{x^2+8x-12}{x(x-3)}$ **23.** [number line: points at $-5, -4, -3, -2$]   **24.** $\dfrac{14}{19}$ **25.** Reals, irrationals

26. -1 **27.** -20 **28.** (a) $1+ax$; (b) $-\dfrac{1}{9}$ **29.** $7xy^3$

problem set 98

1. -18 2. $-5, -4, -3, -2$ 3. 160

4. $R_W T_W = R_D T_D$, $R_W = 60$, $R_D = 50$, $T_W = T_D - 1$, $T_D = 6$, $T_W = 5$; 300 miles

5. $N_F = 700$, $N_T = 550$ 6. $\dfrac{x + 5}{x - 2}$ 7. $\sqrt{41}$ 8. $8\sqrt{2}$ 9. $2\sqrt{5}$ 10. $-\dfrac{5}{4}$

11. $(1, 3)$ 12. 4×10^4 13. (a) $x = -2$; (b) $y = \dfrac{1}{3}x - 2$ 14. $30\sqrt{6}$

15. $30\sqrt{3} - 60\sqrt{2}$ 16. $5, 20$ 17. $x^2 - 2x + 3 - \dfrac{6}{x + 2}$ 18. $\dfrac{20}{3}$ 19. $\dfrac{63}{52}$

20. $\dfrac{k^2 + 8k + 25}{k(k + 5)}$ 21. $\dfrac{p - 4k}{k^2 - 1}$ 22. $\dfrac{8}{75}$ 23. False   24. 0   25. -9

26. (a) $-(1 + x)$; (b) 16   27. $1 - 4x^2 y^2 a^{-4}$

problem set 99

1. -7 2. $28, 30, 32, 34$ 3. 630

4. $R_W T_W + R_J T_J = 332$, $R_W = 60$, $R_J = 46$, $T_W = T_J + 2$; 4 p.m.

5. 40   6. 6   7. $\sqrt{65}$ 8. 10   9. $\dfrac{1}{4}x$ 10. 6×10^{-30} 11. -2

12. $(4, -4)$ 13. (a) $y = 3$; (b) $y = -\dfrac{1}{3}x - 1$ 14. $20\sqrt{2}$ 15. $36\sqrt{3} - 24\sqrt{6}$

16. $2, -7$ 17. $x^2 + x + 1$ 18. $\dfrac{39}{2}$ 19. $-\dfrac{4}{195}$ 20. $\dfrac{4m^2 + 6m + 30}{m(m + 5)}$

21. $\dfrac{5}{58}$ 22. $\dfrac{3xy - 1}{2x - 4y}$ 23. Reals, rationals   24. -36 25. -3

26. (a) $x - 1$; (b) 9   27. Reciprocal   28. $3x^{-7}$ 29.

problem set enrichment 4

1. $R_S T_S + R_P T_P = 490$, $R_S = 20$, $R_P = 35$, $T_S + T_P = 17$; 10 hours

2. $R_M T_M = R_G T_G$, $T_M = 8$, $T_G = 10$, $R_M = R_G + 10$; $R_G = 40$ kilometers per hour,

$R_M = 50$ kilometers per hour   3. $2, 4, 6, 8$ 4. $N_E = 300$, $N_H = 2700$ 5. 4500

6., 7., 8. Proofs are in the lesson.   9. $2\sqrt{6}$ 10. $2\sqrt{10}$ 11. $-x - 3$ 12. $-\dfrac{19}{6}$

13. $(2, -7)$ 14. 2×10^{-40} 15. (a) $y = -\dfrac{7}{2}$; (b) $y = -\dfrac{3}{5}x + 3$

16. $150 + 42\sqrt{2}$ 17. $72 - 16\sqrt{3}$ 18. $-10, 5$ 19. $x^2 - x$ 20. 11

21. $\dfrac{443}{180}$ 22. $\dfrac{10x^2 + 15x + 2}{x^2(x + 1)}$ 23. $\dfrac{x}{y - 1}$ 24. $\dfrac{304}{565}$ 25. Reals, irrationals

26. 79   27. (a) $1 + xm$; (b) 16   28. $1 - 20x^5 y^{-2} a^4$ 29. -36

problem set 100

1. $R_A T_A + 50 = R_P T_P$, $T_A = 7$, $T_P = 5$, $R_A = 20$; $R_P = 38$ kilometers per hour

2. $15 + R_R T_R = R_C T_C$, $R_R = 70$, $R_C = 100$, $T_R = T_C$, $T_C = 1/2$ hour

3. $R_H T_H + 210 = R_J T_J$, $T_H = T_J$, $R_J = 230$, $R_H = 200$, $T_H = 7$ minutes

4. $N_N = 250, N_Q = 200$ **5.** 5000   **6.**   **7.** $\sqrt{65}$ **8.** $3\sqrt{7}$

9. $2\sqrt{10}$ **10.** $\sqrt{34}$ **11.** $-\dfrac{(x+7)(x+2)}{x-3}$ **12.** -2 **13.** $(3, -1)$

14. 5×10^{-9} **15.** (a) $y = 4$; (b) $y = -x - 3$ **16.** $364\sqrt{2}$ **17.** $30 - 24\sqrt{21}$

18. $\pm\dfrac{9}{2}$ **19.** $x^2 + 4x + 16 + \dfrac{60}{x-4}$ **20.** -10 **21.** $-\dfrac{21}{128}$

22. $\dfrac{x^3 + 2x^2 + 6x - 8}{x^2(x+4)}$ **23.** $\dfrac{1 + 4a}{a^3 + 4}$ **24.** $\dfrac{59}{168}$ **25.** True   **26.** -31

27. (a) $2a + 1$; (b) 81   **28.** 31   **29.** $-x^2y^6$

problem set 101

1. $R_E T_E = R_A T_A + 60, T_E = 6, T_A = 4, R_E = 60; R_A = 75$ miles per hour

2. 40   **3.** $T_W R_W = R_B T_B, T_W = 60, T_B = 100, R_W = R_B + 2$; 300 miles   **4.** 7

5. $N_R = 10, N_W = 20$ **6.** $-7, -6, -5, -4$ **7.** $\sqrt{55}$ **8.** $4\sqrt{5}$ **9.** $\dfrac{x-3}{x-4}$

10. $\dfrac{15}{2}$ **11.** $(2, -4)$ **12.** 3×10^{-9} **13.** (a) $x = 4$; (b) $y = \dfrac{2}{3}x + 2$

14. $50\sqrt{3}$ **15.** $100\sqrt{5}$ **16.** $3400\sqrt{5}$ **17.**   **18.** 8

19. $x^3 + 2$ **20.** $-\dfrac{7}{2}$ **21.** 1   **22.** $\dfrac{-2x^3 + 3x^2 + 8x + 4}{x^2(x+1)}$ **23.** $\dfrac{1 - 4x}{yx - 1}$

24. $\dfrac{22}{281}$ **25.** Reals, irrationals   **26.** 26   **27.** $1 - a$ **28.** $-\dfrac{1}{16}$ **29.** -20

30. $1 - x^3 a^{-2}$

problem set 102

1. $R_B T_B + 40 = R_L T_L, T_L = T_B, R_B = 6, R_L = 10$; 10 seconds   **2.** 700

3. $R_R T_R = R_W T_W, R_W = 8, R_R = 6, T_R = T_W + 2$; $T_W = 6$ hours, $T_R = 8$ hours

4. $-11, -9, -7, -5$ **5.** 10 and 50 feet

6. (a) 47,260,000.0; (b) 104.063; (c) 314.03644; (d) 400   **7.** $2\sqrt{5}$ **8.** $2\sqrt{2}$

9. $x - 4$ **10.** $\dfrac{4}{9}$ **11.** $(2, 4)$ **12.** 8×10^{48} **13.** (a) $x = 5$; (b) $y = -\dfrac{4}{3}x - 2$

14. $240\sqrt{3}$ **15.** $48 - 18\sqrt{6}$ **16.** $16\sqrt{3} - 18\sqrt{2}$ **17.**

18. $-7, -5$ **19.** $x^2 - 2x + 16 - \dfrac{27}{x+2}$ **20.** 52   **21.** $-\dfrac{5}{88}$

22. $\dfrac{-9x^3 - 4x^2 + 13x + 16}{4x^2(x+1)}$ **23.** $\dfrac{ay - 4x}{1 + 5x}$ **24.** $\dfrac{17}{121}$ **25.** Reals, irrationals

26. -2 **27.** $1 - a$ **28.** $4xy^{-8}$ **29.** $-\dfrac{1}{27}$ **30.** -634

problem set 103

1. $R_R T_R = R_W T_W, R_R = 6, R_W = 3, T_R + T_W = 6$; 12 miles   **2.** 30

3. $R_R T_R + 500 = R_W T_W, R_W = 40, R_R = 20, T_W = T_R$; 25 seconds

4. 600   **5.** 4, 6, 8, 10   **6.** $(2.03961)(3.16228) \times 10^1$ **7.** $(2.29565)(3.16228) \times 10^{-3}$

8. $(1.80831)(3.16228) \times 10^{-11}$ **9.** $2\sqrt{10}$ **10.** $\sqrt{61}$ **11.** $\dfrac{x+7}{x+10}$ **12.** $-\dfrac{11}{8}$

13. $(-1, 2)$ **14.** 1×10^{-3} **15.** (a) $y = 3$; (b) $y = \dfrac{3}{2}x - 3$ **16.** $400\sqrt{2}$

17. $28\sqrt{6} - 12$ **18.** (number line with points at 2, 3, 4)   **19.** $-10, 7$ **20.** $\dfrac{54}{11}$ **21.** $\dfrac{31}{26}$

22. $\dfrac{-3x^3 + 3x^2 + 9x + 4}{x^2(x+1)}$ **23.** $\dfrac{mx+4}{1+4a}$ **24.** False   **25.** -70 **26.** $1 - x$

27. $\dfrac{1}{16}$ **28.** -16 **29.** $y^{-4} - 2a^{-2}y^{-1}x^{-2}$ **30.** 40.3737374

problem set 104

1. (segment diagram: 40) $R_F T_F + 40 = R_M T_M$, $R_M = 54$, $R_F = 46$, $T_M = T_F$; 5 seconds

2. $-8, -6, -4$

3. (segment diagram) $R_W T_W + R_B T_B = 20$, $R_W = 5$, $R_B = 15$, $T_W + T_B = 2$; 15 miles   **4.** 480

5. $N_O = 750$, $N_T = 600$ **6.** $(1.03923)(3.16228) \times 10^{-3}$ **7.** $4\sqrt{3}$ **8.** $\sqrt{89}$

9. $\dfrac{x+10}{x+7}$ **10.** $\dfrac{3}{4}$ **11.** $(1, -2)$ **12.** 2×10^{-55} **13.** (a) $y = -2$; (b) $y = -2x$

14. $200\sqrt{6}$ **15.** $18\sqrt{15} + 15$ **16.** (number line at $-7, -6, -5$)   **17.** $\pm\dfrac{9}{2}$ **18.** $\dfrac{69}{7}$

19. $\dfrac{8}{11}$ **20.** $\dfrac{6x - px - 3p}{x^2 - x - 12}$ **21.** $\dfrac{2x-5}{x(x-6)(x+1)}$ **22.** $\dfrac{2x^2 + 6x + px}{x^3 - 9x}$

23. Reals, rationals, integers, wholes, naturals   **24.** $1 + 2a$ **25.** $-\dfrac{1}{81}$ **26.** $\dfrac{5p - 4x}{3 - x^2}$

27. -176 **28.** $x^{-2}a^9$ **29.** 60   **30.** 7.18518519

problem set 105

1. (segment diagram) $R_A T_A + R_R T_R = 38$, $R_A = 3$, $R_R = 5$, $T_A = T_R + 2$; 6 p.m.   **2.** 490

3. 4800   **4.** $-10, -9, -8, -7$ **5.** $N_W = 12$, $N_R = 40$

6. $(2.89310)(3.16228) \times 10^{-5}$ **7.** $\sqrt{146}$ **8.** $\sqrt{5}$ **9.** $x + 10$ **10.** $\dfrac{2}{3}$

11. $(2, 5)$ **12.** 2×10^{62} **13.** (a) $x = -4$; (b) $y = \dfrac{5}{2}x - 3$ **14.** $2506\sqrt{6} - 30$

15. $-8, -10$ **16.** (number line at $-2, -1, 0, 1, 2$)   **17.** All real numbers

18. (number line at $-2, -1, 0, 1, 2$)   **19.** $\dfrac{4x + 21}{x^2 - 16}$ **20.** $\dfrac{3x^2 + 3x + 4}{x^2 + 2x + 1}$ **21.** $\dfrac{64}{23}$

22. $\dfrac{65}{56}$ **23.** -13 **24.** $1 + 3x$ **25.** 8   **26.** $\dfrac{4x^2 + y}{x^2 y^2 + y}$ **27.** Reals, rationals

28. $3x^4 y^{-1}a - 12x^{-3}y^{-5}a^{-3}$ **29.** -1

problem set 106

1. 460   **2.** (segment diagram) $R_S T_S + R_F T_F = 52$, $R_S = 3$, $R_F = 4$, $T_S + T_F = 15$; 8 hours

3. (segment diagram) $R_W T_W = R_T T_T$, $R_W = 2$, $R_T = 4$, $T_W = T_T + 2$; 8 miles   **4.** -4

5. $N_D = 10$, $N_C = 15$ **6.** 5   **7.** 2   **8.** 16   **9.** $-\dfrac{8}{3}$

10.

$$-5 \quad -4 \quad -3 \quad -2 \quad -1 \quad 0 \quad 1 \quad 2 \quad 3 \quad 4 \quad 5$$

11. No real numbers

12. $2\sqrt{5}$ **13.** $x - 2$ **14.** $(2, 0)$ **15.** (a) $y = 4$; (b) $y = -\dfrac{1}{3}x - 3$ **16.** $\dfrac{1}{18}$

17. (a) $1 - x$; (b) $\dfrac{1}{4}$ **18.** $\dfrac{a^3 + 1}{a^2 x + b}$ **19.** 0 **20.** $\dfrac{10a + 8}{a^2 - 4}$ **21.** $\dfrac{5x - 13}{x^2 + 2x - 8}$

22. $(2.04450)(3.16228) \times 10^{12}$ **23.** $(2.04450)(3.16228) \times 10^{-28}$ **24.** $-382\sqrt{6} - 6$

25. $x^2 - 7x + 49 - \dfrac{347}{x + 7}$ **26.** $\pm\dfrac{9}{2}$ **27.** $\dfrac{7}{18}$ **28.** Reals, irrationals **29.** 48

30. 0

problem set 107

1. 5 **2.** $-1, 1, 3$ **3.** $270{,}500$ **4.** $R_M T_M = R_H T_H, R_M = 8, R_H = 16,$ $T_M = T_H + 4$; 4 hours

5. $R_T T_T = R_B T_B, T_T = 4, T_B = 48, R_T = R_B + 55$; 240 miles

6. 2 **7.** 3 **8.** $\dfrac{15}{2}$ **9.** -2 **10.** $\dfrac{bx}{cx - 1}$ **11.** $\dfrac{da}{xd - c}$

12. $\dfrac{db}{a - cb}$ **13.** $\dfrac{na}{kn + yn - m}$ **14.** $\dfrac{3x^2 + 6x + 4}{x^2 - 4}$ **15.** $\dfrac{5x - 35}{x^2 - 25}$

16. 2.67208×10^{-5} **17.** $2\sqrt{13}$ **18.** 7 **19.** 2×10^{84} **20.** $-7, -9$

21. $355\sqrt{2} + 24$ **22.** $\dfrac{5x^2 + 22x - 50}{x^3 - 25x}$ **23.** $\dfrac{-x^2 - x}{x^2 + 3x - 10}$ **24.** 44 **25.** y^{-4}

26.

$$-2 \qquad 2$$

27.

$$-7 \quad -6$$

28. (a) $x - 1$; (b) $-\dfrac{1}{4}$

29. $1 - 8x^{-4}a^{-6}$

problem set 108

1.

$R_P T_P + R_F T_F = 440, R_P = 70, R_F = 30, T_P = T_F + 2$; 11 a.m.

2.

$R_S T_S = R_C T_C, R_S = 30, R_C = 50, T_S = T_C + 2$; 3 hours

3. $N_D = 43, N_Q = 21$ **4.** $-20, -18, -16$ **5.** $22{,}500$ **6.** $y = \dfrac{5}{9}x - \dfrac{2}{9}$

7. $y = -\dfrac{6}{7}x + \dfrac{10}{7}$ **8.** $x = 4$ **9.** 6 **10.** 9 **11.** $\dfrac{7}{4}$ **12.** $\dfrac{mx}{pm - 1}$

13. $\dfrac{m}{xm - k}$ **14.** $\dfrac{x}{yx - k}$ **15.** $\dfrac{yc}{xc - yb}$ **16.** $\dfrac{-x^2 - 5x + 4}{x^2 - 25}$ **17.** $\dfrac{-6}{x^2 - x - 6}$

18. $(1.80278)(3.16228) \times 10^{-23}$ **19.** $\sqrt{106}$ **20.** 2 **21.** 1×10^{-38} **22.** $y^9 x^2 a^{-1}$

23. $275\sqrt{3} + 30$ **24.** (a) $x + 1$; (b) $-\dfrac{1}{36}$ **25.** $-8, -7$ **26.** -57

27. No real number **28.** $2x^2 y^3 a^4 - 4x^2 y^7 a^2$ **29.** $4y^{-4} + 12x^3 y^{-4} a^{-4}$ **30.** $50{,}000$

problem set 109

1.

$R_M T_M = R_P T_P, R_M = 2, R_P = 11, T_M + T_P = 15$; 26 miles

2.

$R_S T_S = R_T T_T, R_T = 4, R_S = 2, T_T + T_S = 120$; 160 miles

3. $N_F = 5000, N_H = 100$ **4.** 20 percent **5.** $9, 10, 11, 12$ **6.** $y = -\dfrac{5}{7}x - \dfrac{22}{7}$

7. $y = -2x - 1$ **8.** $y = 2x - 17$ **9.** b, d, e **10.** $D = \{a, 4, 7\}$ **11.** a, c

12. 8 **13.** 44 **14.** $\dfrac{ac}{dc-1}$ **15.** $\dfrac{dx}{ad-bx}$ **16.** $\dfrac{x}{kx-p}$ **17.** $\dfrac{-3x+13}{x^2-9}$

18. $\dfrac{2x+15}{x^2+5x+6}$ **19.** 2.28035×10^{-5} **20.** **21.** $\dfrac{13}{5}$

22. 5×10^{34} **23.** $\dfrac{1}{x}$ **24.** $2\sqrt{5}$ **25.** $25-50\sqrt{5}$ **26.** $\dfrac{xa^2+1}{x+a^2}$ **27.** -1

28. Reals, irrationals **29.** $\dfrac{5}{168}$ **30.** $(2,0)$

31. (a) $x=-3$; (b) $y=-x+6$

problem set 110

1. $R_BT_B+30=R_NT_N$, $T_B=6$, $T_N=6$, $2R_B=R_N$; 5 miles per hour

2. $R_FT_F+20=R_ET_E$, $R_F=40$, $R_E=60$, $T_E+2=T_F$; 4 p.m.

3. 4500, 7500 **4.** $20,000 **5.** $-4,-3,-2$ **6.** ϕ **7.** 20 **8.** ϕ **9.** -1

10. $y=-\dfrac{5}{9}x+\dfrac{7}{9}$ **11.** $y=x$ **12.** $y=-\dfrac{1}{4}x-\dfrac{5}{4}$ **13.** a,b,c,d,e

14. $D=\{a,b,c\}$; $R=\{d,e\}$ **15.** a,b **16.** $\dfrac{3}{20}$ **17.** 16 **18.** $\dfrac{c}{kc-a}$

19. $6\sqrt{2}$ **20.** $\dfrac{x-3}{x^3+8x^2}$ **21.** $(-1,0)$ **22.** (a) $y=-4$; (b) $y=-x$

23. $\dfrac{a^2-x^3}{x^3-a}$ **24.** $x^2+2x+4+\dfrac{13}{x-2}$ **25.** 13 **26.** $\dfrac{20}{9}$ **27.** -28

28. $4ax^{-2}y^4-3y^8$ **29.** Reals, irrationals **30.**

31. $\dfrac{7x^2-20x-6}{x^2-9}$

problem set 111

1. $R_RT_R=R_TT_T$, $R_R=20$, $R_T=8$, $T_R+T_T=14$; 80 miles

2. $R_ST_S+36=R_ET_E$, $T_E=3$, $T_S=3$, $R_E=2R_S$; 24 miles per hour **3.** $50

4. $N_D=500$, $N_Q=100$ **5.** $-11,-9,-7$ **6.** $y=2x+7$ **7.** $y=-\dfrac{4}{3}x-\dfrac{13}{3}$

8. $y=-\dfrac{1}{8}x-\dfrac{19}{8}$ **9.** $y=-\dfrac{1}{5}x$ **10.** 8 **11.** ϕ **12.** a,c,d

13. $D=\{a,b,c\}$; $R=\{4,5\}$ **14.** a,b **15.** $\dfrac{7}{5}$ **16.** $-\dfrac{35}{12}$ **17.** $\dfrac{ax}{bx+1}$

18. **19.** $\sqrt{65}$ **20.** $\dfrac{3}{4}$ **21.** 1×10^{-9} **22.** $\dfrac{1}{4}x^{-4}y^6$

23. $385\sqrt{5}$ **24.** $1+x$ **25.** $-\dfrac{1}{4}$ **26.** 8, 10 **27.** 100 **28.** All real numbers

29. $x^{-6}-3x^{-1}y^{-5}$ **30.** .003

problem set 112

1. $R_R T_R = R_T T_T$, $R_R = 8$, $R_T = 6$, $T_R + T_T = 7$; 24 miles
2. $R_N T_N + R_S T_S = 880$, $T_N = T_S + 2$, $R_N = 40$, $R_S = 60$; 2 p.m.   3. $N_S = 75$, $N_L = 50$
4. 12 percent   5. $-5, -3, -1$ 6. Bees—1885, moths—725   7. $y = \frac{1}{3}x + \frac{11}{3}$
8. $y = -\frac{2}{3}x + \frac{5}{3}$ 9. $y = -\frac{3}{4}x + \frac{25}{4}$ 10. $y = \frac{1}{7}x + \frac{82}{7}$ 11. (a) ϕ; (b) 7
12. b, c, e 13. $R = \{p, 5\}$ 14. $\frac{5}{19}$ 15. $\frac{35}{4}$ 16. $\frac{ym}{xm + ky - pym}$
17. $(1.33791)(3.16228) \times 10^{-9}$ 18. $2\sqrt{10}$ 19. $-\frac{15}{2}$ 20. 7×10^{-25}
21. $a^2 x^4 y^{-4}$ 22. $104\sqrt{15} + 30$ 23. $1 + y$ 24. -3 25. $-12, -10$
26. -94 27. All integers   28. $4y^2 - 12x^{-6}y^2$

problem set 113

1. 1974   2. $R_R T_R = R_W R_W$, $R_W = 2$, $R_R = 10$, $T_W + T_R = 18$; 30 miles
3. $R_B T_B = R_R T_R + 6$, $T_B = 3$, $T_R = 3$, $R_B = 10$; 8 miles per hour   4. 85
5. $N_G = 52$, $N_P = 71$ 6. $y = x - 2$ 7. (a) 1; (b) -6 8. 38   9. ± 4
10. 2   11. 50   12. $-\frac{39}{2}$ 13. 1   14. $\frac{xc}{1 + kc + dc}$
15. $2\sqrt{14}$ 16. $\frac{x + 5}{x - 2}$ 17. $(1, 3)$ 18. (a) $y = -4$; (b) $y = \frac{3}{4}x + 2$
19. $\frac{xz - y}{ayz - 3}$ 20. $2x^2 + 5x + 10 + \frac{14}{x - 1}$ 21. -14 22. $\frac{152}{85}$ 23. 54
24. 0   25. Reals, irrationals   26. $\frac{5}{17}$ 27. $\frac{x^3 y + x^2 y^2 - 3x - 3y - 2x^2 y^3}{y^3 x^2 (x + y)}$
28. $\frac{3x^2 + 12x + 8}{x^2 - 16}$ 29. .0374747

problem set 114

1. 172   2. $R_R T_R = R_J T_J$, $R_R = 14$, $R_J = 7$, $T_R + T_J = 3$; 14 miles
3. $-10, -8, -6, -4$ 4. Short, 10; long, 30   5. 4500   6. $y = -\frac{1}{2}x + \frac{1}{2}$
7. (a) $y = 4$; (b) $y = -\frac{1}{3}x - 1$ 8. a, c, e, f 9. 17   10. 3   11. $\frac{8}{25}$
12. -1 13. $\frac{9}{2}$ 14. $\frac{40}{7}$ 15. $\frac{ncx}{kcn - mc + n}$ 16. 2.29783×10^{18}
17. $\sqrt{145}$ 18. $\frac{20}{3}$ 19. 1×10^{99} 20. x^{-7} 21. $8\sqrt{6} - 24$ 22. $x + 1$
23. $\frac{1}{9}$ 24. $-12, -7$ 25. 80   26. No real number   27. $1 - 3x^2 y^8$
28. False   29. $\frac{ay - x^2}{bc - x^2 y^2}$ 30. $\frac{16}{75}$

problem set 115

1. $R_W T_W + R_R T_R = 66$, $R_W = 3$, $R_R = 15$, $T_R = 2T_W$; 4 hours
2. $R_N T_N + R_S T_S = 880$, $R_S = 20$, $R_N = 60$, $T_S = T_N + 4$; 2 a.m.   3. $N_R = 21$, $N_C = 31$
4. Coons, 605; mongooses, 363   5. 7600   6. $y = -\frac{5}{3}x + \frac{26}{3}$ 7. $\frac{17}{4}$ 8. 8

9. 28 **10.** 25 **11.** Inconsistent **12.** $\dfrac{111}{17}$ **13.** $\dfrac{56}{15}$ **14.** $\dfrac{kc}{mc - pc + 1}$

15. $\sqrt{33}$ **16.** 1 **17.** $(-2, 0)$ **18.** (a) $x = 5$; (b) $y = -\dfrac{1}{2}x - 4$ **19.** $\dfrac{x - a^2}{pay - 3}$

20. $x^2 - x - 1 - \dfrac{6}{x - 1}$ **21.** 46 **22.** 7 **23.** -1 **24.** Reals, rationals, integers

25. $-20, -2$ **26.** $3x^2y + 2x^6y$ **27.** $\dfrac{3xy - 3x - x^2y^2}{y^2x(y - 1)}$ **28.** $\dfrac{4x - 3}{x - 3}$

29. 32.07581582

problem set 116

1. $R_H T_H = R_M T_M + 60$, $R_H = 17$, $T_H = 20$, $T_M = 20$, $R_M = 14$ miles per hour
2. $R_W T_W = R_R T_R$, $R_W = 4$, $R_R = 24$, $T_W + T_R = 14$; 48 miles **3.** 10 **4.** \$135

5. $N_A = 11$, $N_P = 40$ **6.** $y = -\dfrac{1}{7}x + \dfrac{11}{7}$ **7.** a, d, g **8.** 25 **9.** 29

10. **11.**

12. **13.** -76 **14.** $\dfrac{50}{9}$ **15.** $\dfrac{acd}{c + d}$

16. 2.04206×10^{-17} **17.** $\sqrt{145}$ **18.** 1 **19.** 2×10^{38} **20.** $3x^{-4}y^{-5}$

21. $31\sqrt{2}$ **22.** $40 - 16\sqrt{6}$ **23.** $x - 1$ **24.** $-\dfrac{1}{27}$ **25.** $-9, 5$ **26.** -77

27. All real numbers **28.** $1 - 3x^{-4}a^4y^{-2}$ **29.** 478,000

problem set 117

1. 75 percent **2.** $R_N T_N + R_S T_S = 580$, $R_N = 50$, $R_S = 45$, $T_N = T_S + 4$; 4 hours
3. $R_W T_W = R_R T_R$, $R_W = 5$, $R_R = 30$, $T_W + T_R = 21$; 90 kilometers **4.** 165

5. $N_5 = 163$, $N_{20} = 13$ **6.** Yes **7.** $y = -\dfrac{2}{3}x - \dfrac{13}{3}$

8. (a) $y = -2$; (b) $y = -2x$ **9.** ϕ **10.** $\dfrac{49}{4}$ **11.** 20 **12.**

13. **14.** $26 + 13\sqrt{3}$ **15.** $-2 - 6\sqrt{5}$ **16.** $10 + 7\sqrt{2}$

17. $-18 + 2\sqrt{3}$ **18.** $\dfrac{13}{3}$ **19.** $\dfrac{cbx}{pbc + b - ac}$ **20.** $(5, -2)$ **21.** $\dfrac{xy - a}{a + x^2y}$

22. $x^2 - 2x + 2 - \dfrac{8}{x + 2}$ **23.** $\dfrac{21}{176}$ **24.** Reals, irrationals **25.** -9 **26.** -37

27. $-\dfrac{3}{5}$ **28.** $\dfrac{5x^2 + 15x + 4}{x^2 - 9}$ **29.** .003713714 **30.**

31.

problem set 118

1. $R_R T_R = R_B T_B$, $R_R = 8$, $R_B = 20$, $T_R + T_B = 7$; 40 kilometers **2.** 7 liters
3. $200°$K **4.** $N_W = 12$, $N_R = 15$ **5.** 105 kilometers **6.** $y = 2x - 1$

7. a, b, d **8.** $\dfrac{9}{4}$ **9.** 68 **10.** **11.** $-22 - 16\sqrt{3}$

12. $-4 - 8\sqrt{2}$ **13.** $-26 - 28\sqrt{6}$ **14.** $\dfrac{26}{5}$ **15.** $\dfrac{15}{8}$ **16.** $\dfrac{ckx}{mk + c}$

17. $(2.04206)(3.16228) \times 10^{-9}$ **18.** $\sqrt{145}$ **19.** $-\dfrac{4}{3}$ **20.** 7×10^{-56} **21.** x^2y^3

22. $6 + 9\sqrt{2}$ **23.** $10 - 20\sqrt{15}$ **24.** $1 - z$ **25.** $-\dfrac{8}{9}$ **26.** $-1, 8$ **27.** -31

28. No integers **29.** $1 - 4x^3a^{-4}$ **30.** 4.0606

problem set 119

1. 35 liters **2.** 3 hours **3.** 35 **4.** 1500 **5.** 568 acres **6.** $y = -\dfrac{1}{5}x - \dfrac{17}{5}$

7. a, c, d **8.** 36 **9.** 30 **10.**

11. $-28 + 18\sqrt{2}$ **12.** $16 - 7\sqrt{5}$ **13.** $12 - 8\sqrt{3}$ **14.** 12 **15.** $-\dfrac{5}{2}$

16. $\dfrac{cx}{kc + mc - 1}$ **17.** $\sqrt{39}$ **18.** $\dfrac{x^2}{x + 8}$ **19.** $(-4, -4)$

20. (a) $y = 5$; (b) $y = -\dfrac{3}{4}x - 2$ **21.** $\dfrac{ax + a}{1 + a^2x}$ **22.** $x + \dfrac{x - 2}{x^2 - 1}$ **23.** -120

24. **25.** **26.** 1 **27.** $2y^2x^2z^{-1}$

28. $\dfrac{15}{76}$ **29.** $\dfrac{3x^2 + 5x - 11}{x^2 - 9}$

problem set 120

1. 500 liters **2.** 20 **3.** $N_N = 14, N_D = 20$

4. $R_RT_R = R_WT_W, R_R = 7, R_W = 3, T_W + T_R = 20; T_R = 6$ so $D = 42$ miles

5. **6.** **7.** 5 **8.** $\dfrac{25}{4}$ **9.** $\dfrac{41}{4}$

10. **11.** $3 + \sqrt{2}$ **12.** $-4 - 14\sqrt{3}$ **13.** $-2 + 2\sqrt{2}$

14. -50 **15.** $\dfrac{8}{3}$ **16.** $\dfrac{xd}{d^2 + c}$ **17.** $(1.10905)(3.16228) \times 10^{-10}$ **18.** $\sqrt{65}$

19. 4 **20.** 5×10^{-88} **21.** $x^{-4}y^2$ **22.** $2\sqrt{6}$ **23.** -42 **24.** $1 - 4y$

25. $\dfrac{1}{2}$ **26.** $-5, 9$ **27.** 31 **28.** **29.** $1 - x^2y^{-2}$

30. True

problem set 121

1. 40 **2.** $R_ET_E = R_CT_C, R_E = 2, R_C = 10, T_E + T_C = 18$; 30 kilometers **3.** \$6

4. $N_C = 20, N_E = 5$ **5.** 1440 **6.** (a) $y = 3$; (b) $y = 3x$ **7.** a, c, d **8.** 52

9. **10.** $-10 - 8\sqrt{2}$ **11.** $-5 + 4\sqrt{3}$

12. $-14 - 10\sqrt{5}$ **13.** $\dfrac{\sqrt{15}}{3}$ **14.** $\dfrac{\sqrt{10}}{5}$ **15.** $\dfrac{\sqrt{21}}{7}$ **16.** $\dfrac{4\sqrt{2} + \sqrt{6}}{2}$

17. $\dfrac{2\sqrt{5} + 5\sqrt{3}}{5}$ **18.** $\sqrt{2} + 3\sqrt{3}$ **19.** $\dfrac{40}{3}$ **20.** $\dfrac{mx}{a + xb - xk}$ **21.** $\dfrac{ax^2y - x}{p - 3ky}$

22. $4x^2 - 10x + 20 - \dfrac{36}{x + 2}$ **23.** 1152 **24.** -3 **25.** $-\dfrac{25}{144}$

26. $y = -3x + 13$ **27.** $y = -\dfrac{5}{9}x - \dfrac{7}{9}$ **28.** $y = -5x - 23$

29.

problem set 122

1. 1600 tons   **2.** Boys, 1260; Girls, 900

3.   **4.**   **5.**

6. $(2x + 3)(x - 5)$ **7.** $(3x + 2)(x - 1)$ **8.** $(5x + 2)(x - 3)$ **9.** $(3x + 1)(x - 5)$

10. $2(x + 4)(x + 1)$ **11.** $(2x - 3)(x - 6)$ **12.** $(2x - 3)(x + 5)$

13. $2(x + 6)(x - 2)$ **14.** $2(x - 6)(x + 2)$ **15.** $2(x - 2)(x - 1)$

16. $(2x - 3)(x + 6)$ **17.** $2(x + 2)(x + 1)$ **18.** $(3x + 1)(x - 7)$

19. $(3x + 1)(x - 8)$ **20.** $(3x - 4)(x - 1)$ **21.** $16 - 10\sqrt{2}$

22. $30\sqrt{2} - 4\sqrt{3} + 20\sqrt{6} - 8$ **23.** $\dfrac{\sqrt{21}}{7}$ **24.** $\dfrac{\sqrt{10}}{4}$ **25.** $\dfrac{2\sqrt{5} + \sqrt{15}}{5}$

26. $\dfrac{4\sqrt{3} + \sqrt{6}}{3}$ **27.**

problem set 123

1. 30 pounds per square inch   **2.** 4000 pounds per square inch

3. Chickens, 143; pigs, 26   **4.** $R_DT_D + 10 = R_WT_W, R_D = 40, R_W = 70, T_D = T_W + 2$; 2 p.m.

5. $N_N = 100$ **6.**   **7.** $(3x + 2)(x - 7)$

8. $(3x - 2)(x + 7)$ **9.** $(2x - 3)(x + 5)$ **10.** $(2x - 3)(x + 6)$ **11.** $(3x + 2)(x + 7)$

12. $(2x - 3)(x - 7)$ **13.** $(3x - 2)(x - 8)$ **14.** $(2x + 3)(x + 6)$

15. $(3x + 7)(x + 2)$ **16.** $(xy - 4)(a + b)$ **17.** $(a + 2b)(c + 2d)$

18. $(a + b)(c - d)$ **19.** $(2 + a)(b + 4)$ **20.** $(a + x)(b + c)$ **21.** $(m + pc)(2x - 3)$

22. $(k + pc)(4 - xy)$ **23.** $(a + d)(c - xy)$ **24.** $\dfrac{\sqrt{33}}{11}$ **25.** $\dfrac{2\sqrt{15} + 2\sqrt{5}}{5}$

26.

problem set 124

1. 392 meters 2. 1600 feet 3. 36 4. 100 5. Rabbits, 9800; squirrels, 7000

6. 7. $\dfrac{bc}{mb - a}$ 8. $(3x - 2)(x + 9)$

9. $(3x - 4)(x + 1)$ 10. $2(x + 1)(x - 3)$ 11. $(3x - 2)(x + 10)$

12. $(2x + 5)(x + 5)$ 13. $(2x + 5)(x - 5)$ 14. $(5 + b)(a + 3)$ 15. $(y + c)(a + x)$

16. $(3x - 2)(m + p)$ 17. $(k + 3)(x - 5)$ 18. $(pc + 4)(x + c)$ 19. $(ac + 2)(b - k)$

20. [number line: open circles at -2 and 2, shaded between; marks $-2\ -1\ 0\ 1\ 2$]

21. [number line: closed dots at 2, 3, 4]

22. [number line: closed dot at 2 shaded left, open circle at 5; marks $1\ 2\ 3\ 4\ 5\ 6$]

23. 23 24. 67 25. $y = -\dfrac{7}{5}x + \dfrac{11}{5}$

26. $y = -\dfrac{1}{4}x + \dfrac{9}{2}$ 27. $y = -\dfrac{1}{3}x + \dfrac{13}{3}$ 28. $\dfrac{\sqrt{15}}{5}$ 29. $\dfrac{\sqrt{21}}{3}$ 30. 3

problem set 125

1. $R_E T_E + R_F T_F = 420$, $R_E = R_F + 20$, $T_F = 6$, $T_E = 3$; $R_F = 40$ miles per hour, $R_E = 60$ miles per hour

2. Greens, 360; Whites, 1980 3. 24,000 4. $N_O = 530$, $N_T = 237$ 5. 4000

6. 2, 4, 6, 8 7. $\dfrac{-3 \pm \sqrt{29}}{2}$ 8. $\dfrac{-5 \pm \sqrt{57}}{2}$ 9. $-1 \pm \sqrt{5}$ 10. $\dfrac{-3 \pm \sqrt{41}}{2}$

11. $1, -\dfrac{5}{3}$ 12. $\dfrac{-2 \pm 3\sqrt{2}}{2}$ 13. $(3x + 5)(x - 7)$ 14. $(3x - 5)(x + 1)$

15. $(2x + 3)(x - 4)$ 16. $(p^2 - a)(b + c)$ 17. $(x^2 - c)(a + c)$ 18. $(x^3 + y)(m + 2)$

19. $(4 + c)(ab + x)$ 20. [number line shaded between 6 and 10; marks $6\ 7\ 8\ 9\ 10$] 21. [number line: closed dot at 0]

22. [number line: closed dots at $-4, -3, 4, 5$; marks $-4\ -3\ -2\ -1\ 0\ 1\ 2\ 3\ 4\ 5$] 23. $y = -\dfrac{5}{8}x + \dfrac{1}{8}$ 24. 2

25. $\dfrac{6}{5}$ 26. $y = \dfrac{1}{5}x + \dfrac{18}{5}$ 27. 28. $\dfrac{\sqrt{14}}{7}$ 29. $\dfrac{\sqrt{15}}{6}$

30. $\dfrac{4\sqrt{6} + 3\sqrt{2}}{6}$

problem set 126

1. $R_E T_E = R_F T_F$, $T_E = 20$, $T_F = 25$, $R_E = R_F + 10$; $R_F = 40$ miles per hour, $R_E = 50$ miles per hour

2. \$120 3. 125 4. 550 5. $N_G = 42$, $N_S = 40$

6. 23, 25, 27 7. $-2, 3$ 8. $\dfrac{-1 \pm \sqrt{22}}{3}$ 9. $\dfrac{1}{2} \pm \dfrac{\sqrt{29}}{2}$ 10. $\dfrac{1 \pm \sqrt{6}}{5}$

11. $-\dfrac{1}{2} \pm \dfrac{\sqrt{23}}{2}$ 12. $\dfrac{3}{5} \pm \dfrac{\sqrt{29}}{5}$ 13. $-2, 3$ 14. $-\dfrac{1}{3} \pm \dfrac{\sqrt{22}}{3}$ 15. $\dfrac{1}{2} \pm \dfrac{\sqrt{29}}{2}$

16. $(3x - 1)(x + 5)$ 17. $3(x + 9)(x - 1)$ 18. $(2x - 1)(x + 5)$ 19. $(m^2 - c)(k - 2)$

20. $(6 - xy)(a + b)$ **21.** $(ab + c)(x - 2y)$ **22.** $(4x + ab)(n - m)$ **23.** $y = \dfrac{9}{5}x + \dfrac{7}{5}$

24. $y = \dfrac{2}{5}x + \dfrac{29}{5}$ **25.**   **26.** $8\sqrt{2} + 12$ **27.** $\dfrac{\sqrt{6}}{4}$

28. $1 + \dfrac{\sqrt{2}}{2}$

enrichment problems

77-1. Addition **77-2.** Addition and multiplication **78-3.** None

78-4. Addition, multiplication, division

79-5. Addition, subtraction, multiplication, and division

80-6. Addition, subtraction, multiplication **81-7.** Addition **82-8.** Addition

83-9. Multiplication **84-10.** Addition, multiplication

86-11. Addition, subtraction, multiplication **87-12.** None

89-13. Addition, multiplication, division **91-14.** Addition **93-15.** Addition

EL2-16. Addition, multiplication **EL2-17.** Addition, subtraction, multiplication, division

EL2-18. $a \cdot 1 = a, \dfrac{a}{1} = a, a \cdot 0 = 0, a + 0 = a$ **EL2-19.** $a(b + c) = ab + ac$

EL3-20. (a) 2; (b) $-\dfrac{1}{b}$

EL3-21. For both addition and multiplication; identity, inverse, associative property, and commutative property. To this list we add the distributive property.

95-22. (a) $a(bc) = (ab)c$; (b) $a + b = b + a$ **95-23.** 9

95-24. Associative property for addition **96-25.** 0 **96-26.** 1 **96-27.** -9

97-28. $-\dfrac{1}{7}$ **97-29.** Distributive property **98-30.** Addition, multiplication, division

99-31. Commutative property for addition

99-32. For both addition and multiplication, the following are properties of a field: identity, inverse, associative property, and the commutative property. Also, the distributive property is a field property.

100-33.
$a + c + x$	Given
$a + (c + x)$	Associative property
$a + (x + c)$	Commutative property
$(a + x) + c$	Associative property
$(x + a) + c$	Commutative property
$x + (a + c)$	Associative property
$x + (c + a)$	Commutative property
$x + c + a$	Removed parentheses

101-34.
mpa	Given
$m(pa)$	Associative property
$m(ap)$	Commutative property
map	Removed parentheses

102-35. 101 **102-36.** Associative property for addition

103-37.
$abxy$	Given
$ab(xy)$	Associative property
$ab(yx)$	Commutative property
$a(by)x$	Associative property
$a(yb)x$	Commutative property
$(ay)bx$	Associative property
$(ya)bx$	Commutative property
$yabx$	Removed parentheses

107-41.
$abxy$	Given
$a(bx)y$	Associative property
$a(xb)y$	Commutative property
$(ax)by$	Associative property
$(xa)by$	Commutative property
$x(ab)y$	Associative property
$x(ba)y$	Commutative property
$xb(ay)$	Associative property
$xb(ya)$	Commutative property
$xbya$	Removed parentheses

104-38. Addition   **105-39.** Zero   **106-40.** -55 **108-42.** $-\dfrac{3}{2}$

109-43. Multiplication   **110-44.** -105

111-45. Only two numbers can be added at one time

112-46. (a) $(a + b) + c = a + (b + c)$; (b) $ab = ba$ **113-47.** $a(b + c) = ab + ac$

114-48.

$abcd$	Given
$(ab)cd$	Associative property
$(ba)cd$	Commutative property
$ba(cd)$	Associative property
$ba(dc)$	Commutative property
$b(ad)c$	Associative property
$b(da)c$	Commutative property
$bdac$	Removed parentheses

117-51.

$a + x + y + m$	Given
$a + (x + y) + m$	Associative property
$a + (y + x) + m$	Commutative property
$(a + y) + x + m$	Associative property
$(y + a) + x + m$	Commutative property
$y + a + (x + m)$	Associative property
$y + a + (m + x)$	Commutative property
$y + a + m + x$	Removed parentheses

115-49. -5 **116-50.** None